HADRONS AND NUCLEI

Related Titles from AIP Conference Proceedings

589 New Developments in Fundamental Interaction Theories: 37th Karpacz Winterschool of Theoretical Physics
Edited by Jerzy Lukierski and Jakub Rembieliński, October 2001, 0-7354-0029-6

588 Physics with an Electron Polarized Light-Ion Collider: Second Workshop, EPIC 2000
Edited by Richard G. Milner, October 2001, 0-7354-0028-8

576 Application of Accelerators in Research and Industry: Sixteenth International Conf.
Edited by J. L. Duggan and I. L. Morgan, July 2001, 0-7354-0015-6

562 Particles and Fields: Ninth Mexican School
Edited by Gerardo Herrera Corral and Lukas Nellen, April 2001, 1-56396-998-X

561 Tours Symposium on Nuclear Physics IV: Tours 2000
Edited by M. Arnould, M. Lewitowicz, Yu. Ts. Oganessian, H. Akimune, M. Ohta, H. Utsunomiya, T. Wada, and T. Yamagata, April 2001, 1-56396-996-3

549 Intersections of Particle and Nuclear Physics: 7th Conference, CIPANP2000
Edited by Zohreh Parsa and William J. Marciano, December 2000, 1-56396-978-5

541 Theoretical High Energy Physics: MRST 2000
Edited by C. R. Hagen, November 2000, 1-56396-966-1

531 Particles and Fields: Seventh Mexican Workshop
Edited by Alejandro Ayala, Guillermo Contreras, and Gerardo Herrera, July 2000, 1-56396-954-8

508 Hadron Physics: Effective Theories of Low Energy QCD
Edited by A. H. Blin, B. Hiller, M. C. Ruivo, C. A. Sousa, and E. van Beveren, March 2000, 1-56396-927-0

495 Experimental Nuclear Physics in Europe: ENPE 99, Facing the Next Millennium
Edited by Berta Rubio, Manuel Lozano, and William Gelletly, November 1999, 1-56396-907-6

481 Nuclear Structure 98
Edited by C. Baktash, September 1999, 1-56396-858-4

459 Heavy Quarks at Fixed Target
Edited by Harry W. K. Cheung and Joel N. Butler, February 1999, 1-56396-864-9

HADRONS AND NUCLEI

First International Symposium

Seoul, Korea 20–22 February 2001

EDITORS

Il-Tong Cheon
Taekeun Choi
Yonsei University, Seoul, Korea

Seung-Woo Hong
Sungkyunkwan University, Suwon, Korea

Su Houng Lee
Yonsei University, Seoul, Korea

Melville, New York, 2001

AIP CONFERENCE PROCEEDINGS ■ VOLUME 594

Editors:

Il-Tong Cheon, Taekeun Choi, and Su Houng Lee
Department of Physics
Yonsei University
Seoul 120-479
KOREA

E-mail: itcheon@phya.yonsei.ac.kr
tkchoi@phya.yonsei.ac.kr.
suhoung@phya.yonsei.ac.kr

Seung-Woo Hong
Department of Physics
Sungkyunkwan University
Suwon 440-746
KOREA

E-mail: swhong@skku.ac.kr

Illustration on title page: The original painting is part of the collection of the National Museum of Korea.

L.C. Catalog Card No. 2001095833
ISBN 0-7354-0037-7
ISSN 0094-243X
Printed in the United States of America

CONTENTS

STRANGENESS NUCLEAR PHYSICS

HADRONS IN NUCLEAR MATTER

LATTICE QCD AND OTHER RELATED TOPICS

APPENDIX

PREFACE

Research trends in nuclear physics have continuously been changed during the past sixty years. It may well be said that nuclear physics started with the discovery of the pi-meson mediating nuclear forces between the nucleons. Immediately after discovering the nuclear force, researchers embarked on the journey of pursuing the understanding of the structure of nuclei as a many body system. The shell structure and the collective motion of nuclei were revealed. In this journey electron scattering has been one of the keys we have had to unlock the doors to the realm of nuclear and nucleon structure. Meantime, quantum chromodynamics (QCD) was developed and its powerful analytic ability shed light to our understanding and unveiled the substructure of the hadron to some extent. We have obtained a lot of information about the hadron structure through toils. Though solving the QCD is a very formidable task, much progress has been made recently in the lattice calculations. These problems were previously studied in particle physics but they are now hot topics in nuclear physics.

Strangeness is also an important subject in current nuclear physics research. Hyperon production and their interaction with the nucleus have attracted much attention of physicists to understand the roles of strangeness in nuclear dynamics. Hypernuclear structure and strangeness in hadrons are very active research subjects and have given us another key to open a new dimension that was previously unknown.

Physics in dense matter is also a current hot topic. Problems of phase transition in very high energy regions are linked to the characteristics of quark gluon plasma (QGP). The QGP phase has been extensively investigated not only theoretically but also experimentally.

Research activities in these fields are endlessly extended, and tremendous amount of research efforts have been poured into them. Therefore, it seems timely to survey the recent developments made in the field of physics of *Hadrons and Nuclei* by gathering the experts in one place. Thus, we have organized this symposium to discuss some of the recent works and look for future directions in these areas.

It is our pleasure to thank the Institute of Physics and Applied Physics at Yonsei University, Natural Science Research Institute at Yonsei University, the Korean Physical Society, and the Korea Nuclear Society for their sponsorship. We also owe much thanks to Korea Research Foundation, Korean Ministry of Education (BK21 Program) and Cheong Moon Gak Pub. Co. for their financial support.

February, 2001

Il-Tong Cheon
Yonsei University
Seoul, Korea

NOTES FROM THE VOLUME EDITORS

Prof. Walter Greiner composed a poem for Prof. Il-Tong Cheon, the chairman of the symposium, and beautifully presented it during the banquet. With kind permission from Prof. Greiner, we have included the poem in this volume.

We are sorry to add a sad news here that one of the speakers, Prof. Osamu Miyamura at Hiroshima University, passed away very recently. He was hospitalized just after the symposium, so could not submit his manuscript. Thus, his paper is missing in this proceedings. We would like to express our sincere, deep condolence.

WELCOMING ADDRESS

Honorable Guests, Distinguished Physicists, Fellow Scientists, Ladies and Gentlemen!

It is my great honor to give a welcome address to all participants of the International Symposium on Hadrons and Nuclei held here at our University. Yonsei University was founded 116 years ago in 1885 as a missionary school at this site. Since then, it has over the years developed into one of the best universities in the nation. It is still growing rapidly toward a world class academic institute. In order to compete with world major universities, we have established a strategic program to invest funds preferentially and allocate budgets in specific research areas, such as Biotechnology, Clean technology, Environmental technology, Information technology, Nanotechnology. I believe that in the twenty first century, these technologies would be the most important driving forces of revolutionary development of human life.

These technologies cannot and will not be developed without fundamental research in basic sciences. Nuclear Physics has a long history and has contributed greatly to the welfare of mankind. For example, the early counting technique in nuclear physics was developed into modern digital computer and information technology. Modern nuclear power generation would not have existed without discovery of nuclear fission.

In the near future, one of the major challenges for nuclear physicists would be to tackle nuclear waste problems. I believe that in the long run, both information and biotechnology would be greatly benefited by the work you are currently doing.

Through many conversations with Prof. Il-Tong Cheon, I have learned that some of the elementary particles carry strangeness quantum number, just like some people have strange characters. I wonder whether such quantum number is still effective in our body and jumps by one unit by external stimulants to generate strange behaviour for some.

Anyway, it is my pleasure to host all of you at Yonsei University. I hope you will enjoy our beautiful campus and obtain fruitful results through enthusiastic discussions among yourself on your recent research.

Thank you!

February, 2001

Woo Sik Kim
President
Yonsei University, Seoul, Korea

WELCOMING ADDRESS

Ladies and Gentlemen;

On behalf of the Korean Physical Society, I would like to welcome all of you to the International Symposium on Hadrons & Nuclei held at Yonsei University. This is the first major symposium held in Korea in the new millennium. I thank the foreign scholars in particular who will give lectures and make this symposium more enjoyable.

The Korean Physical Society will have its 50th anniversary next year. The Society was established with just a few members in 1952 during the Korean War, but it has now grown to a society with over 6000 members in eleven divisions; particle physics, nuclear physics, astrophysics, condensed matter physics, applied physics, plasma physics, statistical physics, atomic and molecular physics, optics, semiconductor physics and physics education. Nuclear physicists played important roles in the early stage of the Society, and particularly Prof. Il-Tong Cheon, the chairman of the organizers of this symposium, has been a key member of the Society.

We have had a number of international conferences, workshops and symposia held in this country. The largest one among them was the Fourth Asia Pacific Physics Conference held in 1990 at Yonsei University with about 700 participants. Prof. Cheon played the leading role as the chairman of the program committee.

Contribution of Korean physicists to the international physics community is rapidly increasing in the recent years. One of those activities is IUPAP. Eight Korean physicists are serving in the committee of each division. Prof. Cheon is one of the committee members in *C*12 Division. I am sure that the Korean Physical Society will extend its global activities and will strengthen its international collaboration and mutual relations with other countries.

I hope you will have exciting discussions and fruitful results through this Symposium, and since many of you may be visiting Korea for the first time, I hope you will have a chance to explore Korean culture during your stay in this country.

Thank you.

February, 2001 H. S. Song
President of the Korean Physical Society

INTERNATIONAL ADVISORY COMMITTEE

I. R. Afnan (Flinders Univ.)
H. Bhang (Seoul National Univ.)
G. E. Brown (SUNY, Stony Brook)
M. Fujiwara (RCNP, Osaka Univ.)
B. F. Gibson (Los Alamos)
T. Hatsuda (Univ. of Tokyo)
F. Khanna (Univ. of Alberta)
G. Kim (IYF, Tashkent)
T. T. S. Kuo (SUNY, Stony Brook)
Y. Miake (Tsukuba Univ.)
T. Motoba (Osaka E-C Univ.)
Y. Nagai (RCNP, Osaka Univ.)
M. Oka (Tokyo Inst. of Tech.)
M. Rho (Saclay)
K. S. Sim (Korea Univ.)
J. Speth (FZ, Juelich)
H. Toki (RCNP, Osaka Univ.)
S. N. Yang (National Taiwan Univ.)
Z. Y. Zhang (IHEP, Beijing)
Y. Akaishi (KEK)
R. Bertini (INFN, Torino)
H. W. Fearing (TRIUMF, Vancouver)
B. Frois (Saclay)
W. Greiner (Frankfurt Univ.)
W-Y. P. Hwang (National Taiwan Univ.)
B. T. Kim (Sungkyunkwan Univ)
J. C. Kim (Seoul National Univ.)
J. H. Yee (Yonsei Univ.)
D.-P. Min (Seoul National Univ.)
Y. Musakhanov (Uzbekistan National Univ.)
S. Nagamiya (JHF/KEK)
E. Oset (Univ. Valencia)
S. A. Shin (Ewha Women's Univ.)
H. S. Song (Seoul National Univ.)
A. W. Thomas (Univ. of Adelaide)
T. Yamazaki (JSPS, Tokyo)
K. Yazaki (Tokyo Woman's Univ.)

ORGANIZING COMMITTEE

Il-Tong Cheon (Yonsei Univ., *Chairman*)
Myung Ki Cheoun (Seoul National Univ.)
Taekeun Choi (Yonsei Univ., *Scientific Secretary*)
Seung-Woo Hong (Sungkyunkwan Univ., *Scientific Secretary*)
Moon Taeg Jeong (Dongshin Univ.)
Ju Hwan Kang (Yonsei Univ.)
Su Houng Lee (Yonsei Univ.)
Hungchong Kim (Yonsei Univ.)
Yongseok Oh (Yonsei Univ.)
Byung Gil Yu (Hankuk Aviation Univ.)

Organized in celebrating Prof. Il-Tong Cheon's retirement from Yonsei University.

HOST INSTITUTES

Institute of Physics and Applied Physics, Yonsei University
Natural Science Research Institute, Yonsei University
Korean Physical Society
Korea Nuclear Society

SUPPORTING ORGANIZATIONS

Korean Ministry of Education (BK21 program)
Korea Research Foundation
Cheong Moon Gak Publisher

SYMPOSIUM SITE

Yonsei Engineering Research Complex (YERC)
Yonsei University, Seoul, Republic of Korea.

Il-Tong Cheon

Walter Greiner

Traveling to the Far East I had a goal
it was Korea, in particular Seoul,
where professor Il-Tong Cheon
lived, nearly all his life long.

He is a scholar very distinguished,
worked mostly in theoretical physics.
In particular nuclei and the pion
caught the attention of professor Cheon.

But also quantum electrodynamics
with elementary length and without it,
within cavities and within plates,
the Casimir energy and the vacuum's fate

Fascinated this great Korean science teacher,
being broadly interested and acting as a preacher
to generations of students in their desire
to follow modern physics, as much as they could acquire.

He is now sixty five years
and still active as his peers
I wish him all the best,
not only for this Fest,

but for many years to come!
Continue with your work, go on:
It keeps you young and you are winning
together with your daily swimming.

The strength you need
for a long life to exceed
seventy, eighty and even ninety years
All my best wishes, to you, your wife and family, cheers!

Composed and presented by Prof. Walter Greiner during the banquet

International Symposium on Hadrons & Nuclei

INTERACTIONS OF MESONS WITH NUCLEONS AND NUCLEI

Surface Pion Condensation in Finite Nuclei

H. Toki[a,b], K. Ikeda[b] and S. Sugimoto[a,b]

a) Research Center for Nuclear Physics (RCNP), Osaka University, Ibaraki, Osaka 567-0047, Japan
b) RIKEN, 2-1 Hirosawa, Wako, Saitama 351-0105, Japan

Abstract. We review the present relativistic mean field theory from the view point of missing the contributions of pions. We discuss then the role played by pions for hadrons and light nuclei. We introduce the interesting experimental data on pionic states taken at RCNP. These data seem to suggest the occurrence of pion condensation in the nuclear surface. Qualitative discussion is made on the consequence of surface pion condensation on Gamow-Teller transitions and spin response functions and others. We demonstrate that the spin-isospin projection from the surface pion condensed intrinsic state extracts as the dominant term the 2p-2h states connected with 0p-0h state by the tensor force.

RELATIVISTIC MEAN FIELD THEORY

We start with relativistic mean field theory. The relativistic mean field (RMF) theory is quantitatively very successful. This is mainly because RMF includes the strong three body repulsion, which is responsible for saturation, and strong spin-orbit force, which is responsible for magic numbers. However, this theory does not include pions, which should be the most important ingredient in hadron physics.

In fact, the Lagrangian of the relativistic mean field (RMF) theory is

$$\begin{aligned}\mathcal{L} &= \bar{\psi}[i\gamma^\mu\partial_\mu - M - g_\sigma\sigma - g_\omega\gamma^\mu\omega_\mu - g_\rho\gamma_\mu\tau^a\rho^{\mu a} - e\gamma_\mu\frac{(1-\tau_3)}{2}A^\mu]\psi \\ &+\frac{1}{2}\partial_\mu\sigma\partial^\mu\sigma - \frac{1}{2}m_\sigma^2 - \frac{1}{3}g_2\sigma^3 - \frac{1}{4}g_3\sigma^4 \\ &-\frac{1}{4}H_{\mu\nu}H^{\mu\nu} + \frac{1}{2}m_\omega^2\omega_\mu\omega^\mu + \frac{1}{4}c_3(\omega_\mu\omega^\mu)^2 \\ &-\frac{1}{4}G_{\mu\nu}^a G^{a\mu\nu} + \frac{1}{2}m_\rho^2\rho_\mu^a\rho^{a\mu} - \frac{1}{4}F_{\mu\nu}F^{\mu\nu}\,,\end{aligned} \tag{1}$$

where the field tensors H, G and F for the vector fields are defined through

$$\begin{aligned}H_{\mu\nu} &= \partial_\mu\omega_\nu - \partial_\nu\omega_\mu \\ G_{\mu\nu}^a &= \partial_\mu\rho_\nu^a - \partial_\nu\rho_\mu^a - g_\rho\,\varepsilon^{abc}\rho_\mu^b\rho_\nu^c \\ F_{\mu\nu} &= \partial_\mu A_\nu - \partial_\nu A_\mu\,,\end{aligned} \tag{2}$$

and other symbols have their usual meanings. Here, σ denotes the scalar meson, ω the vector meson and ρ the isovector-vector meson. A denotes the photon.

CP594, *Hadrons and Nuclei: First International Symposium,* edited by Il-T. Cheon et al.

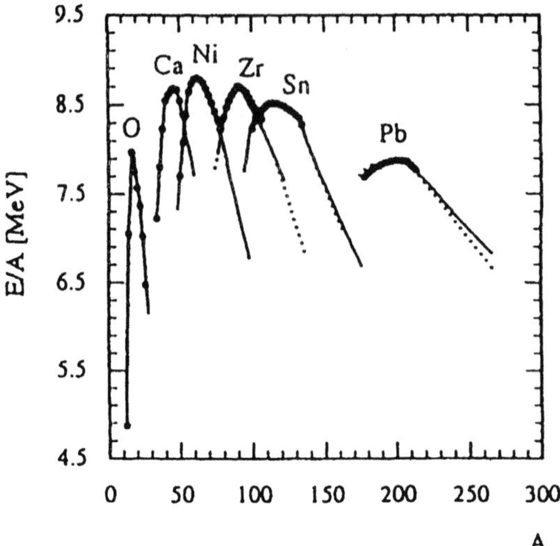

FIGURE 1. The binding energy per particle as a function of the mass number for proton magic number nuclei. The experimental data are shown by dots and the TMA results are shown by solid curves and the NL1 results by dashed curves. Taken from Ref.[1].

This Lagrangian apparently does not include the pion. The pion term should be there, but this term is neglected in the RMF approximation due to the conservation of parity and isospin. There are 6 parameters in the RMF Lagrangian. Under the mean field approximation we can construct coupled differential equations for nucleons and mesons with photon field, which could be easily solved numerically. By adjusting the parameters to the existing data on binding energies and radii of proton magic nuclei, these parameters are fixed by Sugahara and Toki [1]. We find good descriptions of binding energies and radii, out of which the binding energies are shown in Fig.1. We mention here though that we have to take either two parameter sets (TM1 for $A \geq 40$ and TM2 for $A \leq 40$) or TMA with smooth mass dependence in order to describe nuclei in the entire mass regions. The goodness of the parameter sets have been demonstrated by calculating all the even-even mass nuclei in the entire mass region [2].

In addition, we have calculated giant resonances, equation of state of nuclear matter and superheavy elements. We are very much satisfied with the performance of the RMF theory with the TM parameter sets.

If we look, however, the hadron physics, the chiral symmetry and its spontaneous breaking are the essential ingredients for the successful description of the experimental data.

TENSOR FORCE IN LIGHT NUCLEI

The pion exchange interaction is discussed in Nuclear Physics in terms of the tensor force. Let us see this connection first. The pion exchange force in the non-relativistic expression is written as

$$\vec{\sigma}_1 \cdot \vec{q}\vec{\sigma}_2 \cdot \vec{q} = \vec{\sigma}_1 \cdot \vec{\sigma}_2 q^2 + \frac{1}{3} S_{12} \tag{3}$$

where S_{12} is called tensor force. We can then see by direct calculations that the tensor force is much bigger than the central spin-spin force. Therefore, we can say that the pion exchange interaction is the tensor force. Knowing this, let us see what we know about the contribution of the tensor force in light nuclei.

There is a variational calculations on the α particle [3]. According to this calculation about a half of the potential energy is caused by the tensor force. In addition, the wave function of the α particle contains 10% of the 2p-2h components, which consist of 2 particles in p orbit and 2 holes in s orbit even though the energy difference between two orbits is about 20MeV. This large admixture is caused by the tensor force coming from the pion exchange. This essential effect of pions has to be there in the heavy system. Why then the RMF theory without pion works so well. We can say that the use of parameters as g_σ and g_ω mocks up the pion attraction effect. Is there any place for the necessity of pion? The interesting place to look for the answer to this question is the pionic excitation of nuclei.

RCNP (P,N) SPIN EXPERIMENTS

There are two important experiments on pionic excitations performed at RCNP using the (p,n) reactions by Sakai group [4, 5]. They are the zero degree spectra in the (p,n) reactions and the large momentum transfer (p,n) reactions at $E \sim 300MeV$.

We start with the zero degree (p,n) reactions. We take ^{90}Zr as an example [4]. Taking a simple shell model, we expect two states being populated by the Gamow-Teller operator; $g_{7/2}g_{9/2}^{-1}$ particle-hole state and $g_{9/2}g_{9/2}^{-1}$ state. Due to a repulsive interaction between two states, we expect the two states mix and the higher state carries most of the strength, which is called the giant Gamow-Teller state. The experimental data are obtained at forward angles with the incident energy of 300MeV and are analyzed in terms of multipole components. The spectrum is shown in Fig.2 [4]. The large fraction of the GT strength is then found shifted further up to the higher states than the giant GT state by 30 to 40 percent. This shift of the GT strength to higher states were studied theoretically as the coupling of the GT state to 2p-2h states due to the strong tensor force. With the coupling of highly excited states, the perturbative calculations are able to provide the large strengths in the continuum. However, the change of the GT strengths seems too much considering the calculations are done perturbatively.

The GT strength seems to be exhausted by nucleon degrees freedom. This fact suggests that the coupling of the GT states to delta isobars is very small. In terms of g' for delta-hole coupling in the spin-isospin channel, $g'_\Delta \leq 0.2$. This fact indicates that the delta-hole states should contribute largely to pion condensation and its precritical

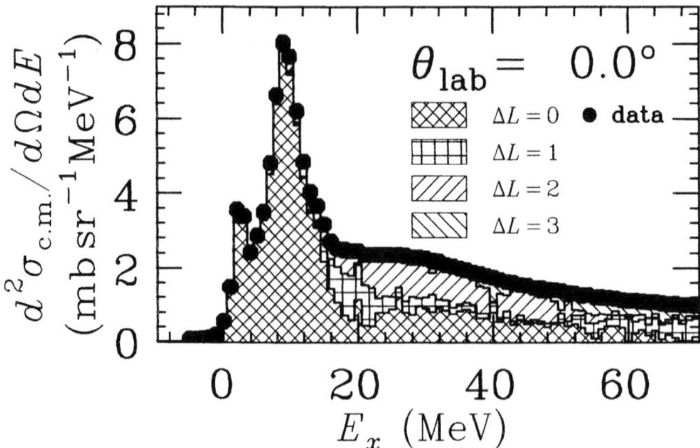

FIGURE 2. The zero degree (p,n) spectrum on ^{90}Zr decomposed into various multipole components. The GT strength is indicated by the hatched area. Taken from Ref.[4].

phenomenon. This possibility is rejected by the experimentally missing precritical phenomenon [6, 7]. The relativity is suggested to reduce the pionic collectivity, but it seems not enough [8]. The precritical phenomenon is expected also in the spin response functions [9]. Hence,we discuss now the second experiment of the Sakai group.

It is the (p,n) reaction with the measurement of polarizations in the initial and final channels at large momentum transfer. In addition, the isovector transfer is identified to 1 by the use of the (p,n) reaction. This spin measurement is able to provide the response functions in the pion and the rho meson channels [5].

In the standard model of spin correlations, we use the interactions in the pion (longitudinal) and rho meson (transverse) channels as

$$V_\pi = \frac{f^2}{m^2}(g' - \frac{\vec{q}^2}{\vec{q}^2 + m_\pi^2})\vec{\sigma}_1 \cdot \vec{q}\vec{\sigma}_2 \cdot \vec{q}\vec{\tau}_1 \cdot \vec{\tau}_2 \quad (4)$$

$$V_\rho = \frac{f^2}{m^2}(g' - C_\rho \frac{\vec{q}^2}{\vec{q}^2 + m_\rho^2})\vec{\sigma}_1 \times \vec{q} \cdot \vec{\sigma}_2 \times \vec{q}\vec{\tau}_1 \cdot \vec{\tau}_2 \quad (5)$$

These interactions provide strong attraction in the pion channel and strong repulsion in the rho meson channel at high momentum region as due to small pion mass, m_π, and large rho meson mass, m_ρ. Naturally then we expect that the pion response is softened from the non-interacting case and the rho meson response is hardened. This tells that the ratios of the spin longitudinal (π) and the spin transverse (ρ) responses are larger than 1 at smaller excitation energy.

The experimental data show surprisingly that the ratios are not larger than one. This is shown in Fig.3 for the case of ^{12}C [5]. This fact is difficult to understand in the present framework. The relativistic description of the response functions is not enough to turn around the ratios unless we use different g' between the pion and the rho meson channels [10]. These two experimental data indicate that there is a serious problem in the present theoretical framework in the pion channel (spin-isospin channel).

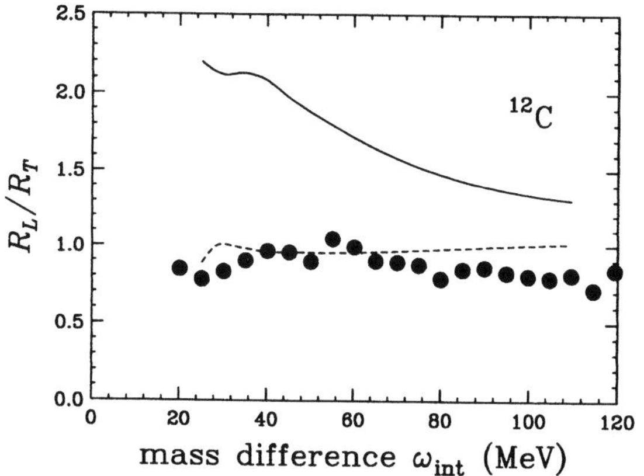

FIGURE 3. The ratio of the longitudinal to transverse spin response functions in (p,n) reaction on ^{12}C as a function of the excitation energy. The experimental data are shown by dots. The DWBA calculation with the pion and rho meson correlations is shown by solid curve, while the one without the correlations is shown by dashed curve. Taken from Ref.[5].

SURFACE PION CONDENSATION

We propose to take the pion terms seriously now, which should be present in the lagrangian. For simplicity of writing, we write explicitly only the sigma and pion terms in the lagrangian density as,

$$\mathcal{L} = \bar{\psi}[i\gamma^\mu\partial_\mu - M - g_\sigma\sigma - g_\pi\gamma_5\gamma^\mu\partial_\mu\tau_a\pi^a]\psi + \mathcal{L}_{meson}. \tag{6}$$

We then assume that the expectation value of the pion field is finite.

We write the equation of motion for nucleons and pions as,

$$[i\gamma^\mu\partial_\mu - M - g_\sigma\sigma - g_\pi\vec{\nabla}\pi^a\gamma_5\vec{\gamma}\tau_a]\psi = 0 \tag{7}$$

and

$$[\vec{\nabla}^2 - m_\pi^2]\pi^a = -g_\pi\vec{\nabla}\langle\bar{\psi}\gamma_5\vec{\gamma}\tau_a\psi\rangle. \tag{8}$$

The sigma meson and others follow the same equations of motion as the standard case. These equations tell the reason why we have not included the pion mean field until now. The source term of the pion Klein-Gordon equation is non-vanishing only when the parity and the isospin are mixed in the single particle state. This violation of the parity and isospin is caused by the pion term in the above Dirac equation for nucleons. Hence, the single particle state can be expressed as

$$\psi_{n,jm}(x) = \sum_{\kappa,t} W^n_{\kappa,t}\phi_{\kappa jm,t} \tag{9}$$

Here, $\phi_{\kappa jm,t}$ denote the eigen functions of nucleons without the pion mean field term. The summation over κ means the parity mixing and the summation over t the isospin mixing.

We would call the case of non-vanishing pion mean field as surface pion condensation, since the pion source term involves derivative of the mean field. The source term is finite only around the nuclear surface.

We do not have yet quantitative numerical results on surface pion condensation. Hence, we discuss here only the qualitative consequence. First, we discuss the Gamow-Teller transitions. Without pion condensation, there exist only two transitions as discussed above for ^{90}Zr. However, the mixing of parity and isospin allow transitions of many neutron states to many proton states. This makes the spectrum of the Gamow-Teller transitions with some GT strengths above the two dominant peaks. Hence, without the strong mixing of two particle two hole states due to large tensor force, we can have large strengths in the continuum in the simple mean field theory as the experiment demands.

The longitudinal spin response functions are largely modified due to pion condensation. The response is caused by the pionic correlations. Since the large pionic strength is used up to construct the nuclear ground state, the pionic fluctuation oughts to be reduced largely. This should make the spin response in the pion channel very weak.

The surface pion condensation provides us the possibility to describe the light mass nuclei using the same parameter set as the one for the heavy mass nuclei. This is because the light mass nuclei have larger proportion of the surface area to the volume as compared to heavy ones and the surface pion condensation is active at the nuclear surface.

There should be many other consequences of surface pion condensation in nuclear phenomena. The pairing correlations and the spin-orbit couplings are all surface phenomena and surface pion condensation would couple with these correlations and provide rich phenomena.

PARITY AND ISOSPIN PROJECTION

We discuss here the role of the parity and isospin projection from the symmetry broken intrinsic state. For simplicity we restrict to the parity projection. We write the single particle state with mixed parity as

$$|\tilde{j}\rangle = \alpha_j |j\rangle + \beta_j |\bar{j}\rangle \tag{10}$$

Here, $|\tilde{j}\rangle$ denotes the mixed single particle state expressed as a linear combination of $|j\rangle$, the normal parity state and $|\bar{j}\rangle$, the non-normal parity state. We can write the intrinsic state with these single particle states as

$$\begin{aligned}
\Psi &= \prod_j (\alpha_j |j\rangle + \beta_j |\bar{j}\rangle) && (11)\\
&= \prod_j \alpha_j |j\rangle + \sum_{j_1} \prod_{j \neq j_1} \alpha_j \beta_{j_1} |j\rangle |\bar{j}_1\rangle + \sum_{j_1,j_2} \prod_{j \neq j_1,j_2} \alpha_j \beta_{j_1} \beta_{j_1} |j\rangle |\bar{j}_1\rangle |\bar{j}_2\rangle + .. && (12)\\
&= A\Psi_+ + B\Psi_- && (13)
\end{aligned}$$

Here Ψ_+ and Ψ_- denote the positive parity and the negative parity states.

Hence the parity projection would provides the state with the definite parity as

$$P_+\Psi = A\Psi_+ = \prod_j \alpha_j|j\rangle + \sum_{j_1,j_2} \prod_{j\neq j_1,j_2} \alpha_j\beta_{j_1}\beta_{j_1}|j\rangle|\tilde{j}_1\rangle|\tilde{j}_2\rangle + .. \quad (14)$$

$$= |0\rangle + |2p-2h\rangle + |4p-4h\rangle + .. \quad (15)$$

This means that the parity projection provides the two particle two hole states as the major correction terms, which are caused by the one pion exchange. Hence, the surface pion condensation together with the parity projection is able to provide the 2p-2h admixture due to the pion exchange interaction as the case of the α particle.

On the other hand, the negative parity projection would provides the negative parity states as

$$P_-\Psi = B\Psi_- = \sum_{j_1} \prod_{j\neq j_1} \alpha_j\beta_{j_1}|j\rangle|\tilde{j}_1\rangle + .. \quad (16)$$

$$= |1p-1h\rangle + |3p-3h\rangle + ..$$

This is the brother state having the quantum number of 0^- to the 0^+ ground state.

CONCLUSION

We have discussed the possible occurrence of surface pion condensation in order to understand the recent (p,n) experimental data taken at RCNP. This suggestion is motivated by the missing pion contribution in the discussion of ground states of finite nuclei, while pions are essential for hadron physics. We have made qualitative discussions on the consequence of surface pion condensation to the Gamow-Teller strengths, the spin response functions and the ground state binding energies.

We would therefore conclude that

1. Relativistic mean field theory is very satisfactory. The RMF theory is however made without the pion contribution.
2. In the hadron physics, however, the pion is essential; the chiral symmetry and its spontaneous breaking.
3. In the few body system, where we can perform a rigorous calculations, we know that the pion plays the major role.
4. There are RCNP spin-isospin experiments; A(p,n) at 0 degree for missing Gamow-Teller strength and at 20 degree for missing pionic enhancement.
5. We assume therefore the surface pion condensation. This condensation makes spin-isospin breaking.
6. This possibility would open up the new development in nuclear physics. The spin observables are particularly interesting.
7. It is further extremely interesting to make unification of hadron and nuclear physics through the role of pion.

This is an invited talk presented by H. Toki in International Symposium on Hadrons and Nuclei held through Feb.20-22, 2001 in Yonsei University. The authors are grateful to Dr. Wakasa for letting them use his figures for presentation and for fruitful discussions. Detailed discussions on the role of tensor force in nucleus with Prof. Akaishi are also greatly acknowledged.

REFERENCES

1. Y. Sugahara and H. Toki, *Phys. Rev.* **D10** (1994) 4262.
2. D. Hirata, K. Sumiyoshi, I. Tanihata, Y. Sugahara, T. Tachibana and H. Toki, *Nucl. Phys.* **A616** (1997) 438c.
3. Y. Akaishi, *Phys. Rev.* **D10** (1994) 4262.
4. T. Wakasa et al., *Phys. Rev.* **C55** (1997) 2909.
5. T. Wakasa et al., *Phys. Rev.* **C59** (1999) 3177.
6. H. Toki and W. Weise, *Phys. Rev. Lett.* **42** (1979) 1034.
7. E. Oset, H. Toki and W. Weise, *Phys. Rep.* **83** (1982) 281.
8. H. Toki and I. Tanihata, *Phys. Rev.* **C59** (1999) 1196.
9. E. Shiino, Y. Saito, M. Ichimura and H. Toki, *Phys. Rev.* **C34** (1986) 1004.
10. K. Yoshida and H. Toki, *Nucl. Phys.* **A648** (1999) 75.

Baryon Resonances in $\pi N \to N\omega(\phi)$ Reactions and Validity of the OZI Rule

A.I. Titov[a], B. Kämpfer[b] and B.L. Reznik[c]

[a] *Bogolyubov Laboratory of Theoretical Physics, JINR, Dubna 141980, Russia*
[b] *Forschungszentrum Rossendorf, PF 510119, 01314 Dresden, Germany*
[c] *Far-Eastern State University, Sukhanova 9, Vladivostok 690090, Russia*

Abstract. Results of a combined analysis are presented for the production of ω and ϕ mesons in πN reactions in the near-threshold region using throughout a conventional "non-strange" dynamics based on such processes which are allowed by the non-ideal $\omega - \phi$ mixing. We find a strong difference between observables in ω and ϕ production reactions caused by the different role of the nucleon and nucleon resonance amplitudes. A series of predictions for the experimental study of this effect is shown.

I. The present interest in a combined study of the ω and ϕ meson production in different elementary reactions is mainly related to the investigation of the hidden strangeness degrees of freedom in the nucleon. Since the ϕ meson is thought to consist mainly of strange quarks, its production should be suppressed according to the OZI rule [1] if the entrance channel does not possess a considerable admixture of strangeness. The standard OZI rule violation is described by the deviation from the ideal $\omega - \phi$ mixing by the angle $\Delta\theta_V \simeq 3.7^0$ [2], which is a measure of the small contribution of light $u, \bar{u}$ and $d, \bar{d}$ quarks in the ϕ meson, or strange $s, \bar{s}$ quarks in the ω meson. Thus, the ratio of ω to ϕ production cross sections is expected to be $R^2_{\omega/\phi} \simeq \mathrm{ctg}^2\Delta\theta_V \simeq 2.4 \times 10^2$. Indeed, the recent experiments on the proton annihilation at rest (cf. [3] for references and a compilation of data) point to a large apparent violation of the OZI rule, which is interpreted as a hint to an intrinsic $s\bar{s}$ component in the proton. A detailed analysis of the current status of the OZI rule in πN and NN reactions has been presented recently in [4], where it is shown that existing data for the ω and ϕ meson production in πN reactions give for the ratio of averaged amplitudes the value of $R_{\omega/\phi} = 8.7 \pm 1.8$, which is much smaller than the standard OZI rule violation value of $R^{\mathrm{OZI}}_{\omega/\phi} = 15.43$ and may be interpreted as a hint to non-zero strangeness components in the nucleon.

Obviously, reliable information on a manifestation of hidden strangeness in the combined study of ϕ and ω production processes can be obtained only when the conventional, i.e. non-exotic, amplitudes have been understood quantitatively. The reaction $\pi N \to VN$ with $V = \omega$, ϕ has the evident advantage to be a simple hadronic reaction representing a subprocess, e.g., in $NN \to VNN$ reactions. The dominant conventional processes are depicted in Fig. 1, where (a) is the t channel meson exchange process, while (b) depicts the s, u nucleon and nucleon resonance channels.

The resonance contribution to vector meson production has its own interest because it might affect significantly in-medium polarization operators and the corresponding

CP594, *Hadrons and Nuclei: First International Symposium,* edited by Il-T. Cheon et al.

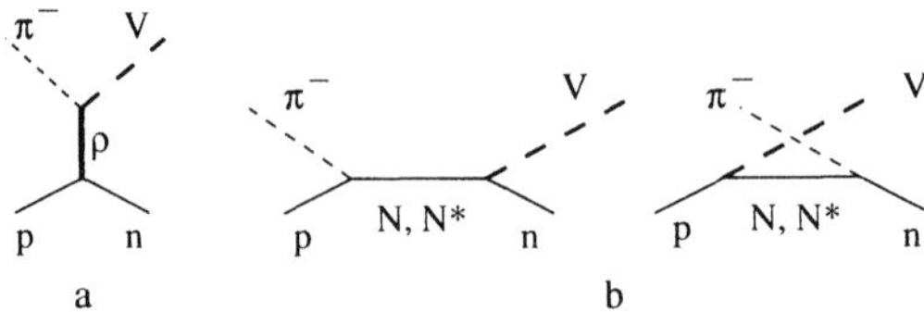

FIGURE 1. Diagrammatic representation of the $\pi^- p \to Vn$ reaction mechanisms with $V = \omega$, ϕ. (a) meson exchange diagram with vector meson emission from the $V\rho\pi$ vertex, (b) nucleon and nucleon-resonance vector-meson production in the VNN and VNN^* vertices.

dilepton emissivity of hadronic matter [6, 7, 8]. In our consideration we use essentially recent results of Riska and Brown [5], where the relevant πNN^*, ωNN^* and ϕNN^* coupling constants are expressed in terms of the corresponding couplings to nucleons using a quark model. Different aspects of the resonance dynamics have been studied in Refs. [9, 10].

Our analysis of the reaction $\pi N \to VN$ is based on calculations of the diagrams in Fig. 1. While the diagrams in Fig. 1 look like usual Feynman diagrams it should be stressed that they give a guidance of how to obtain from an interaction Lagrangian of hadronic fields a covariant parameterization of observables in strict tree level approximation. Additional ingredients are needed to achieve an accurate description of data within such a framework. In particular, the vertices needs to be dressed by form factors. The early theoretical studies [11, 12, 13] show, indeed, that predictions for hadronic observables are sensitive to the parameters of the form factors which can not be fixed unambiguously without adjustments relying on the corresponding experimental data.

II. The differential cross section of the reaction $\pi^- p \to Vn$ with $V = \omega, \phi$ (cf. Fig. 1) has the obvious form in standard notation

$$\frac{d\sigma}{d\Omega} = \frac{1}{64\pi^2 s} \frac{|\mathbf{q}|}{|\mathbf{k}|} |T|^2, \tag{1}$$

where $k = (E_\pi, \mathbf{k})$ and $q = (E_V, \mathbf{q})$ are the four-momenta of the pion and the vector meson in the center of mass system (c.m.s.); the squared invariant amplitude $|T|^2$ includes the average and sum over the initial and final spin states, respectively. We denote the four-momenta of the initial (target) and final (recoil) nucleons by p and p'; Ω and θ are the solid and polar angles of the produced vector meson in the c.m.s.; $s = (p+k)^2$ is the usual Mandelstam variable.

We also consider the spin density matrix $\rho_{rr'}$ which defines the angular distribution in the decays $\omega, \phi \to e^+e^-$, $\omega \to \pi^+\pi^-\pi^0$ and $\phi \to K^+K^-$. It has a simple form in the system where the vector meson is at rest [13]. The decay angles Θ, Φ are defined as polar and azimuthal angles of the direction of the three-momentum of one of the decay particles in the vector meson's rest frame. For the $\omega \to \pi^+\pi^-\pi^0$ decay, Θ is the polar angle of the direction of the vector product $[k_{\pi^+} \times k_{\pi^-}]$, where k_{π^+} and k_{π^-} are the momenta of π^+ and π^- mesons, respectively. The e^+e^- decay distribution integrated over the azimuthal angle Φ, $\mathcal{W}(\cos\Theta)$, depends only on the diagonal matrix elements

ρ_{00}, $\rho_{11} = \rho_{-1-1}$, normalized as $\rho_{00} + 2\rho_{11} = 1$, according to

$$\mathcal{W}^{e^+e^-}(\cos\Theta) \;=\; \frac{3}{4}\left[1 + \rho_{00} + (1 - 3\rho_{00})\cos^2\Theta\right]. \tag{2}$$

The corresponding distributions for the hadronic decays $\phi \to K^+K^-$, $\omega \to \pi^+\pi^-\pi^0$ are

$$\mathcal{W}^h(\cos\Theta) = \frac{3}{2}\left[1 - \rho_{00} - (1 - 3\rho_{00})\cos^2\Theta\right]. \tag{3}$$

In our calculation we choose the quantization axis $\mathbf{z}$ along the beam momentum.

In calculating the invariant amplitudes for the basic processes shown in Fig. 1 we use the following effective interaction Lagrangians.

(i) interactions in the meson exchange process (Fig. 1a):

$$\mathcal{L}_{V\rho\pi} \;=\; g_{V\rho\pi}\,\varepsilon^{\mu\nu\alpha\beta}\,\partial_\mu^{(V)} V_\nu\,\mathrm{Tr}(\partial_\alpha\rho_\beta\pi), \tag{4}$$

$$\mathcal{L}_{\rho NN} \;=\; -\,g_{\rho NN}\bar{\psi}_N\left(\gamma_\mu - \frac{\kappa_{\rho NN}}{2M_N}\sigma_{\mu\nu}\partial^\nu_{(\rho)}\right)\rho^\mu\psi_N, \tag{5}$$

where $\mathrm{Tr}(\rho\pi) = \rho^0\pi^0 + \rho^+\pi^- + \rho^-\pi^+$, and π and ρ^μ denote the pion and rho meson fields.

(ii) interactions in the baryonic channels (Fig. 1b):

$$\mathcal{L}_{MNN}^{N_{\frac{1}{2}^+}(940)\,N} \;=\; \bar{\psi}_N\left[-\frac{f_{\pi NN}}{m_\pi}\gamma_5\gamma_\mu\partial^\mu\pi\cdot\tau - g_{VNN}\left(\gamma_\mu - \frac{\kappa_{VNN}}{2M_N}\sigma_{\mu\nu}\partial^\nu\right)V^\mu\right]\psi_N, \tag{6}$$

$$\mathcal{L}_{MNN^*}^{N_{\frac{1}{2}^+}(1440)\,P_{11}} \;=\; \bar{\psi}_N\left[-\frac{f^{1440}_{\pi NN^*}}{m_\pi}\gamma_5\gamma_\mu\partial^\mu\pi\cdot\tau - g^{1440}_{VNN^*}\,(\gamma_\mu + \partial_\mu\,\partial\!\!\!/\,m_V^{-2})V^\mu\right]\psi_{N^*} + \text{h.c.}, \tag{7}$$

$$\mathcal{L}_{MNN^*}^{N_{\frac{3}{2}^-}(1520)\,D_{13}} \;=\; \bar{\psi}_N\left[i\frac{f^{1520}_{\pi NN^*}}{m_\pi}\gamma_5\,\partial^\alpha\pi\cdot\tau + \frac{g^{1520}_{VNN^*}}{m_V^2}\sigma_{\mu\nu}\partial^\nu\partial^\alpha V^\mu\right]\psi_{N^*\alpha} + \text{h.c.}, \tag{8}$$

$$\mathcal{L}_{MNN^*}^{N_{\frac{1}{2}^-}(1535)\,S_{11}} \;=\; \bar{\psi}_N\left[-\frac{f^{1535}_{\pi NN^*}}{m_\pi}\gamma_\mu\partial^\mu\pi\cdot\tau - g^{1535}_{VNN^*}\gamma_5\,(\gamma_\mu + \partial_\mu\,\partial\!\!\!/\,m_V^{-2})V^\mu\right]\psi_{N^*} + \text{h.c.}, \tag{9}$$

$$\mathcal{L}_{MNN^*}^{N_{\frac{1}{2}^-}(1650)\,S_{11}} \;=\; \bar{\psi}_N\left[-\frac{f^{1650}_{\pi NN^*}}{m_\pi}\gamma_\mu\partial^\mu\pi\cdot\tau - g^{1650}_{VNN^*}\gamma_5\,(\gamma_\mu + \partial_\mu\,\partial\!\!\!/\,m_V^{-2})V^\mu\right]\psi_{N^*} + \text{h.c.}, \tag{10}$$

$$\mathcal{L}_{MNN^*}^{N_{\frac{5}{2}^-}(1675)\,D_{15}} \;=\; \bar{\psi}_N\left[-\frac{f^{1675}_{\pi NN^*}}{m_\pi^2}\partial^\alpha\partial^\beta\pi\cdot\tau\right.$$

$$+ \frac{g^{1675}_{VNN^*}}{m_V^2} \varepsilon^{\alpha\gamma\mu\nu} \gamma_\nu \partial_\gamma \partial^\beta V_\mu \Big] \psi_{N^*\alpha\beta} + \text{h.c.}, \tag{11}$$

$$\mathcal{L}^{N_{\frac{5}{2}^+}(1680)F_{15}}_{MNN^*} = \bar{\psi}_N \Big[-i \frac{f^{1680}_{\pi NN^*}}{m_\pi^2} \gamma_5 \partial^\alpha \partial^\beta \pi \cdot \tau$$

$$+ \frac{g^{1680}_{VNN^*}}{m_V^2} (\gamma_\mu + \partial_\mu \partial\!\!\!/ m_V^{-2}) \partial^\alpha \partial^\beta V^\mu \Big] \psi_{N^*\alpha\beta} + \text{h.c.}, \tag{12}$$

$$\mathcal{L}^{N_{\frac{3}{2}^-}(1700)D_{13}}_{MNN^*} = \bar{\psi}_N \Big[i \frac{f^{1700}_{\pi NN^*}}{m_\pi} \gamma_5 \partial^\alpha \pi \cdot \tau + \frac{g^{1700}_{VNN^*}}{m_V^2} \sigma_{\mu\nu} \partial^\nu \partial^\alpha V^\mu \Big] \psi_{N^*\alpha} + \text{h.c.}, \tag{13}$$

$$\mathcal{L}^{N_{\frac{3}{2}^+}(1720)P_{13}}_{MNN^*} = \bar{\psi}_N \Big[i \frac{f^{1720}_{\pi NN^*}}{m_\pi} \partial^\alpha \pi \cdot \tau$$

$$- \frac{g^{1720}_{VNN^*}}{M_{N^*} + M_N} \gamma_5 \left(\gamma_\mu \partial^\alpha - g^\alpha_\mu \partial\!\!\!/ \right) V^\mu \Big] \psi_{N^*\alpha} + \text{h.c.}, \tag{14}$$

where π, V_μ, ψ_N and ψ_{N^*} are the pion iso-vector, iso-scalar vector meson $V = \omega, \phi$, nucleon and Rarita-Schwinger nucleon resonances field operators, respectively, and the subscript M stands for "meson", τ denotes the Pauli matrix. The notation of the masses is self-explaining. We use the convention of Bjorken and Drell [14] in definitions of γ matrices and the spin matrix $\sigma_{\mu\nu}$. The expressions Eqs. (6 - 14) are based on [5]. As in [5] we include here all three- and four-star resonances up to 1720 MeV (excluding $N_{\frac{1}{2}^+}(1710)P_{13}$, which contribution is very small). All coupling constants with off-shell mesons are dressed by monopole form factors [15] $F_i = (\Lambda_i^2 - m_i^2)/(\Lambda_i^2 - k_i^2)$, where k_i is the four-momentum of the exchanged meson. For the VNN and VNN^* vertices the form factors read

$$F_B(r^2) = \frac{\Lambda_B^4}{\Lambda_B^4 + (r^2 - M_B^2)^2}, \tag{15}$$

where M_B is the baryon mass and r is the four-momentum of the virtual baryons $B = N, N^*$ in Fig. 1b. The invariant amplitude are completely defined by the effective Lagrangians; explicit expressions are is given in Ref. [17]. The final result depends on the parameters of coupling constants and cut-offs.

The coupling constant $g_{\phi\rho\pi}$ is determined by the $\phi \to \rho\pi$ decay. The value $\Gamma_{\phi\to\rho\pi} = 0.69$ MeV [2] results in $|g_{\phi\rho\pi}| = 1.1$ GeV^{-1}. The coupling constant $g_{\omega\rho\pi}$ is determined by the $\omega \to \gamma\pi$ decay. Relying on the vector dominance model one gets $|g_{\omega\rho\pi}| = |12.9 \cdot g_{\phi\rho\pi}| = 14.19$ GeV^{-1} [4]. The sign of $g_{\omega\rho\pi}$ is generally unknown, however, the study of π photoproduction in Ref. [19] shows that it should be positive (cf. Ref. [20]). Therefore, following VDM take $g_{\omega\rho\pi} = 14.19$ GeV^{-1}. The SU(3) symmetry considerations predict a opposite sign for $g_{\phi\rho\pi}$. Thus $g_{\phi\rho\pi} = -1.1$ GeV^{-1}. We will also discuss the case of opposite signs, $g_{\omega\rho\pi} < 0$, $g_{\omega\rho\pi} > 0$.

The remaining parameters of the meson exchange amplitude for the process in Fig. 1a are taken from the Bonn model as listed in Table B.1 (Model II) of Ref. [15]. The nucleon and nucleon resonance amplitudes in Fig. 1b are determined by the couplings $f_{\pi NN(N^*)}$, $g_{VNN(N^*)}$, κ_{VNN}, the resonance widths $\Gamma^0_{N^*}$ the branching ratios $B^\pi_{N^*}$, and the

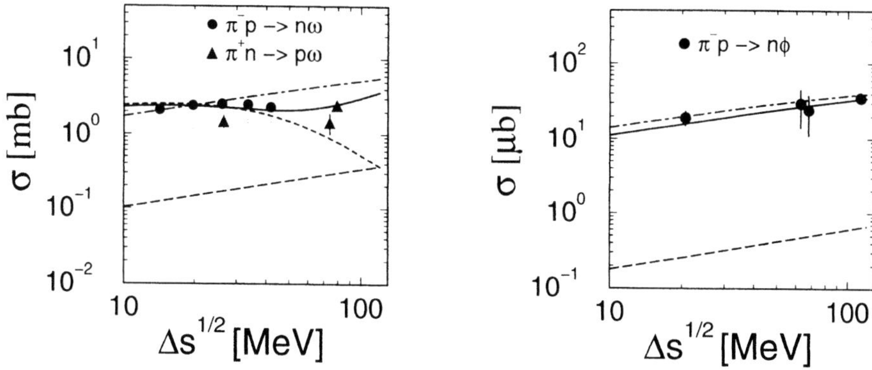

FIGURE 2. Total cross sections for the reactions $\pi N \to N\omega$ (left panel) and $\pi^- p \to n\phi$ (right panel) as a function of the energy excess $\Delta s^{1/2}$. The meaning of the curves is: meson exchange - dot-dashed, direct and crossed nucleon terms - long-dashed, N^* resonances - dashed, full amplitude - solid. Data from [23, 24].

cut-offs Λ_B. For the coupling constant $f_{\pi NN}$ we use the standard value $f_{\pi NN} = 1.0$ [5, 15]. For the ωNN coupling we use the value $g_{\omega NN} = 10.35$ determined recently [21] by fitting the nucleon-nucleon scattering data. This value as well as $\kappa_{\omega NN} = 0$ is close to the one which has been found in a study of πN scattering and the reaction $\gamma N \to \pi N$ [20]. The values of coupling constants $f_{\pi NN^*}$ are determined from a comparison of calculated decay widths $\Gamma_{N^* \to N\pi}$ with the corresponding experimental values [2]. The corresponding signs are taken in accordance with the quark model prediction of Ref. [5]. The values of coupling constants $g_{\omega NN^*}$ are found as $g_{\omega NN^*} = [g_{\omega NN^*}/g_{\omega NN}]g_{\omega NN}$, where the ratio $[g_{\omega NN^*}/g_{\omega NN}]$ is determined by the quark model calculation of Ref. [5]. The couplings ϕNN, ϕNN^* determined by SU(3) symmetry considerations are $g_{\phi NN} = -\mathrm{tg}\Delta\theta_V\, g_{\omega NN}$, $g_{\phi NN^*} = -\mathrm{tg}\Delta\theta_V\, g_{\omega NN^*}$. Similarly, we assume $g_{\phi NN}(\kappa_{\phi NN}) \simeq -\mathrm{tg}\Delta\theta_V\, g_{\omega NN}(\kappa_{\omega NN}) = 0$, or $\kappa_{\phi NN} \simeq 0$, which is consistent with the estimate in [22]. The yet undetermined parameters are: the cut-off parameters $\Lambda^{\rho}_{\phi\rho\pi}$ and $\Lambda^{\rho}_{\omega\rho\pi}$ for the virtual ρ meson in the $V\rho\pi$ vertex, the cut-off Λ_N and the eight cut-offs Λ_{N^*} in Eq. (15). We can reduce the number of parameters by making the natural assumptions

$$\Lambda^{\rho}_{\phi\rho\pi} = \Lambda^{\rho}_{\omega\rho\pi} \equiv \Lambda^{\rho}_{V}, \qquad \Lambda_N = \Lambda_{N^*} \equiv \Lambda_B. \tag{16}$$

Our best fit of total cross sections of existing data is obtained by $\Lambda^{\rho}_{V} = 1.24$ GeV and $\Lambda_B = 0.66$ GeV.

III. The results of our full calculation of the total cross sections as a function of the energy excess $\Delta s^{1/2} = \sqrt{s} - M_N - m_V$, including all amplitudes depicted in Fig. 1, are represented by the solid curves in Fig. 2. We also show separately the contributions of meson exchange, nucleon and nucleon resonance channels. The data for the reaction $\pi^- p \to \phi n$ are taken from Ref. [23], while the data for the reactions $\pi^+ n \to \omega p$ and

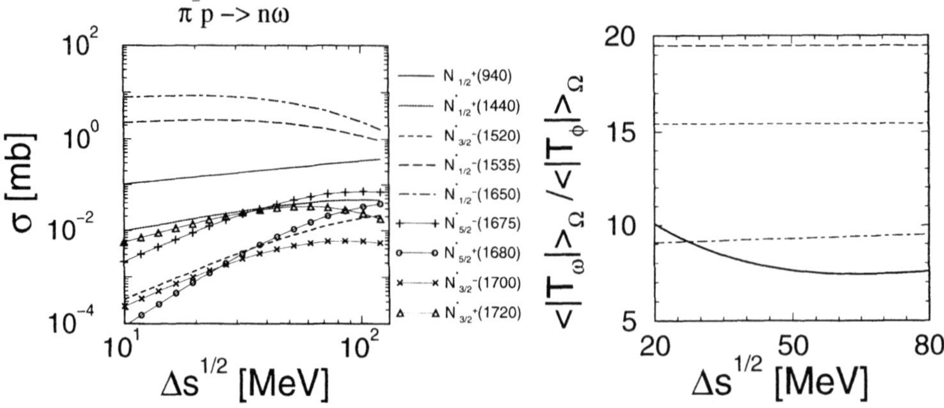

FIGURE 3. Left panel: Individual contributions of nucleon resonances to the total cross section of ω production as a function of $\Delta s^{1/2}$. Right panel: Ratio of averaged amplitudes of ω and φ production as a function of $\Delta s^{1/2}$.

$\pi^- p \to \omega n$ are from Refs. [23, 24], respectively. The data for total cross section of the reaction $\pi^- p \to \omega n$, $\sigma_{\rm tot}$ shown in Fig. 2, differ from the differential cross section $\sigma_{\rm dif}$ in Ref. [24] by a factor [4, 25] included in the phase space of the unstable ω meson. From Fig. 2 (left panel) it is evident that the total amplitude of ω production is a result of strong interferences of all channels: the meson exchange (dot-dashed curve), the nucleon term (long dashed curve) and the resonance contribution (dashed curve) play a comparative role. For the φ production (cf. Fig. 2, right panel) only meson and nucleon terms are important. The resonance contribution is rather small and is therefore not displayed here. Moreover the relative contribution of the nucleon term in φ production is much smaller than for ω production. That is because the initial energy $\sqrt{s} = M_N + m_V + \Delta s^{1/2}$ is greater for the φ production at the same energy excess and, as a consequence, we have two suppression factors: the nucleon/resonance denominators and the form factors F_{N,N^*}. The interference of the meson exchange and nucleon terms is almost destructive, while the contribution of the resonant part is more complicated because the amplitude is complex with different phases for different resonances.

In order to illustrate the structure of the resonant part we show in Fig. 3 (left panel) the contribution of each resonance separately as a function of $\Delta s^{1/2}$. One can see that just near the threshold the resonances with $J = \frac{1}{2}$ are important. Together with the nucleon term they are $N_{\frac{1}{2}^+}(1440)P_{11}$, $N_{\frac{1}{2}^-}(1535)S_{11}$ and $N_{\frac{1}{2}^-}(1650)S_{11}$. It is interesting that the separate contributions of the two latter ones are greater than the nucleon term. But their phases are opposite and, therefore, they cancel each other. The cancellation increases with energy which results in a total decrease of the resonance contribution. On the other hand one can see that the relative role of the higher spin resonances with orbital/radial excitations, being proportional to $\mathbf{q}^2$ and $\mathbf{q}^4$, increases with increasing values of $\Delta s^{1/2}$. This enhancement, however, is smaller than the effect of the strong destructive interference of the resonance amplitudes and, therefore, the total contribution

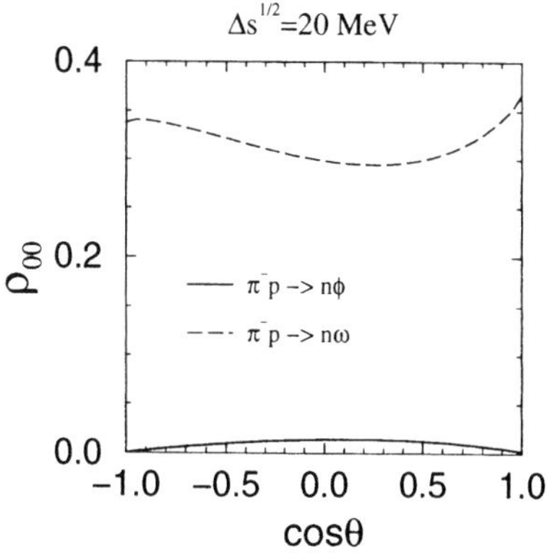

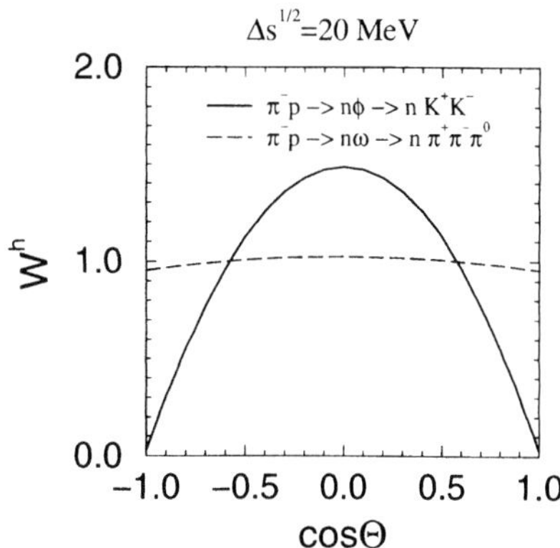

FIGURE 4. Left panel: Spin-density matrix element ρ_{00} for ω and ϕ production as a function of $\cos\theta$. Right panel: Meson angular distributions in the reactions $\pi^- p \to n\phi \to nK^+K^-$ and $\pi^- p \to n\omega \to n\pi^+\pi^-\pi^0$ and electron angular distributions in the reaction $\pi^- p \to nV \to ne^+e^-$ at $\Delta s^{1/2} = 20$ MeV.

of the resonance channel decreases with energy as shown in Fig. 2.

Fig. 3 (right panel) exhibits the ratio of the angular integrated amplitudes

$$< |T_V| >_\Omega = \left[\sum_{m_i, m_f, \lambda} \frac{1}{4\pi} \int d\Omega \, |T^V_{m_f,\lambda;m_i}|^2 \right]^{\frac{1}{2}} \tag{17}$$

of ω and ϕ production as a function of $\Delta s^{1/2}$. The short dashed straight lines in Fig. 3 (right panel) correspond to the standard OZI rule violation value $R^{\rm OZI}_{\omega/\phi} = {\rm ctg}\Delta\theta_V = 15.43$. The long dashed curve corresponds to the ratio of pure nucleon channels taken separately, the dot-dashed curve is result for a pure meson exchange, while the solid line represents the full calculation. Note that the ratio even for pure meson exchange amplitudes $R^M_{\omega/\phi}$ is smaller than $R^{\rm OZI}_{\omega/\phi}$. That is because at the threshold

$$R^M_{\omega/\phi} \simeq \frac{g_{\omega\rho\pi}}{g_{\phi\rho\pi}} \frac{m_\omega}{m_\phi} \frac{f(m_\omega)}{f(m_\phi)} \simeq 9.9 \frac{f(m_\omega)}{f(m_\phi)}, \tag{18}$$

where $f(m)$ is a smooth function of m. The ratio for pure resonance terms is greater than $R^{\rm OZI}_{\omega/\phi}$ by an order of magnitude and more, i.e., $R^{N^*}_{\omega/\phi} \sim 500$ (250) at $\theta = \pi$ (0), because of a strong propagator and form factor suppression for ϕ production. In the absence of the resonant amplitude the destructive interference of meson exchange (M) and nucleon (N) channels results in $R^{M+N}_{\omega/\phi} < R^M_{\omega/\phi}$. The presence of the resonance components leads to $R^M_{\omega/\phi} < R_{\omega/\phi} < R^{\rm OZI}_{\omega/\phi}$. One can see that $R_{\omega/\phi}$ (solid curve) may be slightly above or below the pure $R^M_{\omega/\phi}$ value (dot-dashed curve), however, remaining much smaller than $R^{\rm OZI}_{\omega/\phi}$, namely $R_{\omega/\phi} = 7.5 \cdots 10$, in agreement with the analysis in [4].

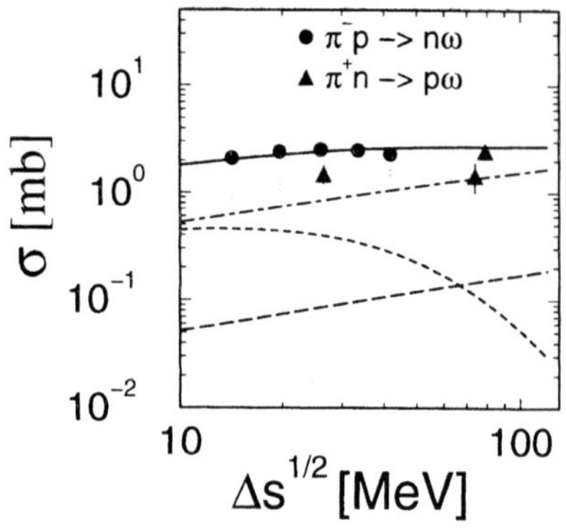

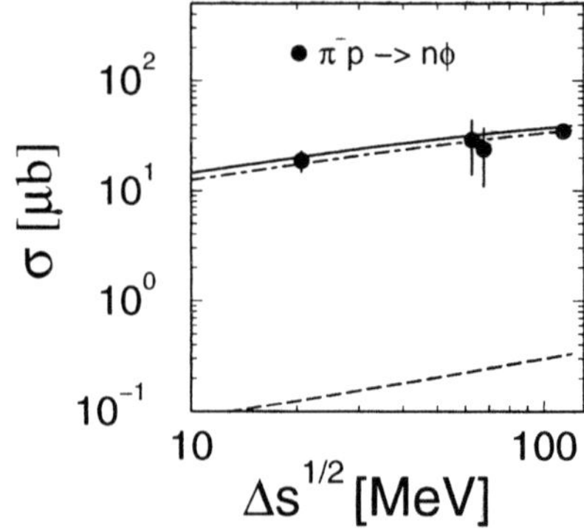

FIGURE 5. Total cross sections for the reactions $\pi N \to N\omega$ (left panel) and $\pi^- p \to n\phi$ (right panel) as a function of the energy excess $\Delta s^{1/2}$ for a constructive interference between meson-exchange and baryonic channels with the parameter set described in the text. Notation as in Fig. 2.

Fig. 4 (left panel) shows the results of our full calculation of the spin density matrix element ρ_{00} at $\Delta s^{1/2} = 20$. Near the threshold, meson exchange amplitude behaves as

$$T_\lambda^{(M)} \sim \mathbf{k} \cdot [\mathbf{k} \times \varepsilon^{*\lambda}]. \tag{19}$$

That means that only polarizations $\lambda = \pm 1$ contribute and, therefore, ρ_{00} is suppressed. The pure nucleon s channel amplitude behaves as

$$T_\lambda^{(N)} \sim \langle f | \sigma \cdot \varepsilon^{*\lambda} | i \rangle, \tag{20}$$

which results in an isotropic spin density, i.e., $\rho_{00} = \rho_{11} = \rho_{-1-1} = 1/3$. The resonance amplitudes have additional terms proportional to $\mathbf{k} \cdot \varepsilon^{*\lambda}$, which also enhance ρ_{00}. This effect is seen clearly in the left panel of Fig. 4 for which our qualitative analysis is valid. For ϕ production, where the main contribution comes from the meson exchange channel (cf. Fig. 2), ρ_{00} is relatively small, $\rho_{00} < 0.05$. But for ω production, where the contribution of the resonance channel is essential, we find $\rho_{00} \sim 0.3$ which is close to an isotropic spin-density distribution with $\rho_{00} \simeq \rho_{11} = \rho_{-1-1} \simeq 1/3$. Fig. 4 (right panel) shows the angular distribution of hadronic decays $\phi \to K^+K^-$ and $\omega \to \pi^+\pi^-\pi^0$ in $\pi^- p \to nV$ reactions at $\Delta s^{1/2} = 20$ MeV. A similar difference is predicted for the angular distribution of electrons in the reaction $\pi^- p \to nV \to ne^+e^-$.

All above calculations have been performed for positive (negative) values of $g_{\omega\pi\rho}$ ($g_{\phi\pi\rho}$), which leads to a destructive interference between meson exchange and baryonic channels. The opposite signs of $g_{V\pi\rho}$ lead to a constructive interference and change our predictions. First of all it is impossible to describe the data with the same set of parameters. In order to do that we have to change (decrease) not only cut-offs Λ_V^ρ, Λ_B, but also decrease $|g_{\omega\pi\rho}|$ and resonance couplings $g_{\omega NN^*}$. The result of the corresponding calculation is shown in Fig. 5, where we use the lowest value for $|g_{\omega\pi\rho}|$ discussed in Ref. [4], $|g_{\omega\pi\rho}| = 8.3$ GeV^{-1}, $\Lambda_V^\rho = 1.2$ GeV, $\Lambda_B = 0.6$ GeV, and the couplings

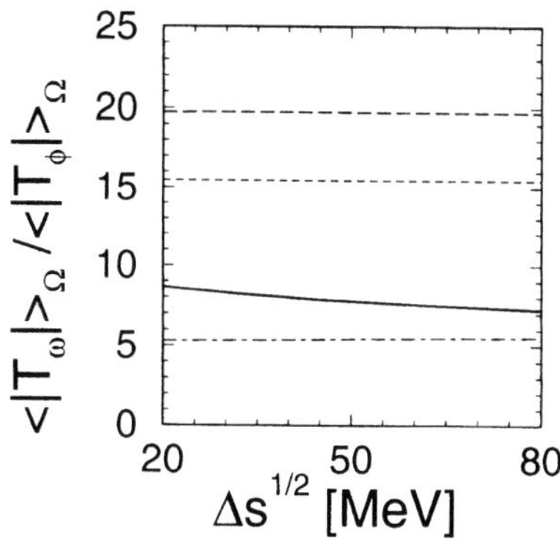

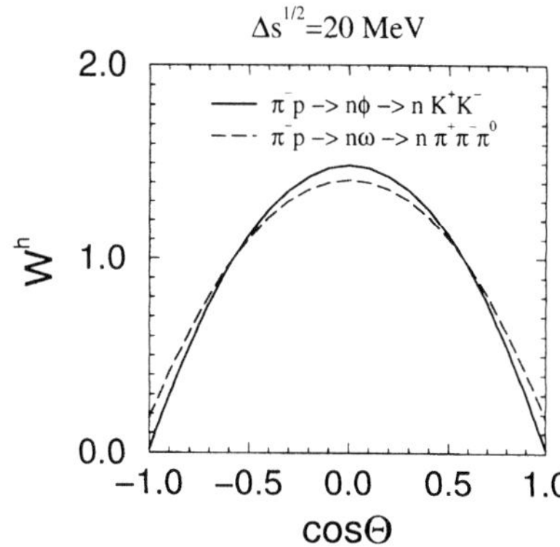

FIGURE 6. Ratio of the amplitudes of ω and ϕ production. averaged over production angle as in Fig.3(right panel) and meson angular distributions in the reactions $\pi^- p \to n\phi \to nK^+K^-$ and $\pi^- p \to n\omega \to n\pi^+\pi^-\pi^0$ as in Fig. 10 but at constructive interference between meson-exchange and baryonic channels.

$g_{\omega NN^*}$ are scaled by a factor 0.44. Since this new parameter set describes the total cross sections we get similar predictions for the ratio of the averaged amplitudes of ω and ϕ productions as shown in Fig. 6, left panel (c.f. Fig. 3, right panel). So, one can not distinguish between the two models only from unpolarized data. But since in this case the meson exchange channel is dominant for both reactions, the prediction for the distribution of ω decay will be different from the previous one. In Fig. 6 (right panel) we show the angular distribution of hadronic decays $\phi \to K^+K^-$ and $\omega \to \pi^+\pi^-\pi^0$ in $\pi^- p \to nV$ reactions at $\Delta s^{1/2} = 20$ MeV. Instead of the isotropic distribution for ω decay shown in Fig.6, now we get an anisotropic distribution. Therefore, this reaction together with a polarization measurement of ω photoproduction can shed light on the phase of $g_{V\rho\pi}$ and the baryon resonance dynamics.

IV. In summary we have performed a combined analysis of ω and ϕ production in πN reactions near the threshold at the same energy excess. We find that the meson-exchange amplitude alone can not describe the existing data, rather the role of the direct nucleon term and the nucleon resonance amplitudes is essential. The latter statement is very important for the ω production, and therefore we investigate the role of nucleon resonances in detail. It is found that the resonance contributions can influence significantly the total and the differential cross sections at small energy excess as well as the ratio of the averaged amplitudes of ω and ϕ production. We have shown that our predictions can essentially be tested by measuring the angular distribution of decay particles in the reactions $\pi N \to N\phi \to NK^+K^-$, $\pi N \to N\omega \to N3\pi$ and $\pi N \to NV \to Ne^+e^-$. Near the thresholds, for the ϕ production we predict an anisotropic distribution, while for the ω production either an almost isotropic distributions or isotropic is obtained, depending on the signs of the coupling strenghts. Experimentally, this prediction can be tested with the pion beam at the HADES spectrometer at GSI/Darmstadt [26].

We would like to thank H.W. Barz, R. Dressler, S.B. Gerasimov, L.P. Kaptari, and J. Ritman for many discussions. This work is supported by BMBF grant 06DR921 and Heisenberg-Landau program. AIT greatfully acknowledges the Organizing Committee of Hadrons and Nuclei 2001 Conference for support and hospitality.

REFERENCES

1. S. Okubo, Phys. Lett. B 5 (1963) 165;
   G. Zweig, CERN report No. 8419/TH 412 (1964);
   I. Iizuka, Prog. Theor. Phys. Suppl. 37/38 (1966) 21.
2. C. Caso et al. (Particle Data Group), Eur. Phys. J. C 3 (1998) 1.
3. J. Ellis, M. Karliner, D.E. Kharzeev, M.G. Sapozhnikov, Nucl. Phys. A 673 (2000) 256.
4. A. Sibirtsev, W. Cassing, Eur. Phys. J. A 7 (2000) 407.
5. D.O. Riska, G.E. Brown, Nucl. Phys. A 679 (2001) 577.
6. B. Friman, H.J. Pirner, Nucl. Phys. A 617 (1997) 496.
7. M. Lutz, G. Wolf, B. Friman, Nucl. Phys. A 661 (1999) 526.
8. M. Post, U. Mosel, nucl-th/0008040.
9. M. Soyeur, M. Lutz, B. Friman, nucl-th/0003013.
10. Y. Oh, A.I. Titov, T.-S.H. Lee, Phys. Rev. C 63 025201 (2001) 02520..
11. K. Nakayama, A. Szczurec, C. Hanhart, J. Haidenbauer, J. Speth, Phys. Rev. C 57 (1998) 1580; K. Nakayama, J.W. Durso, J. Haidenbauer, C.Hanhart, J. Speth, Phys. Rev. C 60 (1999) 055209.
12. A.I. Titov, B. Kämpfer, V.V. Shkyar, Phys. Rev. C 59 (1999) 999.
13. A.I. Titov, B. Kämpfer, B.L. Reznik, Eur. Phys. J. A 7 (2000) 543.
14. J.D. Bjorken, S.D. Drell, *Relativistic quantum mechanics*, McGraw-Hill Inc., 1964.
15. R. Machleidt, Adv. Nucl. Phys. 19 (1989) 189.
16. H. Haberzettl, Phys. Rev. C 56 (1997) 2041;
    H. Haberzettl, C. Bennhold, T. Mart, T. Feuster, Phys. Rev. C 58 (1998) 40.
17. A.I. Titov, B. Kämpfer, B.L. Reznik, nucl-th/0102032.
18. A.I. Titov, T.-S.H. Lee, H. Toki, O. Streltsova, Phys. Rev. C 60 (1999) 035205 .
19. M.T. Jeong and Il-T. Cheon, Nucl. Phys. A 684 (2001) 496;
    B.G. Yu, Il-Y. Cheon and M.T. Jeong, Journ. Phys. Soc. Japan, 63 (1994) 78.
20. T. Sato, T.-S.H. Lee, Phys. Rev. C 54 (1996) 2660.
21. Th.A. Rijken, V.G.J. Stoks, Y. Yamamoto, Phys. Rev. C 59 (1999) 21.
22. U.-G. Meissner, V. Mull, J. Speth, J.W. Van Orden, Phys. Lett. B 408 (1997) 381.
23. Landolt-Börnstein, New Series **I/12**, A. Baldini et al. *Total cross sections of high energy particles*, Springer-Verlag, 1988.
24. H. Karami et al. Nucl. Phys. B 154 (1979) 503.
25. C. Hanhart, A. Kudryavtsev, Eur. Phys. J. A 6 (1999) 325.
26. J. Friese et al. (HADES collaboration), GSI report 97-1, p. 193 (1997).

Antinucleon-nucleus interaction at low energy

F. Iazzi

Dipartimento di Fisica del Politecnico and INFN, Torino, Italy

Abstract. After a brief introduction about the antinucleon-nucleus interaction, the results of the OBELIX experiment concerning the annihilation cross section will be discussed in the frame of the data presently available in literature. Then the preliminary data concerning the final state of the annihilation, in particular the spectra, the multiplicities and the signal of the ρ–meson are presented.

INTRODUCTION

The interaction of the antinucleon ($\bar{N}$) with the nucleus offers the possibility to study several interesting fields. Among the other the $\bar{N} - nucleus$ elastic scattering and polarization measurements can give information about the $\bar{N}$-nucleus mean field, the spin-isospin dependence of $\bar{N} - N$ two body effective interaction and the longitudinal spin response in $(\bar{p}, \bar{n})$ CEX reactions. Also the $\bar{N} - nucleus$ annihilation gives the opportunity to look at a great variety of phenomena: from the simple dependence of the annihilation cross section on the target mass number and on the projectile momentum to the interaction of the annihilation products (pions, kaons) with the residual nucleus, with the related phenomena like the pre–equilibrium, thermalization, coalescence. Moreover, the nature of the annihilation itself can be studied, with the possible effects of the nuclear matter: for instance, it can be investigated whether: a) an eventual QGP formation does enhance the strangeness production or the production of other particles, b) the long-lived mesons (η, η', ρ, ω ...) do change their properties inside the nuclear matter c) do form quasi–stable $nucleon - meson$ or $nucleon - nucleon - meson$ states, d) the nuclear matter is color transparent [1].

The annihilation cross section and the annihilation products have been studied as soon as the $\bar{p}$ machines (AGS, KEK, LEAR) were operating and a brief summary of the results is reported in Chapters 2 and 5 respectively.

$\bar{N}$-NUCLEUS ANNIHILATION CROSS SECTION

The $\bar{p}$ nucleus annihilation cross section was studied first at momenta above 1 GeV/c and a scaling law of the kind:

$$\sigma_{abs}(p,A) \sim \sigma_0(p_{\bar{p}})A^{\nu} \quad (1)$$

with ν close to 2/3 was found. Such a value was consistent with a black sphere picture or, in terms of optical model, the nucleus should be black to the $\bar{p}$'s. The whole

CP594, *Hadrons and Nuclei: First International Symposium,* edited by Il-T. Cheon et al.

momentum dependence should be contained in $\sigma_0(p_{\bar{p}})$.

A number of annihilation cross section measurements were performed in the $80's$ and $90's$ in order to check also at lower momenta this behavior and the results are summarized in the following.

The absorption cross section was measured at KEK and at AGS on C, Al, Cu, Pb in the range 470–880 MeV/c [2, 3, 4] and the observed behavior as a function of A is quite consistent with the scaling law (1). Also consistent with (1) is the measurement at LEAR [5] of the annihilation cross section at 200 and 300 MeV/c on Ne, while recent measurements at very low momentum ($\leq 70\ MeV/c$, [6, 7]) on D, ^{3}He and Ne showed a quite different trend. It must be noticed that in these last cases the nuclei are light and the statistics very poor.

Obviously the Coulomb interaction of the charged projectile plays an important role in the distortion of the $\bar{p}$-nucleus annihilation measurements, due to the multiple scattering in the targets and to the Coulomb scattering on the nucleus, particularly at the lowest momenta. A way for overcoming this difficulty is to use the antineutrons ($\bar{n}$) instead of the $\bar{p}$'s as projectiles but the difficulty of producing and managing $\bar{n}$'s is much higher than for $\bar{p}$'s. In fact the $\bar{n}-nucleus$ data are very scarce: apart the few points of the pionieering work of Gunderson et al. [8], only 3 sets of data were produced at LEAR on 6 nuclei, Fe, C, Al, Cu, Sn and Pb from 150 up to 780 MeV/c [9, 10, 11]. The results confirmed the power law $A^{2/3}$ and showed a dependence on the $\bar{n}$ momentum of the kind:

$$\sigma_{ann}(p,A) \sim a + b/p_{\bar{n}} \qquad (2)$$

similar to the elementary $\bar{p}-p$ one. Nevertheless the momentum range was still quite high and the statistics of the data still poor.

ANNIHILATION CROSS SECTION: OBELIX RESULTS

Some of the $\bar{N}-nucleus$ data previously mentioned were produced by the OBELIX experiment with both $\bar{p}$'s and $\bar{n}$'s at the LEAR machine during the first data taking of the experiment, in parasiting mode. The good quality of the results with the $\bar{n}$'s suggested to use the $\bar{n}$ beam facility, built in the apparatus, in order to increase the statistics of the "distortion free" $\bar{n}-nucleus$ annihilation reaction and this was done in 1994 and 1995, always in parasiting mode, with more nuclei and in a larger momentum range.

A complete description of the OBELIX apparatus, mainly devoted to the meson spectroscopy measurements, is very long and can be found elsewhere [12]; therefore only the main performances are recalled hereafter:

- a facility for producing $\bar{n}$'s (via CEX from the $\bar{p}$'s) to be used as projectiles in the range 50 - 400 MeV/c is incorporated in the apparatus, with the momentum of each $\bar{n}$ measured,
- the annihilation vertex of the $\bar{n}$ in the target is fully reconstructed,
- a disk shaped solid target (NT) located downstream the central LH_2 target (RT) could take advantage of the $\bar{n}$'s transmitted through RT

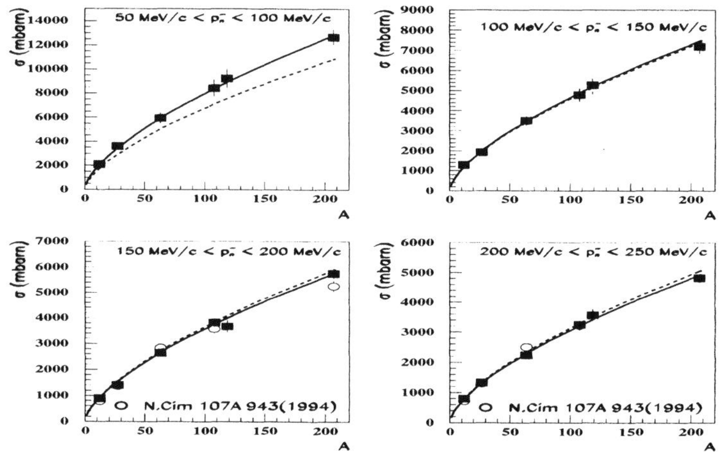

FIGURE 1. $\bar{n}-nucleus$ annihilation cross sections vs A for different $p_{\bar{n}}$;

- the momentum of each charged annihilation product is measured in an angular acceptance of $\sim 3\pi$ sr.

The 6 nuclear targets put in the NT position were C, Al, Cu, Ag, Sn and Pb: from 13000 to 40000 annihilation events were collected per nucleus at momenta from 50 up to 400 MeV/c. The annihilation cross sections in each momentum interval were evaluated from the following formula:

$$\sigma_{ann}(p_{\bar{n}},A) = \frac{A}{\rho_A\ \mathcal{N}_{Av}\ l_A\ \varepsilon_A}\ \frac{N_{ann}(p_{\bar{n}},A)}{N_{\bar{n}}^{inc}(p_{\bar{n}},A)} \tag{3}$$

where A is the target nucleus mass number, ρ_A the density, $\mathcal{N}_{Av}$ the Avogadro number, l_A the thickness, ε_A the overall efficiency for the specific target, $N_{ann}(p_{\bar{n}},A)$ the number of annihilation events in each momentum bin and $N_{\bar{n}}^{inc}(p_{\bar{n}},A)$ the corresponding number of incoming $\bar{n}$'s deduced from the annihilations in LH_2. The absolute cross section was evaluated only for Pb: due to the equal efficiency ε_A of every target the relative cross section was calculated from $N_{ann}(p_{\bar{n}},A)$ and $N_{\bar{n}}^{inc}(p_{\bar{n}},A)$ for the other nuclei.

For each nucleus, the annihilation events have been divided in 7 momentum intervals of 50 MeV/c: for the momenta below 250 MeV/c the annihilation cross section is reported in Fig.1 as a function of A, with the statistical error bars only. A systematic error due to the counting of the annihilations in LH_2 and a normalization error due to the CEX reaction in the OBELIX production system have been estimated around $\sim 3\%$ and $\sim 8\%$ respectively.

All the data have been fitted simultaneously in $p_{\bar{n}}$ and A with the function:

$$\sigma_{ann}(p_{\bar{n}},A)\ \sim\ (a+b/p_{\bar{n}}+c/p_{\bar{n}}^2)\cdot A^{\nu} \tag{4}$$

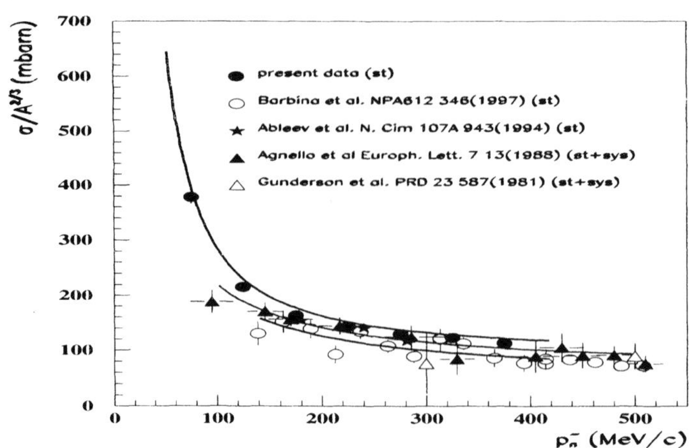

FIGURE 2. $\bar{n}-nucleus$ annihilation cross section, divided by A, vs $p_{\bar{n}}$;

obtaining the values for the free parameters: $a \sim 87.4 \pm 4.2$ [mb], $b \sim (9.9 \pm 1.7) \cdot 10^3$)[mb/(MeV/c)], $c \sim (9.8 \pm 1.5) \cdot 10^5$ [mb/(MeV/c)]^2, $\nu \sim 0.65$ and $\chi^2 \sim 0.93$. The very good agreement between the points and the fitting curve (continuous line) in Fig. 1 remains unchanged also at the higher momenta. A function like in eq. (2) (dashed line in Fig. 1) succeeds to fit the data above 100 MeV/c but the term with the power $p_{\bar{n}}^{-2}$ is necessary below.

In order to check this behavior in more detail, the data have been also divided in smaller bins of 10 MeV/c and the value of ν remained the same with, of course, an increased statistical error: the scaling law has been confirmed down to 50 MeV/c for the light nuclei down to C.

Therefore the above mentioned anomalous behavior observed with $\bar{p}$'s in Ne by Bianconi et al. [7] below 70 MeV/c is not consistent with the present results: a possible explanation of the discrepancy should be probably searched for in the Coulomb distortion due to the $\bar{p}$ projectile and in the low statistics.

A comparison with the $\bar{n}$-nucleus data existing in literature is shown in Fig. 2, where the annihilation cross section divided by $A^{2/3}$ is reported as a function of $p_{\bar{n}}$. In the overlapping region above 100 MeV/c the trend is the same for all the sets of measurements but a little shift, within the errors, appears in the overall normalization. Below 100 MeV/c the present data are unique and show a quite steep rise which requires the power $p_{\bar{n}}^{-2}$ term in eq. (4) for the best fit.

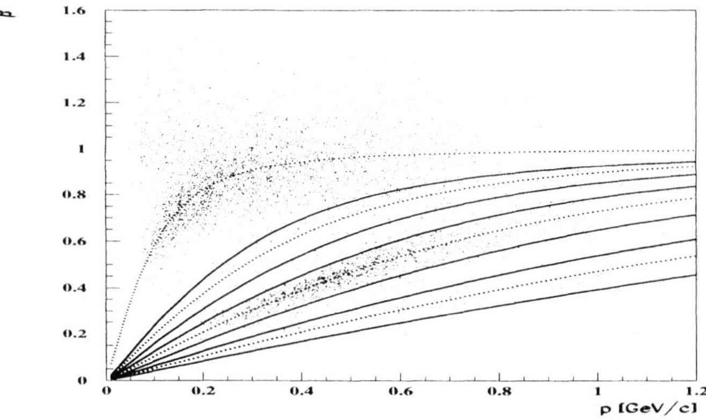

FIGURE 3. β vs momentum of the charged annihilation products

FINAL STATE OF $\bar{N}$-NUCLEUS ANNIHILATION

The annihilation products in nuclei can be distinguished in primary products, which are the mesons produced in the elementary $\bar{N} - N$ annihilation and the secondary products, which are the emitted protons (p), deuterons (d) and the other nuclear fragments coming from the interaction of the primary products with the residual nucleus. Several works, almost all with $\bar{p}$'s unless one, have been produced since the 80's and a selection of them can be found in Refs.[13, 14, 15, 16, 17, 18, 19, 20, 21, 22]. The main results, in terms of inclusive spectra and multiplicity, can be summarized as follows: 1) the π multiplicity is nearly indipendent on the nucleus (suggesting a very low π absorption), 2) no K excess has been observed up to now, 3) the p, d spectra reveal a component of the $\pi - nucleus$ interaction comparable with the other processes (pre-equilibrium, equilibrium, coalescence) following the annihilation.

ANNIHILATION PRODUCTS: OBELIX RESULTS

In the OBELIX experiment the charged products of the annihilation can be identified by the relation between the velocity β (detected from the time of flight) and the momentum. As an example in Fig. 3 the plot of the velocity vs the momentum is shown and 4 different regions corresponding to the masses of the π's, K's, p's and d's (from upside to downside respectively) are clearly distinguishable.

>From these data the particle multiplicity and the momentum distribution are easily extracted. Further improvements in the identification are under study as well as the corrections to be taken into account for the particle absorption and scattering inside the nuclei.

	C	Al	Cu	Ag	Pb
m_ρ (MeV)	742 $\pm$ 102	707 $\pm$ 153	644 $\pm$ 144	701 $\pm$ 137	726 $\pm$ 133
Γ_ρ (MeV)	241 $\pm$ 68	247 $\pm$ 102	194 $\pm$ 96	254 $\pm$ 169	285 $\pm$ 189

Even though affected by these distortions, estimated less than 10%, some preliminary results can be pointed out.

The ratio of the K's and π's to the number of annihilations gives the K and π "overall" multiplicity m_k and m_π: a ratio $m_k/m_\pi \sim 3\%$ has been found, indipendent on A (within the errors), which indicates no enhancement of the strangeness production with respect to the free annihilation. On the contrary the p and d "overall" multiplicity show an increasing trend with saturation for the baryon production with A. The ratio $m_d/m_p \sim 3-4\%$ is a bit lower than the average value in literature.

Also the multiplicity per annihilation, n_k, n_π, n_p, n_d is measured by the OBELIX apparatus identifying and counting the prongs of each event: this multiplicity has been found to be constant with A for π, k, and d and slightly increasing for p. By comparing the behavior of m_p with n_p and m_d with n_d one can deduce that the probability of p and d emission after the annihilation increases with A but in different ways: d's are always single emitted, p's are sometimes emitted more than 1 per annihilation.

Finally, the observed p momentum distribution in the present data shows a decrease of the average kinetic energy (related to the slope parameter of the distribution) with increasing A, probably due to the increasing probability of the scattering inside the nucleus.

RESONANCES IN THE $\bar{N}$-NUCLEUS ANNIHILATION

The possibility of a variation in the vector meson mass inside the hot and dense nuclear matter is subject of a recent debate and the still scarce data seem indicate a mass shift toward the lower energy, the so called "dropping mass" effect [23].

Among the OBELIX performances there is the possibility of searching for mesonic resonances produced in the annihilation in nuclei looking at the invariant mass of the $\pi\pi$ pairs. A first trial has been recently done for ρ_0 decaying into $\pi^+\pi^-$ dealing with the problems of the overwhelming combinatorial background and of the efficiency evaluation in the following simplified way.

The background subtraction has been performed by using the $\pi^+\pi^+$ and $\pi^-\pi^-$ invariant mass spectra, properly normalized, between $0.2-1$ GeV/c, where no $I=2$ resonance exists. The selection of the $\pi^+\pi^-$ pairs has been done keeping only those events containing an exclusive $\pi^+\pi^-$ pair: in this way the cut was severe but uniform for all nuclei. After normalization and subtraction the values reported in the above Table have been obtained for the ρ_0 mass m_ρ and width Γ_ρ for various nuclei. It appears that the masses are systematically below the "free" ρ_0 mass (770 MeV/c) but also that the errors are bigger or of the same order of magnitude of the shift. No conclusion can be drawn about the dropping mass effect but great improvements in the analysis are still possible.

CONCLUSIONS

The $\bar{N} - nucleus$ annihilation cross section has been measured by OBELIX for the first time from 50 to 400 MeV/c, for C, Al, Cu, Ag, Sn, Pb. The simultaneous fit as a function of A and $p_{\bar{n}}$ confirms the behavior $A^{2/3}$, while the momentum dependence shows a quite steep rise at low values, never seen up to now.

Concerning the particle production, the multiplicities of the primary products do not show any dependence on A nor any strangeness enhancement. The multiplicities of the secondary products increase with A; for protons also the emission frequency per annihilation increases. >From the proton spectra a small decrease of the average kinetic energy with A has been observed.

Finally, the ρ production in nuclear matter is clearly visible in the OBELIX experiment but the weak indication of a "dropping mass" effect has to be confirmed (or ruled out) by further improved analyses.

REFERENCES

1. Dover, C. B., "Antinucleon-nucleus physics: theoretical overview" in *Proceedings of the First Workshop on Intense Hadron Facilities and Antiproton Physics*, SIF Conf. Proc. vol.26, Edited by T. Bressani, F. Iazzi and G. Pauli., Editrice Compositori, Bologna,1989,pp.171-189
2. Aihara, H., et al., *Nucl. Phys.*, **A360**, 291-296 (1981)
3. Nakamura, K., et al., *Phys.. Rev. Lett.*, **52**, 731-734 (1984)
4. Ashford, V., et al., *Phys. Rev.*, **C31**, 663-665 (1985)
5. Balestra, F., et al., *Nucl. Phys.*, **A452**, 573-590 (1986)
6. Zenoni, A., et al., *Phys. Lett.*, **B461**, 413-417 (1999)
7. Bianconi, A., et al., *Phys. Lett.*, **B481**, 194-198 (2000)
8. Gunderson, B., et al., *Phys. Rev.*, **D23**, 587-594 (1981)
9. Agnello, M., et al., *Europh. Lett.*, **7**, 13-17 (1988)
10. Ableev, V., et al., *N. Cim.*, **107A**, 943-953 (1994)
11. Barbina, C., et al., *Nucl. Phys.* **A612**, 346-352 (1997)
12. Agnello, M., et al., *N.I.M.* **A399**, 11-26 (1997)
13. Mc Gaughey, P. L., *et al.*, *Phys.. Rev. Lett.* **56**, 2156-2159 (1986)
14. Angelopoulos, A. et al., *Phys. Lett.* **B205**, 590-596 (1988)
15. Armstrong, T., et al., *Z. Phys.* **A332**, 467-475 (1989)
16. Riedlberger, J., et al., *Phys. Rev.* **C40**, 2717-2731 (1989)
17. Balestra, F., et al., *Nucl. Phys.* **A491**, 541-571 (1989)
18. Hofmann, P., et al., *Nucl. Phys.* **A512**, 669-673 (1990)
19. Minor, E. D., et al., *Z. Phys.* **A342**, 447-454 (1992)
20. Sudov, A. S., et al., *Nucl. Phys.* **A554**, 223-245 (1993)
21. Polster, D., et al., *Phys. Rev.* **C51**, 1167-1172 (1995)
22. Agnello, M., et al., *Nucl. Phys.* **A516**, 662-672 (1990)
23. Lenkeit, B., et al., *Nucl. Phys.* **A654**, 236-244 (1999)

Mesonic Nuclear Collective States

Haruki Kurasawa* and Toshio Suzuki†

*Department of Physics, Faculty of Science, Chiba University, Chiba 263-8522, Japan
†Department of Applied Physics, Fukui University, Fukui 910-8507, Japan
RIKEN, 2-1 Hirosawa, Wako-shi, Saitama 351-0198, Japan

Abstract. The new giant dipole state excited by exchange current is investigated. While the well-known giant dipole resonance state is described as an oscillation of protons against the neutrons, the new one is composed of two giant resonance states as $[0^+(T=1)\times 1^-(T=1)]1^-(T=1,T_z=0)$ or $[1^-(T=1)\times 2^+(T=1)]1^-(T=1,T_z=0)$. These double giant resonance states may be observed experimentally in $(\gamma,pn),(e,e'pn)$, or (π,π).

The dipole state(GDR), which exhausts the TRK sum-rule value, is described as an oscillation of protons against the neutrons. Among isospin-dependent two-body nuclear interactions $V(r_{ij})\vec{\tau}_i\cdot\vec{\tau}_j$, the main part which is responsible for the oscillation is $V(r_{ij})\tau_{0i}\tau_{0j}$, where τ_0 denotes the third component of the isospin operator. This part does not change the value of the TRK sum rule[1].

The dipole states correspond to the polarization of the charge distribution. There is another way for the charge distribution to be polarized. The proton becomes a neutron, by transferring its charge to another neutron, which becomes a proton. In this process, the proton does not carry the current, so that we may call this new dipole state the mesonic dipole state (MDS). The nuclear interaction which is responsible for this state is the exchange force $V(r_{ij})(\tau_{+i}\tau_{-j}+\tau_{-i}\tau_{+j})$. Since this part does not commute with the dipole operator, the value of the TRK sum rule is increased.

The excitation of the MDS is considered in two ways. The one way is to use the exchange current itself. The main component of the MDS should be 2 particle-2hole(2p-2h) state, since two nucleons contribute to it, satisfying the Pauli principle. The 2p-2h states are excited through the two-body exchange current given by

$$\begin{aligned}\vec{\nabla}\cdot\vec{j}_{\rm ex} &= -i\,[V,\rho_{\rm v}]\\ &= \frac{i}{4}\sum_{i,j}V(r_{i,j})\{\delta(\vec{r}-\vec{r}_i)-\delta(\vec{r}-\vec{r}_j)\}(\tau_{-i}\tau_{+j}-\tau_{+j}\tau_{-j}),\end{aligned}\qquad (1)$$

where the nuclear force V and the isovector density $\rho_{\rm v}$ are written as

$$V=\frac{1}{2}\sum_{i,j}^{A}V(r_{i,j})\vec{\tau}_i\cdot\vec{\tau}_j,\qquad \rho_{\rm v}=-\frac{e}{2}\sum_{i=1}^{A}\delta(\vec{r}-\vec{r}_i)\tau_{0i}.\qquad (2)$$

Then the total photo-absorption cross section is given by

$$\sigma_\gamma=(2\pi)^3\frac{e^2}{k}\frac{2}{3}|\langle f|\int d\vec{r}rY_{10}(\hat{r})(\vec{\nabla}\cdot\vec{j}_{\rm N}+\vec{\nabla}\cdot\vec{j}_{\rm ex})|i\rangle|^2,\qquad (3)$$

CP594, *Hadrons and Nuclei: First International Symposium,* edited by Il-T. Cheon et al.

where k denotes the photon energy and $\vec{j}_{\rm N}$ the isovector one-body current.

The another way to estimate the cross section is by rewriting Eq.(3) by using the continuity equation,

$$\vec{\nabla}\cdot\vec{j}_{\rm N}+\vec{\nabla}\cdot\vec{j}_{\rm ex}=-i[H,\rho_{\rm v}]. \tag{4}$$

We have

$$\sigma_\gamma=(2\pi)^3\frac{e^2}{k}\frac{2}{3}|\langle f|\int d\vec{r}rY_{10}(\hat{\vec{r}})[H,\rho_{\rm v}]|i\rangle|^2. \tag{5}$$

What we need to calculate in the second way is the one-body matrix element, as well known,

$$\langle f|[H,D]|i\rangle=(E_f-E_i)\langle f|D|i\rangle. \tag{6}$$

It should be noticed that in the l.h.s. in the above equation we have the two-body operator, $[V,D]$, while in the r.h.s. the one-body operator only. This implies that the MDS should couple with 1p-1h states through the exchange forces. We note that MDS can not be described within the framework of RPA, since it yields the continuity equation as[2]

$$(E_f-E_0)\langle f|\rho_{\rm v}|0\rangle=\sum_{m,i}\left(\langle mi^{-1}|[H,\rho_{\rm v}]|\ \rangle Y^*_{mi}+\langle\ |[H,\rho_{\rm v}]|mi^{-1}\rangle Z^*_{mi}\right), \tag{7}$$

with

$$\langle mi^{-1}|[H,\rho_{\rm v}]|\ \rangle=\sum_{j<F}\left(\langle mj|[H,\rho_{\rm v}]|ij\rangle-\langle mj|[H,\rho_{\rm v}]|ji\rangle\right). \tag{8}$$

Thus the two-body current excites only 1p-1h states in RPA.

We estimate the photo-absorption cross section for MDS. In order to discuss the "collective" MDS, we assume that the isospin-dependent force is given by Bohr-Mottelson model,

$$V=\frac{1}{2}\chi\sum_{\mu}\sum_{i,j}r_iY^*_{1\mu}(\hat{\vec{r}}_i)\vec{\tau}_i\cdot r_jY_{1\mu}(\hat{\vec{r}}_j)\vec{\tau}_j, \tag{9}$$

where the strength, χ, is determined so as to reproduce the excitation energy of GDR, $\omega_1=82/A^{1/3}$ MeV,

$$\chi=4\pi m\omega^2/A,\quad \omega=41/A^{1/3}\ \text{MeV}. \tag{10}$$

By calculating $[V,D]$, we find that the exchange current excites the MDS's which are composed of the two giant resonance states,

$$[0^+(T=1)\times1^-(T=1)]1^-(T=1,T_z=0)\quad(\text{E0}\times\text{E1}),$$
$$[1^-(T=1)\times2^+(T=1)]1^-(T=1,T_z=0)\quad(\text{E1}\times\text{E2}).$$

These are the collective 2p - 2h ($T=1, T_z=0$) states with the lowest spin excited by the exchange current.

In this paper we discuss MDS composed of E1×E2. As mentioned before, the MDS should couple with the 1p-1h states owing to the continuity equation. In the first order perturbation, we finally obtain a set of the nuclear states which are relevant to the present discussions,

$$|\text{MDS}\rangle = |[1^- \times 2^+]1^-\rangle - \alpha|1^-\rangle, \tag{11}$$
$$|\text{DGDR},0^+\rangle = |[1^- \times 1^-]0^+\rangle - \beta|\ \rangle, \tag{12}$$
$$|\text{DGDR},2^+\rangle = |[1^- \times 1^-]2^+\rangle, \tag{13}$$
$$|\text{GDR}\rangle = |1^-\rangle + \alpha|[1^- \times 2^+]1^-\rangle, \tag{14}$$
$$|0\rangle = |\ \rangle + \beta|[1^- \times 1^-]0^+\rangle, \tag{15}$$

where α and β denote the mixing amplitudes. We can show that the above equations satisfy the continuity equations,

$$(E_{\text{MDS}} - E_0)\langle \text{MDS}|D|0\rangle = \langle \text{MDS}|[H,D]|0\rangle, \tag{16}$$
$$(E_{\text{GDR}} - E_0)\langle \text{GDR}|D|0\rangle = \langle \text{GDR}|[H,D]|0\rangle, \tag{17}$$

and the sum of the dipole strength for GDR is given by

$$\begin{aligned} S_1^{(1)} &= S_1^{(1)}(0) + \frac{1}{2}\langle\ |[D,[V,D]]|\ \rangle \\ &= S_1^{(1)}(0)\ (1+\kappa), \end{aligned} \tag{18}$$

where $S_1^{(1)}(0)$ represents the value of the TRK sum rule. Thus, the sum value for GDR is enhanced by κ in first order of V. This enhancement is already taken into account in RPA for GDR. The excitation strength of MDS is given by second order of V. We note that the Skyrme forces contribute to κ, but can not provide the exchange current which excites MDS.

When we use the observed excitation energies of GDR and isovector GQR as

$$\omega_1 = 2\omega, \quad \omega_{\text{GQR}} = 3\omega, \tag{19}$$

the photo-absorption cross section for MDS is estimated to be

$$\begin{aligned} \sigma_\gamma(\text{MDS}) &= \frac{5.96}{A^{2/3}}\sigma_\gamma^0 \\ &= 0.510\,\sigma_\gamma^0 \qquad (^{40}\text{Ca}), \\ &= 0.170\,\sigma_\gamma^0 \qquad (^{208}\text{Pb}), \end{aligned} \tag{20}$$

where σ_γ^0 denotes the photo-absorption cross section exhausting the TRK sum-rule value. Thus, MDS seems to have enough strength to be observed. The excitation energy and width(variance) of MDS may be given by those of GDR and the isovector GQR,

$$\omega_{\text{MDS}} = \omega_1 + \omega_{\text{GQR}}, \quad \sqrt{\sigma_{\text{MDS}}} = \sqrt{\sigma_1 + \sigma_{\text{GQR}}}. \tag{21}$$

Recently the double giant resonance states corresponding to Eqs. (12) and (13) have been observed through Coulomb excitation in relativistic heavy ion reactions[3]. This fact shows that the giant resonance states are well-defined elementary modes of nuclei. It is interesting to investigate in more detail the new double giant resonance states discussed here. These are expected to be excited through $(\gamma, pn), (e, e'pn)$, or (π, π) reactions.

REFERENCES

1. T. Suzuki, Ann. de Phys. **9**, 535 (1984).
2. H. Kurasawa and Suzuki, Prog. Theor. Phys. **99**, 145 (1998).
3. T. Aumann, P. F. Bortignon and H. Emling, Annu. Rev. Nucl. Part. Sci. **48**, 351 (1998) and references therein.

Deeply Bound Pionic Atoms

S. Hirenzaki

Department of Physics, Nara Women's University, Nara 630-8506, Japan

Abstract. We study the structure and formation of the deeply bound pionic states theoretically. These states have attracted much attention since they provide us with valuable information on the behavior of "real" pions in the interior of the nucleus. In this paper, we briefly summarize the history of the research and, as one of the current activities, we show the isotope dependence of deeply bound pionic states in Sn and Pb isotopes. It is shown the isotope effects can be observed from the data of (d,^{3}He) reactions with accessible energy resolution. The isotope dependence are expected to provide us new information on the pion-nucleus isovector interaction and/or the neutron density distributions of nuclei.

INTRODUCTION

Pionic atoms have been studied for a long time to investigate the behavior of pions in nuclei. Recently, deeply bound pionic atoms have attracted much attention, since they provide us with valuable information on the behavior of "real" pions in the interior of the nucleus. These states, however, cannot be observed in conventional pionic X-ray experiments due to the strong absorption of pions. In the conventional method, pionic atoms have been formed by the capture and subsequent electromagnetic deexcitation of stopped π^-. X-rays emitted in the electromagnetic cascade give experimental access to determine the pionic binding energies [1]. In heavy atoms, the electromagnetic cascade stops before reaching the lowest states due to the absorptive part of the strong interaction. States inside the "last" orbital, where the X-ray cascade terminates, is referred to as "deeply bound pionic states" in this paper. For deeply bound pionic states, assuming a pure Coulomb potential and neglecting the strong interaction, the calculated probability for the pion being inside the nucleus is comparable or even higher than for being outside. Thus we expect to have large absorption effect due to strong interaction.

Despite of this, the repulsive pion-nucleus optical potential pushes the pion wave function outwards [2, 3] and deeply bound pionic states are predicted to have rather narrow widths. The pion mainly resides in close vicinity of the nucleus in the potential pocket, which is formed by the repulsive strong interaction together with the attractive Coulomb interaction. In the corresponding halo-like states of the π^-, the pion absorption by the nucleus is comparatively low due to the strongly reduced overlap of the pion wave function and the imaginary part of the strong interaction. Resulting level widths are large compared to the electromagnetic widths but small compared to the level distances of the pionic states.

Because of this, there is a possibility of observing the deeply bound pionic states with appropriate experimental method. Obviously new methods were required. There have been much theoretical and experimental effort to find out proper reactions for the

CP594, *Hadrons and Nuclei: First International Symposium,* edited by Il-T. Cheon et al.

formation of deeply bound pionic states. Toki and Yamazaki pointed out the possibility of observing deeply bound pionic atoms using pion-transfer reactions such as (n,p) and (d,^{2}He) [2, 3]. Following their suggestions, the (n,p) reaction at T_n = 420 MeV at TRIUMF [4] and the (d,^{2}He) reaction at T_d = 1000 MeV at SATURNE [5] were carried out but no positive evidence was observed. After these experiments, it was pointed out that the charge-exchange pion-transfer reactions at large momentum transfer are quite sensitive to the initial- and final-state interactions, and the distortion effects reduce the cross section by about two orders of magnitudes [4, 6] compared with the predictions of plane-wave approximation [2, 3]. Thus the pion-transfer reactions of this type turned out to be not suited for the formation of deeply bound pionic atoms [4, 6].

Other kinds of reactions, such as the single nucleon pick-up pion-transfer reactions (n,d) [7] and (d,^{3}He) [8] were studied theoretically for the formation of deeply bound pionic states. Since the angular momentum matching condition is satisfied in these reactions, we can expect to have smaller distortion effects than the (n,p) and the (d,^{2}He) cases. According to these theoretical expectations, there were experimental attempts using (n,d) [9] and (p,pp) [10] reactions. In both cases, some extra strength was observed in the pion subthreshold region which suggests the existence of deeply bound pionic states. However, due to weak neutron intensity or difficulty in observing the two unbound protons, clear evidence of deeply bound pionic atoms was not found from these experiments.

In 1996, deeply bound pionic atoms were observed experimentally in the (d,^{3}He) reactions with ^{208}Pb [11, 12]. The observed spectrum agrees with the theoretical predictions made before the experiment [8]. In this experiment, the peak structure in the observed spectrum was attributed to the pionic $2p$ state contribution. The pionic $1s$ contribution can be found only as the skewed shape of the largest peak since the strength due to the $[(2p)_\pi \otimes (3p_{1/2})_n^{-1}]$ configuration, which exists between the contributions from $[(2p)_\pi \otimes (3p_{3/2})_n^{-1}]$ and $[(1s)_\pi \otimes (3p_{3/2}, 2f_{5/2})_n^{-1}]$ configurations, prevents us from observing the $(1s)_\pi$ contribution as a separate peak.

In order to observe the pionic $1s$ state more clearly, the (d,^{3}He) reaction with a ^{206}Pb target was performed very recently [13]. The theoretical prediction was also made before the experiment [14] . In this reaction, due to the absence of the contributions from $3p_{1/2}$ neutron component, the pionic $1s$ contribution is identified as a peak structure as predicted in the theoretical investigation [14]. This is clear evidence for the existence of quasi-stable pionic $1s$ state. The binding energy and width of the $1s$ state will be determined from the data [13]. These successful observations convince the reliability of the theoretical model and the experimental feasibility of the spectroscopic studies with the (d,^{3}He) reaction.

We would like to mention here the theoretical activities by J. Nieves and E. Oset in Valencia University, who have studied theoretically the pionic atom formation reactions extensively. Their works are well summarized in ref. [15].

These research activities on the deeply bound pionic states stimulated the studies of other meson bound states in nucleus [16, 17, 18, 19]. As in the case of the pionic atoms, the eigen energies of the bound states provide us important information on the meson properties at a finite nuclear density[20], which is connected to the prediction of the QCD-inspired models [21] and is one of the most interesting subjects in contemporary nuclear physics [22].

Concerning the spectra of ^{206}Pb(d,^{3}He) reaction mentioned above, it was pointed out that the observed $(1s)_\pi$ contribution consists of two subcomponents, $[(1s)_\pi \otimes (3p_{3/2})_n^{-1}]$ and $[(1s)_\pi \otimes (2f_{5/2})_n^{-1}]$ which cannot be well separated due to the natural width of the $1s$ state. Thus, the deduced $1s$ binding energy and width are somewhat ambiguous due to the uncertainties of the relative strengths of these contributions. It is, therefore, desirable to find suitable candidates to observe pionic $1s$ state more clearly and to extract its binding energy and width precisely.

In order to observe the peak structure consisting of a single subcomponent $[(1s)_\pi \otimes j_n^{-1}]$ in the formation cross section, it is advantageous to study target nuclei with the valence $s_{1/2}$ neutron orbit. As such a candidate, we select ^{136}Xe, ^{116}Sn and ^{112}Cd targets, since they are heavy nuclei with $3s_{1/2}$ neutron states in the valence orbits. We can expect to have peak structures in the cross section dominated by a single quasi-substitutional $[(1s)_\pi \otimes (s_{1/2})_n^{-1}]$ configuration. It is shown qunatatively that the (d,^{3}He) reaction around T_d = 500 MeV with ^{136}Xe and ^{116}Sn targets are best suited to observe the deepest $1s$ pionic states. In these reactions, the pionic $1s$ components are clearly seen as a separate peak in the reaction spectrum. [23]

The isotope dependence of the deeply bound pionic states is also studied in Sn and Pb. We made use of the long isotope chain of Sn, which is expected to be suitable for the observation of the isotope dependence of the $1s$ pionic state in the (d,^{3}He) reactions [23]. For the Pb isotopes, we can compare the results with the experimental data of (d,^{3}He) reactions for ^{206,208}Pb [11, 12, 13].

The isotope dependence of the pionic atoms are expected to be influenced by the isovector term in the pion-nucleus optical potential [24] and we expect to obtain important information on the following two points by studying the isotope dependence of the deeply bound pionic states [25, 26]: (i) The neutron density distribution, which is very interesting in the context of the β-unstable nuclear physics [27]. The pionic atom spectroscopy in the (d,^{3}He) reactions could be applied to the unstable nuclear targets using the inverse kinematics technique in future [28, 29]; and (ii) The medium effects on the pion decay constant f_π. The f_π is related to the isovector pion-nucleon scattering length by the Tomozawa-Weinberg relation [30, 31] and was suggested to be changed by the medium effects [22]. The pion decay constant f_π is one of the most important parameters in the nuclear chiral dynamics and thus it is interesting to see the medium effects on f_π.

We show the isotope dependence of deeply bound pionic states in Sn and Pb isotopes in this paper. It will be shown the isotope effects can be observed from the data of (d,^{3}He) reactions with accessible energy resolution.

EFFECTIVE NUMBER APPROACH FOR (D,^{3}HE) REACTIONS

We calculate the pionic-atom formation cross sections using the effective number approach. The effective number approach is often used to calculate a reaction cross section involving complex nuclei by using the cross section of elementary process. This approach is justified from the recoilless kinematics, in which a transferred particle (pion in this case) carries little momentum in the target frame. In this approximation, the (d,^{3}He) reaction cross section in the laboratory frame is expressed as [7, 8]

$$\left(\frac{d\sigma}{d\Omega}\right)_{dA\to^3He(A-1)\pi} = \left(\frac{d\sigma}{d\Omega}\right)_{dn\to^3He\pi} \times \sum_{[l_\pi\otimes j_n^{-1}]} N_{\text{eff}}, \tag{1}$$

with

$$N_{\text{eff}} = \sum_{JMm_s} \left| \int d^3r \chi_f^*(\mathbf{r}) \xi^*_{1/2,m_s} [\phi^*_{l_\pi}(\mathbf{r}) \otimes \psi_{j_n}(\mathbf{r})]_{JM} \chi_i(\mathbf{r}) \right|^2. \tag{2}$$

Here, $\left(\frac{d\sigma}{d\Omega}\right)_{dn\to^3He\pi}$ indicates the elementary differential cross section at forward angles for the $d+n \to ^3$He+π^- reaction in the laboratory system, which is extracted from the experimental data of the $p+d \to \pi^+ + t$ reaction assuming the charge symmetry [32, 33]. The neutron and the pion wave functions are denoted as ψ_{j_n} and ϕ_{l_π}. We adopt the harmonic-oscillator wave function for ψ_{j_n}. In all cases, we have used the oscillator parameter given by $\hbar\omega = 40\ \mathrm{A}^{-\frac{1}{3}}$ MeV, with A the nuclear mass number. The pion wave functions were obtained by solving the Klein-Gordon equation. The spin wave function is denoted as $\xi_{1/2,m_s}$ and we take the spin average with respect to m_s so as to take into account the possible spin direction of the neutrons in the target nucleus. χ_i and χ_f are the initial and the final distorted waves of the projectile and the ejectile, respectively. We use the Eikonal approximation and replace χ_f and χ_i according to

$$\chi_f^*(\mathbf{r})\chi_i(\mathbf{r}) = \exp(i\mathbf{q}\cdot\mathbf{r})D(z,\mathbf{b}), \tag{3}$$

where distortion factor $D(z,\mathbf{b})$ is defined as

$$D(z,\mathbf{b}) = \exp\left[-\frac{1}{2}\sigma_{dN}\int_{-\infty}^{z} dz'\rho_A(z',\mathbf{b}) - \frac{1}{2}\sigma_{hN}\int_{z}^{\infty} dz'\rho_{A-1}(z',\mathbf{b})\right]. \tag{4}$$

Here σ_{dN} and σ_{hN} are the deuteron-nucleon and ^{3}He-nucleon total cross sections. The functions $\rho_A(z,\mathbf{b})$ and $\rho_{A-1}(z,\mathbf{b})$ are the density distribution of the target and daughter nucleus at beam-direction coordinate z with an impact parameter $\mathbf{b}$.

In order to predict the spectrum of the (d,^{3}He) reactions, we need to take into account the configuration mixing effect in the target nuclei, the excitation energies, and the reaction strengths leading to the daughter nuclei [14, 23]. To predict the absolute value of a total strength for the neutron pickup from each orbital, we need to normalize the calculated effective numbers using the neutron occupation probabilities in the ground state of target nucleus. The occupation probabilities are obtained from the analyses of the A(p,d)A-1 and the A(d,t)A-1 reaction data, and are less than 1 in general.

As for the excited levels of the daughter nuclei, we use experimental excitation energies and strengths obtained from the A(p,d)A-1, the A(d,t)A-1 and A(d,p)A+1 reactions whenever the experimental data are available. Since single neutron pickup reaction from a certain orbital in the target can couple to several excited states of the daughter nuclei, we need to distribute the effective numbers among these excited levels of the daughter nuclei in proportion to the experimental strengths. Thus, the effective

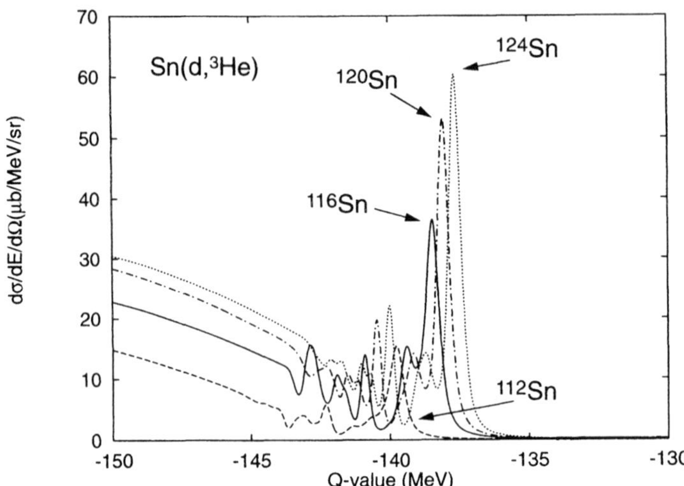

FIGURE 1. The Sn isotope dependence of total (d,^{3}He) spectrum for the pionic atom formation at T_d=500MeV with 300keV experimental resolution. The target nucleus is indicated in the figure.

number for the pionic state (ℓ_π) formation with the N-th nuclear excited state coupled to single neutron pickup from a certain neutron orbit j_n is written as

$$N_{\rm eff}(\ell_\pi \otimes (j_n^{-1})_N) = N_{\rm eff}(\ell_\pi \otimes j_n^{-1}) \times F_O(j_n) \times F_R((j_n^{-1})_N), \tag{5}$$

where $N_{\rm eff}(\ell_\pi \otimes j_n^{-1})$ is the effective number defined in Eq. (2), F_O is the normalization factor due to the occupation probabilities of the neutron states j_n in the target nucleus, and F_R is the relative strength of the N-th excited states in the daughter nucleus coupled to the single neutron pickup from the state j_n [25].

We have also evaluated negative pion quasi-free production contributions, as described in ref. [34], by taking into account the normalization coming from the occupation probability of neutron orbits and the relative strength of excited levels.

NUMERICAL RESULTS

It is shown that the (d,^{3}He) spectrum for the ^{116}Sn target is expected to have peak structure with a single subcomponent [23]. Hence, we can expect that the Sn isotpe is suited for the observation of the π atom isotope shifts. Figure 1 shows the isotope dependence of the total (d,^{3}He) spectra. The shape of these spectra varies significantly depending on the targets. The position and the height of the largest peak exhibit the isotope dependence. Though the shift of the peak position mainly comes from the difference of the neutron separation energy for these isotopes, it is possible to deduce the isotope dependence of the pionic states from the data with realistic energy resolution and hence the observation of the isotopic dependence of the pionic-atom formation spectrum will give us better insight for the behavior of pions in the nuclear medium [25].

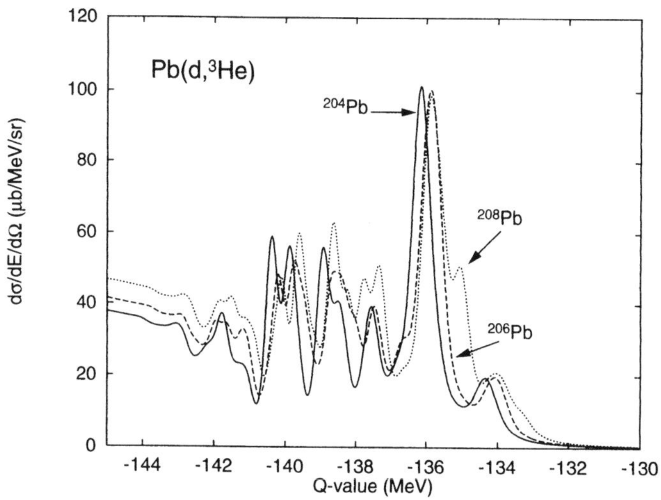

FIGURE 2. The isotope dependence of total Pb(d,^{3}He) spectrum for the pionic atom formation at T_d=600MeV with 300keV experimental resolution. The target nucleus is indicated in the figure.

We also consider the case of Pb isotopes. The theoretical results have already been reported in refs. [8, 14] for ^{206,208}Pb(d,^{3}He). Then, the results for the case of ^{204}Pb(d,^{3}He) are mainly shown in this section. The experimental data are found in refs. [11, 12, 13] for the cases of ^{206,208}Pb(d,^{3}He). At T_d = 600 MeV, the theoretical and the experimental results are available for ^{206,208}Pb. For ^{204}Pb, separated peaks may be observed corresponding to the pionic $1s$ and $2p$ contributions as in the case of ^{206}Pb. The peak corresponding to the $1s$ pionic state, however, comprises two subcomponents, and it could be difficult to determine the $1s$ eigen energy precisely. On the other hand, the highest peak mainly consists of a single $2p$ subcomponent and thus, for the Pb isotopes, the isotope dependence of the $2p$ pionic states should be determined easily from the data. The isotope dependence of the total spectrum is shown in Figure 2. For the case of ^{204}Pb, the largest peak of the pionic $2p$ state is expected to be even higher than the heavier Pb isotopes since the separation energies of the $(p_{1/2})_n$ and $(p_{3/2})_n$ states are almost the same, and their contributions are added to make the peak higher. The spectra corresponding to the shallower pionic states also demonstrate the isotope dependence. Due to the small contribution of the $p_{1/2}$ neutron state, the minor subcomponents, which are hidden in the ^{208}Pb(d,^{3}He) spectrum, appear clearly for the ^{204,206}Pb cases.

SUMMARY

In this paper, we briefly summarize the history of our research for the deeply bound pionic states using the (d, ^{3}He) reactions. And we show the isotope dependence of the pionic atom formation cross sections by the reactions which was studied recently, as an example of the current activities of theoretical investigations. The calculated spectrum is, however, known to have the significant discrepancies to experimental results for the ^{206}Pb case. Thus, in addition to extract physical quantities on the pionic states, we need to improve the theoretical model for the calculations of the (d,^{3}He) spectra in order to perform precise spectroscopic studies. The experimental results for the isotope shift, studied in this paper, are expected to be obtained soon. We believe that the data will provide new insights both for pion behaviors and for the reaction mechanisms which must stimulate the further investigations.

REFERENCES

1. T. E. O. Ericson and W. Weise, 'Pions and Nuclei', Oxford Univ. press (1988).
2. H. Toki and T. Yamazaki, Phys. Lett. **B213**(1988) 129.
3. H. Toki, S. Hirenzaki, T. Yamazaki and R. S. Hayano, Nucl. Phys. **A501** (1989) 653.
4. M. Iwasaki et al, Phys. Rev. **C 43** (1991) 1099.
5. R. S. Hayano, Proc. Int. Workshop on Pions in Nuclei, Penyscola, Spain (World Scientific, 1991) p. 330.
6. J. Nieves and E. Oset. Nucl. Phys. **A 518** (1990) 617.
7. H. Toki, S. Hirenzaki, and T. Yamazaki, Nucl. Phys. **A530** (1991) 679.
8. S. Hirenzaki, H. Toki and T. Yamazaki, Phys. Rev. **C44** (1991) 2472.
9. A. Trudel *et al.*, TRIUMF E628 and TRIUMF progress report (1991).
10. N. Matsuoka *et al*, Phys. Lett. B. **359** (1995) 39.
11. T. Yamazaki, *et al.*, Z. Phys. **A355** (1996) 219.
12. H. Gilg *et al.*, Phys. Rev. **C62** (2000) 025201; K. Itahashi *et al.*, Phys. Rev. **C62** (2000) 025202.
13. A. Gillitzer *et al.*, Proc. of PANIC (1999) Uppsala, Sweden, Nucl.Phys. **A663, 664** (2000) 206c.
14. S. Hirenzaki and H. Toki, Phys. Rev. **C55** (1997) 2719.
15. J. Nieves and E. Oset, Phys. Lett. **B282** (1992) 24.
16. R.S. Hayano, S. Hirenzaki and A. Gillitzer, Euro. Phys. J. **A6** (1999) 99.
17. K. Tsushima, D. H. Lu, A. W. Thomas, and K. Saito, Phys. Lett. **B443** (1998) 26.
18. K. Saito, K. Tsushima, D. H. Lu, and A. W. Thomas, Phys. Rev. **C59** (1999) 1203.
19. F. Klingle, T. Waas, and W. Weise, Nucl. Phys. **A650** (1999) 299.
20. T. Yamazaki *et al.*, Phys. Lett. **B418** (1998) 246.
21. T. Hatsuda, Nucl. Phys. **A544** (1992) 27c.
22. G. E. Brown and M. Rho, Phys. Rev. Lett. **66** (1991) 2720.
23. Y. Umemoto, S. Hirenzaki, K. Kume and H. Toki, Prog. Theor. Phys. **103** (2000) 337.
24. R. Seki and K. Masutani, Phys.Rev. **C27** (1983) 2799.
25. Y. Umemoto, S. Hirenzaki, K. Kume and H. Toki, Phys. Rev. **C62** (2000) 024606.
26. A. Gillitzer *et al.*, An experiment at GSI, S236.
27. H. Toki, S. Hirenzaki and T. Yamazaki, Phys. Lett. **B249** (1990) 391.
28. T. Yamazaki, R. S. Hayano, and H. Toki, Nucl. Inst. Meth. Phys. Res. **A305** (1991) 406.; T. Yamazaki, R. S. Hayano, H. Toki and P. Kienle, Nucl. Instrum. Methods **A292** (1990) 619.; Y. Umemoto, S. Hirenzaki, K. Kume, H. Toki, I. Tanihata, Nucl. Phys. **A679** (2001) 549.
29. I. Tanihata, Nucl. Phys. **A616** (1997) 56c.
30. Y. Tomozawa, Nuovo Cim. **46A** (1966) 707.
31. S. Weinberg, Phys. Rev. Lett. **17** (1966) 616.
32. E. Aslanides *et al.*, Phys. Rev. Lett. **39** (1977) 1654.
33. H. W. Fearing, Phys. Rev. **C16** (1977) 313
34. S. Hirenzaki and H. Toki, Nucl. Phys. **A628** (1998) 403.

Chiral Symmetry of Baryons [1]

Daisuke Jido*, Atsushi Hosaka† and Makoto Oka**

*Consejo Superior de Investigaciones Científicas, Universitat de Valencia, IFIC, Institutos de Investigación de Peterna, Aptdo. Correos 2085, 46071, Valencia, Spain
†Research Center for Nuclear Physics (RCNP) Osaka University
**Department of Physics, Tokyo Institute of Technology, Meguro, Tokyo 152-8551 Japan

Abstract. Chiral symmetry of baryons is studied by identifying them with linear representations of the chiral symmetry group. We show that two distinctive chiral assignments are possible. Based on these two assignments, linear sigma models are constructed and their physical implications are studied. Then we propose experiments to observe the relative sign of axial charges of the positive and negative parity nucleons through the π and η productions at the threshold region. This can be used as a probe to distinguish the two chiral assignments.

INTRODUCTION

Chiral symmetry is one of the most important concepts for low energy hadron physics. Due to spontaneous breaking, almost massless Nambu-Goldstone bosons appear, which makes the dynamics of hadrons very rich and interesting. Recent interests of chiral symmetry includes, for instance, the mechanism of the dynamical symmetry breaking and the realization of chiral symmetry in hadrons. The latter is important, since it is one way to understand the compositeness of hadrons.

Naively hadrons would respect chiral symmetry and form representations of the chiral symmetry group. In reality, however, due to spontaneous breaking, such symmetry pattern may not necessarily be realized. Still, one can imagine that observed hadrons would form representations when chiral symmetry is restored. This is suggested by a common pattern of the mass splitting between the positive and negative parity states as shown in Fig. 1. In particular, all the mass splittings are at the same order of about half GeV, which is generated by the spontaneous breaking. As usual, we expect simple representations for low lying hadrons. For instance, the ground state nucleon could be in a linear representation, $(\frac{1}{2},0)\oplus(0,\frac{1}{2})$, of the isospin chiral group $SU(2)_R\times SU(2)_L$. The pions are assigned to $(\frac{1}{2},\frac{1}{2})$ which is supplemented by the still controversial particle, sigma. Another example is a representation for ρ and a_1, which can be identified with $(1,0)\oplus(0,1)$.

In this report, we would like to discuss particle classifications based on chiral symmetry, especially for baryons. So far, such investigation has not been performed very much, although it is a very fundamental question in hadron physics.

[1] Presented by AH.

CP594, *Hadrons and Nuclei: First International Symposium,* edited by Il-T. Cheon et al.

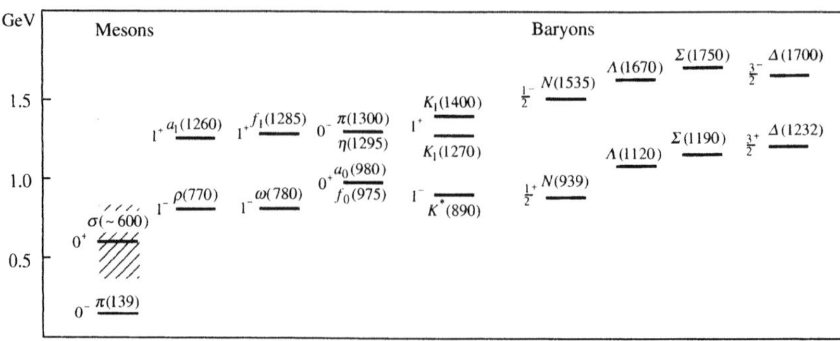

FIGURE 1. Mass splittings of positive and negative parity hadrons in various channels. The uncertain mass of sigma (σ) is hatched.

In our previous works[1], it was pointed out that there are two distinctive assignments for the chiral symmetry representations of baryons. We have called them naive and mirror. In the following sections, we introduce these assignments and discuss physical implications for positive and negative parity nucleons. Then we propose meson production reactions to study the two chiral assignments.

CHIRAL ASSIGNMENTS FOR THE NUCLEON AND LINEAR SIGMA MODELS

Chiral symmetry is a flavor symmetry for the right and left handed fermions. In QCD, the current quarks possess this symmetry. The right and left handed components (of the Lorentz group) are defined in term of a four component nucleon field N by

$$N = N_r + N_l; \quad N_r = \frac{1+\gamma_5}{2} N, \; N_l = \frac{1-\gamma_5}{2} N. \tag{1}$$

For $SU(2)_R \times SU(2)_L$, chiral (symmetry) transformations are defined as flavor transformations for the right and left handed components: $N_r \to g_R N_r, N_l \to g_L N_l$, where g_R and g_L are elements of $SU(2)_R$ and $SU(2)_L$, respectively.

Now consider two fermions and their transformation rules. Naively, one would expect the same transformations for both N_1 and N_2,

$$N_{1r} \to g_R N_{1r}, \; N_{1l} \to g_L N_{1l}; \quad N_{2r} \to g_R N_{2r}, \; N_{2l} \to g_L N_{2l}. \tag{2}$$

As was first pointed out by Lee[2], it is also possible to assign the opposite transformation for the second nucleon,

$$N_{1r} \to g_R N_{1r}, \; N_{1l} \to g_L N_{1l}; \quad N_{2r} \to g_L N_{2r}, \; N_{2l} \to g_R N_{2l}. \tag{3}$$

The point is that it is sufficient to assign different internal symmetries for the right and left handed components. It does not matter whether which of $SU(2)_R$ or $SU(2)_L$

is assigned. Since the transformations of the right and left handed components are interchanged, the second assignment of (3) is called mirror, while the first assignment of (2) is called naive.

A distinguished feature of the mirror assignment is that it is possible to introduce the chiral invariant mass term in the form of interaction between N_1 and N_2: $m_0(\bar{N}_1 N_2 + \bar{N}_2 N_1)$, where m_0 is a parameter which can not be determined within the present theoretical arguments. Such a mass term was used by DeTar and Kunihiro[3] in order to explain a possiblity of finite mass of baryons when chiral symmetry is restored.

Now we consider linear sigma models based on the naive and mirror assignments. To do this, meson fields are introduced as components of the representation $(1/2, 1/2)$ of the chiral group, which are subject to the transformation rule: $\sigma + i\vec{\tau}\cdot\vec{\pi} \to g_L(\sigma + i\vec{\tau}\cdot\vec{\pi})g_R^\dagger$.

In the naive assignment, the chiral invariant lagrangian up to order (mass)^4 is given by:

$$\begin{aligned} L_{\text{naive}} &= \bar{N}_1 i\partial\!\!\!/ N_1 - g_1\bar{N}_1(\sigma + i\gamma_5\vec{\tau}\cdot\vec{\pi})N_1 + \bar{N}_2 i\partial\!\!\!/ N_2 - g_2\bar{N}_2(\sigma + i\gamma_5\vec{\tau}\cdot\vec{\pi})N_2 \\ &\quad - g_{12}\{\bar{N}_1(\gamma_5\sigma + i\vec{\tau}\cdot\vec{\pi})N_2 - \bar{N}_2(\gamma_5\sigma + i\vec{\tau}\cdot\vec{\pi})N_1\} + L_{\text{mes}}, \end{aligned} \tag{4}$$

where g_1, g_2 and g_{12} are free parameters. The terms of g_1 and g_2 are ordinary chiral invariant coupling terms of the linear sigma model. The term of g_{12} is the mixing of N_1 and N_2. Since the two nucleons have opposite parities, γ_5 appears in the coupling with σ, while it does not in the coupling with π. The meson lagrangian L_{mes} in (4) is not important in the following discussion.

Chiral symmetry breaks down spontaneously when the sigma meson acquires a finite vacuum expectation value, $\sigma_0 \equiv \langle 0|\sigma|0\rangle$. This generates masses of the nucleons. From (4), the mass can be expressed by a 2×2 matrix in the space of N_1 and N_2. The mass matrix can be diagonalized by the rotated states,

$$\begin{pmatrix} N_+ \\ N_- \end{pmatrix} = \begin{pmatrix} \cos\theta & \gamma_5\sin\theta \\ -\gamma_5\sin\theta & \cos\theta \end{pmatrix} \begin{pmatrix} N_1 \\ N_2 \end{pmatrix}, \tag{5}$$

where the mixing angle and mass eigenvalues are given by

$$\tan 2\theta = \frac{2g_1}{g_1+g_2}, \quad m_\pm = \sigma_0 g_\pm = \frac{\sigma_0}{2}\left(\sqrt{(g_1+g_2)^2 + 4g_{12}^2} \pm (g_1 - g_2)\right). \tag{6}$$

In the naive model, since the interaction and mass matrices takes the same form, the physical states, N_+ and N_-, decouple exactly; the lagrangian becomes a sum of the N_+ and N_- parts. [2] Therefore, chiral symmetry imposes no constraint on the relation between N_+ and N_-. The role of chiral symmetry is just the mass generation due to its spontaneous breaking. We present a schematic plot of $m_\pm$ as functions of σ_0 in Fig.2. When chiral symmetry is restored and $\sigma_0 \to 0$, both N_+ and N_- become massless and degenerate. However, the degeneracy is trivial as they are independent; they no

[2] Small chiral symmetry breaking might induce a small coupling $g_{\pi N_+ N_-}$.

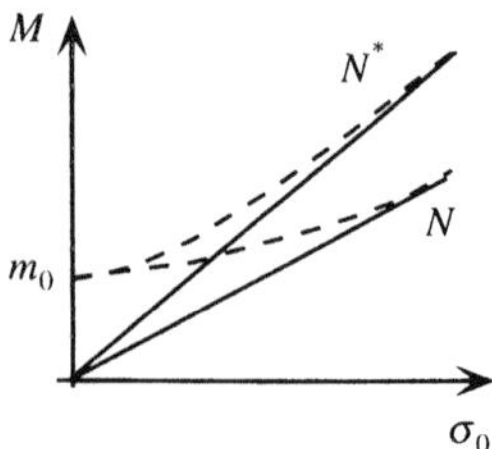

FIGURE 2. Masses of N and N^* as functions of σ_0. Solid and dashed lines are for the naive and mirror assignments.

longer transform among themselves. The decoupling of N_+ and N_- implies that the off-diagonal Yukawa coupling $g_{\pi N_+ N_-}$ vanishes. This is a rigorous statement up to the order we considered.

Now we turn to the mirror assignment. It is rather straightforward to write down the chiral invariant lagrangian compatible to the mirror transformations[3, 1]:

$$\begin{aligned} L_{\rm mirror} &= \bar{N}_1 i\partial\!\!\!/ N_1 - g_1 \bar{N}_1(\sigma + i\gamma_5 \vec{\tau}\cdot\vec{\pi})N_1 + \bar{N}_2 i\partial\!\!\!/ N_2 - g_2 \bar{N}_2(\sigma - i\gamma_5\vec{\tau}\cdot\vec{\pi})N_2 \\ &- m_0(\bar{N}_1\gamma_5 N_2 - \bar{N}_2\gamma_5 N_1) + L_{\rm mes}\,. \end{aligned} \tag{7}$$

Here the chiral invariant mass term has been added. Note that in the g_2 term, the sign of the pion field is opposite to that of the g_1 term. This compensates the mirror transformation of N_2. The lagrangian (7) was first formulated by DeTar and Kunihiro.

When chiral symmetry is spontaneously broken, the mass matrix of the lagrangian (7) can be diagonalized by a linear combination similar to (5). The mixing angle and mass eigenvalues are given by

$$\tan 2\theta = \frac{2m_0}{\sigma_0(g_1+g_2)}, \quad m_\pm = \frac{1}{2}\left(\sqrt{(g_1+g_2)^2\sigma_0^2 + 4m_0^2} \pm (g_1 - g_2)\sigma_0\right). \tag{8}$$

In the mirror model, the interaction term is not diagonalized in the physical basis, unlike the naive model.

In Fig.2, a typical behavior of masses of $m_\pm$ are shown as functions of σ_0. When chiral symmetry is restored, the two nucleons are degenerate with a finite mass m_0.

It is worth noting the axial coupling constants g_A in the mirror model. They can be extracted fro the commutation relations between the axial charge operators Q_5^a and the nucleon fields,

$$\begin{aligned} [Q_5^a, N_+] &= \frac{\tau^a}{2}\gamma_5(\cos 2\theta N_+ - \sin 2\theta\gamma_5 N_-) \\ [Q_5^a, N_-] &= \frac{\tau^a}{2}\gamma_5(-\sin 2\theta\gamma_5 N_+ - \cos 2\theta N_-)\,. \end{aligned} \tag{9}$$

This implys that g_A's are expressed by a 2×2 matrix whose elements are given by the coefficients of (9). From this we see that the signs of the diagonal axial charges g_A^{++} and g_A^{--} are opposite. The absolute value is, however, smaller than one in contradiction

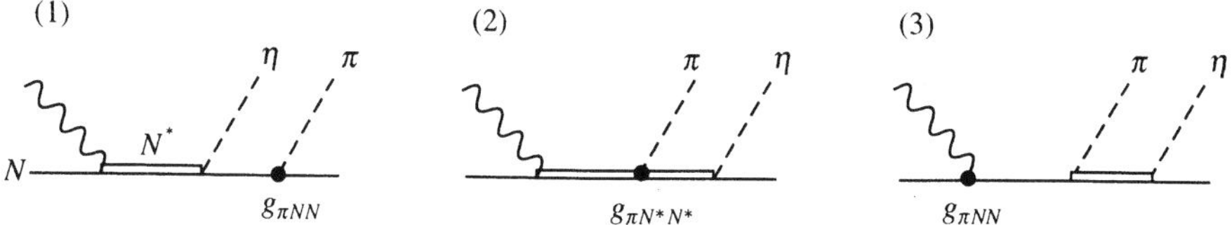

FIGURE 3. Dominant three diagrams for π and η productions. The incident wavy line is either a pion of photon.

with experimental value $g_A \sim 1.25$. In the present model, the physical states $N_\pm$ are the superpositions of $N_{1,2}$, whose axial charges are ± 1. This explains why $|g_A^{++}|, |g_A^{--}| < 1$. The correction to this small g_A might be supplemented by delta contributions as Adler-Weisberger sum rule and the Weinberg's algebraic method implies[4, 5, 6].

π AND η PRODUCTIONS AT THRESHOLD REGION

In this section, we propose experimental method to study the two chiral assignments. As discussed in the preceding sections, one of the differences between the naive and mirror assignments is the relative sign of the axial coupling constants of the positive and negative parity nucleons. In the following discussions, we identify $N_+ \sim N(939)$ and $N_- \sim N(1535) \equiv N^*$. Strictly, the identification of the negative parity nucleon with the first excited state $N(1535)$ is no more than an assumption. From experimental point of view, however, $N(1535)$ has a distinguished feature that it has a strong coupling with an η meson, which can be used as a filter to observe the resonance. In practice, we observe the pion couplings which are related to the axial couplings through the Goldberger-Treiman relation $g_{\pi N_\pm N_\pm} f_\pi = g_A M_\pm$.

Let us consider π and η productions induced by a pion or photon. Suppose that the two diagrams of (1) and (2) as shown in Fig. 3 are dominant in these process. Modulo energy denominator, the only difference of these processes is due to the coupling constants $g_{\pi NN}$ and $g_{\pi N^*N^*}$. Therefore, depending on their relative sign, cross sections are either enhanced or suppressed. In the pion induced process, due to the p-wave coupling nature, another diagram (3) also contributes substantially.

In actual computation, we take the interaction lagrangians:

$$\begin{aligned} L_{\pi NN} &= g_{\pi NN} \bar{N} i\gamma_5 \vec{\tau}\cdot\vec{\pi} N, \quad L_{\eta NN^*} = g_{\eta NN^*}(\bar{N}\eta N^* + \bar{N}^*\eta N), \\ L_{\pi NN^*} &= g_{\pi NN^*}(\bar{N}\tau\cdot\pi N^* + \bar{N}^*\tau\cdot\pi N), \quad L_{\pi N^*N^*} = g_{\pi N^*N^*}(\bar{N}^* i\gamma_5 \tau\cdot\pi N^*). \end{aligned} \tag{10}$$

We use these interactions both for the naive and mirror cases with empirical coupling constants for $g_{\pi NN} \sim 13$, $g_{\pi NN^*} \sim 0.7$ [3] and $g_{\eta NN^*} \sim 2$. The coupling constants $g_{\pi NN^*} \sim 0.7$ and $g_{\eta NN^*} \sim 2$ are determined from the partial decay widths, $\Gamma_{N^*(1535)\to\pi N} \approx \Gamma_{N^*(1535)\to\eta N} \sim 70$ MeV, although large uncertainties for the width

[3] As pointred out in the previous footnote, a small coupling of $g_{\pi NN^*}$ could be induced also in the naive model.

TABLE 1. Parameters used in our calculation.

m_N	m_{N^*}	Γ_{N^*}	$g_{\pi NN}$	$g_{\pi NN^*}$	$g_{\eta NN^*}$	$g_{\pi N^* N^*}$
938 (MeV)	1535 (MeV)	140 (MeV)	13	0.7	2.0	13 (naive) −13 (mirror)

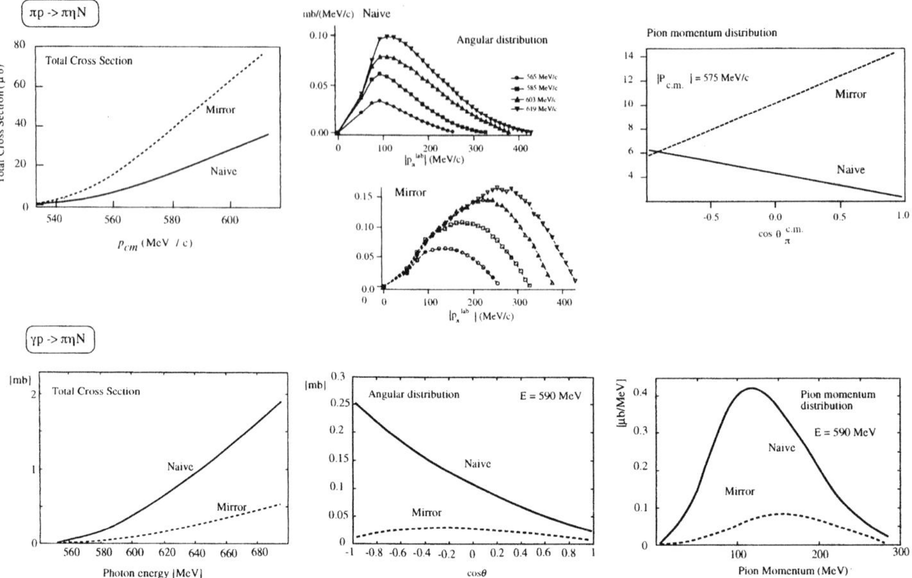

FIGURE 4. Various cross sections for π and η productions.

have been reported[7]. The unknown parameter is the $g_{\pi N^* N^*}$ coupling. One can estimate it by using the theoretical value of the axial charge g_A^* and the Goldberger-Treiman relation for N^*. When $g_A^* = \pm 1$ for the naive and mirror assignments, we find $g_{\pi N^* N^*} = g_A^* m_{N^*}/f_\pi \sim \pm 17$. Here, just for simplicity, we use the same absolute value as $g_{\pi NN}$. The coupling values used in our computations are summarized in Table 1.

Several remarks follow:

- We assume resonance (N^*) pole dominance. This is considered to be good particularly for the η production at the threshold region, since η is dominantly produced by N^*.
- There are altogether twelve resonance dominant diagrams. Due to energy denominator, the three diagrams in Fig. 3 are dominant.
- Background contributions, in which two meson (seagull) or three meson vertices appear, are suppressed due to G-parity conservation.

Hence, the processes are indeed dominated by the N^* resonance diagrams.

We show various cross sections for the pion and photon induced processes in Fig. 4. We briefly discuss the results:

1. Total cross sections are of order of micro barn, which are well accessible by the present experiments. In the photon induced process, the interference between the diagram (1) and (2) occurs and the destruction in the mirror assignment is observed. In the pion induced case, due to the momentum dependence of the initial vertex the third term (3) becomes dominant and the mirror assignment is rather enhanced.
2. In the pion induced reaction, the momentum distribution of the final state pion differs for the naive and mirror assignments; the peak position varies differently as function of the incident energy.
3. In the pion induced reaction, the angular distribution of the final state pion differs crucially depending on the sign of the πNN and πN^*N^* couplings.
4. In the photon induced reaction, the distinction of the momentum and angular distributions are not very well recognized in the mirror assignments, since the cross sections are suppressed.

SUMMARY

In this report, we have discussed chiral symmetry of nucleons. We have shown that there are two symmetry assignments, naive and mirror, which were implemented by the fundamental representations $(\frac{1}{2},0)$ and $(0,\frac{1}{2})$. Inclusion of higher dimensional representations such as $(1,\frac{1}{2})$ and $(\frac{1}{2},1)$ is also possible and is an interesting direction[6, 8]. In order to make these theoretical discussions realistic, we proposed to study reactions of π and η production. Various differntial cross sections were computed which can be measured at existing meson and photon facilities. Chiral symmetry of the baryon sector is an interesting subject which should be studied both in theory and experiment.

REFERENCES

1. D. Jido et al., Nucl. Phys. A671 (2000) 471, hep-ph/9805306,
2. B.W. Lee, Chiral Dynamics (Academic Press, 1970).
3. C. DeTar and T. Kunihiro, Phys. Rev. D39 (1989) 2805,
4. S.L. Adler, Phys. Rev. 140 (1965) B736,
5. W.I. Weisberger, Phys. Rev. 143 (1966) 1302,
6. S. Weinberg, Phys. Rev. 177 (1969) 2604,
7. Particle Data Group, Euro. Phys. J. C15 (2000) 1.
8. D. Jido, T. Hatsuda and T. Kunihiro, Phys. Rev. Lett. 84 (2000) 3252, hep-ph/9910375,

Discriminating between effective theories of $U_A(1)$ symmetry breaking

V. Dmitrašinović

Research Center for Nuclear Physics (RCNP), Osaka University;
Vinča Institute of Nuclear Sciences, P.O.Box 522, Belgrad, Yugoslavia

Abstract. We address the question if one can empirically distinguish between the two proposed solutions to the "$U_A(1)$ problem": the 't Hooft, and the Veneziano-Witten $U_A(1)$ symmetry breaking effective interactions. Two hadronic observables are offered as discriminants: (1) The scalar (0^+) meson spectrum; (2) Weinberg's second spectral sum rule. Their present experimental status is discussed.

WHAT IS THE "$U_A(1)$ PROBLEM"?

The $U_A(1)$ problem [1, 2] consists of two parts: (i) the discrepancy between the left- and right-hand side in the "Gell-Mann–Okubo" type mass relation:

$$m_{\eta'}^2 + m_\eta^2 = (1111\ \mathrm{MeV})^2 \neq 2m_K^2 = (700\ \mathrm{MeV})^2,$$

and (ii) the $\eta - \eta'$ mixing angle being negative and far from the ideal one:

$$\theta_{\mathrm{ps}} \simeq -20^o \neq 35.3^o\ .$$

The presently accepted solution postulates an explicit breaking of the $U_A(1)$ symmetry, i.e., a new interaction $\mathcal{H}_{\mathrm{U}(1)}$ that satisfies

$$\begin{aligned} \left(f m_{\mathrm{U}(1)}^2 f\right)_{ab} &= -\langle 0|[Q_a^5,[Q_b^5,\mathcal{H}_{\mathrm{U}(1)}(0)]]|0\rangle \\ &= 0, \quad a,b = 1,\ldots 8\ ; \\ &\neq 0, \quad a = b = 0\ . \end{aligned} \tag{1}$$

Here a, b are the flavour indices of the axial charges corresponding to the appropriate pseudoscalar (ps) meson(s). This interaction raises the mass of the SU(3) flavour-singlet and thus provides for the mass difference

$$m_{\mathrm{U}(1)}^2 = m_{\eta'}^2 + m_\eta^2 - 2m_K^2 \simeq (855\mathrm{MeV})^2.$$

The $U_A(1)$ symmetry-breaking mass $m_{\mathrm{U}(1)}$ is 855 MeV, provided all pseudoscalar decay constants are equal. Inclusion of the variation in ps decay constants leads to $830 \pm$

CP594, *Hadrons and Nuclei: First International Symposium,* edited by Il-T. Cheon et al.

60 MeV. The same interaction $\mathcal{H}_{\mathrm{U}(1)}$ solves the ps mixing angle problem:

$$\tan 2\theta_{\mathrm{ps}} = \frac{\left(2\sqrt{2}/3\right)\Delta_{\mathrm{ps}}^2}{(1/3)\Delta_{\mathrm{ps}}^2 - f_0^2 m_{\mathrm{U}(1)}^2}, \tag{2}$$

where

$$\Delta_{\mathrm{ps}}^2 = f_K^2\left(m_{K^0}^2 + m_{K^+}^2\right) - f_\pi^2\left(m_{\pi^0}^2 + m_{\pi^+}^2\right) \tag{3}$$

$$f_0^2 m_{\mathrm{U}(1)}^2 = f_{\eta'}^2 m_{\eta'}^2 + f_\eta^2 m_\eta^2 - f_K^2\left(m_{K^+}^2 + m_{K^0}^2\right) + f_\pi^2\left(m_{\pi^+}^2 - m_{\pi^0}^2\right), \tag{4}$$

which leads to a negative ps. mixing angle $\theta_{\mathrm{ps}} = -(25 \pm 10)^o$.

TWO SOLUTIONS

Two $U_A(1)$ symmetry breaking effective operators are discussed in the literature: (1) the't Hooft-Kobayashi-Kondo-Maskawa ("'t Hooft", for short) effective interaction [3, 4], and (2) the "Veneziano-Witten" effective interaction [5, 6].

The 't Hooft interaction. This interaction is believed to be induced by instantons in QCD. It reads

$$\mathcal{L}_{\mathrm{tH}}^{(N_f=3)} = -K_{\mathrm{tH}}\left[\det_f\left(\bar{\psi}(1+\gamma_5)\psi\right) + \det_f\left(\bar{\psi}(1-\gamma_5)\psi\right)\right] \tag{5}$$

where, $\det_f\left(\bar{\psi}(1 \pm \gamma_5)\psi\right)$ is a determinant of the flavour-space matrix. Eq. (1) then leads to

$$\begin{aligned} m_{00}^2(\mathrm{tH}) f_0^2 &= 6\langle 0|\mathcal{L}_{\mathrm{tH}}^{(3)}(0)|0\rangle + O(1/N_C) \\ &= -12 K_{\mathrm{tH}}\langle\bar{q}q\rangle_0^3 + O(1/N_C), \end{aligned} \tag{6}$$

The symbol $O(1/N_C)$ remind us that we have neglected $1/N_C$ suppressed terms. So long as the respective coupling constant is sufficiently large, the U(1) problem is solved.

The Veneziano-Witten interaction. This interaction

$$\mathcal{L}_{\mathrm{VW}}^{(N_f=3)} = K_{\mathrm{VW}}\left[\det_f\left(\bar{\psi}(1+\gamma_5)\psi\right) - \det_f\left(\bar{\psi}(1-\gamma_5)\psi\right)\right]^2 \tag{7}$$

is *not* induced by instantons, but rather by $1/N_C$ effects in QCD. Again from Eq. (1) we find

$$m_{00}^2(\mathrm{VW}) f_0^2 = 48 K_{\mathrm{VW}}\langle\bar{q}q\rangle_0^6 + O(1/N_C). \tag{8}$$

The same comments about the coupling constant hold as for the 't Hooft interaction. Once again, the flavour-singlet pseudoscalar mass moves up above the octet one.

DISCRIMINATING BETWEEN THE SOLUTIONS

Manifestly, no study of the pseudoscalar η, η' mesons' properties alone can resolve this issue. We offer two new tests discriminating between the two effective interactions in a chiral quark model. Differences between models with the 't Hooft- and the Veneziano-Witten (VW) $U_A(1)$ symmetry-breaking interactions arise in: (i) the *scalar* mesons spectra, [7, 8, 9]. The 't Hooft interaction leads to a mass shift within the scalar nonet that is identical in size, but opposite in sign to that found in pseudoscalars, whereas the VW one does not shift the scalar meson masses at all. (ii) the second spectral (Weinberg) sum rule [10]. The 't Hooft interaction strongly modifies this sum rule, whereas the VW one does not change it at all.

Scalar meson spectrum

In the following we use an effective chiral field theory of quarks with a non-trivial ground state characterized by a finite quark condensate and an effective $U_A(1)$ symmetry-breaking interaction, following Nambu and Jona-Lasinio (NJL) [11]. The $U_L(3) \times U_R(3)$ symmetric gluon-exchange interaction (the first line) is modelled by a quartic quark selfinteraction

$$\mathcal{L}_{\mathrm{NJL}}^{(3)} = \bar{\psi}[i\partial\!\!\!/ - m^0]\psi \quad + \quad G\sum_{i=0}^{8}[(\bar{\psi}\lambda_i\psi)^2 + (\bar{\psi}i\gamma_5\lambda_i\psi)^2] + \mathcal{L}_{\mathrm{U(1)}}\,, \tag{9}$$

ammended by the $U(1)_A$ symmetry breaking effective interaction. There are at present no easily applicable nonperturbative methods for a direct approach to the 6- or 12-point operators in Eq. (9). Therefore one has to construct an "effective mean-field quartic self-interaction Lagrangian" $\mathcal{L}_{\mathrm{eff}}^{(4)}$ from Eq. (5,7) following the procedure employed in Ref. [7]. This leads to consistent chiral dynamics in the sense that the Goldstone theorem and other chiral Ward-Takahashi identities pertaining to the ps octet remain intact in the chiral limit. Mathematically that procedure is equivalent to taking a quark and an antiquark external line and closing them into a loop using Feynman rules for the Lagrangian (9) in all possible ways while taking into account the proper symmetry number of the diagram. The meson masses are read off from the poles of their propagators, which in turn are constrained by the gap equation. This model has turned out to be a reliable laboratory for calculating light spinless meson mass relations induced by $U_A(1)$ symmetry-breaking, as can be seen from the comparison between the NJL model results [7] and a confining potential model's predictions [8]. The close agreement of the spectra is the best justification of the NJL model.

't Hooft interaction. The ps meson flavour singlet - octet mass shift due to the 't Hooft interaction in the NJL model has been established to be in exact agreement with the general result (6), see Ref. [7]. One finds, however, that the singlet - octet mass splitting in the scalar (0^+) channel is just as large as, though of opposite sign to the ps one. This statement is embodied in the $N_f = 3$ scalar – pseudoscalar meson mass sum

rule

$$m_\eta^2 + m_{\eta'}^2 - m_{K^+}^2 - m_{K^0}^2 = m_{K_0^{*+}}^2 + m_{K_0^{*0}}^2 - m_{f_0}^2 - m_{f_0'}^2 . \tag{10}$$

Equivalent results were found in a confining chiral quark model, Ref. [8], also as an effect of the 't Hooft $U_A(1)$ symmetry breaking interaction. For two flavours the 't Hooft interaction predicts a mass splitting between the isoscalar (f_0^*) and isovector (a_0) scalar (0^+) states that has been observed on the lattice, see Fig. 6 in Ref. [12]. The corresponding pseudoscalar mass splitting with two flavours has not been calculated, so direct comparison with the $N_f = 2$ sum rule

$$m_\pi^2 - m_{\eta^*}^2 = m_{f_0^*}^2 - m_{a_0}^2 \tag{11}$$

is not possible at the moment. The sum rule (10) shifts the masses of the physical isosinglet scalar states f_0, f_0' from their simple quark model values, see Fig. 1.

Assuming that the $f_0(1500)$ is one of the two isoscalar scalar states, the sum rule (10) predicts the mass of the other scalar state as 1000±50 MeV. As there are *two* isosinglet scalar states f_0 in the Particle Data tables with mass(es) very close to 1 GeV, the $f_0(980)$ and the ("σ") $f_0(\varepsilon(1000))$, one is left with an ambiguity. The $f_0(1500)$ is in better shape: Ritter et al. [13] have explained the puzzling absence of $K\bar{K}$ pairs from the $f_0(1500)$ two-body decay products as a consequence of the 't Hooft interaction. This explanation depends critically on the scalar mixing angle θ_s being small and *positive*, which follows from

$$\tan 2\theta_s = \frac{\left(4\sqrt{2}/3\right)\left(m_{K_0^*}^2 - m_{a_0}^2\right)}{m_{U(1)}^2 + (2/3)\left(m_{K_0^*}^2 - m_{a_0}^2\right)} . \tag{12}$$

One must say that higher-order corrections are known to modify our sum rule (10). For example, upon taking into account of vector- and axial-vector mesons, the r.h.s. (scalar masses) of Eqs. (10),(11) are reduced by a multiplicative factor equal to the flavour-singlet axial coupling constant of the constituent quarks [14].

Veneziano-Witten interaction. To leading order in N_C we find again that the general result Eq. (8) holds for the meson masses, see Ref. [9]. This time scalar mesons are unaffected by the VW interaction. In the absence of flavour singlet-octet mass splitting in the scalar sector, the flavour-singlet scalar mesons mix ideally (see Eq. (12), but with $m_{U(1)}^2$ omitted from the denominator) and one finds one $u\bar{u}, d\bar{d}$ and one $s\bar{s}$ state, with a mass splitting of about 300 MeV. The lower-lying state is degenerate with the isovector scalar mesons, i.e., around 1320 MeV in this model. Curiously, there is an f_0 state at 1370 MeV. Then the heavy scalar meson ought to be near 1600 MeV. There are two candidates in the vicinity: (a) the familiar $f_0(1500)$, and (b) the new $f_0(1720)$. The former has a puzzling absence, for an $s\bar{s}$ state, of the $K\bar{K}$ decay mode. This has prompted suggestions that it is not an ordinary $q\bar{q}$ octet member, as the Veneziano-Witten model predicts. This evidence and the apparent success of the 't Hooft model at explaining the $f_0(1500)$ decay pattern [13] seem to rule against the Veneziano-Witten model.

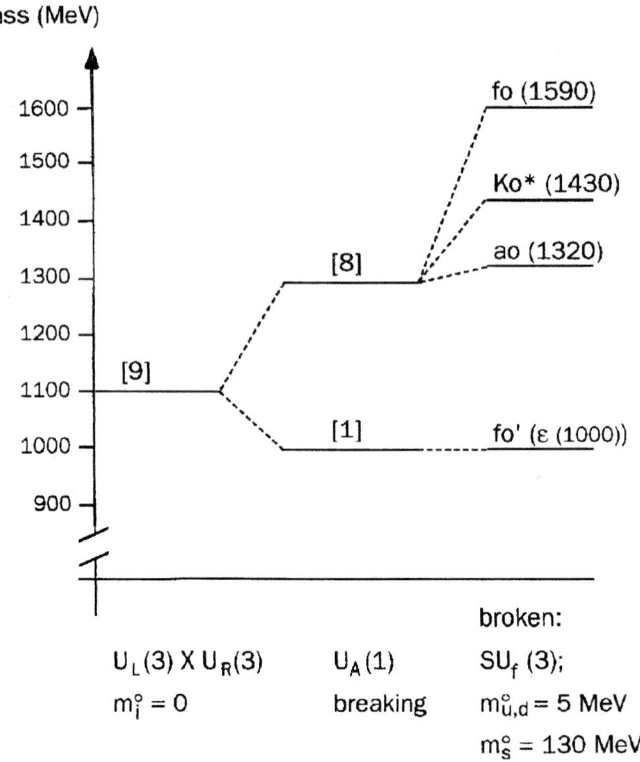

FIGURE 1. The scalar meson spectrum in the three flavour NJL model with 't Hooft interaction.

The second spectral sum rule

Weinberg's second sum rule (Wsr II) [15]

$$\int_0^\infty ds \left(\rho^V(s) - \rho^A(s)\right) = 0\,, \tag{13}$$

for the (difference of) vector and axial vector spectral functions is a statement about the chiral symmetry of the underlying theory at asymptotically large momenta. Here $\rho^{ab}_{V,A}$ are the spectral functions as defined by

$$\begin{aligned}\langle 0| \left[A^a_\mu(x), A^b_\nu(y)\right] |0\rangle &= -i \int d\mu^2 \rho^{ab}_A(\mu^2) \left(g_{\mu\nu} - \frac{\partial^x_\mu \partial^y_\nu}{\mu^2}\right) \Delta(x-y;\mu^2) \\ &\quad + i \int d\mu^2 \rho^{ab}_{PS}(\mu^2) \partial^x_\mu \partial^y_\nu \Delta(x-y;\mu^2)\end{aligned} \tag{14}$$

$$\langle 0| \left[V_\mu^a(x), V_\nu^b(y)\right] |0\rangle = -i\int d\mu^2 \rho_V^{ab}(\mu^2)\left(g_{\mu\nu} - \frac{\partial_\mu^x \partial_\nu^y}{\mu^2}\right)\Delta(x-y;\mu^2) \ , \quad (15)$$

V_μ^a, A_ν^b are the vector- and axial currents and Δ is the commutator of the free scalar fields at two space-time points.

The two Weinberg spectral sum rules have been examined in an effective field theoretical model of QCD, Ref. [16], with the result that the first sum rule is exactly satisfied, and the second one is broken *even in the exact $SU_L(2)\times SU_R(2)$ chiral limit*, i.e., with current quark masses $m_u^0 = m_d^0 = 0$.

Violations of the second Weinberg sum rule. Nieh [17] gave a critical assessment of this sum rule very early, though it passed largely unnoticed. One can express the violation of the Wsr II due to the Hamiltonian H as

$$\begin{aligned}\delta_{ij}\int_0^\infty ds\left(\rho_V^{ab}(s) - \rho_A^{ab}(s)\right) &= \int d^3x \langle 0| \left[[H, A_i^a(0,\mathbf{x})], A_j^b(0,\mathbf{y})\right] |0\rangle \\ &- \int d^3x \langle 0| \left[[H, V_i^a(0,\mathbf{x})], V_j^b(0,\mathbf{y})\right] |0\rangle \ , \quad (16)\end{aligned}$$

where $(i,j = 1,2,3)$ are the spatial dimension indices. Thus, Eq. (16) shows that the second sum rule actually tests the commutators of the *spatial current components* and the Hamiltonian, i.e., the invariance of the theory under $U_L(6)\times U_R(6)$ current algebra symmetry transformations [18], rather than merely the usual $U_L(3)\times U_R(3)$ chiral charge algebra, which is its subalgebra. Manifestly, any term in the Hamiltonian that breaks the $U_L(3)\times U_R(3)$ symmetry will also break the $U_L(6)\times U_R(6)$ symmetry. There are three sources of $U_L(3)\times U_R(3)$ symmetry breaking in QCD: (i) the current quark masses; (ii) $U_A(1)$ symmetry-breaking effective interaction; and (iii) the electroweak (EW) interactions. In the following we shall examine only the first two.

The second spectral sum rule Eq. (16), upon Nieh's correction [17], is also satisfied in the effective model of Ref. [16]. That model already contains the $U_A(1)$ symmetry breaking 't Hooft interaction in its two-flavor version. Thus there is a large violation of the Wsr II even in the chiral limit, due the $U_A(1)$ symmetry breaking 't Hooft interaction in this model [10].

Current quark mass terms. Inserting the current quark mass Hamiltonian $\mathcal{H}_{\chi SB}(0) = \bar{\Psi}(0)M_q^0\Psi(0)$ into Eq. (16) we find

$$\begin{aligned}\int_0^\infty ds\left(\rho_V^{ab}(s) - \rho_A^{ab}(s)\right) &= \int d^3x \langle 0| \left[[H_{\chi SB}, A_i^a(0,\mathbf{x})], A_j^b(0,\mathbf{y})\right] |0\rangle \\ &- \int d^3x \langle 0| \left[[H_{\chi SB}, V_i^a(0,\mathbf{x})], V_j^b(0,\mathbf{y})\right] |0\rangle \\ &= -\langle 0|\bar{\Psi}\left\{\left\{M_q^0, \frac{\lambda^a}{2}\right\}, \frac{\lambda^b}{2}\right\}\Psi|0\rangle \\ &= -\langle 0\left|\left[Q_5^a, \left[Q_5^b, \bar{\Psi}M_q^0\Psi\right]\right]\right|0\rangle \\ &= \left(f_{\rm ps} m_{\rm ps}^2({\rm mech}) f_{\rm ps}\right)_{ab} \ . \quad (17)\end{aligned}$$

The expression on the right-hand side of (17) is the same as the one entering the Gell-Mann-Oakes-Renner (GMOR) formula relating the pseudoscalar (ps) meson masses and decay constants to the above vacuum matrix elements. Eq. (17) is in agreement with the ITEP sum rule results [19].

$U_A(1)$ symmetry-breaking effective interactions. Insert the effective 't Hooft quark self-interaction Eq. (5) into the double commutators in Eq. (16) to find

$$\int d^3x \langle 0| \left[\left[H_{\rm tH}^{(3)}, A_i^a(0,\mathbf{x}) \right], A_j^b(0,\mathbf{y}) \right] |0\rangle - \int d^3x \langle 0| \left[\left[H_{\rm tH}^{(3)}, V_i^a(0,\mathbf{x}) \right], V_j^b(0,\mathbf{y}) \right] |0\rangle = -6\langle 0|\mathcal{H}_{\rm tH}^{(3)}|0\rangle \delta_{ij}\delta^{ab} . \quad (18)$$

The interacting ground state ("vacuum") expectation value of the 't Hooft interaction is related to the 't Hooft mass with three light flavours *via* Eqs. (6,1), see Ref. [7, 9]. The empirical value of $f_0^2 m_{\rm tH}^2$ was determined above as $(300{\rm MeV})^4$ from the definition Eq. (6). This leads to

$$\int d^3x \langle 0| \left[\left[H_{\rm tH}^{(3)}, A_i^a(0,\mathbf{x}) \right], A_j^b(0,\mathbf{y}) \right] |0\rangle - \int d^3x \langle 0| \left[\left[H_{\rm tH}^{(3)}, V_i^a(0,\mathbf{x}) \right], V_j^b(0,\mathbf{y}) \right] |0\rangle = \delta_{ij}\delta^{ab} f_0^2 m_{\rm U(1)}^2 . \quad (19)$$

Note that this result holds for *all* $a,b = 0,...,8$, *i.e.*, not only in the flavour-singlet channel $(a,b=0)$. This is something of a surprise, as we have come to expect its influence only in the flavour-singlet ps and scalar channels. Here one is sensitive to the violation of the $U_L(6) \times U_R(6)$ *current* algebra, rather than that of the (usual) $SU_L(3) \times SU_R(3)$ algebra of chiral *charges*, and that the 't Hooft interaction violates the $U_L(6) \times U_R(6)$ symmetry. Adding now the current quark mass contribution to the right-hand side of Eq. (19) we find

$$\int_0^\infty ds \left(\rho_V^{ab}(s) - \rho_A^{ab}(s) \right) = \delta_{ab} f_0^2 m_{\rm U(1)}^2 + f_a m_{ab}^2({\rm mech.}) f_b . \quad (20)$$

There is, however, another way to effectively break the $U_A(1)$ symmetry with quark degrees of freedom: the Veneziano-Witten effective quark interaction Eq. (7) Insert this into the double commutators in Eq. (16); direct calculation yields *zero*, to leading order in $1/N_C$,

$$\int d^3x \langle 0| \left[\left[H_{\rm VW}^{(3)}, A_i^a(0,\mathbf{x}) \right], A_j^b(0,\mathbf{y}) \right] |0\rangle - \int d^3x \langle 0| \left[\left[H_{\rm VW}^{(3)}, V_i^a(0,\mathbf{x}) \right], V_j^b(0,\mathbf{y}) \right] |0\rangle = 0 + O(1/N_C) , \quad (21)$$

thus leaving only the current quark mass induced corrections to the Wsr II

$$\int_0^\infty ds \left(\rho_V^{ab}(s) - \rho_A^{ab}(s) \right) = f_a m_{ab}^2({\rm mech.}) f_b , \quad (22)$$

in the Veneziano-Witten model. Hence there is an order of magnitude difference between these two models of $U_A(1)$ symmetry breaking in all the flavour channels $(a,b = 0,1,\ldots 8)$.

Just one precise measurement, say in the isovector channel, should discriminate between the two models. Some data at low energies already exist, see Fig. 1. in Ref. [10], but the range of energy integration is limited by the τ lepton mass, and saturation of the sum rule is not achieved. Thus far, the results are inconclusive. New kinds of experiments seem necessary. There is hope, however, that the methods described by Hatsuda in these proceedings [20] will allow an "exact" calculation of the second spectral sum rule in lattice QCD.

CONCLUSIONS

1. There are two effective $U_A(1)$ symmetry breaking interactions: 't Hooft and Veneziano-Witten.

2. Scalar meson spectrum and the second spectral sum rule discriminate between them.

3. Flavour-singlet scalar mesons can be identified in accord with either the 't Hooft or Veneziano-Witten interactions. Decay properties seem to slightly prefer 't Hooft, but more work is necessary.

4. Present (τ lepton decay) data on the second Weinberg sum rule does not extend high enough in energy to differentiate between these two interactions.

REFERENCES

1. Nambu, Y., *Symmetry Principles and Fundamental Particles*, W.H. Freeman and Co., San Francisco, 1967.
2. Weinberg, S., *Phys. Rev. D*, **11**, 3583 (1975).
3. 't Hooft, G., *Phys. Rev. D*, **14**, 3432 (1976).
4. Kobayashi, M., Kondo, H., and Maskawa, T., *Prog. Theor. Phys.*, **45**, 1955 (1971).
5. Veneziano, G., *Nucl. Phys. B*, **159**, 213 (1979).
6. Witten, E., *Nucl. Phys. B*, **156**, 269 (1979).
7. Dmitrašinović, V., *Phys. Rev. C*, **53**, 1383 (1996).
8. Klempt, E., Metsch, B. C., Münz, C. R., and Petry, H. R., *Phys. Lett. B*, **361**, 160 (1995).
9. Dmitrašinović, V., *Phys. Rev. D*, **56**, 247 (1997).
10. Dmitrašinović, V., *Phys. Rev. D*, **57**, 7019 (1998).
11. Nambu, Y., and Jona-Lasinio, G., *Phys. Rev.*, **122**, 345 (1961).
12. Karsch, F., *(Proc. Suppl.) Nucl. Phys. B*, **83–84**, 14 (2000).
13. Ritter, C., Metsch, B. C., Münz, C. R., and Petry, H. R., *Phys. Lett. B*, **380**, 431 (1996).
14. Dmitrašinović, V., *Nucl. Phys. A*, **686**, 379 (2001).
15. Weinberg, S., *Phys. Rev. Lett.*, **18**, 507 (1967).
16. Dmitrašinović, V., Klevansky, S., and Lemmer, R. H., *Phys. Lett. B*, **386**, 45 (1996).
17. Nieh, H. T., *Phys. Rev.*, **163**, 1769 (1967).
18. Feynman, R. P., Gell-Mann, M., and Zweig, G., *Phys. Rev. Lett.*, **13**, 678 (1964).
19. Pascual, P., and de Rafael, E., *Z. Phys. C*, **12**, 127 (1982).
20. Hatsuda, T., "", in *these proceedings*, 2001.

ELECTROMAGNETIC PROBES OF NUCLEONS AND NUCLEI

From known to undiscovered resonances

Bijan Saghai

Service de Physique Nucléaire, DAPNIA, CEA/Saclay, 91191 Gif-sur-Yvette, France,
email: bsaghai@cea.fr

Abstract. Electromagnetic meson production formalisms are reviewed, with emphasise placed on their ability in search for new baryon resonances *via* $\gamma p \rightarrow K^+\Lambda$ and $\gamma p \rightarrow \eta p$ processes. The relevant studies, aiming to deepen our insights to hadron spectroscopy, constitute strong tests of the QCD inspired theoretical developements.

1. INTRODUCTION

presently, our knowledge on the baryon resonances comes [1, 2] mainly from partial wave analysis of the "pionic" processes $\pi N \rightarrow \pi N$, ηN, $\gamma N \rightarrow \pi N$, and to less extent, from two pion final states.

The advent of new facilities offering high quality electron and photon beams and sophisticated detectors, has stimulated intensive experimental and theoretical study of the mesons photo- and electro-production. One of the exciting topics here is the search for new baryon resonances which do not couple or couple too weakly to the πN channel. Several such resonances have been predicted [3] by different QCD inspired approaches, offering strong test of the underlying concepts.

In this note, we concentrate on the interpretation of pseudoscalar mesons photoproduction recent data [4, 5, 6, 7, 8], where manifestations of new resonances were reported [9, 10]. The processes under consideration, $\gamma p \rightarrow K^+\Lambda$ and $\gamma p \rightarrow \eta p$, are basically studied *via* two families of formalisms:

- Effective Lagrangian approach, where the amplitudes are in general expressed as Feynman diagrams at tree level.
- Constituent quark approach based on the broken $SU(6) \otimes O(3)$ symmetry.

Below, we summarize the basis of these formalisms and examin their findings with respect to the reported new resonances.

2. THEORETICAL FRAMES

For several decades, effective Lagrangian family approaches have been extensively developped and applied to the electromagnetic production of pseudoscalar mesons.

The most studied channel is by far, the single pion photoproduction where, in the investigated kinematic regions, the reaction mechanism is dominated by the $\Delta(1234)$ resonance. Such a feature has also been observed, although not fully understood, in the

CP594, *Hadrons and Nuclei: First International Symposium,* edited by Il-T. Cheon et al.

TABLE 1. Isospin-1/2 baryon resonances [1] with mass $M_{N^*} \leq 2.5$ GeV. Notations are $L_{2I\ 2J}(mass)$ and $L_{I\ 2J}(mass)$ for N^* and Y^*, respectively.

Baryon	Three and four star resonances	One and two star resonances
N^*	$S_{11}(1535), S_{11}(1650)$, $P_{11}(1440), P_{11}(1710), P_{13}(1720)$, $D_{13}(1520), D_{13}(1700), D_{15}(1675)$, $F_{15}(1680)$, $G_{17}(2190), G_{19}(2250)$, $H_{19}(2220)$,	$S_{11}(2090)$, $P_{11}(2100), P_{13}(1900)$, $D_{13}(2080), D_{15}(2200)$, $F_{15}(2000), F_{17}(1990)$,
Λ^*	$S_{01}(1405), S_{01}(1670), S_{01}(1800)$, $P_{01}(1600), P_{01}(1810), P_{03}(1890)$, $D_{03}(1520), D_{03}(1690), D_{05}(1830)$, $F_{05}(1820), F_{05}(2110)$, $G_{07}(2100)$, $H_{09}(2350)$,	$D_{03}(2325)$, $F_{07}(2020)$,
Σ^*	$S_{11}(1750)$, $P_{11}(1660), P_{11}(1880), P_{13}(1385)$, $D_{13}(1670), D_{13}(1940), D_{15}(1775)$, $F_{15}(1915), F_{17}(2030)$.	$S_{11}(1620), S_{11}(2000)$, $P_{11}(1770), P_{11}(1880), P_{13}(1840)$, $P_{13}(2080)$, $D_{13}(1580)$, $F_{15}(2070)$, $G_{17}(2100)$.

case of the η meson production, where the $S_{11}(1535)$ resonance plays a dominant role, at least up to $\approx$ 100 MeV above threshold. However, the associated strangeness production channel has not shown any strong preference for a given resonance.

The recent data from high duty cycle accelerators allow a real break through in this field and extend the measured (measurable) domains well above threshold and give acces to polarization observables. Then, the relevant formalisms need to have the ability of incorporating a large number of resonances summarized in Table 1. This requirement becomes crucial in searching for new baryon resonances, on which this note focuses.

In this Section, we concentrate on two of the most commonly used formalisms, namely, tree level diagrammatic effective Lagrangians [9, 11, 12, 13, 14, 15, 16, 17, 18, 19, 20], and constituent quark approaches [10, 21, 22, 23]. For other relevant approaches, the reader is refered to Refs. [10, 24].

2.1 Meso-baryonic effective Lagrangian approach

In lines with single photoproduction formalisms, the effective Lagrangian approaches have been extended to the KY [9, 11, 12, 13, 14, 15, 16, 17, 18] and the ηN [20] final states. In this Section, we limit ourseleves to the former channel.

The history of strangeness physics studies *via* electromagnetic probes can be divided into two periods (see, e.g., Refs. [15, 24]). The early works started in the late 50's and went on for about 15 years. Then, in the early 80's, several experimental projects restored this dormant field and gave it a promising future, and due to several foreseen facilities with high quality polarized electron and/or photon beams, revived theoretical investigations in this realm. The starting point of these studies is the effective hadronic Lagrangian approach, using diagrammatic techniques, developed in the old days by Thom, Renard and Renard [11]. However, these works led to a confusing situation [15] on the ingredients of the elementary operator. These pioneering attempts were followed by more extensive investigations [12, 13, 14, 15, 16, 17, 18] of the elementary reactions: $\gamma p \rightarrow K^+\Lambda$, $K^+\Sigma^\circ$, $K^\circ\Sigma^+$, with real and virtual photons, as well as the crossing symmetry channels [13, 16, 17] $K^- p \rightarrow \gamma\Lambda$, $\gamma\Sigma^\circ$.

In this note, we wish to comment on the capability of different formalisms in handling the exchanged resonances in the elementary reaction mechanism.

The most widely used effective Lagrangians are based on the tree approximation, allowing the inclusion of a large number of possible exchanged particles in the *s*-, *u*-, and *t*-channels *via* the relevant Feynman diagrams. Within such phenomenological approaches, *a priori* more than 30 exchanged baryonic resonances (Table 1) can intervene. This uncomfortable situation, where no dominante resonances could be identified, is due to our lack of knowledge [1] on the photo-excitation couplings and/or on the branching ratios of these resonances to the relevant KY final states.

This raises a *crucial question*: does a given formalism allow us to introduce in a model, baryon resonances with spin 1/2, 3/2, and 5/2? Capabilities of the most commonly used formalisms which deal with the above question are summarized in Table 2.

One of the main sources of the level of success of the models built within these formalisms, is the inclusion of the *t*-channel resonances. Actually, we know [25] from the duality hypothesis that there is a close relationship between a dynamical model's content with respect to the included baryon resonances and the corresponding strength of the *t*-channel exchanges needed to fit the data. However, this artifact does not provide fine enough insights into the reaction mechanism: in a given model, we cannot say which baryon resonances, absent in the model, are mimiked by the *t*-channel contributions. However, within the formalisms discussed in this Section, serious technical difficulties prevent us from introducing all the known baryon resonances and hence to discard *t*-channel contributions.

A major problems in the formalisms handling baryonic resonances with spin higher than 1/2 is related to the adopted propagators. As shown by the RPI group [19], the most

TABLE 2. Capabilities of the most commonly used formalisms for the process $\gamma p \rightarrow K^+\Lambda$ in handling nucleonic resonances with spin 1/2, 3/2, and 5/2, and hyperonic resonances with spin 1/2 and 3/2. All the models issued from these formalisms include, besides the extended Born terms, the $K^*(892)$ and $K1(1270)$ exchanges in the *t*-channel.

Resonance (spin) → Group [Ref.] ↓	$N^*(1/2)$	$N^*(3/2)$	$N^*(5/2)$	$Y^*(1/2)$	$Y^*(3/2)$	Off-shell treatment
North Carolina [13]	Y			Y		
Ohio - GWU [9,14]	Y	Y		Y		
Saclay - Lyon [16]	Y	Y	Y	Y		
VPI - Lyon - Saclay [17]	Y	Y		Y	Y	Y
Yonsei [18]	Y	Y	Y	Y	Y	

commonly used propagator [14, 16] for spin 3/2 nucleonic resonances has no inverse. Moreover, this undesirable situation prevents those formalisms from introducing spin 3/2 hyperonic resonances, which otherwise would lead to an unwanted singularity in the u-channel. To overcome this serious shortcoming, the RPI group, investigating the pion and η photoproduction reactions, has applied [19] the Rarita-Schwinger approach [26]. Along the same lines, the authors of Ref. [17] have produced a formalism allowing a proper treatment of *both* nucleonic and hyperonic spin 3/2 resonances. For recent discussions on various aspects of this topic see, e.g. Refs [17, 19, 27, 28].

In Sec. III.A, the results of the formalism discussed above, will be compared to the $\gamma p \to K^+\Lambda$ data.

2.2 Constituent quark effective Lagrangian approach

The starting point of the meson electromagnetic production in the chiral quark model is the low energy QCD Lagrangian [29]. The baryon resonances in the s- and u-channels are treated as three quark systems. The transition matrix elements based on the low energy QCD Lagrangian include the s-, u-, and t-channel contributions $\mathcal{M}_{if} = \mathcal{M}_s + \mathcal{M}_u + \mathcal{M}_t$. The contributions from the s-channel resonances can be written as

$$\mathcal{M}_{N^*} = \frac{2M_{N^*}}{s - M_{N^*}(M_{N^*} - i\Gamma(q))} e^{-\frac{k^2+q^2}{6\alpha_{ho}^2}} \mathcal{A}_{N^*}, \tag{1}$$

where k and q represent the momenta of the incoming photon and the outgoing meson respectively, $\sqrt{s} \equiv W$ is the total c.m. energy of the system, $e^{-(k^2+q^2)/6\alpha_{ho}^2}$ is a form factor in the harmonic oscillator basis with the parameter α_{ho}^2 related to the harmonic oscillator strength in the wave-function, and M_{N^*} and $\Gamma(q)$ are the mass and the total width of the resonance, respectively. The amplitudes $\mathcal{A}_{N^*}$ are split into two parts [21]: the contribution from each resonance below 2 GeV, the transition amplitudes of which have been translated into the standard CGLN amplitudes in the harmonic oscillator basis, and the contributions from the resonances above 2 GeV treated as degenerate, since little experimental information is available on those resonances.

The u-channel contributions are divided into the nucleon Born term and the contributions from the excited resonances. The matrix elements for the nucleon Born term is derived explicitly, while the contributions from the excited resonances above 2 GeV for a given parity are assumed to be degenerate so that their contributions could be written in a compact form.

The t-channel contribution contains two parts: *i)* charged meson exchanges which are proportional to the charge of outgoing mesons and thus do not contribute to the process $\gamma N \to \eta N$; *ii)* ρ- and ω-exchange in the η production which are excluded here due to the duality hypotheses; as discussed in Ref. [10].

Within the exact $SU(6) \otimes O(3)$ symmetry the $S_{11}(1650)$ and $D_{13}(1700)$ do not contribute to the investigated reaction mechanism. However, the breaking of this symmetry leads to the configuration mixings. Here, the most relevant configuration mixings are [10] those of the two S_{11} and the two D_{13} states around 1.5 to 1.7 GeV. The configuration mixings, generated by the gluon exchange interactions in the quark model [30], can be expressed in terms of the mixing angles, Θ_S and Θ_D, between the two $SU(6) \otimes O(3)$ states $|N(^2P_M)>$ and $|N(^4P_M)>$, with the total quark spin 1/2 and 3/2. Results of this approach will be compared to the data for $K^+\Lambda$ and ηp channels in Sections III.A and III.B, respectively.

3. EVIDENCE FOR NEW RESONANCES?

3.1 Associated strangeness production

Recent SAPHIR data [4] for the $\gamma p \rightarrow K^+\Lambda$ process has been claimed [9] to provide *evidence* for a missing D_{13} resonance [31]. In their work based on a meso-baryonic effective Lagrangian approach (see Sec. II.A), Mart and Bennhold (MB) produce a model which contains contributions from;

- Extended Born terms.
- Two kaonic resonances in the t-channel: $K^*(892)$ and $K1(1270)$.
- Three established nucleonic resonances in the s-channel: $S_{11}(1650)$, $P_{11}(1710)$, and $P_{13}(1720)$; hereafter referred to as $N4$, $N6$, and $N7$, respectively. Notice that (Table 1) the two first resonances have spin-1/2, while the third one is a *spin-3/2* resonance. The propagators are in line with those in Ref. [14] and *do not* embody off-shell treatments.

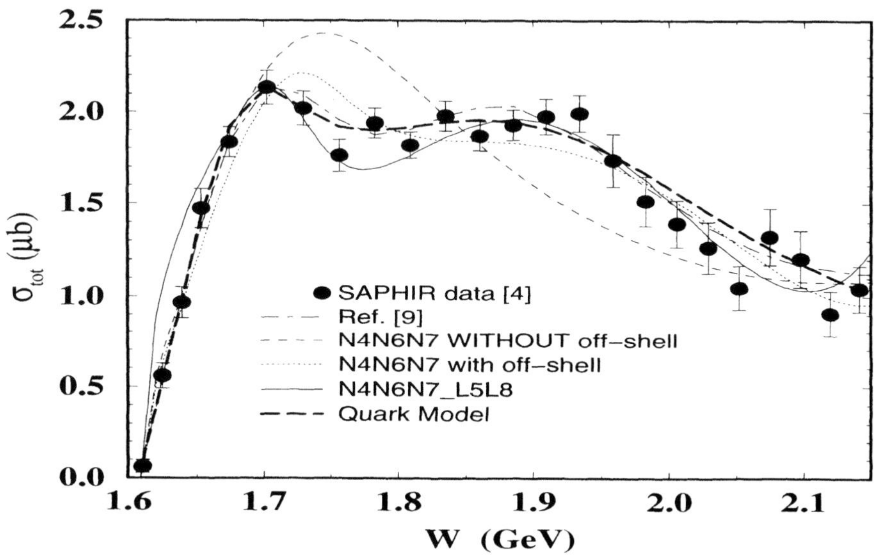

FIGURE 1. Total cross section for the process $\gamma p \rightarrow K^+\Lambda$ as a function of total center-of-mass energy.

- One unknown (or missing) *spin-3/2* nucleonic resonance, that the authors determine as $D_{13}(1895)$.

It is to be noted that this model has no hyperonic resonances.

Figure 1 shows the total cross section data and the MB complete model (dash-dotted curve). MB report [9] also results with a model containing all the above ingredients, except the missing resonance. The main feature of this latter model is that it does not produce any structure around W=1.9 GeV, as required by the (fitted) data.

Using the Saclay-Lyon model [16], we [24] have fitted the same data base (including differential and total cross-sections from SAPHIR, old recoil Λ asymmetry [32], and recent JLab electroproduction data [33]) as MB. Limiting the reaction mechanism to the above ingredients without the missing D_{13} resonance, we obtain the same features as MB. Our fit for the [N4, N6, N7] set is shown in Fig. 1 (dashed curve) and decreases monotonically beyond a maximum around W=1.72 GeV. As a next step, and for the reason explained in Section II.A, we include the off-shell effect treatments in line with Ref [17]. We then get the dotted curve which shows a significant enhancement in the cross section above 1.85 GeV. By introducing two hyperonic resonances $P_{01}(1810)$ and $P_{03}(1890)$ (hereafter called *L5* and *L8*, respectively), the data are well reproduced (full curve). This set of resonances [N4, N6, N7, L5, L8] reproduces reasonably also the other fitted data. Our results therefore show that there is *no need* for a missing resonance.

The above considerations can not be taken as an attempt to produce a new model, but just as an *illustration* as how cautious we have to be in using the existing formalisms when searching for new resonances. A more reliable approach in this respect, should allow us to embody all known resonances.

Such an opportunity is offered to us by the constituent quark formalims presented in Section II.B. We [23] have included all known nucleon and hyperon resonances given in Table 1, and have fitted the photoproduction data base. The result is given in Fig. 1 (heavy dashed curve). The agreement between this latter curve and the data endorses our conclusions that the SAPHIR data does not show any manifestations of a new resonance.

3.2 η-meson production

Using a constituent quark model (Section II.B), we [10] have fitted the following sets of the η-photoproduction data: differential cross-sections from MAMI/Mainz [5] and Graal [8], as well as the polarized beam asymmetry from Graal [7]. Then we have predicted [10] the total cross-section and the polarized target asymmetry. This latter observable has been measured at ELSA [6].

In Fig. 2, we show comparison for the total cross-section, for the models I and II. Both models include all 3 and 4 star known nucleon resonances (Table 1). They also satisfy the configuration mixing requirements and for both models, the extracted mixing angles are in agreement with the Isgur-Karl [30] predictions.

The model I reproduces fairly well the total cross-section data up to $W \approx 1.61$ GeV. Between this latter energy and ≈ 1.68 GeV, the model overestimates the data, and above 1.68 GeV, the predictions underestimate the experimental results, missing the total cross-section increase.

In summary, results of the model I show clearly that an approach containing a correct treatment of the Born terms and including *all known resonances* in the s- and u-channels *does not* lead to an acceptable model, even within broken $SU(6) \otimes O(3)$ symmetry scheme. To go further, one possible scenario is to investigate manifestations of yet undiscovered resonances. As already mentioned, rather large number of such resonances has been predicted by several authors. To find out which ones could be considered as relevant candidates, we examined the available data.

The differential cross-sections [10, 34], show clearly that this mismatch is due to the forward angle peaking of the differential cross-section for $W \geq 1.68$ GeV ($E_\gamma^{lab} \geq$ 1. GeV). Such a behaviour might likely arise from missing strength in the S-waves. This latter conclusion is endorsed by the role played by the E_0^+ in the multipole structure of the differential cross-section and the single polarization observables. If there is indeed an additional S-wave resonance in this mass region, its dependence on incoming photon and outgoing meson momenta would be qualitatively similar to that of the $S_{11}(1535)$, even though the form factor might be very different. Thus, for this new resonance, we use the same CGLN amplitude expressions as for the $S_{11}(1535)$. We have hence introduced [1] a third S_{11} resonance and refitted the same data base as for the model I, leaving it's mass and width as free parameters. The results of this model, depicted in Fig. 1 (full curve), reproduce nicely the data. This is also the case [10] for the polarized beam and polarized target asymmetries. For this latter observable, our predictions yield a good agreement with the data.

For the new S_{11} resonance, we find M=1.729 GeV and Γ=183 MeV. These values are amazingly close to those of a predicted [35] third S_{11} resonance, with M=1.712 GeV

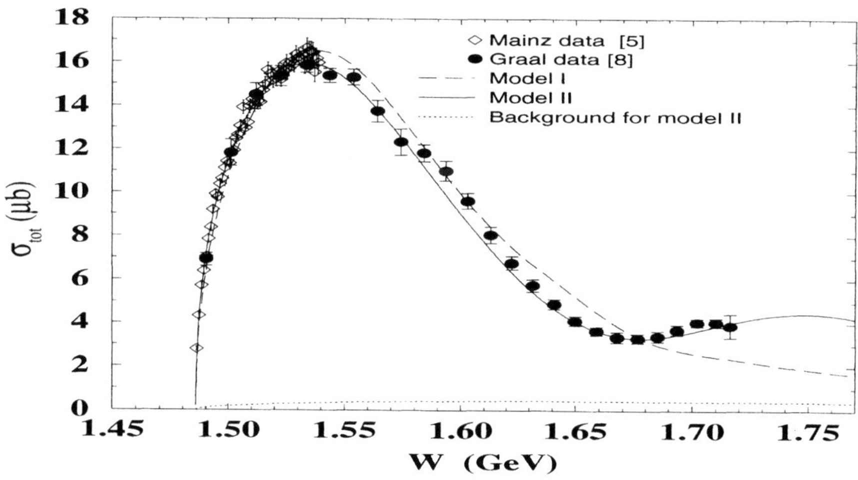

FIGURE 2. Total cross section for the reaction $\gamma p \to \eta p$ as a function of total center-of-mass energy. The curves come from the models I (dashed), II (full). The dotted curve shows the background terms contribution in the model II.

and Γ_T=184 MeV.

4. SUMMARY AND CONCLUDING REMARKS

In this note we concentrated on the search for new resonances *via* the two processes $\gamma p \to K^+\Lambda$, ηp for which recent data have become available.

The effective Lagrangian approaches, using Feynman diagrams at tree level, applied to the above channels, allow to study some specific aspects of the reaction mechanism. However, they are not suitable in looking for new resonances. The reason for this shortcoming is that they do not allow the inclusion of *all* relevant known resonances. Although the introduction of spin-1/2 resonances is straightforward, higher spin resonances are more complicated to be handlded. The main difficulty comes from the incorporation of the so called *off-shell effects*, inherent to the fermions with spin $\geq 3/2$. Presently, these effects can be embodied for spin-3/2 resonances, but no conclusive attempt has been made for higher spins. Another limitation of these approaches is due to the number of free parameters: one for each spin-1/2, two for each higher spin, plus 3 off-shell parameters per resonance. In other words, even if we were able to treat all higher spin resonances correctly, the very large number of parameters would not allow to reach any clear conclsions on the possible manifestations of new resonances. Notice that such resonances are expected above the first resonance region, where higher spin resonances are expected to play significant roles.

The advantage of the quark model for the meson photoproduction is the ability to introduce all known resonances. Moreover, the number of adjustable parameters, one per resonance in the broken $SU(6) \otimes O(3)$ limit, stays much smaller than in the case of the above formalism. Contrary to the former approache, the quark model adjustable parameters measure the extent to which the $SU(6) \otimes O(3)$ symmetry is broken. Hence, they should stay rather close to their $SU(6) \otimes O(3)$ symmetry values, while in the case of effective Lagrangians, apart for a few exceptions, there are no constraints on the range of the fitted parameters. Besides, the constituent quark models allow us to relate the data directly to the internal structure of the baryon resonances.

The main conclusion here is therefore: *the appropriate framework in search for new baryonic resonances is constituent quark approaches.*

The above conclusion was illustrated in this note by two examples and the findings are:

- Recent $\gamma p \to K^+\Lambda$ SAPHIR data can be understood by taking into account the known resonances within an effective hadronic Lagrangian approach embodying off-shell effects for spin-3/2 baryon resonances, as well as within a constituent quark model. There is hence no need for introducing unkown resonances.
- Investigation of the recent $\gamma p \to \eta p$ Graal data within a chiral constituent quark approach based on the broken $SU(6) \otimes O(3)$ symmetry, shows clear need for a new S_{11} resonance, with mass $M \approx 1.730$ GeV and total width $\Gamma \approx$180 MeV.

To gain insights to the nature of this resonance, an extension of the η electroproduction studies above the first resonance region is in progress [36]. Investigation of vector meson electromagnetic production within constituent quark models [37, 38] appears also very promising in baryon spectroscopy.

ACKNOWLEDGEMENTS

It is a pleasure to thank the organizers for their kind invitation to this very stimulating Memorial meeting. I am indebted to my collaborators: C. Fayard, G.H. Lamot, Z. Li, T. Mizutani, P. Oswald, F. Tabakin, T. Ye, and Q. Zhao.

REFERENCES

1. Particle Data Group, *Eur. Phys. J.* **15**, 1 (2000).
2. T.P. Vrana, S.A. Dytman, and T.S.H. Lee, *Phys. Rep.* **328**, 181 (2000); R. Workman, *nucl-th/0104028*; D.M. Manley and E.M. Saleski, *Phys. Rev. D* **45**, 4002 (1992).
3. S. Capstick and W. Roberts, *Prog. Part. Nucl. Phys.*, **45** (Suppl. 2), 5241 (2000); R. Bijker, F. Iachello, and A. Leviatan, *Ann. Phys.* **284**, 89 (2000); L.Ya. Glozman and D.O. Riska, *Phys. Rep.* **268**, 263 (1996); M. Anselmino *et al.*, *Rev. Mod. Phys.* **65**, 1199 (1993).
4. M.Q. Tran *et al.*, *Phys. Lett.* **B445**, 20 (1998).
5. B. Krusche et al., *Phys. Rev. Lett.* **74**, 3736 (1995).
6. A. Bock *et al.*, *Phys. Rev. Lett.* **81** 534, (1998).
7. J. Ajaka *et al.*, *Phys. Rev. Lett.* **81**, 1797 (1998).
8. F. Renard *et al.*, *hep-ex/0011098.*
9. T. Mart and C. Bennhold, *Phys. Rev. C* **61**, 012201 (2000).
10. B. Saghai and Z. Li, *nucl-th/0104084*, submitted to *Eur. Phys. J.*
11. H. Thom, *Phys. Rev.* **151**, 1322 (1966); F.M. Renard and Y. Renard, *Nucl. Phys.* **B25**, 490 (1971); Y. Renard, *ibid.* **B40**, 499 (1972).
12. J. Cohen, *Int. J. Mod. Phys.* **A4**, 1 (1989).
13. R.A. Williams, C.R. Ji, and S.R. Cotanch, *Phys. Rev. C* **46**, 1617 (1992).
14. R.A. Adelseck, C. Bennhold, and L.E. Wright, *Phys. Rev. C* **32**, 1681 (1985); C. Bennhold and L.E. Wright, *ibid C* **39**, 927 (1989); T. Mart, C. Bennhold, and C.E. Hyde-Wright, *ibid C* **51**, R1074 (1995).
15. R.A. Adelseck and B. Saghai, *Phys. Rev. C* **42**, 108 (1990); *ibid C* **45**, 2030 (1992).
16. J.C. David, C. Fayard, G.H. Lamot, and B. Saghai, *Phys. Rev. C* **53**, 2613 (1996).
17. T. Mizutani, C. Fayard, G.H. Lamot, and B. Saghai, *Phys. Rev. C* **58**, 75 (1998).
18. M.K. Cheoun, B.S. Han, I.T. Cheon, and B.G. Yu, *Phys. Rev. C* **54**, 1811 (1996); B.S. Han, M. K. Cheoun, K.S. Kim, and I-T. Cheon, *nucl-th/9912011.*
19. M. Benmerrouche *et al.*, *Phys. Rev. C* **39**, 2339 (1989).
20. M. Benmerrouche, N.C. Mukhopadhyay, and J. F. Zhang, *Phys. Rev. D* **51**, 3237 (1995); B. Krusche *et al.*, *Phys. Lett.* **B397**, 171 (1997); N.C. Mukhopadhyay and N. Mathur, *ibid* **B444**, 7 (1998); R.M. Davidson *et al.*, *Phys. Rev. C* **62**, 058201 (2000).
21. Z. Li, H. Ye, and M. Lu, *Phys. Rev. C* **56**, 1099 (1997).
22. Z. Li and B. Saghai, *Nucl. Phys.* **A644**, 345 (1998).
23. Z. Li, B. Saghai, T. Ye, and Q. Zhao, *in progress.*
24. B. Saghai, *Nucl. Phys.* **A639**, 217c (1998).
25. B. Saghai and F. Tabakin, *Phys. Rev. C* **53**, 66 (1996); *ibid C* **55**, 917 (1997).
26. W. Rarita and J. Schwinger, *Phys. Rev.* **60**, 61 (1941).
27. V. Pascalutsa, *Phys. Lett.* **B503**, 85 (2001); and references therein.
28. T.H. Hemmert, B.R. Holstein, and J. Kambor, *J. Phys. G* **24**, 1831 (1998).
29. A. Manohar and H. Georgi, *Nucl. Phys.* **B234**, 189 (1984).
30. N. Isgur and G. Karl, *Phys. Lett.* **B72**, 109 (1977); N. Isgur, G. Karl, and R. Koniuk, *Phys. Rev. Lett.* **41**, 1269 (1978).

31. S. Capstick, *Phys. Rev. D* **46**, 2864 (1992); S. Capstick and W. Roberts, *ibid D* **49**, 4570 (1994); *ibid D* **58**, 074011 (1998).
32. B. Borgia *et al.*, *Nuovo Cimento* **32**, 218 (1964); M. Grilli *et al.*, *ibid* **38**, 1467 (1965); D.E. Groom and J.H. Marshall, *Phys. Rev.* **159**, 1213 (1967); R. Hass *et al.*, *Nucl. Phys.* **B137**, 261 (1978).
33. G. Niculescu *et al.*, *Phys. Rev. Lett.* **81**, 1805 (1998).
34. B. Saghai, *nucl-th/0104086*.
35. Z. Li and R. Workman, *Phys. Rev. C* **53**, R549 (1996).
36. Q. Zhao, B. Saghai, and Z.-P. Li, *nucl-th/0011069*, and *in progress*.
37. Q. Zhao, J.-P. Didelez, M. Guidal, and B. Saghai, *Nucl. Phys.* **A660**, 323 (1999); Q. Zhao, B. Saghai, and J.S. Al-Khalili, *nucl-th/0102025*, to appear in *Phys. Lett.* **B**; Y. Oh and H.C. Bhang, *nucl-th/0104068*.
38. Y. Oh, A.I. Titov, and T.-S. H. Lee, *Phys. Rev. C* **63**, 025201 (2001); *nucl-th/0104046*; Q. Zhao, *Phys. Rev. C* **63**, 025203 (2001).

The CEBAF Electron Accelerator, Physics Program and First Results

Bernhard A. Mecking

Thomas Jefferson National Accelerator Facility, 12000 Jefferson Avenue, Newport News, U.S.A.

Abstract. The Continuous Electron Accelerator Facility, CEBAF, is devoted to the investigation of the electromagnetic structure of mesons, nucleons, and nuclei using high-energy and high duty-cycle electron and photon beams. The physics program and the experimental equipment at CEBAF will be described. Selected examples for the results obtained will be given. The motivation for upgrading the accelerator to higher energy will be discussed.

INTRODUCTION

The Southeastern University Research Association (SURA), a university consortium, operates the Thomas Jefferson National Accelerator Facility, also called Jefferson Lab (or JLab), for the U.S. Department of Energy. JLab's main instrument is the Continuous Electron Beam Accelerator Facility, CEBAF, a continuous-wave electron accelerator capable of delivering precise electron beams for experiments in three experimental areas. CEBAF's user community consists of about 1,500 physicists, roughly 800 of them are actively involved in the experimental program. In addition to its main mission, JLab makes contributions to the development of Free Electron Lasers, to medical imaging, and to community outreach programs.

THE CEBAF ACCELERATOR

CEBAF's central instrument is a superconducting electron accelerator with a maximum design energy of 4 GeV and 100 % duty-cycle. Three electron beams with a maximum total current of 200 μA can be used for simultaneous electron scattering experiments in three experimental areas, Halls A, B, and C.

The accelerator design is based on two parallel superconducting continuous-wave linear accelerators joined by magnetic recirculation arcs. The accelerating structures are 338 five-cell superconducting niobium cavities operating at a temperature of $2K$ with an average design energy gain of 5 MeV/m.

The isochronous magnetic recirculation arcs allow the electron beam to be recirculated up to five times. After each of the first four passes, a transverse radio frequency separator can be activated to extract every third bunch, and to deliver an electron beam with an energy of $\frac{1}{5}, \frac{2}{5}, \frac{3}{5}$, or $\frac{4}{5}$ of $E_{\max}$ to one (and only one) of the three experimental areas. At the end of the fifth pass, a last separator can split the beam up to three ways;

CP594, *Hadrons and Nuclei: First International Symposium,* edited by Il-T. Cheon et al.

this allows the highest energy beam to be sent to all three halls. Typical beam parameters are $\delta E/E \leq 10^{-4}$ and a beam size of a few hundred μm.

A very important component of the accelerator is the polarized electron source. Using three independent pulsed lasers illuminating a photocathode makes it possible to create up to three polarized electron beams that can be varied individually in intensity by varying the intensities of the lasers. A common Wien filter compensates the spin precession in the recirculation arcs and in the beam lines leading to the halls to give the desired longitudinal polarization on target. All halls monitor the electron polarization with Møller polarimeters; in addition, Hall A uses a Compton backscattering polarimeter.

EXPERIMENTAL HALLS

Three experimental halls are available for simultaneous experiments. Since the beam intensity can be varied over a large dynamic range, the only important restriction for multi-hall operation is that the beam energies have to be multiples of the single-pass energy. To cover a wide range of physics problems the halls contain complementary equipment: Hall A has two high resolution magnetic spectrometers, Hall B houses a large acceptance spectrometer, and Hall C uses a combination of a high momentum and a short orbit spectrometer. The hall equipment will be discussed in more detail in the following sections.

Hall A

The Hall A instrumentation is designed for experiments that require the precise reconstruction of the mass of the unobserved hadronic final state via missing-mass techniques; a typical example is the excitation energy of the unobserved nucleus A^* in the $A(e,e'p)A^*$ reaction. The instrumentation consists of two high resolution magnetic spectrometers (HRS) that can be rotated around a common pivot. Design requirements are 10^{-4} momentum resolution in the 10% momentum bite and the solid angle of 7 msr. The spectrometers are based on a QQDQ design with superconducting magnetic elements. The tracking part of the detector systems is identical for both spectrometers; the particle identification part is specialized for electrons in one spectrometer, and for hadrons in the other. The hadron arm also contains a focal plane polarimeter to measure the transverse polarization components of the detected protons.

Hall B

Hall B is equipped with a large acceptance detector (CEBAF Large Acceptance Spectrometer, CLAS). Its main mission is to carry out experiments that require the detection of several, only loosely correlated particles in the hadronic final state, and measurements at limited luminosity (e.g. due to the use of a polarized target or the use of a tagged bremsstrahlung photon beam).

CLAS is a toroidal multi-gap spectrometer; its magnetic field is generated by six iron-free superconducting coils. The particle detection system consists of drift chambers to determine the trajectories of charged particles, Cerenkov counters for the identification of electrons, scintillation counters for the trigger and for time-of-flight measurements, and electromagnetic calorimeters to identify electrons and to detect photons and neutrons. CLAS has been designed for both electron and photon induced reactions. For real photon experiments, a bremsstrahlung tagging spectrometer, located at the entrance of the hall, is used to determine photon energy and flux.

Hall C

The main instruments in Hall C are two focusing magnetic spectrometers: the High Momentum Spectrometer (HMS) and the Short Orbit Spectrometer (SOS). Typical values for the momentum resolution are 0.1%. The HMS is a QQQD spectrometer with iron-dominated quadrupoles and a homogeneous field dipole. All HMS magnets use superconducting coils to reach momenta up to 7 GeV/c. The SOS is a normal-conducting QDD spectrometer with a maximum central momentum of 1.5 GeV/c. Its short path length makes it ideal for the detection of decaying particles, such as pions and kaons.

ACCELERATOR AND HALL STATUS

Due to the excellent performance of the superconducting cavities, the accelerator has by far exceeded the 4 GeV design goal. Experiments have been conducted at 5.7 GeV, and 6 GeV has been reached in a short test run. Three-hall operation has become routine, with intensity ratios up to $10^5 : 1$ (note that Hall B is typically operating at a few nA, only).

The polarized source using a strained GaAs crystal has reached a CW current of $140 \mu A$ at about 75% polarization. Helicity correlated changes in the beam parameters are small enough to make parity violation experiments possible.

Hall C was the first hall to become operational. Seventeen experiments have been completed, including several that required specialized setups, e.g. deuteron polarimeter, polarized target, and hypernuclear spectroscopy.

The Hall A HRS spectrometers have been fully operational since 1997. Nineteen experiments have been completed. Special efforts were required to measure small asymmetries caused by parity violation effects, and to carry out experiments using a polarized ^{3}He target.

In Hall B, CLAS has been in routine use since 1998. In electron scattering experiments the design luminosity of $10^{34} cm^{-2} s^{-1}$ has been reached; up to 3,000 events/s can be recorded and stored for later analysis. Special efforts were required to conduct experiments with a dynamically polarized hydrogen/deuterium target.

EXPERIMENTAL RESULTS (EXAMPLES)

Proposals submitted by CEBAF's user community are reviewed by the Program Advisory Committee (PAC) which meets twice a year. As of the summer of 2000, the PAC has recommended approval of a total of 121 experiments proposed by 869 authors from 156 institutions located in 24 countries. Thirty experiments have been completed, partial data has been taken for 47 additional experiments.

In the following sections, some representative examples for completed CEBAF experiments will be given. The experiments have been chosen not only for their physics interest, but also to highlight the capabilities of the accelerator and the halls.

Parity Violation in $e-p$ Elastic Scattering

The proton is a composite object containing constituent u and d quarks, and potentially a sea of $q\bar{q}$–pairs. Little is known about the $q\bar{q}$–contributions, especially the strange quark content of the nucleon. Combining the form factors of proton and neutron, and invoking isospin symmetry, allows the separation of the contributions of u and d quarks. The additional information required to determine the $s\bar{s}$ content can be obtained from measuring parity violation in electron–nucleon scattering. The parity violating effect arises from the interference between one-photon and Z^o exchange. The parity violating asymmetry in elastic $e-p$ scattering – in combination with the electromagnetic form factors – makes it possible to separate the contributions of u, d, and s quarks.

Measuring parity violating effects is conceptually straightforward: all that is required is to flip the helicity of the incident electrons and look for a change in the reaction cross section. However, due to the smallness of the parity violating effect (a few ppm or less), these experiments have been notoriously difficult in the past. The main reason is that beam parameters like intensity, energy, halo, etc. are not perfectly stable when the helicity is flipped. It had been hoped that these problems would be easier to handle with the new generation of high duty-cycle electron accelerators.

An experiment was carried out in Hall A [1] to measure parity violation in elastic $e-p$ scattering. A $100\mu A$ longitudinally polarized electron beam with an energy of 3.3 GeV hit a high–power liquid hydrogen target. The beam helicity was changed in a pseudo–random fashion with 30 Hz frequency. Both HRS spectrometers were set up at small angles ($\approx 12^o$), elastically scattered electrons were detected in lead-lucite sandwich counters. To handle the high counting rates, the electronic output signals of the counters were integrated over 30 Hz periods.

The performance of the accelerator during the experiment exceeded the high expectations: helicity correlated systematic changes in beam position were on the order of a few nm, changes in the beam energy at the 10^{-8} level. Therefore, the corrections to the raw asymmetry to account for helicity correlated effects were negligible, a first for parity violation experiments.

The experimental parity violating asymmetry has been compared to what is expected on the basis of the assumption that there is no strange quark contribution. The measured difference is plotted in Fig. 1 as a function of Q^2, and compared to various theoretical

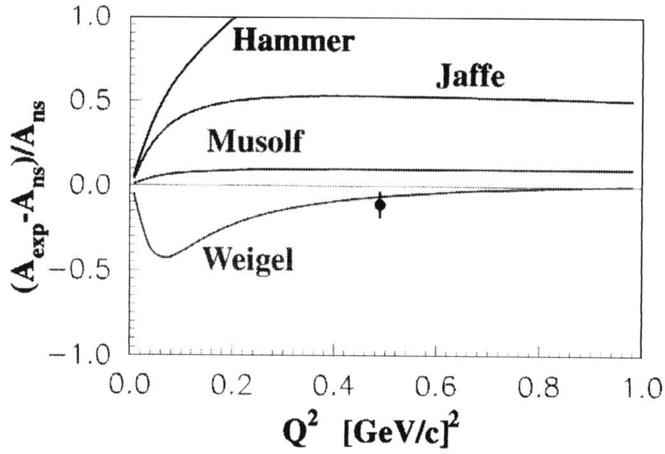

FIGURE 1. Parity Violating Asymmetry in Elastic e–p Scattering.

predictions. The measurement does not support a large contribution of strange quarks to the proton.

The Hall A spectrometers will continue to be used for parity experiments. In addition, a specialized toroidal spectrometer with eight gaps is being constructed for Hall C. This device will be able to cover both forward and backward electron scattering (in separate runs), and will allow separation of the strange electric and magnetic form factors.

Measurement of t_{20} in Elastic $e-d$ Scattering

The electromagnetic properties of the deuteron (a spin–1 nucleus) are described by three independent form factors, characterizing the spatial distribution of charge, magnetization, and quadrupole strength. The standard Rosenbluth technique (varying the electron scattering angle while keeping the momentum transfer constant) determines only two combinations of these form factors. An additional polarization measurement is required to isolate all three form factors.

A custom setup [2] was used in Hall C to determine the tensor polarization, t_{20}, of the recoiling deuteron in elastic $e-d$ scattering. An electron beam with currents up to $120\mu A$ hit a high–power liquid deuterium target, resulting in an average luminosity of $3 \cdot 10^{38} cm^{-2} s^{-1}$. Elastically scattered electrons were detected in the HMS. Deuterons were transported from the production target to a secondary scattering target in a specially constructed achromatic magnetic channel. The tensor polarization of the recoiling deuterons was determined from the angular distribution of the protons in the $p(\vec{d},pp)n$

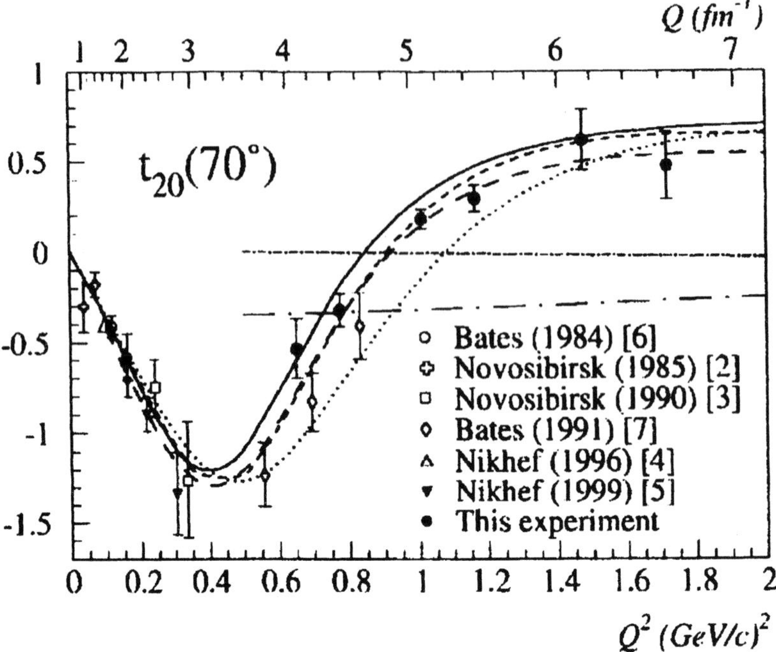

FIGURE 2. Tensor Analyzing Power t_{20} in Elastic $e-d$ Scattering.

reaction. The analyzing power of the polarimeter had been determined previously at SATURNE using a deuteron beam of known polarization.

The experiment covered a range in momentum transfer $Q^2 \approx (0.65-1.7)(GeV/c)^2$. The results are displayed in Fig. 2. In this Q^2 range there are no indications yet for the onset of the scaling effects predicted by perturbative QCD. In fact, the results are well described by conventional models assuming that the deuteron is a bound proton-neutron system (with appropriate corrections for meson–exchange currents and relativistic effects).

Measurement of the $N \to \Delta$ Transition

Exploring the electromagnetic structure of the excited states of the nucleon is one of the major experimental programs at CEBAF. These studies are of fundamental importance since they determine charge and current distribution as well as the transition currents. Measurements of these quantities may be used to test microscopic models, such as QCD based quark models, chiral bag models, and Skyrme models.

Electromagnetic excitation of the nucleon resonances on the free nucleon yields information on the γNN^* vertex. The broad width of the resonances makes it impossible

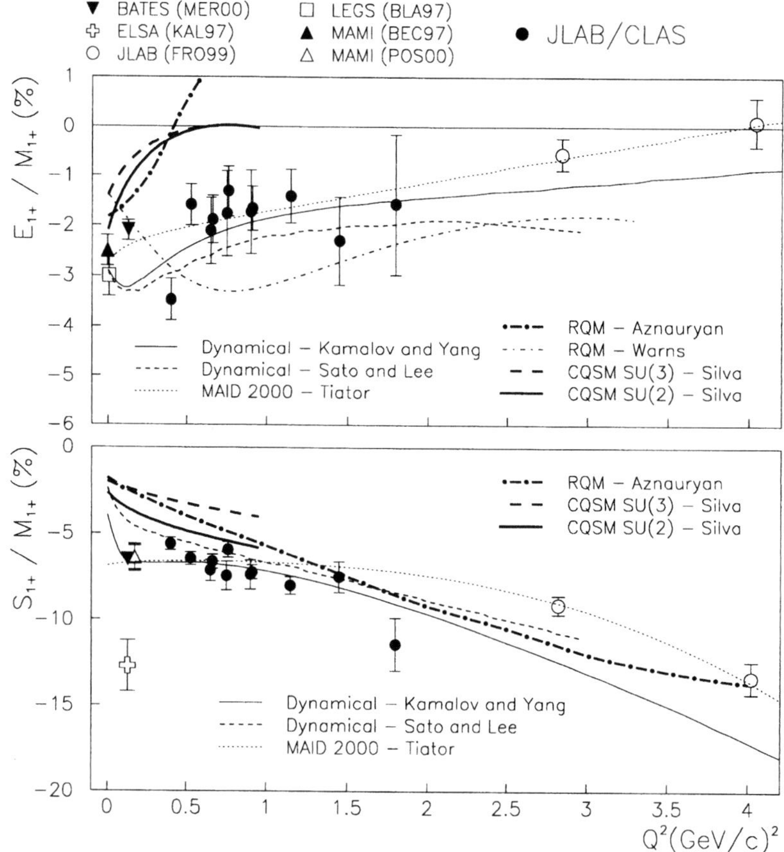

FIGURE 3. Electric and Scalar Quadrupole Amplitudes for the $N \to \Delta(1232)$ Transition.

to separate them in inclusive reactions. By determining in coincidence the angular distribution of the N^* decay products, such as πN, ηN, ρN, and $\pi\Delta$, it is possible to separate the contributions of the individual resonances.

There is a broad program in Hall B using the CLAS detector to determine these transition amplitudes from a measurement of differential cross sections and polarization asymmetries for the photo- and electro-production of mesons. A prominent example is the quadrupole strength in the $N \to \Delta$ transition.

In SU(6) symmetric quark models, this transition is described by a simple quark spin-flip in the $L_{3Q} = 0$ ground state, corresponding to a magnetic dipole transition M_{1+}. The electric and scalar quadrupole transitions are predicted to be vanishing: $E_{1+} = S_{1+} = 0$. In more elaborate QCD based models which include color-magnetic interactions due to one-gluon exchange at small distances, the $\Delta(1232)$ acquires an $L_{3Q} = 2$ component, leading to small electric and scalar contributions. A deformation of the Δ can also be created by a pion cloud surrounding the three-quark core.

The values of the electric and scalar quadrupole amplitudes can be derived from an analysis of the angular distribution (in both polar and azimuthal angles) of the pion in the $p(e,e'p)\pi^o$ reaction. The CLAS detector in Hall B was used to cover a large range in Q^2, mass of the hadronic final state, W, and pion angles simultaneously [3]. Figure 3 shows the electric $Re(E_{1+}M^*_{1+})/M^2_{1+}$ and the scalar $Re(S_{1+}M^*_{1+})/M^2_{1+}$ amplitudes at the position of the $\Delta(1232)$ as a function of Q^2. The results demonstrate the importance of taking the coupling of the photon to the pion cloud explicitly into account.

THE CEBAF ENERGY UPGRADE

There is a broad class of important physics questions that can not be addressed at the present time due to the lack of an electron accelerator with the combination of high energy, high duty-cycle, and the high beam power required to study rare phenomena. Of particular importance are to study:

- the origin of quark confinement by searching for mesons with exotic quantum numbers. These objects can be formed by exciting the flux tube that binds $q\bar{q}$-systems.
- the quark-gluon structure of the nucleon by mapping out the longitudinal quark momentum distributions in the valence quark region, and by studying quark–quark correlations via exclusive processes (by exploiting the newly discovered Generalized Parton Distributions).

In addition to these new programs, the present program can be extended to higher momentum transfers.

The optimization of the physics reach, and practical considerations have led to the development of the present 12 GeV upgrade proposal [4]. Higher energy than 12 GeV would allow to get to higher momentum transfer, Q^2, and thus make access to the deep-inelastic regime easier. However, with increasing energy the requirements on detector resolution and luminosity (to get sufficient count rate at the higher Q^2) will become increasingly difficult and costly to meet.

Upgrading the CEBAF accelerator to 12 GeV will require to increase the energy of the linear accelerator from presently 1200 MeV to 2200 MeV. This will be accomplished by adding additional accelerating structures, and by replacing some of the original cavities by newer, higher gradient versions. All of the recirculating arcs will have to be upgraded to higher integral magnetic field.

The highest energy of 12 GeV will be available in a new experimental area at the end of the north linac, Hall D. This area will be devoted to the photoproduction of exotic mesons using a linearly polarized photon beam and a large acceptance detector. The three existing areas will be able to use any multiple of 2200 MeV (or a lower energy), up to 11 GeV.

The hall equipment will be upgraded to take advantage of the higher energy. In Hall A, a broad-range spectrometer (Medium Acceptance Spectrometer, MAD) will be added. The CLAS in Hall B will be upgraded to higher luminosity and improved forward coverage. In Hall C, the SOS will be replaced by a new Super High Momentum

Spectrometer, SHMS, to cover very small angles and high momenta.

SUMMARY

The physics motivation and the experimental instrumentation of the Continuous Electron Beam Accelerator have been discussed. The unique combination of high energy, high duty-cycle, high intensity and small phase space beams from CEBAF's superconducting accelerator, in combination with complementary equipment in three experimental halls, allows performing experiments that determine the internal structure of mesons, nucleons, and nuclei with unprecedented precision.

Many important physics issues that are addressed in CEBAF experiments require isolating small components of the cross section via the measurement of interference terms. The use of polarized electrons, polarized targets and polarimeters has become increasingly important.

The energy of the CEBAF accelerator has exceeded the initial design energy, and is now approaching 6 GeV. Exciting new physics questions require the energy of the accelerator to be increased to 12 GeV. An economical plan has been developed to achieve this increase by small changes to the present accelerator. Concurrently, the experimental equipment needs to be upgraded to handle the challenges at the higher energy.

REFERENCES

1. K.A. Aniol *et al.*, PRL 82, 1096 (1999).
2. D. Abbott *et al.*, PRL 84, 5053 (2000).
3. K. Joo *et al.*, to be published
4. L.S. Cardman *et al.*, White Paper 'The Science Driving the 12 GeV Upgrade of CEBAF', 2001

First Results from SPring-8/LEPS and Prospects

J. K. Ahn[*], H. Akimune[†], Y. Asano[**], W. C. Chang[‡], S. Daté[§], M. Fujiwara[¶], K. Hicks[||], T. Hotta[¶], K. Imai[††], T. Ishikawa[††], T. Iwata[‡‡], H. Kawai[§§], H. Kohri[¶], N. Kumagai[§], S. Makino[¶¶], T. Matsumura[¶], T. Matsuoka[¶], T. Mibe[¶], M. Miyabe[††], Y. Miyachi[‡‡], N. Muramatsu[**], T. Nakano[¶], M. Nomachi[***], Y. Ohashi[§], H. Ohkuma[§], T. Ooba[§§], A. Sakaguchi[***], Y. Shiino[§§], H. Shimizu[†††], Y. Sugaya[***], M. Sumihama[***], H. Toyokawa[§], C. Rangacharyulu[‡‡‡], A. Wakai[‡‡], C. W. Wang[‡], S. C. Wang[‡], K. Yonehara[†], M. Yosoi[††], T. Yorita[¶] and R. Zegers[**]

[*]*Department of Physics, Pusan National University, Pusan 609-735, Korea*
[†]*Department of Physics, Konan University, Kobe 658-8501, Japan*
[**]*Advanced Science Research Center, JAERI, Tokai, Ibaraki 319-1195, Japan*
[‡]*Institute of Physics, Academia Sinica, Taipei 11529, Taiwan*
[§]*JASRI, SPring-8, Sayou, Hyogo 679-5198, Japan*
[¶]*RCNP, Osaka University, Ibaraki, Osaka 567-0047, Japan*
[||]*Department of Physics, Ohio University, Athens, Ohio 45701, USA*
[††]*Department of Physics, Kyoto University, Kyoto 606-8502, Japan*
[‡‡]*Department of Physics, Nagoya University, Nagoya 464-8602, Japan*
[§§]*Department of Physics, Chiba University, Chiba 263-8522, Japan*
[¶¶]*Wakayama Medical College, Wakayama 641-0012, Japan*
[***]*Department of Physics, Osaka University, Toyanaka, Osaka 560-0043, Japan*
[†††]*Department of Physics, Yamagata University, Yamagata 990-8560, Japan*
[‡‡‡]*Department of Physics, University of Saskatchewan, Saskatoon, SK S7N 5E2, Canada*

Abstract. The Laser-Electron-Photon facility at SPring-8 (LEPS) is now taking data for all hadronic reactions at compton-backscattered photon energies between 1.5 GeV and 2.4 GeV. A linearly polarized photon beam is incident on a liquid hydrogen target or nuclear targets. A large dipole spectrometer analyzes charged particles produced at forward angles. First results from SPring-8/LEPS will be presented with special emphasis on the study of photoproduction of ϕ mesons near threshold. Some prospects for the near future will also be presented and discussed in close connection with photoproduction of hyperons.

INTRODUCTION

Photoproduction of the ϕ vector meson from the proton is of fundamental interest since the process is OZI suppressed in the absence of meson-exchange process due to nearly pure $\bar{s}s$ state of the ϕ meson. The photon mixes into a neutral vector meson and then diffractively scatters off the target in the context of the vector dominance model. The process is dominated by diffractive scattering via Pomeron exchange, as shown in Fig 1. Diffractive photoproduction processes involve exchange of vacuum quantum numbers in t-channel and a characteristic t-dependence in an exponential parameterization; $d\sigma/dt \sim \exp(-b\,|\,t\,|)$, where t is the Lorentz-invariant four-momentum

CP594, *Hadrons and Nuclei: First International Symposium*, edited by Il-T. Cheon et al.

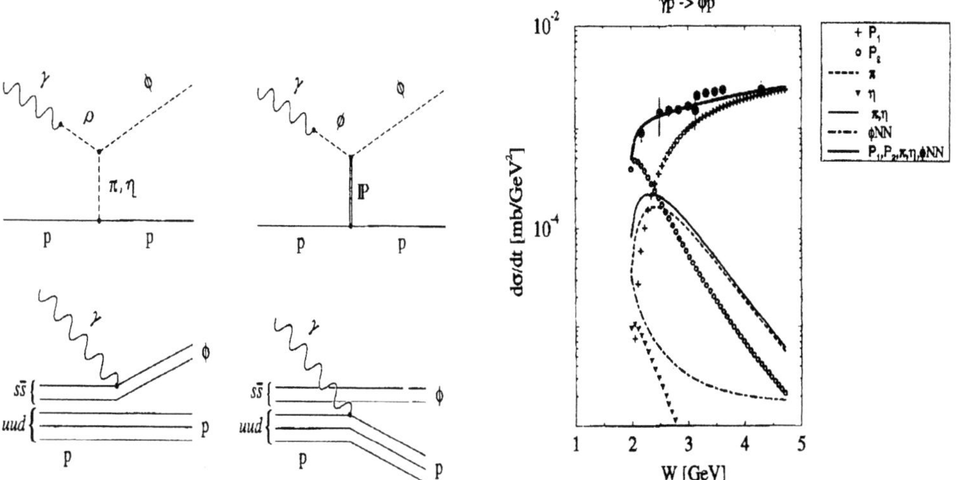

FIGURE 1. Left panel: the diagrams describing production mechanism of the ϕ mesons with photons: pseudoscalar meson exchange (top-left), pomeron exchange (top-right), and direct knock-out processes of $s\bar{s}$ pair into the ϕ (bottom). Right panel: the energy dependence of the optical point ($d\sigma/dt\,|_{t=0}$) taken from Ref. [1] and theoretical estimates from Ref. [3] .

transfer squared. Existing cross-section data indicate enhancement near threshold above levels predicted by a Pomeron-exchange model, as shown in Fig.1[1]. It is suggestive of the existence of another gluon-exchange contribution (0^{++} glueball trajectory) [2]. The new trajectory is called a second Pomeron or a super-soft Pomeron trajectory. It has a negative intercept in the spin J and mass-squared m^2 plot. It yields enhancement in $d\sigma/dt\,|_{t=0}$ near the threshold, while its contribution decreases as the energy E_γ increases. However, recent theoretical works of Titov *et al.* [3, 4] show that pseudoscalar meson exchange processes have a quite similar energy dependence in $d\sigma/dt\,|_{t=0}$. Therefore, one should resolve these two competing processes, in order to confirm the existence of a new Pomeron or 0^{++} glueball trajectory.

Measurement of spin observables is then crucial to enhance small meson-exchange amplitudes hidden behind predominant processes by the Pomeron exchange. Since the ϕ meson and the photon have a negative parity, the Pomeron ($J = 0^+$) carries a natural parity ($P = P_\gamma P_\phi(-1)^L$), while the pseudoscalar mesons a unnatural parity. Production amplitudes of the vector mesons with circularly polarized photons are represented as $A^N \pm A^U$. The negative sign indicates left circular polarization ($|\,L\rangle$), while the positive sign is right circular polarization ($|\,R\rangle$, helicity $= +1$). Linearly polarized photon states are given by a linear combination of the two different circularly polarized photon states; $|\,x\rangle = (|\,R\rangle + |\,L\rangle)/\sqrt{2}$ and $|\,y\rangle = -i(|\,R\rangle - |\,L\rangle)/\sqrt{2}$. The horizontal polarization selects only the amplitude (A^N) via natural parity exchange, whereas the vertical polarization the amplitude (A^U) via unnatural parity exchange. It can be seen in the azimuthal angle distribution for the $\phi \to K^+K^-$ decay, as depicted in Figure 2. It therefore emphasizes the importance of photoproduction of the ϕ mesons with linearly polarized photons. It should be stressed that at present the SPring-8/LEPS facility is the only place to produce

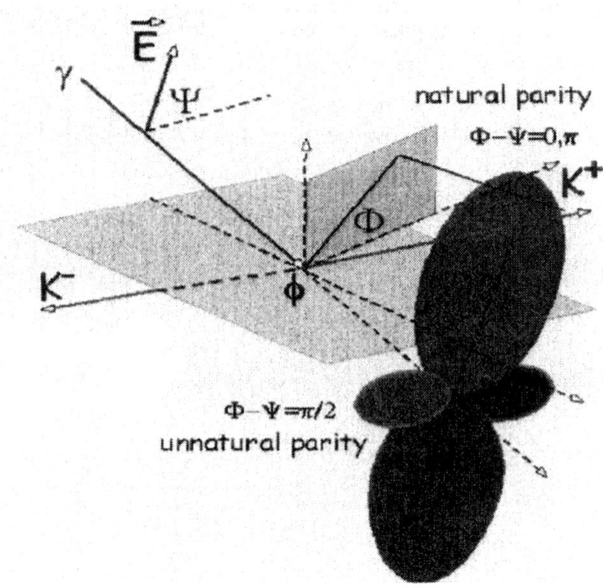

FIGURE 2. Schematic drawing describing distribution of K^-K^+ decay azimuthal angles which indicates the parity of the exchange particle.

ϕ mesons with linearly polarized photons in the world.

The ϕ decay asymmetry is defined as $\Sigma_\phi = (\sigma_\parallel - \sigma_\perp)/(\sigma_\parallel + \sigma_\perp)$, where $\sigma_\parallel(\sigma_\perp)$ are the cross sections for parallel and perpendicular alignments of the photon polarization vector and ϕ production $(\gamma - \phi)$ plane. Direct knock-out of $s\bar{s}$ pair into the ϕ can contribute to the ϕ photoproduction in the presence of strange-quark content in the nucleon. One may have to observe double-polarization observables to see the pre-existing strange-quark content in the nucleon [4].

SPRING-8 LEPS FACILITY

A large multi-GeV photon facility, Laser-Electron-Photon at SPring-8 (LEPS), started physics runs since last December, 2000. The SPring-8 is one of the most powerful third-generation synchrotron radiation facilities in the world. The large forward acceptance spectrometer was built in the beamline BL33LEP for extensive studies in quark nuclear physics. Linearly or circularly polarized laser lights are focused on a 7.7 m long straight section of the 8-GeV electron storage ring. Compton backscattered photons of $\sim 10^7$ s^{-1} flux are then transported into the large aperture and forward spectrometer apparatus through an approximately 60 m long beam line. The beam energy is determined by tagging scattered electrons using two planes of the silicon strip arrays and plastic scintillation counters. The tagging system is located at the exit of the first bending magnet

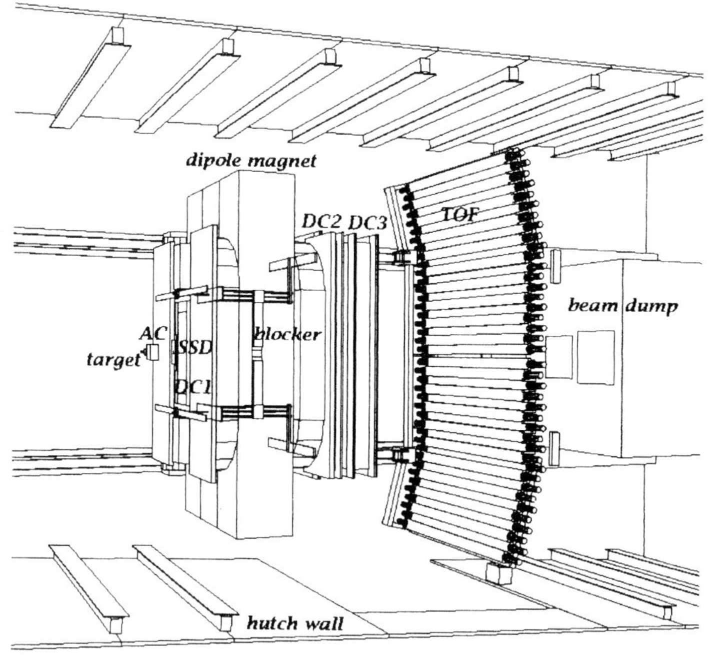

FIGURE 3. Layout of the LEPS spectrometer apparatus in the experimental hutch.

downstream of the interaction region. Tagged photon energies range from about 1.5 GeV to the end-point energy of Compton backscattering. It is worth noting that earlier commissioning runs yielded great success in producing the world-highest real photon beams at 2.42 GeV maximum energy. Figure 3 shows a layout of the LEPS spectrometer housed in a radiation shielding hutch. The LEPS spectromter, approximately 4 m long and 3 m wide, consists of a large-aperture dipole magnet, three drift chambers (DC1–DC3), a silica aerogel counter (AC), two planes of the silicon strip detectors (SSD) and a big array of 40 plastic scintillation counters (TOF) for timing and triggering purpose. The maximum magnetic field strength is 1.1 T near the center of the pole and currently operated at 0.7 T. The aperture of the magnet is 135 cm wide and 55 cm high, and the iron yoke is 60 cm long. Two planes of the single-sided silicon-strip detector are placed 70 cm upstream from the magnetic center for vertexing and tracking purpose. Multiple scattering due to the 300 μm thick SSD plane results in 1 mrad angular spread (σ), thereby yielding 0.3 MeV/c^2 and 2×10^3 GeV/c^2 spreads in K^-K^+ invariant mass and momentum transfer plots [5]. The multi-wire drift chamber DC1 is positioned right after the SSD planes. It consists of five wire planes with three different wire orientations, to resolve the left-right ambiguity. Two other drift chambers, DC2 and DC3, are placed downstream of the magnet, each having the same geometry and wire configuration. The effective size of the DC2 and the DC3 is 200 cm wide and 80 cm high.

Particle identification relies largely on the combination of measured momenta and time-of-flight (TOF) from the target to the respective TOF slats. The start time is determined by the accelerator RF signal from the storage ring. The RF beam buckets comes in every 1.9685 ns with 12 ps spread (σ). In last physics runs we used a thin plastic scintillation counter as for the start timing. The start counter was placed right after the target. Forty plastic scintillation counters are placed 3 m away from the magnetic center. Each plastic scintillator is 4 cm thick, 12 cm wide and 200 cm long. The TOF slat is aligned in a manner that its position is staggered to make two adjacent slats overlapped by 1 cm. Half the TOF slats are positioned in the central plane and the other half in sidewards planes titled by 15 degree. The time resolution of TOF slats was measured to be less than 100 ps.

Two pieces of the lead blocker bar are placed inside the magnet, so as to suppress e^+e^- triggers online and also to avoid hitting of the e^+e^- pairs directly on the experimental hutch walls. The latter case is strongly related to radiation shielding problem. Each blocker bar is 4 cm high, 10 cm thick and separated by about 15 cm from one another. The optimum location and the dimension of the blocker pieces were determined by a Monte Carlo simulation. Non-magnetic support structures are movable along the beam axis. This blocker also stops other charged particles scattered out in a nearly median plane. Such e^+e^- online suppression was made in last runs using a silica aerogel Cerenkov counter with n =1.03. The Cerenkov counter was placed between the start counter and the silicon-strip vertex detector. With this system the e^+e^- triggers went considerably suppressed to a level of 0.1%. It enabled us to trigger effectively hadronic events without the need for higher-level trigger scheme.

PRELIMINARY RESULTS AND OUTLOOK

We report very preliminary results from a set of data taken since last December, 2000. A linearly polarized beam of compton backscattered photons is incident on a liquid hydrogen target with the effective length of 5 cm. We selected a 351 nm wavelength of the argon-ion laser and operated in a multi-line mode. The trigger was made using fast coincidences between the Cerenkov counter and the TOF counters.

Figure 4 shows the distribution of masses for all reconstructed charged particles without any kinematical cuts. We track all the charged particle trajectories and do magnetic analysis by using a numerical Runge-Kutta integration method. Precise calibration is now underway. Nevertheless, the mass peaks for $\pi^{\pm}$, $K^{\pm}$, p and d are clearly visible. The mass spectrum for the momentum region lower than 1 GeV/c is displayed as an overlaid histogram. The K^+K^- invariant mass spectrum in Fig. 4 indicates a clear peak corresponding to the ϕ.

The integrated photon flux is approximately 2.5×10^6 s^{-1} in the energy region from 1.5 GeV to 2.4 GeV. The data analysis is now underway. The statistics is already large enough to obtain the decay asymmetry of the ϕ mesons. Owing to the forward spectrometer, we are able to measure the ϕ photoproduction at the minimum value of momentum transfer. We also see the A-dependence of the ϕ photoproduction cross sections.

In our near-term physics program photoproduction of the $\Lambda(1405)$ is of great interest since it is a unique candidate for a meson-baryon bound state and for testing a reliability

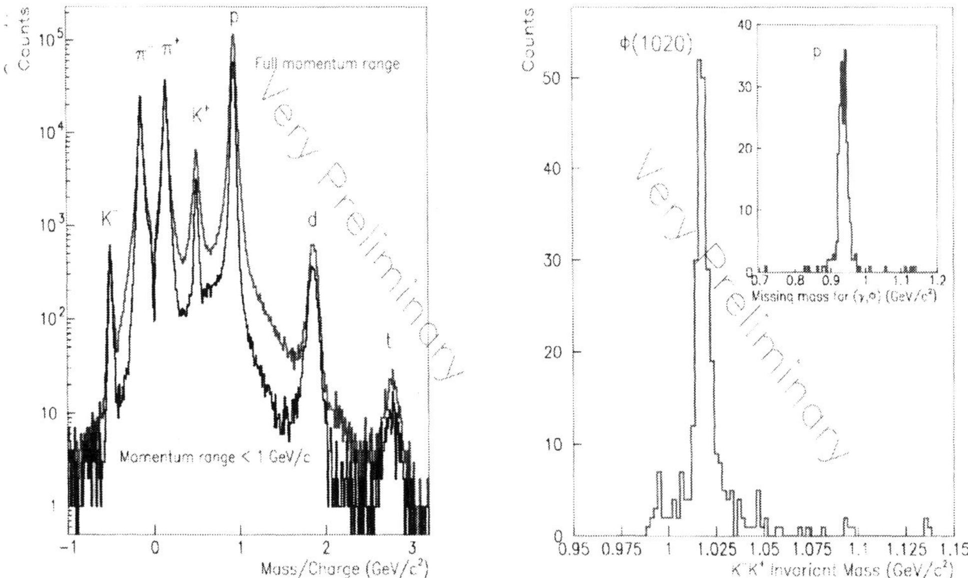

FIGURE 4. Left panel: reconstructed masses of charged particles. Right panel: distribution of K^+K^- invariant masses.

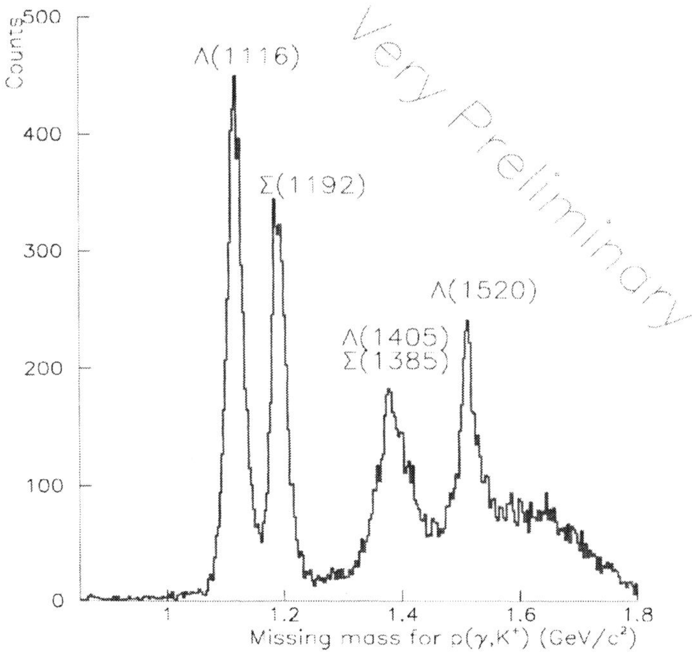

FIGURE 5. Distribution of missing masses for (γ, K^+) reactions.

of the χPT. The well-established $J^P = \frac{1}{2}^- \Lambda(1405)$ hyperon vies for the distinction between an unstable $\bar{K}N$ bound state and an $L = 1$ $SU(3)$-singlet q^3 state coupled with the S-wave $\bar{K}N$ channels [6]. If the $\Lambda(1405)$ is a strongly bound $\bar{K}N$ state, there must exist another undetected $J^P = \frac{1}{2}^- \Lambda$ resonance lying close to the $\Lambda(1520)$. If, instead, the $\Lambda(1405)$ is the $SU(3)$-singlet *uds* state, it must be mass-degenerate with the $J^P = \frac{3}{2}^- \Lambda(1520)$ state. The large mass splitting must then be explained in the quark models with an anomalously large spin-orbit coupling. In the $J^P = \frac{1}{2}^-$ octet, the $\Lambda(1405)$ is, however, 130 MeV/c^2 lighter than the $J^P = \frac{1}{2}^- N^*(1535)$ even though it contains an s quark which is 100–150 MeV/c^2 heavier than the d quark. There is also a suggestion of the $\Lambda(1405)$ being a hybrid baryon $(uds)g$ where uds is in $\frac{1}{2}^+$ and g the gluon. However, if it costs only 300 MeV for a hybrid excitation, we would have seen several hybrid states in the non-strange baryon spectrum [7].

Experimental efforts have been far behind theoretical efforts. Only two old experiments had enough statistics for a detailed analysis. Thomas *et al.* [8] reported 470 $\Sigma^+\pi-$ and $\Sigma^-\pi^+$ events which contained no more than 50 $\Sigma^0(1385)$ background events in $\pi^- p \to K^0\Sigma^\pm\pi^\mp$ at 1.69 GeV/c. Later, Hemingway *et al.* [9] reported 766 $\Sigma^+\pi^-$ and 1106 $\Sigma^-\pi^+$ events from $K^- p \to (\Sigma\pi\pi)^+\pi^-$ at 4.2 GeV/c. In the context of the effective range approximation, these data are consistent with a $\Lambda(1405)$ mass of approximately 1405 MeV/c^2 and a width of 45-55 MeV. Due to its asymmetric lineshape of the $\Lambda(1405)$ mass spectrum, Hemingway's mass of 1391 ± 1 MeV/c^2 from Breit-Wigner fit is unacceptably very poor. The values quoted in Particle Data Book are $m = 1407 \pm 4$ MeV/c^2 and $\Gamma = 50.0 \pm 2.0$ MeV from Dalitz's M-matrix fit [10].

The Byer-Fenster tests on these data suggest a $J = \frac{1}{2}$ state, but cannot rule out other possibility completely. No $\Lambda(1405)$ parity determination was possible because of absence of the significant polarization: Neither J nor P has yet been determined directly. The striking S-shape cusp behavior (the lineshape dropping dramatically at the approach of the $\bar{K}N$ threshold) is characteristic of S-wave coupling; the other below-threshold hyperon, the $\Sigma(1385)$, has no such threshold distortion because of its $\bar{K}N$ coupling is P-wave. For the $\Lambda(1405)$, this asymmtry is the sole direct evidence that $J^P = \frac{1}{2}^-$ [6].

Therefore, the $\Lambda(1405)$ still remains as the object of sustained curiosity. Moreover, much attention is recently attracted to experimentalists for testing the reliability of χPT because it is a unique candidate for a meson-baryon bound state. Large and attractive S-wave scattering lengths of ηN and $\eta\Lambda$ make up a new η octet of $J^P = \frac{1}{2}^-$ ($N(1535)$ and $\Lambda(1670)$, respectively) associated with S-wave threshold interaction. It is also interesting to study the properties of $\Sigma(1750)(=\eta + \Sigma)$ and to search for undetected $\eta\Xi$ members.

Recent works of Oset *et al.* [11] predict cross sections for the lineshape of the $\Lambda(1405)$ into different channels using a chiral unitary model. In the model, the resonance $\Lambda(1405)$ is generated dynamically from the 10 coupled channels, K^-p, $\bar{K}^0n$, $\pi^0\Lambda$, $\pi^0\Sigma^0$, $\pi^+\Sigma^-$, $\pi^-\Sigma^+$, $\eta\Lambda$, $\eta\Sigma^0$, $K^+\Xi^-$, $K^0\Xi^0$. It is interesting to see the different shapes of the three $\pi\Sigma$ channels, which can be understood in terms of the isospin decomposition. Without negligible $I = 2$ contribution, they shows that

$$3\frac{d\sigma}{dM_{\pi^0\Sigma^0}} \simeq \frac{d\sigma}{dM_{I=0}}$$

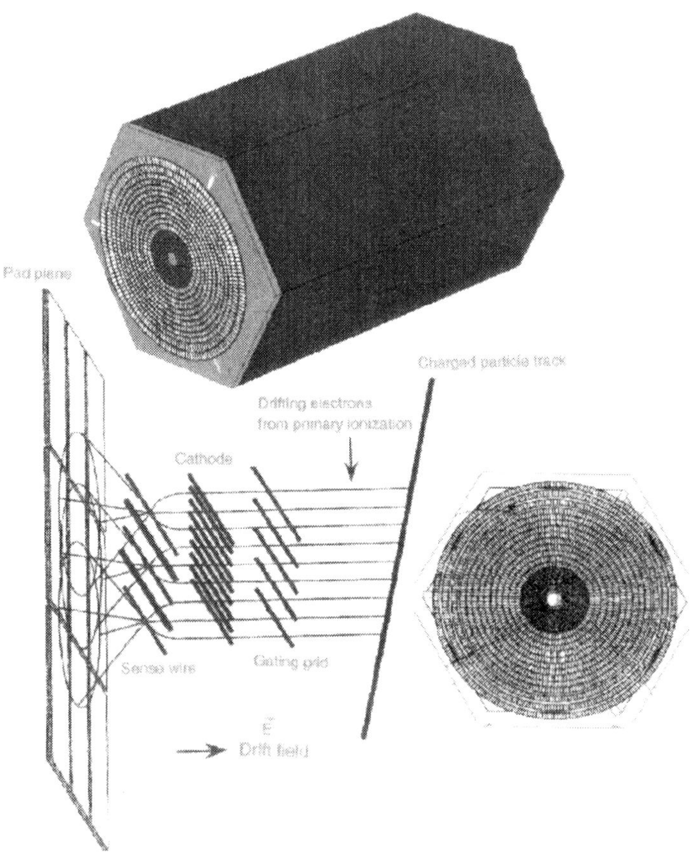

FIGURE 6. Schematic view of the time projection chamber (TPC) under construction (top); Principle of charged particle detection with the TPC (bottom-left); Simulated charged particle tracks from a $\Lambda(1405)$ photoproduction event.

$$\frac{d\sigma}{dM_{\pi^0\Sigma^0}} + \frac{d\sigma}{dM_{\pi^+\Sigma^-}} + \frac{d\sigma}{dM_{\pi^-\Sigma^+}} \simeq \frac{d\sigma}{dM_{I=0}} + \frac{d\sigma}{dM_{I=1}}.$$

This means that the real shape of the $\Lambda(1405)$ can be seen in either the $\pi^0\Sigma^0$ channel or the sum of the three $\pi\Sigma$ channels. For photoproduction of the $\Lambda(1405)$, it is suggestive of the photon energy of $E_\gamma^{\text{lab}} \sim 1.7$ GeV in order to suppress a P-wave contribution like the $\Sigma^0(1385)$ [11]. The compton-scattered photons come out with energy-dependent angular distribution. The highest energy photons emerge at 0 degree, and the scattering angle becomes larger as the energy goes down. Incident 1.7 GeV photons thus impinge on the target region slightly off the beam axis. Then decay particles of the $\Lambda(1405)$ would undergo less energy-loss.

To observe a real shape of the $\Lambda(1405)$, we should measure the $\Sigma^0\pi^0$ channel and the sum of all $\Sigma\pi$ channels. Measurement of missing mass spectrum in $p(\gamma, K^+)X$ permits observation of the sum of the $\Sigma^0\pi^0$, $\Sigma^-\pi^+$, and $\Sigma^+\pi^-$ channels because of little contribution from the $\Lambda\pi^0$ channel. However, direct measurement of the line-shape from the decay $\Lambda(1405) \to (\Sigma\pi)^0$ is rather difficult in several points of view; three photons from $\Sigma^0\pi^0$ and a neutron from $\Sigma^-\pi^+$, and short-lived $\Sigma^\pm$ particles leaving only several cm tracks. It is therefore necessary to develop a novel high-resolution vertex detector [13, 14].

Figure 5 shows distribution of missing masses for $p(\gamma, K^+)$ reaction which is associated with hyperon photoproduction. Not only ground states of Λ and Σ^0 hyperons but also their low-lying excited states $\Sigma^0(1385)$, $\Lambda(1405)$, and $\Lambda(1520)$ are clearly seen in the missing mass spectrum. The central tracking device of the SPring-8/LEPS facility will be a Time Projection Chamber (TPC) in Fig. 6 with an active volume of 70 cm in length and 40 cm in diameter. It is now under construction. Our primary goal with the TPC is to achieve excellent two-body invariant mass resolution and particle identification with a large solid angle coverage. Photoproduction of $\Lambda(1405)$ and ϕ meson off nuclei is of considerable interest in the current TPC physics program. We attempt to measure the decay of charged Σ hyperons, thereby reconstructing the $\Lambda(1405) \to \Sigma\pi$ decay. Success of SPring-8/LEPS physics programs with the TPC will open a new era of low-energy regime for hadron physics.

ACKNOWLEDGMENTS

The work of JKA was supported by the JSPS Postdoctoral Fellowship for Foreign Researchers.

REFERENCES

1. H.J. Behrend *et al.*, Nucl. Phys. **B144**, 22 (1978).
2. T. Nakano and H. Toki, in Proceedings of EXPAF 97, Aioi, Japan, 11-13 Mar 1997.
3. A.I. Titov *et al.*, Phys. Rev. **C59**, 2993 (1999).
4. A.I. Titov *et al.*, Phys. Rev. Lett. **79**, 1634 (1997); E.M. Henley *et al.*, Phys. Lett. **B281**, 178 (1992).
5. T. Nakano *et al.*, Nucl. Phys. **A629**, 559c (1998).
6. R.H. Dalitz, "Note on the Λ(1405)" in C. Caro *et al. Euro. Phys. J.* **C3** 676 (1998).
7. S. Pakvasa and S.F. Tuan, *Phys. Lett.* **B459** 301 (1999).
8. D.W. Thomas, A. Engler, H.E. Fisk, R.W. Kraemer *Nucl. Phys.* **B56** 15 (1973).
9. R.J. Hemingway, *Nucl. Phys.* **B253** 742 (1985).
10. R.H. Dalitz and A. Deloff *J. Phys. G: Nucl. Part. Phys.* **17** 289 (1991).
11. J.C. Nacher, E. Oset, H. Toki, A. Ramos *Phys. Lett.* **B455** 55 (1999).
12. T. Nakano *et al.* LEPS is an acronym standing for Laser-Electron Photons at SPring-8. It will soon be ready for publication of physics results under international LEPS collaboration led by T. Nakano.
13. K. Imai *et al.* "Photoproduction of Hyperon Resonances" *Letter of Intent* at SPring-8
14. J.K. Ahn, "Photoproduction of the Λ(1405)" KEK Proceedings 2000-5, Workshop on Chiral Dynamics, December 9-10, Tokyo, 1999.

PROTON STRUCTURE FUNCTIONS AT HERA

Bruno Stella
on behalf of H1 and ZEUS Collaborations

III Università degli Studi di Roma and INFN Sez. Roma 3 (Italy)
V. d. Vasca Navale 89 , I-00146 Roma (E.mail: bruno.stella@roma1.infn.it)

Abstract. The electron-proton collider HERA, like an electron-mycroscope, explores the structure of the proton down to 10^{-16} cm and up to the situation of very high parton densities. The proton energy was upgraded from 820 to 920 GeV in the Fall of '98 and the luminosity has also substantially improved, with another factor of 3 upgrade expected to follow this year.

Inclusive proton structure functions have been studied with incident e^+ and e^- of 27 GeV in the neutral (NC) and charged (CC) current interactions as functions of the squared four-momentum transfer, Q^2, and of the fractional proton momentum carried by partons, x. The structure function F_2, as well as the $\gamma - Z^0$ interference term xF_3, have been measured in a range of Q^2 and $1/x$ that extends by orders of magnitude that reached by fixed target experiments. The DGLAP evolution equations [1] allow for a perturbative NLO QCD fit of the measured non-perturbative structure functions in the available kinematic range: α_S and the gluon density at low x are fitted at the same time with good precision. The longitudinal structure function, F_L, can be determined within the DGLAP formalism.

With CC, the electroweak unification has been tested; at high x, a first flavour decomposition of the light quarks is achieved. The contribution to F_2 of the charm quark has been measured and results to be relevant. Bounds on the radius of quarks and on compositeness are derived from the data at the highest Q^2, $100 < Q^2 \leq 40,000\ GeV^2$.

INTRODUCTION

HERA is the 27.6 GeV electron - 920 GeV proton storage ring accelerator built by DESY (Hamburg) with a substantial contribution of other nations. The proton structure is explored in the colliding mode in two of the straight interaction sections by H1 and ZEUS experiments and in the fixed (polarised) target mode by polarised electron beam in a third section (by the experiment HERMES, wose results will not be reported here). After starting its operation at the end of '92, in 1994-'97 HERA has been operated with e^+'s and 820 GeV protons ($\sqrt{s} = 300$ GeV, s being the cms energy squared), accumulating a luminosity of $\approx 40pb^{-1}$ per experiment. The proton energy has been upgraded to 920 GeV in '98 ($\sqrt{s} = 318$ GeV) and the machine has been operated first with e^-'s ($L \approx 15pb^{-1}$ per experiment) and then with e^+'s ($L \approx 60 - 70pb^{-1}$ per experiment). The rate of integrated luminosity has increased every year, achieving recently order of 100 pb^{-1} per year. The upgrade of the machine going on this year will provide an other factor 3 improvement.

H1 and ZEUS apparata are very general and have common features: a central tracking system in a uniform magnetic field (along the beam axis); a huge electromagnetic and hadronic calorimeter system (more detailed in the proton direction, due to the

CP594, *Hadrons and Nuclei: First International Symposium,* edited by Il-T. Cheon et al.

asymmetric momenta of the beams); instrumented iron for muon tracking; "forward" (at 0° respect to the outgoing proton) proton and neutron detectors and "backward" (at 180° respect to the outgoing proton) electron and photon taggers. Their sensitivities and performances are different enough: to be short, H1 has been optimised for leptons ($\sigma(E)/\sqrt{E} = 12\%$ for electrons, $\approx 50\%$ for hadrons) and ZEUS has a better resolution for hadronic calorimetry ($\sigma(E)/\sqrt{E} = 18\%$ for electrons and 35% for hadrons) .

HERA is a world unique colliding facility for probing the proton at very small distances ($\sim 10^{-16}$cm). The main diagram, the t-channel exchange of a highly virtual gauge boson (γ, Z neutral current, W charged current) in DIS, with the definition of the relevant kinematic variables, is depicted in Fig. 1: Q^2 is the squared 4-momentum transfer, x is the fraction of proton momentum carried by the struck quark, y is the inelasticity. In the quasi-two body kinematics, only two variables are independent: $Q^2 = xys$.

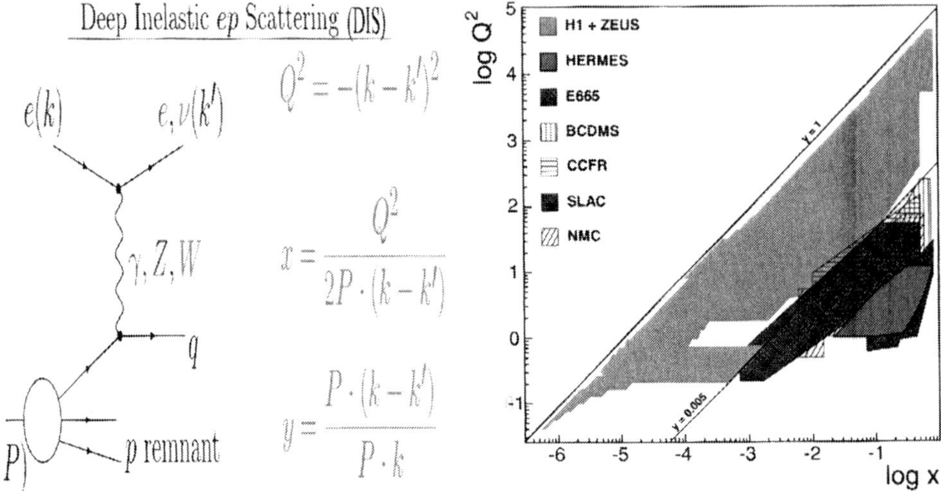

Figure 1. *(1a) Main diagram of DIS at HERA and definition of the kinematic variables. (1b) Q^2, x regions explored in various experiments, including fixed target ones.*

Fig. 1b shows the kinematic range (in $\log Q^2$, $\log x$) explored by HERA experiments (including HERMES) and by previous fixed target experiments; they overlap at small inelasticity (y$\sim$ 0.005). HERA has extended the explored range by almost three orders of magnitude, both in Q^2 and x.

QCD describes the proton as a dynamic fog of partons (quarks and gluons), u and d quarks at high fractional momentum x (valence quarks), irradiating gluons cascading in lower momentum gluons and quarks (sea quarks, of all possible flavours, including antiquarks) with degrading fractional momentum.

The high Q^2, high x limit (hard interactions) is the region of probing the proton at high resolution ($\equiv 1/\sqrt{Q^2_{max}}$), to check the pointlike nature of partons and their possible compositeness. The weak effects become sizable in NC , and EM and weak charged interactions give comparable cross sections for the first time. At lower x ($0.1 < x < 1.$) the quark densities can here be separated for the first time according to their flavour.

At very low x ($10^{-6} < x < 10^{-3}$), a region again explored for the first time by HERA, Q^2 is low (soft interactions), the parton density goes like a negative power of x and is than very high (gluons, sea quarks) and α_S is large; here the perturbative expansion of QCD (pQCD) is not valid anymore and the mechanism of confinement could be studied. Below $Q^2 \approx 1$ GeV^2 the physics is going to be hadronic (Regge limit) instead than partonic.

The neutral current cross section.

The "neutral current" events are identified inclusively by the presence in the final state of an high energy isolated electron, coplanar with the beam and the hadronic jet of the struck quark. An example of these NC events is shown in Fig. 2a with the ZEUS detector. They have very clear signatures: the proton remnant is visible in the proton (forward) direction , togeteher with a very energetic hadronic jet balanced in p_t (see transverse view) by the scattered electron.

The Born cross section for $e^{\pm}p \to e^{\pm}X$ is a linear combination of the "structure functions" F's:

$e^{\pm}p : (\frac{d^2\sigma^{NC}}{dxdQ^2})_{Born} = \frac{2\pi\alpha^2}{xQ^4}(Y_+F_2^{NC} \mp Y_- xF_3^{NC} - y^2F_L^{NC})$ with $Y_{\pm} = 1 \pm (1-y)^2$ (elicity term). By dividing by the coefficient, one gets the so called "reduced" cross section $\tilde{\sigma}$. xF_3 is the γ/Z interference term; $F_L(x,Q^2)$ is the longitudinal structure function, sensitive to gluons (it is zero with fermion partons only and at first order QCD).

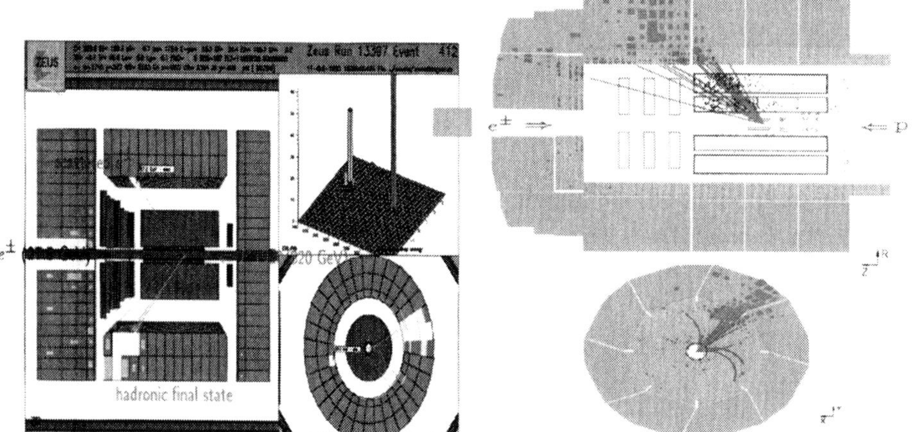

Figure 2. *(2a) A typical NC event detected by ZEUS experiment. (2b) A typical charged current event as detected by H1 (central detector). In both cases the projections along the beams and transverse to the beams are shown.*

In leading order QCD, the structure functions of x , Q^2 are combinations of the quark and antiquarks probability densities: $F_2(x,Q^2) = x\Sigma_{Quarks}A_f[q(x,Q^2) + \bar{q}(x,Q^2)]$. The coefficients A_{flavor} are the quark and anti-quark electric and weak charges (squared). The derivative $\frac{\delta F_2}{\delta log(Q^2)}$ is proportional to the gluon density $g(x,Q^2)$. The function

$xF_3(x,Q^2) = x\Sigma_{Quarks}B_f(q(x,Q^2) - \bar{q}(x,Q^2))$. The coefficients B_{flavor} are products of electric and weak charges of the different flavours.

The charged current cross section.

The charged current events ($e^{\pm}p \to \nu X$) have an undetected neutrino in the final state and are identified by a pure hadronic jet with high transverse momentum (together with remnants of the proton in the proton direction). A typical event is shown in Fig. 2b. In this case the recostruction is based on hadronic calorimeter data.

The CC cross section is suppressed at $Q^2 << M_W^2$ due to the high $W^{\pm}$ mass M_W:
$\frac{d^2\sigma^{CC}}{dxdQ^2} = \frac{G_F^2}{2\pi x}\left(\frac{M_W^2}{Q^2+M_W^2}\right)^2 \cdot \tilde{\sigma}_{CC}^{\pm}$ $\quad(\tilde{\sigma}_{CC}^{\pm} = Y_+F_2^{CC} \mp Y_-xF_3^{CC} - y^2F_L^{CC})$. The leading order cross sections are (for e^+ and e^-): $\quad \tilde{\sigma}_{CC}^{e^+} = x \cdot [(\bar{u}+\bar{c}) + (1-y)^2(d+s)]$ (as valence quark, only d is probed, but is suppressed by the $(1-y)^2$ factor).
$\tilde{\sigma}_{CC}^{e^-} = x \cdot [(u+c) + (1-y)^2(\bar{d}+\bar{s})]$ (as valence quark, only u is probed and is not suppressed). The two beams give important different informations with CC.

Outlook

In the following, in section 2, the proton structure is studied inclusively, both because the NC structure functions come out from the inclusive cross sections and because the proton is parameterised as an inclusive sum of partons. The data, analysed in the (non unique) framework of pQCD, as "evoluted" (functions of Q^2) according to DGLAP equations [1], give all the three structure functions, the gluon density and α_s. In section 3, CC cross sections are studied, with first evidence for the electroweak unification at high Q^2; the first attempts to measure flavour discriminated quark densities are reported, both for light (valence) quarks and for heavy quarks (charm). In section 4 the present limits for quark structure (quark radius, compositeness) are reported. A summary concludes the talk. [2]

INCLUSIVE PROTON STRUCTURE (NC).

The structure function F_2 is dominant at HERA for $Q^2 << M_Z^2$ and $y < 0.7$. At high x , by definition of "valence" only the three valence quarks are expected to contribute: the momentum of the three quarks would be statistically distributed, giving a gaussian distribution peaking at $x = 0.3$. At lower x , the structure function cannot be predicted by pQCD. The pre-HERA measurements (Fig. 3a) almost agreed with parton model, requiring little "sea quarks" contribution. Instead, as shown in Fig. 3a, the HERA experiments produced a structure function F_2 rising with decreasing x : the quark densities are rapidly increasing at low fractional momenta. This can be interpreted as due to gluon emission and subsequent splitting of the gluons into a quark-antiquark pair (or two gluons).

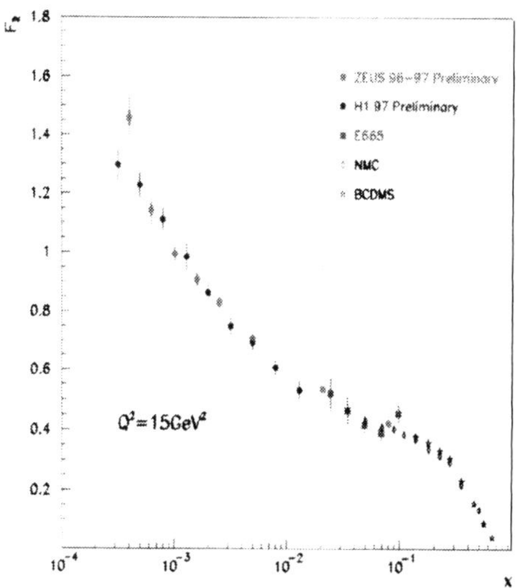

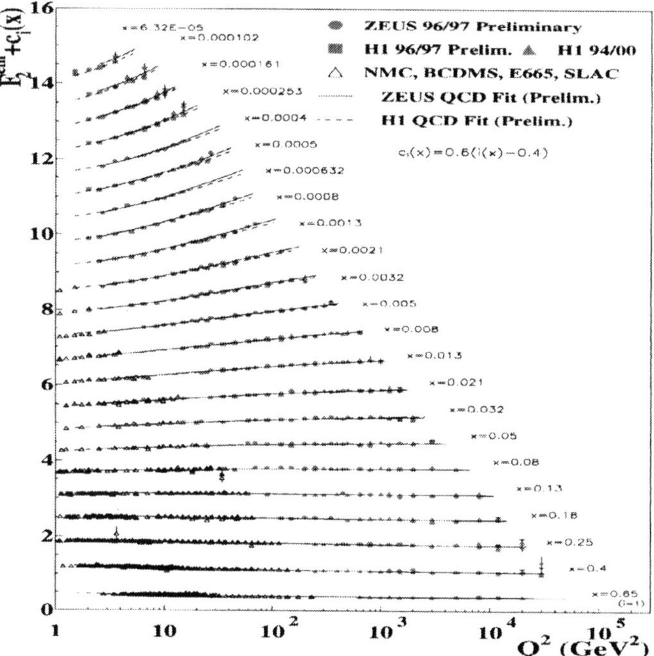

Figure 3. *(3a) The proton structure function* F_2 *as a function of* x *at* $Q^2 = 15$ GeV^2 *(H1, ZEUS). (3b)* F_2 *as function of* Q^2*, for fixed* x, $3.2\ 10^{-5} < x < 0.85$.

Scaling violations of F_2 ('94-'97 data).

As function of Q^2, in the parton model F_2 is expected to stay constant because the quarks are considered to be pointlike and the photon can interact always with the same number of partons. In fact an almost constant F_2 had been observed by fixed target experiments (see in Fig. 3b the triangular points at low Q^2).

By extending very much the range in Q^2 and x , HERA experiments have evidentiated large scaling violations at low x (Fig. 3b, top lines) and also at high x (Fig. 4, bottom lines).

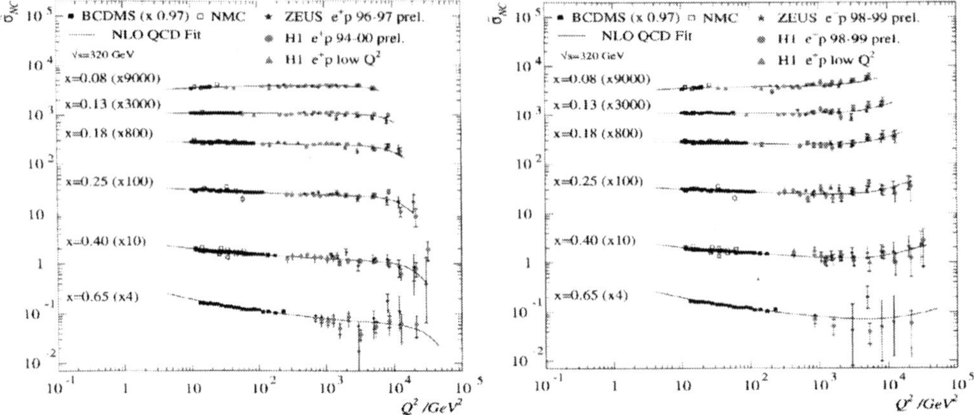

Figure 4. NC reduced cross section $\tilde{\sigma}_{NC}$ at high x, Q^2 (a test of the standard model (SM) over 4 orders of magnitude in Q^2; note the effect of the different sign of the γ/Z^0 interference in e^-(right) and e^+(left)).

It is remarkable that in the four (and more) decades of Q^2 explored by HERA the Q^2 dependence can be explained by gluon radiation by quarks (at high x) and quark emission by gluons (at low x), with a perfect description due to DGLAP evolution equations (see Next Leading Order (NLO) QCD fit lines in Fig. 3b, 4). Moreover, not only ZEUS and H1 agree with each other, but also with the low Q^2 experiments.

Recent results at high Q^2 with e^-p and e^+p at $E_p = 920$ GeV ($\sqrt{s} = 318$ GeV)

After the machine upgrade in '98, with protons accelerated at 920 GeV, the comparison of the NC data with e^- and e^+ beam make now possible to test the Standard Model over 4 orders of magnitude in Q^2, including the small electroweak effects. The γ/Z interference is expected to be positive with e^- and negative with e^+. This is clearly visible experimentally in Fig. 4, where the reduced cross sections as function of Q^2 are presented, and Fig. 5 (the same as function of x) . Again the flat Q^2' and the steep x dependences are perfectly fitted by pQCD. [It must be said that the fit needs a large number ($\sim$ 25) of parameters, with technical choices; the choices and the parameters are not unique, even within the same evolution equations (DGLAP, [1]); other choices of evolution equation (CCFM, [3]) could reproduce as well the data at present]. The difference between e^- and e^+ is used to measure the interference term (next par.).

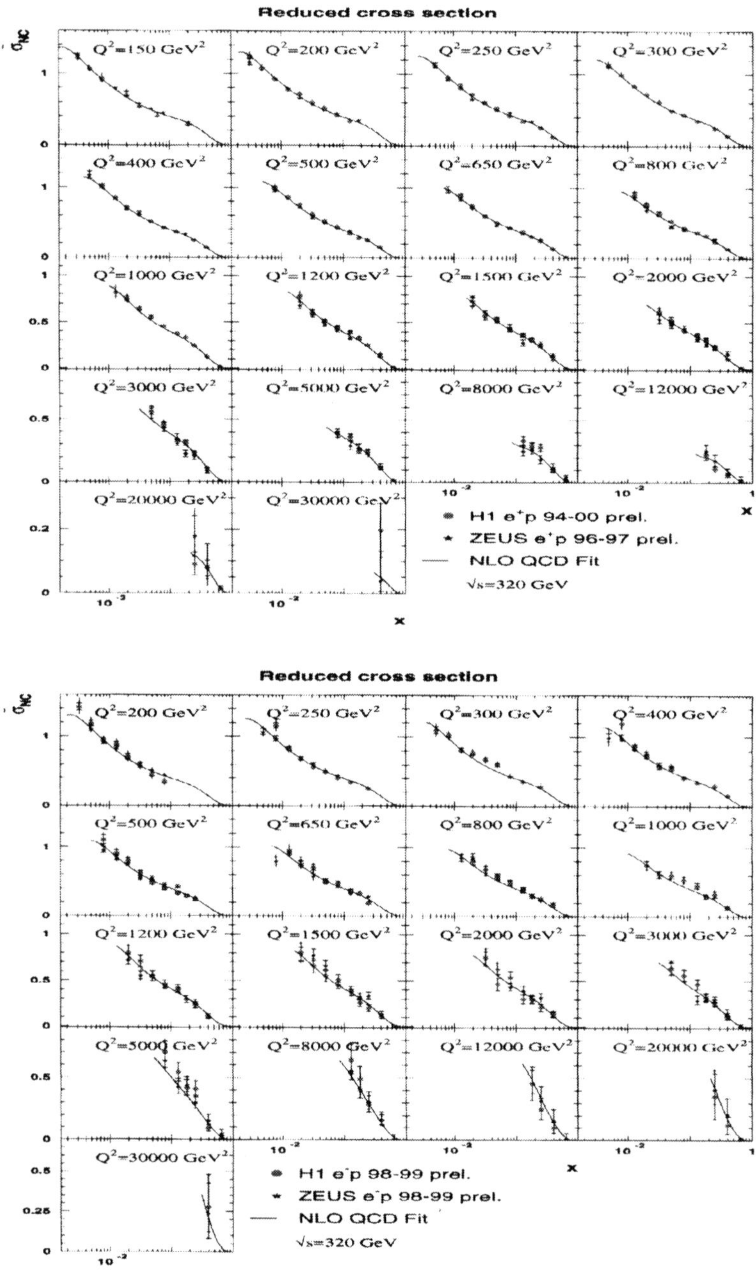

Figure 5. *NC reduced cross section $\tilde{\sigma}_{NC}$ for e^+ (upper) and e^- (lower): a stringent test of the pQCD + DGLAP (lines).*

Extraction of xF_3.

$\tilde{\sigma}^- - \tilde{\sigma}^+ = 2Y_- xF_3 = 2[1-(1-y)^2]xF_3$ (In case of different energies, kinematical corrections and F_L contributions are to be accounted for.) Fig. 6 shows the first measurements of xF_3. The preliminary results, for high x and high Q^2 (up to 30,000 GeV^2) compare well with the SM expectations.

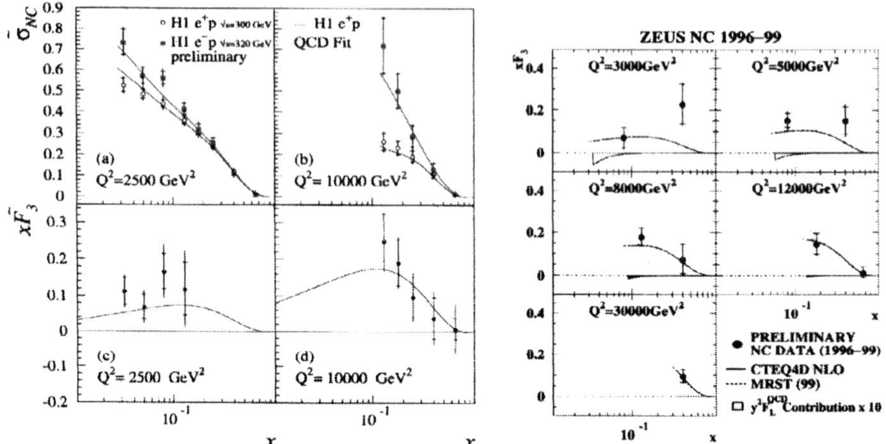

Figure 6. *First measurements of xF_3^{NC} at high Q^2 (sensitive to valence quark densities) by H1 (left) and ZEUS. [The F_L effect ($\approx 1\%$) is taken here from the QCD fit and lowest x]*

Determination of α_S and gluon density.

The most important parameters of pQCD are the coupling constant $\alpha_S(Q^2)$ and the gluon density g(x ,Q^2). Up to now one of them was fixed by the world average and the other one was fitted. Now a simultaneous global fit is performed, which accounts for the anti-correlation of the two parametric functions. The results are shown in Fig. 7a (here in case of H1). The gluon density rises steeply at low x and high Q^2.

The corresponding α_S is: $\alpha_S(Z^0) = 0.1150 \pm 0.0017(exp) \pm 0.0012(model) \pm 0.005$ due to renormalisation scale uncertainty. Similar results are obtained by ZEUS. The values are in good agreement with the ones obtained from the DIS jet cross sections by the same experiments and with the world average, but show the running of α_S in a single experiment.

Longitudinal structure function F_L.

Assuming the validity of the pQCD theory used for the fit, the F_L structure function can be extracted. F_L shows up only at low x , when the partons are not only fermions, as a negative contribution to the cross section (see Fig. 7b). By subtracting $\tilde{\sigma}$ from F_2 (or

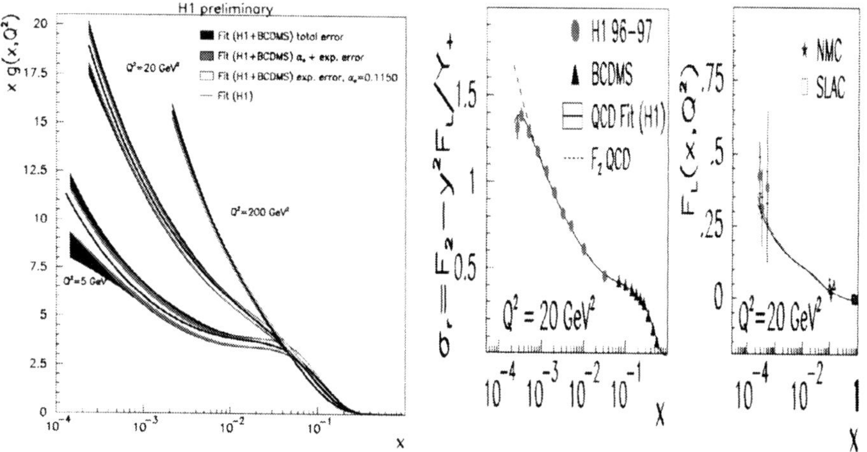

Figure 7. *(7a) Fit results on gluon density by H1. (7b) Extraction of F_L at low x ; left: $\tilde{\sigma}$ (the dashed line corresponds to $F_L = 0$.), right: resulting F_L; both at $Q^2 = 20 GeV^2$.*

by derivating $\tilde{\sigma}$ respect to y) H1 has got F_L within the DGLAP formalism (an example is shown in Fig. 7b), at x values 1000 times smaller than previous measurements.

Low Q^2, Low x: pQCD stops to work.

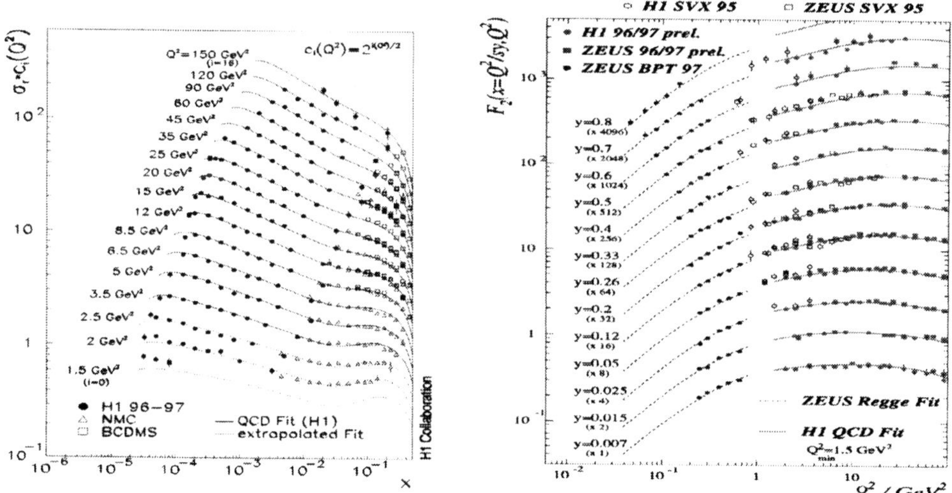

Figure 8. *(8a) Low Q^2, low x : behaviour of $\tilde{\sigma}$. (8b) F_2 at very low Q^2: hadronic $\leftrightarrow$ partonic junction.*

The pQCD fits could be performed technically even at very low Q^2, close to λ^2_{QCD}, when α_S^2 is not anymore much smaller than α_S; but they would loose theoretical meaning. This is why the fits must stop at some Q^2_{min} of the order of 1 GeV^2. Fig. 8a shows how nicely the fit describes the x dependence at various Q^2's, but stops to work at low x and low Q^2 (bottom lines).

At very low Q^2 we expect a transition between perturbative (partonic) and non-perturbative ($\equiv$ hadronic) QCD. This is experimentally evidentiated in Fig. 8b, where above $Q^2 \approx 1$ GeV^2 the H1 QCD fit is used to describe the data, and below $Q^2 \approx 1$ GeV^2, ZEUS data are well described by a fit inspired by Regge model. The approximate continuity between the two regimes is remarkable. Will we achieve a unified (partonic-hadronic) description of the proton structure function ?

CC AND FLAVOUR DISCRIMINATED QUARK DENSITIES

CC differential cross section: testing the electro-weak sector (NC and CC, e^- and e^+).

The results for the CC cross section as function of Q^2 are shown in Fig. 9a. We see that HERA data agree with SM on wide Q^2 range. It is evident the different coupling of the exchanged W (differently charged in the e^- and e^+ cases) with partons of the proton.

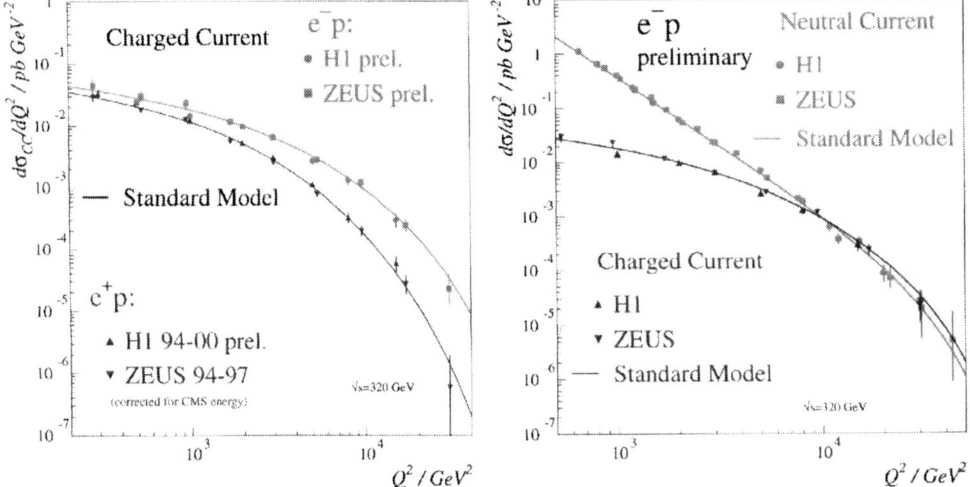

Figure 9. *(9a) CC cross section $d\sigma/dQ^2$ (e^- and e^+). (9b) Comparison of NC and CC cross section $d\sigma/dQ^2$ in case of e^-: electro-weak unification.*

If we compare NC (mostly electromagnetic) and CC (weak), we test that beyond $Q^2 \simeq M_W^2 \simeq M_Z^2$ the effect of the W propagator becomes small and the cross section are now similar. This is shown in Fig. 9b in case of e^-. So, we have encountered the electroweak unification !

Flavor specific parton density functions (PDF).

The NC structure functions are linear combinations of quark densities weighted with their (flavor specific) effective charges (squared). In case of CC, the different beam charges select (couple to) different combinations of flavors (as shown in the introduction). This can be used to measure the flavor discriminated PDF's.

Light quarks

At high x only the light quarks contribute to $\tilde{\sigma}_{NC}$ and $\tilde{\sigma}_{CC}$:
$\tilde{\sigma}_{NC}(e^+) \propto (1+(1-y)^2)\Sigma e_f^2 x q_f$
$\tilde{\sigma}_{CC}(e^+) \propto x\bar{u} + (1-y)^2 xd$; $\tilde{\sigma}_{CC}(e^-) \propto xu + (1-y)^2 x\bar{d}$.

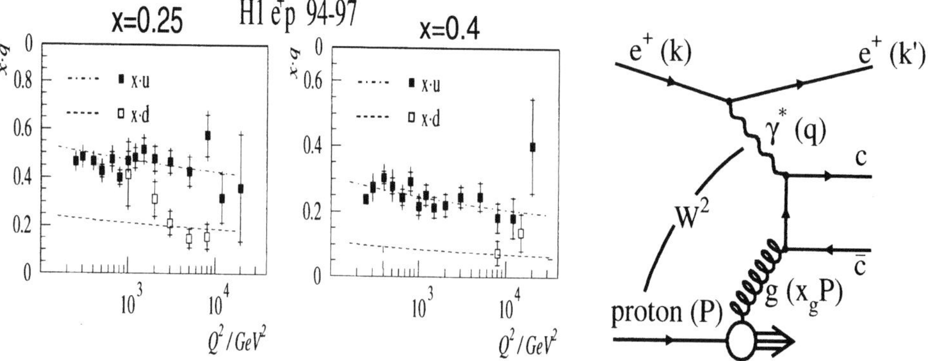

Figure 10. *(10a) Valence quark densities from high Q^2/high x, CC/NC (H1 e^+p data 94-97). (10b) Main charm-anticharm production diagram (boson-gluon fusion).*

H1 has used only the (high statistics) e^+ data to extract $u(x,Q^2)$ and $d(x,Q^2)$. The results at $x = 0.25$ and $x = 0.4$ are shown in Fig. 10a. The proton composition at high fractional momenta is here experimentally measured, without QCD inputs: u/d is slightly larger than 2. (More results have been presented at Osaka conference last year.) We don't have here the large uncertainties present in case of nuclei (fixed target experiments).

Heavy quark contribution to F_2.

Heavy quarks are produced in deep inelastic scattering mainly by the "boson-gluon fusion" process (Fig. 10b), which is sensitive to the gluon density in the proton. The diagram can also be interpreted as a probe to the charm (sea) quark content in the proton. The NC events are selected by requesting a $D^* \to K\pi\pi$ in the final state. The visible cross section has to be extrapolated to find the total one, by using xg(x) from QCD fit to inclusive F_2. ZEUS and H1 achieve good agreement in the $ep \to eD^*X$ cross section.

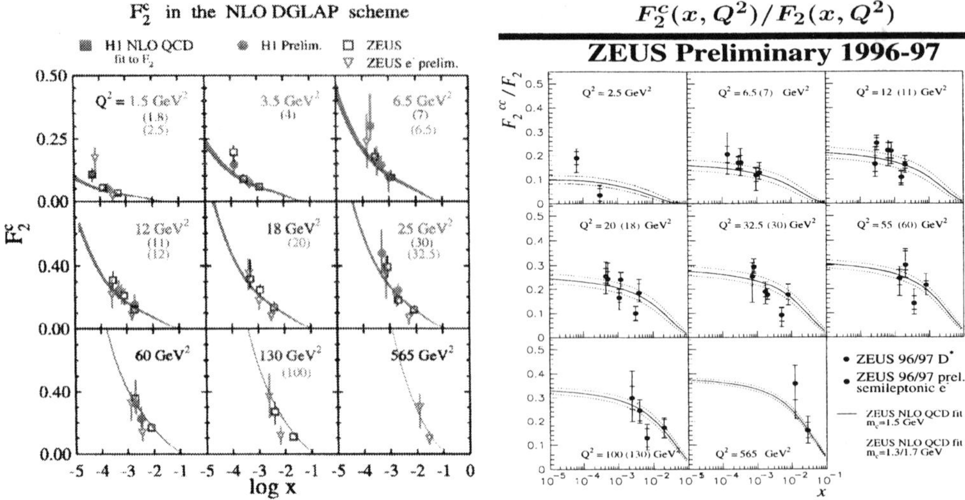

Figure 11. *Charm contribution to F_2 in the NLO DGLAP scheme: left, absolute; right, fractional.*

The $F_2^{\bar{c}c}(x,Q^2)$ results are shown in Fig. 11a. Again a strong rise towards low x is observed, as $F_2^{\bar{c}c} \propto xg(x)$. Fig. 11b presents the ratio $\frac{F_2^{\bar{c}c}}{F_2}$, as measured by ZEUS. It is evident that the charm content of the proton is non negligible, arriving even at 30% at high Q^2 and low x .

QUARK STRUCTURE

At high Q^2 the resolution is such that we might expect to probe the structure of the single quark in the proton.

Quark radius

The simplest (over-simplified: static) interpretation of negative deviations from Q^2 spectrum at highest Q^2 is a spheric space charge distribution of the target quark. Parametrization of form factor of quarks: $f(Q^2) = 1 - 1/6 < R_q^2 > Q^2$ where R_q is the radius of the (uniform) quark charge distribution, to be fitted from the Q^2 dependence of the ratio of the experimental cross section with the SM expectation multiplied by the form factor. Fig. 12a shows the upper limits achieved by H1 with the low statistics e^-data and by ZEUS with the full available data.

HERA has then achieved a limit of $1/1000$ of the proton radius for quarks. With only the valence quarks, the proton would then be almost empty, as the atom is!

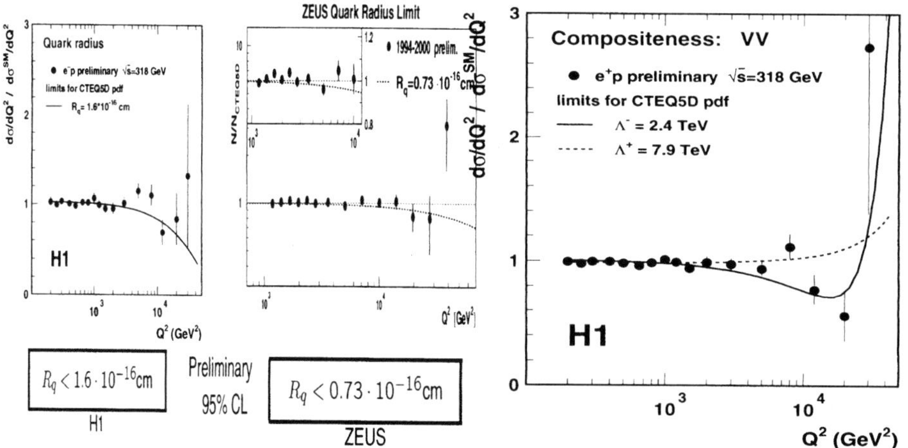

Figure 12. *(12a) Quark radius. Limits of H1 (e^-p data only) and ZEUS (global data). (12b) Fits on contact interaction models (VV case).*

Compositeness: contact interactions.

The structure of quarks can be studied in terms of compositeness by sub-elementary constituents, bound by a new interaction with caracteristic mass scale much larger than then the present center of mass scale: $\Lambda >> \sqrt{s}$. This can be parameterised as an effective 4 Fermion (eeqq) contact interaction with effective coupling $\eta^q_{ij} = \varepsilon^q_{ij}(\frac{g}{\Lambda^q_{ij}})^2$ with i,j= left/right-handed particles (helicities); q= u,d quark flavors; $\varepsilon^q_{ij} = \pm 1$: positive or negative interference with SM. Only vector coupling is assumed for HERA (scalar and tensor coupling terms are beyond the sensitivity of HERA). An example of the fits which give the lower limit of the coupling Λ is shown in Fig. 12b. The limits in both experiments achieve a few TeV in the various cases and are determined by the presence of fluctuations at the end of the Q^2 spectrum.

CONCLUSIONS

HERA experiments H1 and ZEUS have built up a wealth of new knowledge on structure functions and QCD in a very wide range of phase space ($0.11 < Q^2 < 46000\ GeV^2$; $\sim 10^{-6} < x < 0.7$). We have now at HERA all 4 double differential cross sections: NC, CC for e^+ and e^-. This enables us to extract special contributions for the first time, providing a sound phenomenological construction of strong interaction physics in the proton.

xF_3^{NC}, the $\gamma - Z^0$ term, has started to be measured at large Q^2. The Q^2 evolution equations named DGLAP allow for a NLO pQCD fit of the measured non-perturbative structure functions in the range $1 < Q^2 < 30,000$ GeV^2, $10^{-6} <$ x < 1, with an impressive good description of the x , Q^2 behaviour. $xg(x,Q^2)$ and α_S have been fitted at the same time; the rapid rise of $g(x,Q^2)$ at low x has been confirmed and α_S has been measured with high precision.

The CC cross section shows the unification of electromagnetic and weak interactions at HERA energies. CC (alone and with NC) at high x allow for the measurement of flavour separated parton density functions (for valence quarks) with sensitivity to the d/u ratio, without nuclear corrections. The contribution of the charm quark to F_2 has been measured: it follows the $g(x,Q^2)$ rise at low x and is relevant (up to $\approx 30\%$). At the highest Q^2 the quark structure begins to be explored. The quark radius is measured to be smaller than 0.1% of the proton radius. Limits for compositeness (contact interactions) have been inproved.

HERA is facing a luminosity and detectors upgrade. In the years 2001-2005 HERA will have potential for significant and competitive improvements of existing limits on new phenomena (like leptoquarks, leptogluons, R_p-violating SUSY particles, excited electron, excited neutrino, contact interactions eeqq) and of course the opportunity to improve in details its precise measurements of the structure functions.-

REFERENCES

1. Yu. L. Dokshitzer, Sov. Phys, JETP 46 (1977) 641;
V.N. Gribov and L.N. Lipatov, Sov. J. Nucl. Phys. 15 ('72) 438 and 675;
G. Altarelli and G. Parisi, Nucl. Phys. B126 (1977) 298.
2. Due to the lack of space, no detailed references can be given. The papers concerning the experimental results can be found in the WEB pages of the two experiments:
$http://www-zeus.desy.de/$
$http://www-h1.desy.de/h1/www/general/home/intra\_home.html$
3. M. Ciafaloni, Nucl. Phys. B296 (1988) 49;
S. Catani, F. Fiorani and G. Marchesini, Nucl. Phys. B336 ('90) 18;
G. Marchesini, Nucl. Phys. B445 ('95) 45.

Coulomb Distortion Effects for Electron or Positron Induced (e,e') Reactions in the Quasielastic Region

K.S. Kim*, L.E. Wright† and D.A.Resler†

*Department of Physics and Institute of Basic Science, Sung Kyun Kwan University, Suwan, 440-746, Korea email

†Institute of Nuclear and Particle Physics, Department of Physics and Astronomy, Ohio University, Athens, Ohio 45701

Abstract. In response to recent experimental studies we investigate Coulomb distortion effects on (e,e') reactions from medium and heavy nuclei for the case of electrons and positrons. We extend our previously reported full DWBA treatment of Coulomb distortions to the case of positrons for the $^{208}Pb(e,e')$ reaction in the quasielastic region for a particular nuclear model. In addition, we use previously reported successful approaches to treating Coulomb corrections in an approximate way to calculate the Coulomb distortion effects for (e,e') reactions for both electrons and positrons for the case of a simple nuclear model for quasielastic knock-out of nucleons. With these results in hand we develop a simple *ad-hoc* approximation for use in analyzing experiments, and discuss methods of extracting the "longitudinal structure function" which enters into evaluation of the Coulomb sum rule. These techniques are generally valid for lepton induced reactions on nuclei with momentum transfers greater than approximately 300 MeV/c.

INTRODUCTION

A persistant problem in using electron scattering for investigating nuclear structure and nuclear properties, especially in the quasielastic region, is the large static Coulomb field of medium and heavy nuclei. The presence of the static Coulomb potential (of order 25 MeV at the surface of the ^{208}Pb nucleus) invalidates one of the primary attributes of electron scattering as usually presented. Namely that in the electron plane-wave Born approximation, the cross section can be written as a sum of terms each with a characteristic dependence on electron kinematics and containing various bi-linear products of the Fourier transform of charge and current matrix elements. That is, various structure functions for the process can be extracted from the measured data by so-called Rosenbluth separation methods. The trouble with this picture is that when Coulomb distortion of the electron (or positron) wavefunctions arising from the static Coulomb field of the target nucleus is included exactly by partial wave methods, the structure functions can no longer be extracted from the cross section, even in principle.

In the early 90's, Coulomb distortion for the reactions (e,e') and $(e,e'p)$ in quasielastic kinematics was treated exactly by the Ohio University group[1, 2, 3, 4, 5] using partial wave expansions of the electron wavefunctions. Such partial wave treatments are referred to as the distorted wave Born approximation (DWBA) since the static Coulomb distortion is included exactly by numerically solving the radial Dirac equation contain-

CP594, *Hadrons and Nuclei: First International Symposium*, edited by Il-T. Cheon et al.

ing the Coulomb potential for a finite nuclear charge distribution to obtain the distorted electron wave functions. While this calculation permits the comparison of nuclear models to measured cross sections and provides an invaluable check on various approximate techniques of including Coulomb distortion effects, it is numerically challenging and computation time increases rapidly with higher incident electron energy. It was not possible to separate the cross section into various terms containing the structure functions and develop insights into the role of various terms in the transition charge and current distributions.

To avoid the numerical difficulties associated with DWBA analyses at higher electron energies and to look for a way to still define structure functions, our group [6, 7, 8] developed an approximate treatment of the Coulomb distortion based on the work of Knoll[9] and the work of Lenz and Rosenfelder[10]. We were able to greatly improve some previous attempts along this line[11, 12] where various additional approximations were made which turned out not to be valid. The essence of the approximation is to calculate the four potential A_μ arising from the lepton four current in the presence of the static Coulomb field of the nucleus. This is possible for momentum transfers greater than approximately 300 MeV/c in a limited spatial region which we take to be of order $3R$ where R is the nuclear charge radius. The Coulomb distortion is included in the four potential A_μ by the elastic scattering lepton phase shifts and by letting the magnitude of the lepton momentum include the effect of the static Coulomb potential. This last step leads to an r-dependent momentum. A key result of our approximation method is that the separation of the cross section into a "longitudinal" term and a "transverse" term is still possible.

We compared our approximate treatment of Coulomb distortion (which we will designate as *approximate DW*) to the exact DWBA results for the reaction $(e,e'p)$ and found good agreement (at about the 1-2% level) near the peaks of cross sections even for heavy nuclei such as ^{208}Pb. With an improved parametrization of the elastic scattering electron phase shifts [8], we achieve quite good agreement away from the peaks in the cross sections. Using this approximate DW treatment of Coulomb distortions for the inclusive (e,e') reaction in much more difficult numerically since the direction of the outgoing nucleon has to be integrated over, and all the nucleons in the nucleus have to be knocked out. Therefore, we sought even more severe approximations in order to obtain a simple *ad-hoc* method of calculating the structure functions for (e,e') reactions. In this paper we will use a simple non-relativistic *toy* model to calculate the Coulomb corrections to the longitudinal structure function with our *approximate DW* methods that we applied to $(e,e'p)$ and then investigate the *ad-hoc* treatment of the longitudinal structure function which is a key ingredient in investigating the Coulomb sum rule. After developing an improved *ad-hoc* procedure using our *toy* model we compare it the the full DWBA calculation which we have now extended to include positron induced reactions.

APPROXIMATE TREATMENT OF COULOMB DISTORTION

Our approximate method of including the static Coulomb distortion in the electron wavefunctions is to write the wave functions in a plane-wave-like form[7];

$$\Psi^{\pm}(\mathbf{r}) = \frac{p'(r)}{p} e^{\pm i\delta(\mathbf{L}^2)} e^{i\Delta} e^{i\mathbf{p}'(r)\cdot\mathbf{r}} u_p , \tag{1}$$

where the phase factor $\delta(\mathbf{L}^2)$ is a function of the square of the orbital angular momentum, u_p denotes the Dirac spinor, and the local effective momentum $\mathbf{p}'(\mathbf{r})$ is given in terms of the Coulomb potential of the target nucleus by

$$\mathbf{p}'(\mathbf{r}) = \left(p - \frac{1}{r}\int_0^r V(r)dr \right) \hat{\mathbf{p}} . \tag{2}$$

The $ad-hoc$ term $\Delta = a[\hat{\mathbf{p}}'(r)\cdot\hat{r}]\mathbf{L}^2$ denotes a small higher order correction to the electron wave number which we have written in terms of the parameter $a = -\alpha Z(\frac{16MeV/c}{p})^2$. The value of 16 MeV/c was determined by comparison with the exact radial wave functions in a partial wave expansion. We have examined the positron case ($Z \mapsto -Z$) and find that this parametrization works equally well when compared to the exact radial positron wave functions.

Initially [6] we fitted the phases δ_κ to a quadratic function of κ^2 which worked reasonably well for lower electron energies, but with the prospect of new higher energy electron accelerators, we needed a fit to the phases that will work at higher energies. We were able to find a parametrization of the elastic scattering phases shifts in terms of κ^2 which has the correct large κ^2 behaviour and becomes linear in κ^2 at low angular momentum, and since we have the correct large κ behaviour, we need only calculate the exact scattering phase shifts for κ values up to order pr. After some investigation [8], we found that the following parametrization of elastic scattering phase shift describes the exact phase shifts very well:

$$\delta(\kappa) = [a_0 + a_2\frac{\kappa^2}{(pR)^2}]e^{-\frac{1.4\kappa^2}{(pR)^2}} - \frac{\alpha Z}{2}(1 - e^{-\frac{\kappa^2}{(pR)^2}}) \times \ln(1+\kappa^2) \tag{3}$$

where p is the electron momentum and we take the nuclear radius to be given by $R = 1.2A^{1/3} - 0.86/A^{1/3}$. We fit the two constants a_0 and a_2 to two of the elastic scattering phase shifts ($\kappa = 1$ and $\kappa = Int(pR)+5$). To a very good approximation, $a_0 = \delta(1)$ and $a_2 = 4\delta(Int(pR)+5) + \alpha Z ln(2pR)$, where $Int(pR)$ replaces pR by the integer just less than pR. Note that this parametrization only requires the value of the exact scattering phase shift for $\kappa = 1$ and $\kappa = Int(pR)+5$. For this paper we have confirmed that this same parametrization works equally well for the positron phase shifts.

Using the new phase shift parametrization and the local effective momentum approximation, we construct plane-wave-like wave functions for the incoming and outgoing electrons. Since the only spinor dependence is in the Dirac spinor all of the Dirac algebra goes through as usual and we end up with a Møller-like potential given by,

$$A_{\mu}^{appro.DW}(\mathbf{r}) = \frac{4\pi e}{q^2-\omega^2} e^{i[\delta_i((\mathbf{r}\times\mathbf{p}_i'(r))^2)+\delta_f((\mathbf{r}\times\mathbf{p}_f'(r))^2)]} e^{i(\Delta_i-\Delta_f)} e^{i\mathbf{q}'(r)\cdot\mathbf{r}} \bar{u}_f \gamma_\mu u_i \tag{4}$$

where the phase shift parametrization is given in Eq. 3 with κ^2 being replaced by $(\mathbf{r}\times\mathbf{p})^2$, the parameter Δ is given following Eq. 2, and the r-dependent momentum transfer is given by $\mathbf{q}'(r) = \mathbf{p}_i'(r) - \mathbf{p}_f'(r)$.

APPLICATION TO THE INCLUSIVE PROCESS

In the plane wave Born approximation (PWBA), where electrons or positrons are described as Dirac plane waves, the cross section for inclusive quasielastic (e,e') processes can be written simply as

$$\frac{d^2\sigma}{d\Omega_e d\omega} = \sigma_M\{\frac{q_\mu^4}{q^4}S_L(q,w) + [\tan^2\frac{\theta_e}{2} - \frac{q_\mu^2}{2q^2}]S_T(q,w)\} \tag{5}$$

where $q_\mu^2 = \omega^2 - \mathbf{q}^2$ is the four-momentum transfer, σ_M is the Mott cross section given by $\sigma_M = (\frac{\alpha}{2E})^2\frac{\cos^2\frac{\theta}{2}}{\sin^4\frac{\theta}{2}}$, and S_L and S_T are the longitudinal and transverse structure functions which depend only on the momentum transfer q and the energy transfer ω. As is well known, by keeping the momentum and energy transfers fixed while varying the electron energy E and scattering angle θ_e, it is possible to extract the two structure functions with two measurements. As we will summarize below, our approximate treatment of Coulomb distortions still permit *Rosenbluth-like* separations but with Coulomb corrections which require the use of models.

For the inclusive cross section (e,e'), the longitudinal and transverse structure functions in Eq. (5) are bi-linear products of the Fourier transform of the components of the nuclear transition current density integrated over outgoing nucleon angles. Explicitly, the structure functions for knocking out nucleons from a shell with angular momentum j_b are given by

$$S_L(q,\omega) = \sum_{\mu_b s_p} \frac{\rho_p}{2(2j_b+1)}\int |N_0|^2 d\Omega_p \tag{6}$$

$$S_T(q,\omega) = \sum_{\mu_b s_p} \frac{\rho_p}{2(2j_b+1)}\int (|N_x|^2+|N_y|^2) d\Omega_p \tag{7}$$

where the nucleon density of states $\rho_p = \frac{pE_p}{(2\pi)^2}$, the z-axis is taken to be along $\mathbf{q}$, and μ_b and s_p are the z-components of the angular momentum of the bound and continuum state particles. The Fourier transfer of the nuclear current $J^\mu(\mathbf{r})$ is simply,

$$N^\mu = \int J^\mu(\mathbf{r}) e^{i\mathbf{q}\cdot\mathbf{r}} d^3r. \tag{8}$$

ans the continuity equation has been used to eliminate the z-component (N_z) via the equation $N_z = -\frac{\omega}{q} N_0$.

We created such an *ad-hoc* procedure in a previous paper [6], but we were comparing our *ad-hoc* procedures to the exact DWBA calculation which was largely dominated by the transverse terms. Hence, our *ad-hoc* procedures for the longitudinal term were not very well determined. In addition, our full DWBA calculation was only set up for electrons, so we could not check the *ad-hoc* approximation for positrons. In order to address this matter, we created a simple *toy* model which assumes harmonic oscillator bound state protons and takes the outgoing continuum proton wavefunction to be a plane wave. Using this simple model to calculate the transition charge distributions allows us to calculate the longitudinal contribution to the cross section using the approximate DW expression for N_0 in Ref. [6] and to compare this result to various *ad-hoc* proscriptions. Based on this investigation, coupled with our previous investigation of the transverse contributions which dominate the cross section at large electron scattering angles, we propose the following *ad-hoc* expressions for the longitudinal and transverse structure functions:

$$N_0^{ad-hoc} = \int (\frac{q'_\mu(r)}{q_\mu})^2 (\frac{q}{q'(r)})^2 e^{i<\delta(\kappa_i^2)+\delta(\kappa_f^2)>} e^{i\mathbf{q}'(r)\cdot\mathbf{r}} J_0(\mathbf{r}) d^3r \quad (9)$$

$$\mathbf{N}_T^{ad-hoc} = (\frac{p'_i(0)}{p_i}) \int e^{i\mathbf{q}'(r)\cdot\mathbf{r}} \mathbf{J}_T(\mathbf{r}) d^3r. \quad (10)$$

where $< \delta(\kappa_{i,f}^2) >$ denotes an average over the angles of the vector $\mathbf{r}$. That is, $< \kappa_{i,f}^2 > = < (\mathbf{r} \times \mathbf{p}_{i,f})^2 > = r^2 p_{i,f}^2 (3 - cos^2\theta_{p_{i,f}})/4$. Note that under this averaging, the Δ term goes to zero. This removes the angular dependence in the phase factors, and thus permits a multipole treatment of the matrix element as usual.

In the following figures for the longitudinal parts of the cross sections based on our simple model we will compare our new recommended longitudinal *ad-hoc* result given in Eq. (9) and to our previous *ad-hoc* results called LEMA$'$ which we give below for convenience [6]:

$$N_0^{LEMA'} = (\frac{p'_i(0)}{p_i} \int e^{i\mathbf{q}''(r)\cdot\mathbf{r}} J_0(\mathbf{r}) d^3r \quad (11)$$

where $\mathbf{q}'' = \mathbf{p}''_i(r) - \mathbf{p}''_f(r)$, $p''(r) = p - \frac{\lambda}{r} \int_0^r V(r')dr'$ and the factor λ, which depends on the energy transfer ω, is given by $\lambda = (\omega/\omega_o)^2$ with $\omega_o = \frac{q^2}{1.4M}$.

In Fig. 1 we compare the two approximate calculations with the DW approximation for the longitudinal contribution to the cross section for knocking protons out of various shells at a forward angle in ^{208}Pb by electrons or positrons. Note that while we use harmonic oscillator wavefunctions for all orbitals, we do use the binding energies of the orbitals that correspond to the values we find for our relativistic $\sigma - \omega$ model for ^{208}Pb. While the *ad-hoc* result is not in perfect agreement with the full DW result, it clearly is in better agreement that the LEMA$'$ result and, for cases where the electron incident and final energy exceed 300 MeV is in reasonable agreement, particularly near the maxima. Note that the positron results are not very sensitive to which approximation is used.

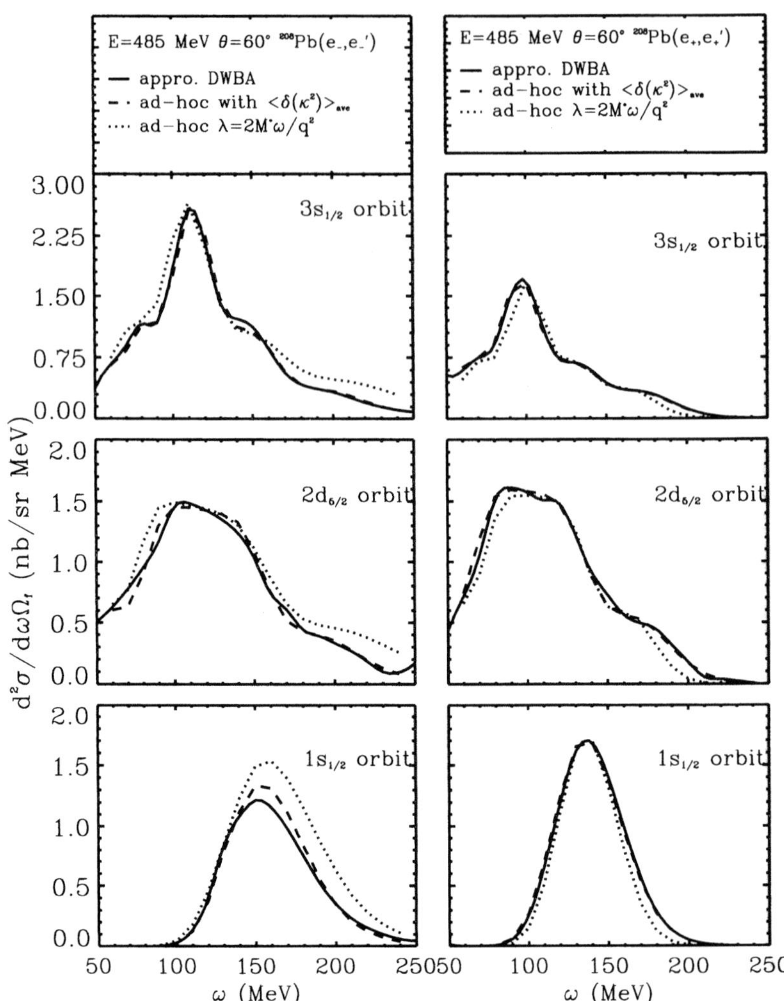

FIGURE 1. Longitudinal contributions to the differential cross sections at a forward scattering angle for $^{208}Pb(e_{\pm},e'_{\pm})$ for different bound state orbitals. The solid line is the approximate DW result, the dashed line is our *ad-hoc* result and the dotted line is our previous LEMA$'$ approximation.

In Fig. 2 we show similar results at a backward angle. We note that our *ad-hoc* DWBA results for positrons tend to be in much better agreement with the DW result than the electron case. We again find that our new *ad-hoc* approximation for the longitudinal contribution is considerably better that our previous LEMA$'$ result. We note that while the agreement between our *ad-hoc* calculation and the DW calculation for knocking out protons from individual orbitals is not excellent, the discrepancies do not seem have a systematic tendency to be either low or high and we have reason to hope that when all the orbitals are added together as in the case of (e,e') reactions from nuclei that these

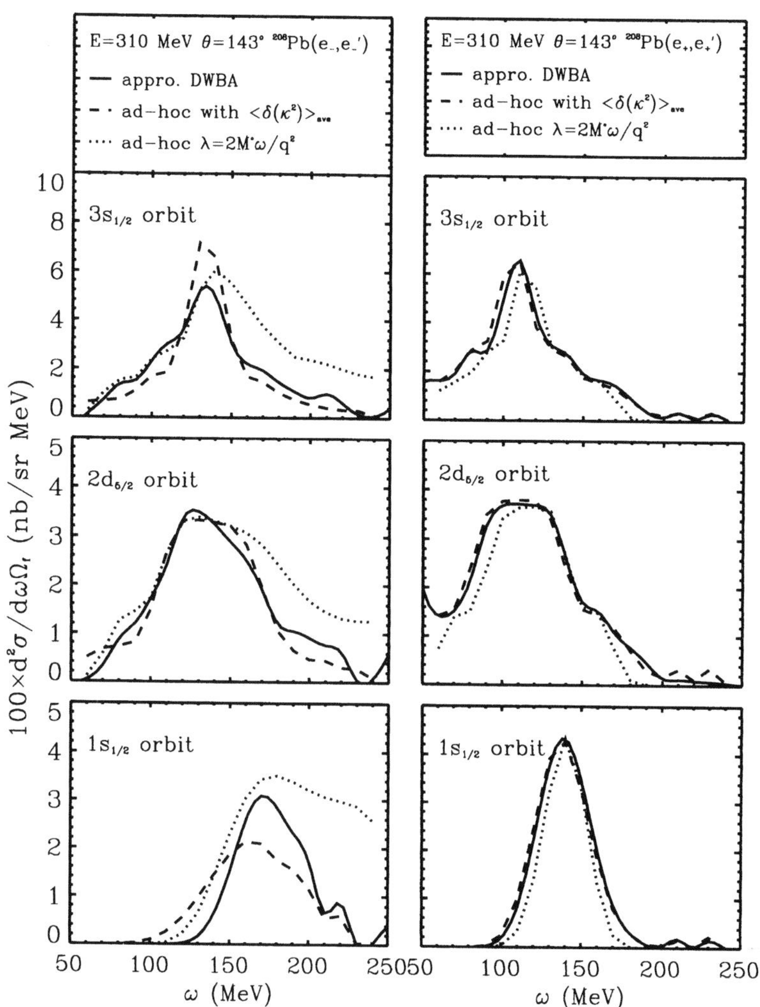

FIGURE 2. Longitudinal contributions to the differential cross sections for $^{208}Pb(e_{\pm},e'_{\pm})$ at a backward angle for different bound state orbitals. The solid line is the approximate DW result, the dashed line is our *ad-hoc* DWBA result and the dotted line is our previous LEMA$'$ approximation.

discrepancies will tend to average out.

COMPARISON TO EXPERIMENT AND CONCLUSIONS

Based of our investigation of this simple *toy* model, we adopt our new *ad-hoc* model for the longitudinal structure functions and return to our full nuclear model for investigating Coulomb corrections for $^{208}Pb(e,e')$ in the quasielastic region where the lepton can be

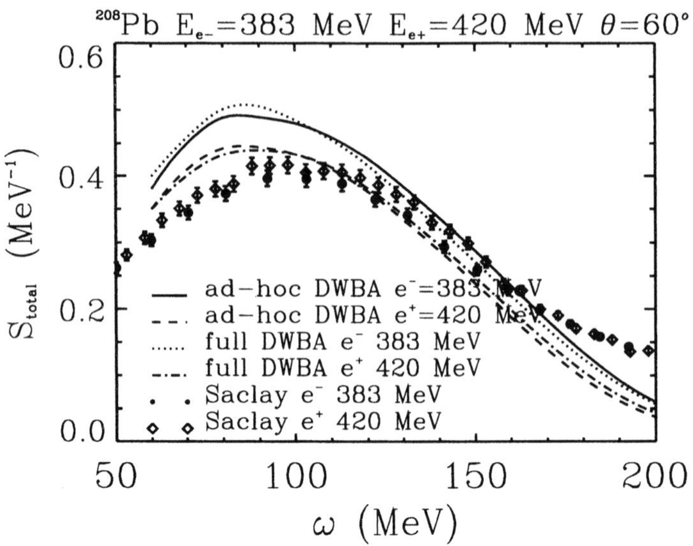

FIGURE 3. The total structure function S_{total} generated by dividing the differential cross section by σ_M for $^{208}Pb(e_{\pm},e'_{\pm})$ at a forward scattering angle of 60° with electrons of energy 383 MeV and positrons with energy 420 MeV. The theoretical curves correspond to the full DWBA calculation and to our *ad-hoc* DWBA calculation. The data were taken at Saclay [13, 14]. The bound state and continuum neutron and proton orbitals are solutions to relativistic Hartree potential based on the $\sigma - \omega$ model.

electrons or positrons. Our first step is to re-examine our full DWBA calculation [3] and modify the code for the case of positrons. We were successful in doing this and can now compare the full DWBA calculation for electrons and positrons based on a realistic relativisitic nuclear model to our *ad-hoc* treatment of Coulomb corrections which still permit a separation into longitudinal and transverse terms.

With our capability of examining Coulomb distortion of both positrons and leptons with the full DWBA calculation and with our improved *ad-hoc* procedure we can compare our model predictions to experiment. In Fig. 3, we compare our model calculations with Coulomb distortion included exactly and with our *ad-hoc* method for quasielastic scattering of electrons of energy 383 MeV and positrons of energy 420 MeV both at a scattering angle of $\theta = 60^o$ from ^{208}Pb to the experimental data from Saclay [13, 14]. Note that in this and the following figure, we are plotting the total structure function $S_{total} = \frac{d^2\sigma}{d\omega d\Omega_f}/\sigma_M(E_i)$.

We first note that our *ad-hoc* and exact DWBA results are in reasonable agreement although the lepton energy is somewhat low for our approximate result, and further that the positron and electron total structure functions have approximately the same shape as a function of the energy transfer ω. However, they do not have the same magnitude as do the data from Saclay. The positron theory result is in reasonable agreement with the experimental data, but the electron result is approximately 15%-20% larger than the data.

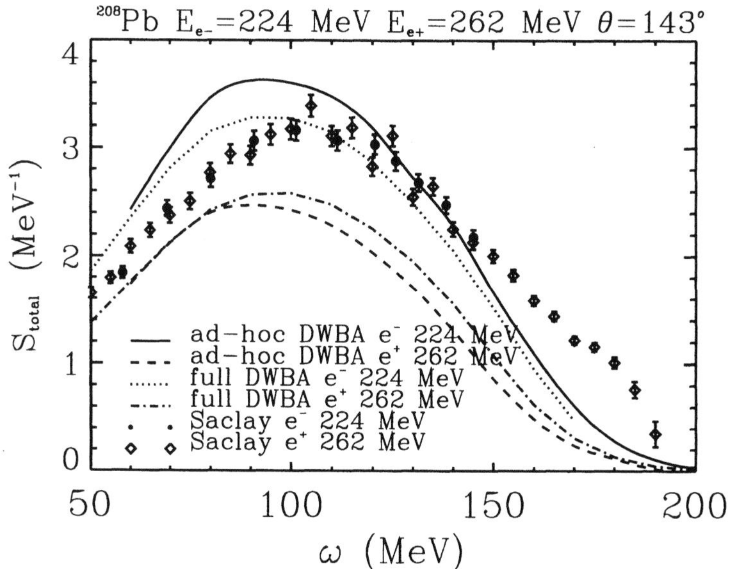

FIGURE 4. The total structure function S_{total} generated by dividing the differential cross section by σ_M for $^{208}Pb(e_{\pm},e'_{\pm})$ at a backward scattering angle of 143° with electrons of energy 224 MeV and positrons with energy 262 MeV. The theoretical curves correspond to the full DWBA calculation and to our *ad-hoc* DWBA calculation. The data were taken at Saclay [13, 14]. The bound state and continuum neutron and proton orbitals are solutions to relativistic Hartree potential based on the $\sigma-\omega$ model.

In Fig. 4 we make a similar comparison except that now the scattering angle is $\theta = 143^o$, and the electron incident energy is 224 MeV while the positron incident energy is 262 MeV. Again, when S_{total} is plotted the positron and electron shapes as a function of energy transfer ω are very similar, but again, unlike the experimental data, the magnitudes are quite different. At this backward scattering angle case, our electron result (DWBA) is in quite good agreement with the data. At these much lower energies, clearly our *ad-hoc* approximation is beginning to fail, particularly for the electron case.

There is considerable interest in extracting the longitudinal contributions from (e,e') reactions from medium and heavy nuclei in order to investigate the Coulomb sum rule. Clearly, Coulomb distortion effects have to be handled properly. Our results indicate that we could use a *Rosenbluth-like* procedure in order to separate our "longitudinal" and "transverse" contributions to the cross section. However, these contributions depend on a modified (by Coulomb distortion) Fourier transform of the transition charge and current distributions. It is necessary to use a nuclear model to extract the longitudinal and/or tranverse structure functions from the data. It is not clear to us that a *Rosenbluth-like* procedure is the best way to proceed, since our *ad-hoc* procedure is not accurate in the wings of the cross section distributions and in many cases, some of the Rosenbluth points fall on either the low ω or high ω side of the quasielastic peak. It seems that a better procedure might be to choose some semi-realistic nuclear model for the process in question. Use Eqs. (9) and (10) to calculate the structure functions and then fit the calculations to the available data using a least squares procedure to determine nomalization factors N_L and N_T in front of the appropriate terms. The nuclear model should have the overall correct spatial and kinematic dependence, but the longitudinal or transverse strength will be determined by fitting these normalization factors. Once these factors are determined, one can use the same nuclear model weighted with these factors to calcuate the plane wave structure functions, thereby having "measured" the nuclear longitudinal and transverse response. Furthermore, we notice in our investigations a general tendency that Coulomb distortion effects for positrons tend to be smaller than Coulomb distortion effects for electrons. This corresponds to an observation made many years ago when looking at inelastic lepton scattering from nuclei [15], where we noted that Coulomb distortion for positrons tends to saturate. As electrons pass near the nucleus, the attactive Coulomb interaction pulls them into regions of stronger potential which increases the Coulomb distortion effects, while positrons are pushed away from the region with a stronger potential.

REFERENCES

1. Yanhe Jin, D. S. Onley, and L. E. Wright, Phys. Rev. **C45**, 1311(1992).
2. C.F. Williamson, T.C. Yates, W. M. Schmitt, M. Osborn, M. Deady, Peter, D. Zimmerman, C.C. Blatchley, Kamal K. Seth, M. Sarmiento, B. Parker, Yahne Jin, L.E. Wright and D.S. Onley, Phys. Rev. **C56**, 3152(1997).
3. Yanhe Jin, D. S. Onley, and L. E. Wright, Phys. Rev. **C45**, 1333(1992).
4. Yanhe Jin, J.K. Zhang, D.S. Onley and L.E. Wright, Phys. Rev. C**47**, 2024(1993).
5. Yanhe Jin, D.S. Onley and L.E. Wright, Phys. Rev. C**50**, 168(1994).
6. K. S. Kim, L. E. Wright, Yanhe Jin, and D. W. Kosik, Phys. Rev. **C54**, 2515(1996).
7. K. S. Kim and L. E. Wright, Phys. Rev. **C56**, 302(1997).
8. K. S. Kim and L. E. Wright, Phys. Rev.**C**60, 067604 (1999).
9. J. Knoll, Nucl. Phys. **A201**, 289 (1973): **A223**, 462(1974).
10. F. Lenz and R. Rosenfelder, Nucl. Phys. **A176**, 513(1971); F. Lenz, thesis, Freiburg (1971).
11. C. Giusti, and F. D. Pacati, Nucl. Phys. **A473**, 717(1987).
12. M. Trani, S. Turck-Chieze and A. Zghiche, Phys. Rev. **C38**, 2799 (1988).
13. P. Guèye, et al., Phys. Rev. **C60**, 044308 (1999).
14. J. Morgenstern, Private Communication.
15. D.S. Onley, J.T. Reynolds and L.E. Wright, Phys. Rev. **134**, B945 (1965).

Pion and Kaon Electromagnetic Form Factors and Their Related Decays in an $SU_L(3) \otimes SU_R(3)$ Effective Lagrangian

Yun Chang Shin[a], Myung Ki Cheoun[a,b] , K. S. Kim[a] and T. K. Choi [a]

a) Department of Physics, Yonsei University, Seoul, 120-749, Korea
b) IUCNSF, Seoul National University, Seoul 151-742, Korea

Abstract. An $SU(2)$ effective lagrangian is extended to an $SU_L(3) \otimes SU_R(3)$ by including the vector and axial vector meson. With this effective lagrangian, electromagnetic form factors of charged pion and kaon are calculated in both time- and space-like regions. Good agreement with experimental data is obtained for those form factors. Decay widths of $\rho \to \pi\pi$ and $\phi \to K^+K^-$ are also calculated and shown to agree with experimental data very well.

INTRODUCTION

At energy below 1 GeV, the vector meson plays an important role in electromagnetic interactions of the hadron. The vector meson dominance model (VMD) has been proved remarkably successful in the description of electromagnetic form-factors and decays of the hadron, although it is a phenomenological approach. Many approaches, such as a hidden gauge symmetry approach (HGS)[1], a massive Yang-Mills approach (MYM)[2], and so on, have been developed to include the vector meson in a fundamental manner. By taking higher order terms into account, redefining suitable field and adjusting parameters, all of the model can be shown to be equivalent[3]. However, a simple addition of higher order terms is not a convenient method for those calculations. In our previous papers[4, 5], we have proposed an effective chiral Lagrangian for the description of vector and axial-vector mesons by considering all the relevant symmetries and the low-energy constraints from chiral perturbation theory(ChPT). The spin-1 mesons are introduced in a non-linear realization of chiral symmetry, with which it is easy to check consistency with chiral perturbation theory. In constructing our model Lagrangian, we have stressed simplicity. Only mass terms and kinetic terms of spin-1 meson fields are necessary to meet experimental results. In some approach, such as $O(p^4)$ order expansion of ChPT, $\mathcal{L}_2$ lagrangian gives loop contribution as well known[6], which helps a good phenomenological description. But our effective lagrangian theory, which is aimed for large energy process, uses $O(p^2)$ expansion because most of the higher order contributions in other approaches are incorporated by a single change in the kinetic terms of vector field with only one parameter in our model. Since full reviews concerning effective theories and their relationships to other approaches can be found in other papers [2, 3], we skip them here.

In this talk our lagrangian is briefly summarized in section 2. In section 3, the

CP594, *Hadrons and Nuclei: First International Symposium,* edited by Il-T. Cheon et al.

electromagnetic pion and kaon form factors and some related decays with this lagrangian are presented with detailed discussions. A summary is done with a remark concerning a role of axial vector mesons on heavy ion collision in the final section.

LAGRANGIAN

Our lagrangian consists of a pseudoscalar meson sector $\mathcal{L}(\pi)$, a spin-1 vector and axial vector meson sector $\mathcal{L}(V,A)$, and a term of interactions with scalar particles $\mathcal{L}_S$, which comes from mass splittings in SU(3) extension, i.e.,

$$\mathcal{L} = \mathcal{L}(\pi) + \mathcal{L}(V,A) + \mathcal{L}_S \, . \tag{1}$$

The lagrangian for the pseudoscalar meson sector, which is a leading Lagrangian of the ChPT, is given as

$$\begin{aligned} \mathcal{L}(\pi) &= \frac{f^2}{4}\langle D^\mu U^\dagger D_\mu U\rangle + \frac{f^2}{4}\langle U^\dagger \chi + \chi^\dagger U\rangle \, , && (2) \\ D_\mu U &= \partial_\mu U - i(v_\mu + a_\mu)U + iU(v_\mu - a_\mu) \, , && (3) \end{aligned}$$

where bracket denotes a trace in flavor space, f is a pseudoscalar meson decay constant, chiral field $U = \exp(i2\pi/f)$ with $\pi = T^a\pi^a$, $T^a = \lambda^a/2(a = 1,2,...8)$. External gauge fields are introduced via v_μ and a_μ. The χ is defined by $\chi = 2B_0(\mathcal{S} + i\mathcal{P})$. Explicit chiral symmetry breaking due to current quark masses can be introduced by treating those masses as if they were uniform external scalar field $\mathcal{S}$[3]. The non-linear realization of chiral symmetry is expressed in terms of $u = \sqrt{U}$ and $h = h(u, g_R, g_L)$ defined from $u \to g_R u h^\dagger = h u g_L^\dagger$. In this realization, we naturally have the following covariant quantities

$$\begin{aligned} i\Gamma_\mu &= \frac{i}{2}(u^\dagger\partial_\mu u + u\partial_\mu u^\dagger) + \frac{1}{2}u^\dagger(v_\mu + a_\mu)u + \frac{1}{2}u(v_\mu - a_\mu)u^\dagger \, , \\ i\Delta_\mu &= \frac{i}{2}(u^\dagger\partial_\mu u - u\partial_\mu u^\dagger) + \frac{1}{2}u^\dagger(v_\mu + a_\mu)u - \frac{1}{2}u(v_\mu - a_\mu)u^\dagger \, , \\ \chi_+ &= u^\dagger\chi u^\dagger + u\chi u \, , && (4) \end{aligned}$$

whose transformations are carried out in terms of h, i.e., $\Gamma_\mu \to h\Gamma_\mu h^\dagger - \partial_\mu h \cdot h^\dagger$, $\Delta_\mu \to h\Delta_\mu h^\dagger$, and $\chi_+ \to h\chi_+ h^\dagger$. With these quantities, the Lagrangian in eq.(2) can be rewritten as

$$\mathcal{L}(\pi) = f^2\langle i\Delta_\mu i\Delta^\mu\rangle + \frac{f^2}{4}\langle\chi_+\rangle. \tag{5}$$

As for the massive spin-1 mesons, we include only the mass and kinetic terms[4]

$$\mathcal{L}(V,A) = m_V^2\langle(V_\mu - \frac{i\Gamma_\mu}{g})^2\rangle + m_A^2\langle(A_\mu - \frac{ir\Delta_\mu}{g})^2\rangle - \frac{1}{2}\langle({}^G V_{\mu\nu})^2\rangle - \frac{1}{2}\langle(A_{\mu\nu})^2\rangle \tag{6}$$

with

$$\begin{aligned} {}^G V_{\mu\nu} &= \partial_\mu V_\nu - \partial_\nu V_\mu - ig[V_\mu, V_\nu] - iG[A_\mu, A_\nu] \, , \\ A_{\mu\nu} &= \partial_\mu A_\nu - \partial_\nu A_\mu - ig[V_\mu, A_\nu] - ig[A_\mu, V_\nu] \, , && (7) \end{aligned}$$

where $V_\mu = T^a V^a_\mu (A_\mu = T^a A^a_\mu)$ denotes spin-1 vector (axial-vector) meson field and g denotes a $V\pi\pi$ coupling constant. Note that we have introduced a new form of ${}^G V_{\mu\nu}$. The chiral symmetry is preserved for any value of G at chiral limit in ${}^G V_{\mu\nu}$, so that the value of G cannot be determined from the chiral symmetry. If G is equal to g as in the HGS approach, the result may reproduce experimental data by including other higher order terms.

The introduction of the $\mathcal{L}_S$ term can be found in ref. [5]. The resulting Lagrangian is given as

$$\mathcal{L}_S \sim -\frac{1}{2}(\frac{s_m}{f})^2(\tilde{M})^2_a(\pi^a)^2 + \frac{1}{2}s_m M_a j^a, \tag{8}$$

where $\tilde{M}^2_a = \frac{1}{6}(2B_0\alpha)^2\delta_{8a} + M^2_a$. In expanding our lagrangian, we choose only photon, pseudoscalar meson and vector meson parts. The lagrangian is, then, summarized as

$$\begin{aligned}
\mathcal{L} \;=\; & \frac{1}{2}m^2_{Va}V_\mu V^\mu + \frac{1}{2}m^2_{Aa}A_\mu A^\mu \\
& + \frac{m^2_{Va}}{2gf^2_a}(1 - \frac{Gr^2}{g}) f_{abc} V^a_\mu \pi^b \partial^\mu \pi^c \\
& + eQ f_{abc} B^a_\mu \pi^b \partial^\mu \pi^c \\
& - \frac{1}{4}(\partial_\mu V_\nu - \partial_\nu V_\mu)^2 - \frac{1}{4}(\frac{e}{g})^2(\partial_\mu B_\nu - \partial_\nu B_\mu)^2 \\
& - \frac{e}{2g}(\partial_\mu V_\nu - \partial_\nu V_\mu)(\partial^\mu B^\nu - \partial^\nu B^\mu) \\
& + \frac{1}{2}\frac{Gr}{gf} f_{abc}(\partial_\mu V^a_\nu - \partial_\nu V^a_\mu)(A^{\mu b}\partial^\nu \pi^c + \partial^\mu \pi^b A^{\nu c}) \\
& + \frac{1}{2}\frac{Gr}{gf}(\frac{e}{g}) Q f_{abc}(\partial_\mu B^a_\nu - \partial_\nu B^a_\mu)(A^{\mu b}\partial^\nu \pi^c + \partial^\mu \pi^b A^{\nu c}) \\
& - \frac{1}{4}(\partial_\mu A_\nu - \partial_\nu A_\mu)^2 \\
& + \frac{1}{2}\frac{r}{f} f_{abc}(\partial_\mu A^a_\nu - \partial_\nu A^a_\mu)(V^{\mu b}\partial^\nu \pi^c + \partial^\mu \pi^b V^{\nu c}) \\
& - \frac{1}{2}\frac{r}{f}\frac{e}{g} Q f_{abc}(\partial_\mu A^a_\nu - \partial_\nu A^a_\mu)\pi^b(\partial^\nu B^c - \partial^\mu B^c) \\
& - \frac{1}{2}m^2_{\pi a}\pi^a\pi^a + \frac{1}{2}\partial_\mu \pi^a \partial^\mu \pi^a,
\end{aligned} \tag{9}$$

where $m^2_{Va} = g^2(f^2_V + s_m s_\nu M_a)$ and V_μ and A_μ stand for redefined fields $V^{'}_\mu$ and $A^{em'}_\mu$. In order to determine pseudoscalar meson mass and decay constants, we exploit the following covariant quantities

$$\begin{aligned}
m^2_{\pi a} \;&=\; (M_a + (\frac{s_m}{f})^2 \tilde{M_a}^2)/Z^2_{\pi a}\,, \; f_a = Z_{\pi a} f \\
& with \;\; Z^2_{\pi a} = (1 + s_m s_d \frac{M_a}{f^2})\,.
\end{aligned} \tag{10}$$

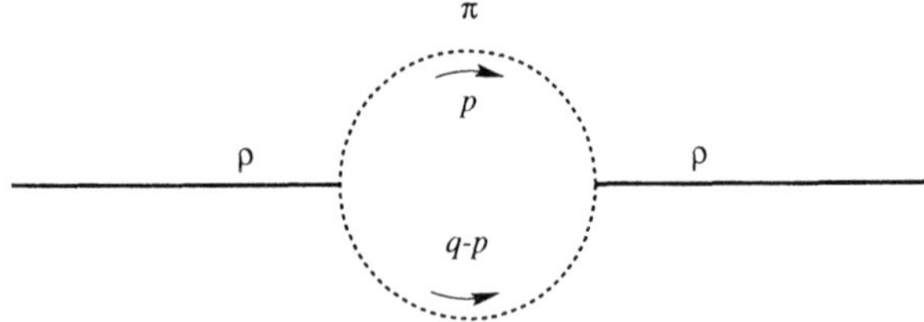

FIGURE 1. ρ meson self energy

Mass splitting between non-strange particles and strange particles is generated from interaction theses fields with scalar field.

PION AND KAON ELECTROMAGNETIC FORM FACTOR

The pion form-factor in the time-like region is dominated by the ρ-meson resonance. Likewise to the pion the kaon form-factor is influenced mainly by the φ-meson. But the contribution of ρ-ω meson mixing is also important. With the effective lagrangian in section 2, we improve the analysis of both form-factors. The pseudoscalar meson loops are also taken into account. From the effective lagrangian, V-π interaction term in eq.(9) generates a vector current of pion as

$$J^{\mu}_{had} = i(\pi^{+}\partial^{\mu}\pi^{-} - \pi^{-}\partial^{\mu}\pi^{+}). \tag{11}$$

This coupling to the ρ-meson field produces the self energy as shown in Fig. 1 , which is calculated as

$$-i\Pi^{\mu\nu} = g^{2}_{\rho\pi\pi}\int \frac{d^{4}p}{(2\pi)^{4}} \frac{(2p-q)^{\mu}(2p-q)^{\nu}}{(p^{2}-m_{\pi}^{2}+i\varepsilon)((p-q)^{2}-m_{\pi}^{2}+i\varepsilon)} \,. \tag{12}$$

The ρ meson coupling to a conserved current implies that the self energy is transverse, i.e.,

$$q_{\mu}\Pi^{\mu\nu}(q) = q_{\nu}\Pi^{\mu\nu}(q) = 0. \tag{13}$$

This property, combined with Lorentz invariance, uniquely determines the tensor structure of the self energy

$$\Pi^{\mu\nu} = (-g^{\mu\nu} + \frac{q^{\mu}q^{\nu}}{q^{2}})\Pi(q^{2}). \tag{14}$$

The full propagator of ρ meson is then given as

$$D^{\mu\nu} = \frac{1}{q^{2} - \dot{m}_{\rho}^{2} - \Pi_{\rho}}(-g^{\mu\nu} + \frac{q^{\mu}q^{\nu}}{q^{2}}) + \frac{1}{\dot{m}_{\rho}^{2}}\frac{q^{\mu}q^{\nu}}{q^{2}} \,. \tag{15}$$

Here, the bare ρ meson mass, $\dot{m}_{\rho}$, is introduced so that its physical mass is given by

$$m_{\rho P}^{2} = \dot{m}_{\rho}^{2} + \mathrm{Re}[\Pi_{\rho}(q^{2} = m_{\rho}^{2})]. \tag{16}$$

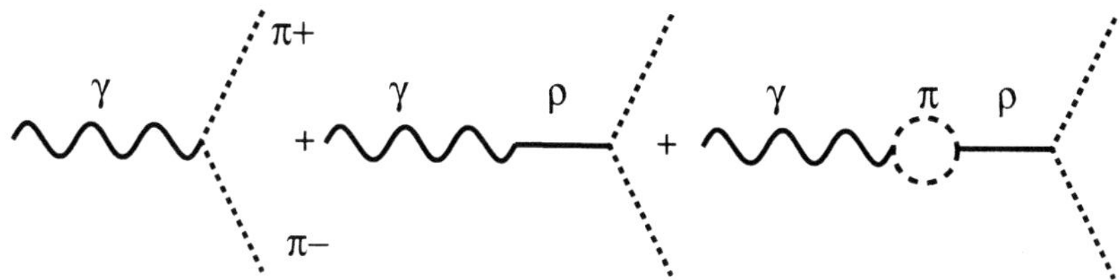

FIGURE 2. Feynmann Diagram of Pion Form-factor with self energy

Since the full propagator of ρ meson is given as eq.(16), the $\rho \to \pi\pi$ decay width at resonance is given as

$$\Gamma_{\rho\to\pi\pi} = -\mathrm{Im}\Pi_\rho(q^2 = m_\rho^2)/m_\rho. \tag{17}$$

Detailed calculation for the self-energy is given in ref. [5]. The electromagnetic pion form-factor is defined by the following matrix element

$$\langle \pi^\pm(k')|J_\mu^{em}(0)|\pi^\pm(k)\rangle = \pm(k+k')_\mu F_\pi(q^2). \tag{18}$$

The leading term of $F_\pi(q^2)$ obtained from $\mathcal{L}_{\gamma\pi}$, $\mathcal{L}_{\gamma V}$ is expressed in the following way

$$F_\pi^{(o)}(q^2) = 1 - \frac{g_{\rho\pi\pi}}{g}\frac{q^2}{q^2 - m_\rho^2 + i m_\rho \Gamma_\rho}, \tag{19}$$

where g is a bare coupling constant which does not consider the loop effect. m_ρ and $\dot{m}_\rho$ are means of ρ meson and bare ρ meson, respectively. The relation of both masses is given by eq.(16).

Introducing the $\rho\pi\pi$ self energy in Fig. 2, we obtained

$$\begin{aligned} F_\pi(q^2) &= 1 - \frac{g_{\rho\pi\pi}}{g}\frac{q^2}{q^2 - \dot{m}_\rho^2 - \Pi_\rho} + \frac{\Pi_\rho}{q^2 - \dot{m}_\rho^2 - \Pi_\rho} \\ &= 1 - \frac{g_{\rho\pi\pi}}{g(q^2)}\frac{q^2}{q^2 - \dot{m}_\rho^2 - \Pi_\rho}. \end{aligned} \tag{20}$$

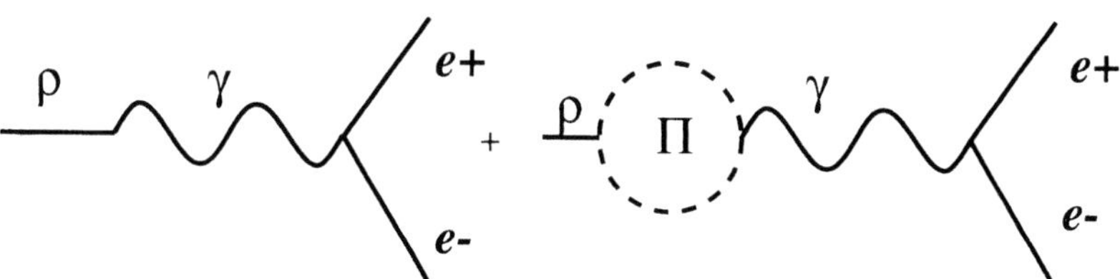

FIGURE 3. ρ meson decay

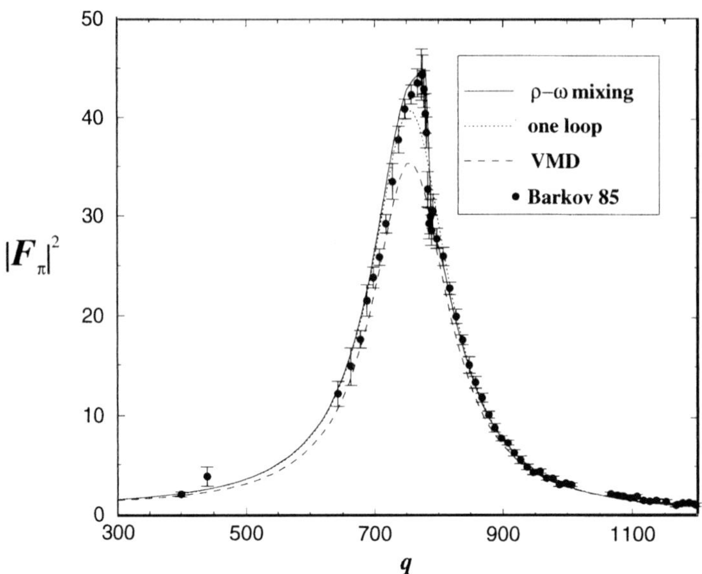

FIGURE 4. Pion Electromagnetic Form-Factor in time like region: solid, dotted and dashed lines represent eqs.(45), (42) and (41), respectively. Here q means $\sqrt{q^2}$.

Note that not only the ρ meson propagator, but also $\gamma-\rho$ coupling is modified by the pion loop as shown in Fig. 3 as follows

$$-\frac{eq^2}{g(q^2)} = -\frac{eq^2}{g} + \frac{e\Pi_\rho}{g_{\rho\pi\pi}}. \tag{21}$$

The constant g determined from the experimental $\rho \to e^+e^-$ decay width should be compared with $\mathrm{Re}[g(q^2)]_{q^2=m_\rho^2}$. Using $\mathrm{Re}[g(q^2)]_{q^2=m_\rho^2}$ and experimental results of $\Gamma_{\rho\to\pi\pi}$, we find g and $\beta = Gr^2$ values. When g is 5.36 and β is 0.32, $g_{\rho\pi\pi}$ goes to 6.037. Under universality ($g_{\rho\pi\pi} = g_{\rho\gamma}$) used in VMD model, the prediction of the pion form factor is underestimated compared to the experimental values. Brown et.al [14] allows its violation i.e. $\beta = g_{\rho\pi\pi}/g_{\rho\gamma} = 1.2$ by considering the intrinsic size due to the vector meson. In the papers, for example of Brown[14] and Klingl[13], this contribution is attributed to those of vector mesons, which are incorporated by the gauge fields in the hidden gauge symmetry approach, while in this talk these fields are introduced explicitly using the non-linear realization of the chiral symmetry exploited originally by Weinberg.

Using eqs. (16) and (17), we also find physical mass and decay width:

$$m_{\rho P} \approx 771\mathrm{MeV}\ ,\ \Gamma_{\rho\to\pi\pi} \approx 150\mathrm{MeV}\ ,\ \mathring{m}_\rho \approx 808\mathrm{MeV}\ . \tag{22}$$

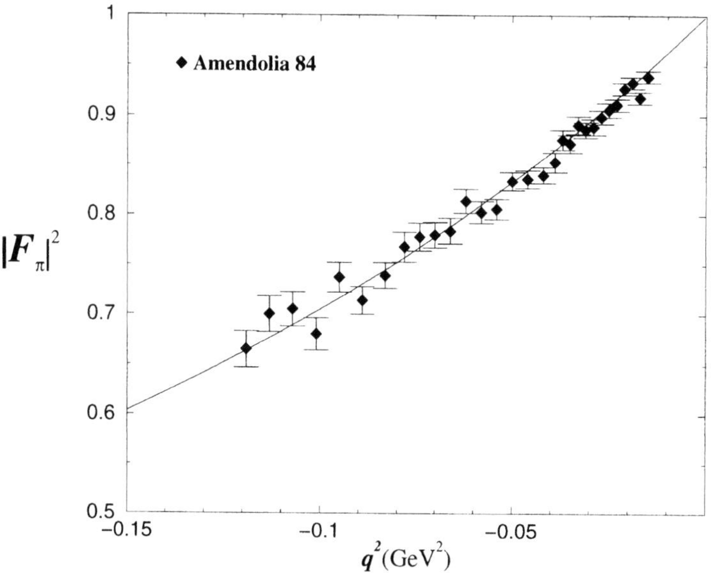

FIGURE 5. Pion Electromagnetic Form-Factor in space like region.

Finally, the inclusion of $\rho-\omega$ mixing turned out to give another factor to eq.(42) in the following way

$$F_\pi(q^2) = (1 - \frac{g_{\rho\pi\pi}}{g(q^2)} \frac{q^2}{q^2 - \dot{m}_\rho^2 - \Pi_\rho})(1 + \frac{g(q^2)}{g_\omega} \frac{z_{\rho\omega}}{q^2 - m_\omega^2 - im_\omega\Gamma_\omega}). \tag{23}$$

The ω meson width Γ_ω=8.4MeV is used and the mixing parameter $z_{\rho\omega} = -4.52 \times 10^{-3}\mathrm{GeV}^2$ from Ref. [13] is also exploited. The corresponding optimal result for $F_\pi(q^2)$ compared with experimental data[16] is shown in Fig. 4. Dashed line corresponds to VMD model prediction. Dotted line includes one loop correction while solid line takes $\rho-\omega$ mixing contribution into account.

The form factor in the space like region ($q^2 < 0$) is also given in Fig. 5. Our approach gives a good agreement with experimental result [15]. The electromagnetic form-factor of charged kaon is also defined by

$$\langle \mathrm{K}^\pm(k')|J_\mu^{em}(0)|\mathrm{K}^\pm(k)\rangle = \pm(k+k')_\mu F_K(q^2). \tag{24}$$

The leading behavior of $F_K(q^2)$ is obtained just by transcribing the previous formalism developed for $F_\pi(q^2)$ and replacing the ρ meson by the ϕ meson and the pion loop by the kaon loop. It leads to yield the following result obtained

$$F_K(q^2) \quad = \quad 1 + \frac{\sqrt{2}}{3} \frac{g_{\phi K^+K^-}}{g} \frac{q^2}{q^2 - \dot{m}_\phi^2 - \Pi_\phi} + \frac{\Pi_{\phi\to K^+K^-}}{q^2 - \dot{m}_\phi^2 - \Pi_\phi}$$

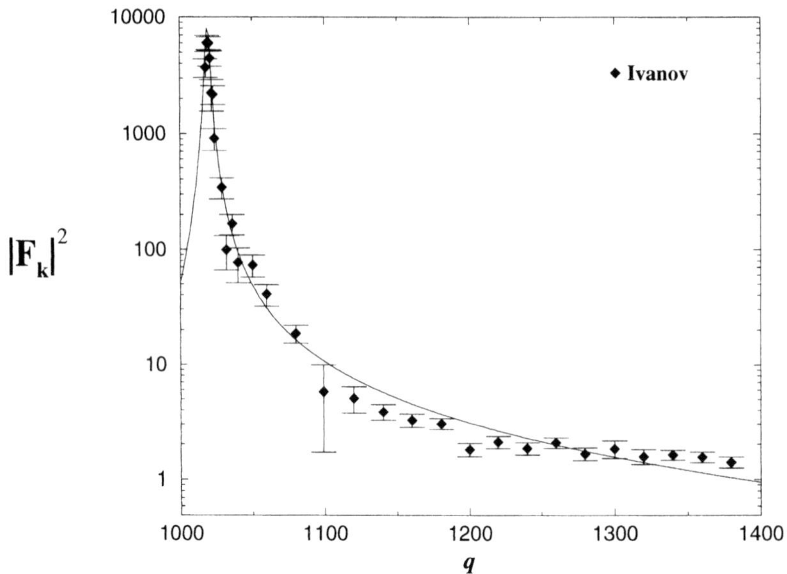

FIGURE 6. Kaon Electromagnetic Form-Factor in time like region : Here q means $\sqrt{q^2}$.

$$= \quad 1+\frac{\sqrt{2}}{3}\frac{g_{\phi K^+K^-}}{g(q^2)}\frac{q^2}{q^2-\mathring{m}_\phi^2-\Pi_\phi}, \tag{25}$$

where the ϕ meson self energy has the contributions not only from K^+K^- but also from $K_L^oK_S^o$, i.e.,

$$\Pi_\phi = \Pi_{\phi\to K^+K^-} + \Pi_{\phi\to K_L^oK_S^o}\ . \tag{26}$$

The photon coupling of the ϕ meson is also modified by means of the charged kaon loop including the renormalization

$$\frac{1}{g_\phi(q^2)} = \frac{1}{g} + \frac{3}{\sqrt{2}}\frac{\Pi_{\phi\to K^+K^-}}{g_{\phi K^+K^-}q^2}. \tag{27}$$

Considering the additional contributions of both ρ meson and ω meson we obtain the final form of the charged kaon form-factor

$$\begin{aligned} F_K(q^2) \quad &= \quad 1+\frac{\sqrt{2}}{3}\frac{g_{\phi K^+K^-}}{g_\phi(q^2)}\frac{q^2}{q^2-\mathring{m}_\phi^2-\Pi_\phi} \\ &\quad -\frac{g_{\rho K^+K^-}}{g_\rho(q^2)}\frac{q^2}{q^2-\mathring{m}_\rho^2-\Pi_\rho}-\frac{4}{3}\frac{g_{\omega K^+K^-}}{g_\omega(q^2)}\frac{q^2}{q^2-\mathring{m}_\omega^2-\Pi_\omega}. \end{aligned} \tag{28}$$

Fig. 6 shows the charged kaon form-factor at the time-like region compared with experimental data [17]. Here the coupling constant $\mathrm{Re}[g_\phi(q^2)]$ is approximately 6.5.

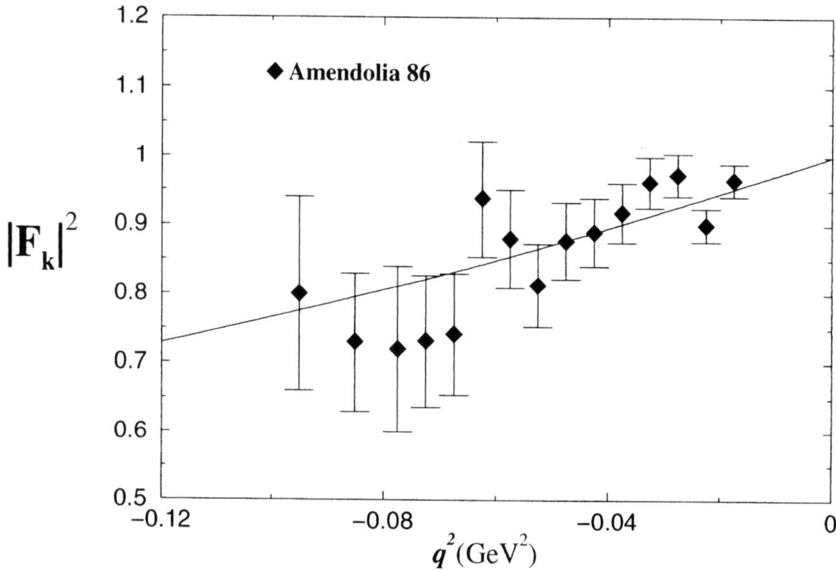

FIGURE 7. Kaon Electromagnetic Form-Factor in space like region

And physical ϕ meson mass and $\phi \to K^+K^-$, $\phi \to K_S^0 K_L^0$ decay width are given by

$$m_{\phi P} \approx 1019\text{MeV} \ , \ \dot{m}_\phi \approx 940\text{MeV}$$
$$\Gamma_{\phi \to K^+K^-} \approx 2.32\text{MeV} \ , \ \Gamma_{\phi \to K_S^0 K_L^0} \approx 1.517\text{MeV}.$$

The kaon form factor at space like region $q^2 < 0$ is shown in Fig. 7. The final inclusion of the ρ and ω meson contributions reproduce successfully experimental results[18].

CONCLUSION

We extended a chiral effective lagrangian by including the vector and the axial-vector mesons as well as pions to $SU_R(3) \otimes SU_L(3)$. The meson fields are introduced through the non-linear realization of chiral symmetry, which provides an easy way of imposing consistency with the ChPT. In order to have mass splitting of strange and non-strange particles, the interactions between scalar mesons and each meson i.e, vector, axial-vector, and pseudoscalar mesons are taken into account.

For phenomenological side of this lagrangian, pion and kaon electromagnetic form-factors and some related decays are calculated. In the process of calculating, our effective Lagrangian is shown to give a good agreement with experimental data without considering the effects from the higher orders in other effective theories.

Recently, a role of the axial vector mesons, such as, K_1 and a_1, is emphasized regarding the emission spectrum of the photon in the heavy ion collision. But the recipes for both axial vector mesons are not consistent with each other. Consistent description about the role in the heavy ion collision is under progress by using the Lagrangian suggested in this report.

REFERENCES

1. M.Bando, T.Kugo and K.Yamawaki, Phys.Rep. **164** (1988) 217
2. U.G.Meissner, Phys.Rep. **161** (1988) 213
3. M. C. Birse, Z. Phys. **A355** (1996) 231
4. T. S. Park, B. S. Han, M. K. Cheoun, and S. H. Lee, Jour. Kor. Phys. Soc., **29** (1996) S381
5. Y. C. Shin, M. K. Cheoun, B. S. Han, K. S. Kim, and Il-T. Cheon, Eur. Phys. J. **A 9**, (2000) 269.
6. J. Gasser, H. Leutwyler, Nucl. Phys. **B250** (1985) 517
7. G. Ecker, J. Gasser, A. Pich and E. de Rafael, Nucl. Phys. **B321** (1989) 311
8. M. Bando, T. Kugo and K. Yamawaki, Nucl. Phys. **B259** (1985) 493
9. H. B. O'Connell et. al, Prog. Nucl. Part. Phys 39(1997) 201
10. R. K. Bhaduri, *Models of Nucleon*, (Addison-Wesley Publishing Company, Inc., Redwood city, 1988)
11. M. Herrmann, B. L. Friman and W. Norenberg, Nucl. Phys. **A560** (1993) 411
12. Fayyazuddin and Riazuddin, Phys. Rev. **D36** (1987) 2768
13. F. Klingl, N. Kaiser and W. Weise, Z. Phys **A356** (1996) 193
14. G.E. Brown, M. Rho and W. Weise, Nucl. Phys **A454** (1986) 669
15. S.R Amendolia et al., Phys. Lett. **B146** (1984) 116
16. L. M. Barkov et.al, Nucl.Phys. **B259** (1985) 365
17. P. M. Ivanov, Phys. Lett. **B107** (1981)
18. S.R Amendolia et al., Phys. Lett. **B178** (1986) 435
19. L.S.Celenza, X.D.Li and C.M.Shakin, Phys.Rev. **C56** (1997) 3326
20. W.Molzon et al., Phys. Rev. Lett. **41** (1978) 1213.

Spin of a relativistic composite system and nucleon magnetic moment

Miho Takayama* and Hiroshi Toki*

*Research Center for Nuclear Physics (RCNP) Osaka University, Japan

Abstract. Magnetic moment of a relativistic three-body system of Dirac particles is studied. Special attention is paid to the Lorentz boost of the center of mass coordinate, which allows us to take the correct non-relativistic limit of the matrix element of the electromagnetic interaction.

INTRODUCTION

The non-relativistic constituent quark model (NRCQM) has been employed widely for the study of baryon properties, such as spectrum, form factors, and transition amplitudes. A typical success of NRCQM is the magnetic moment of baryons. The magnetic moment in NRCQM is just a sum of the magnetic moments of each constituent quark. Even in the simplest approach, where wave function has spin-flavor SU(6) symmetry and loosely binding limit (the nucleon mass M_N is three times of the constituent quark mass m_q), one could obtain magnetic moment of proton $\mu_p = 3\mu_N$ ($\mu_N = \frac{e}{2M_N}$) and of neutron $\mu_n = -2\mu_N$ which are consistent with the empirical data within 15 % error (see Table 1).

Since magnetic moment is closely related to spin, whose origin is spinor structure of fermion in relativistic formulation, it is natural to think that relativistic constituent quark model (RCQM) can also describe the nucleon magnetic moment automatically. However, it is reported that RCQM cannot reproduce the nucleon magnetic moment, the value is too small, only 55% [1, 2]. These results seem to suggest that the success of NRCQM is just an accident and baryon is not provided as a simple three-quark composite system but more complicated object. However, on the other hand, since NRCQM can be regarded as one particular form of RCQM with non-relativistic limit, we expect that static properties in RCQM coincide with that of NRCQM.

The purpose of the present study is to solve the discrepancy between RCQM and NRCQM. To do this, we investigate the relativistic three-body system including boost effect for low momentum region [3]. First we write the wave function in center of mass (CM) frame, then obtain the wave function boosted to a desired momentum frame. In this procedure, we will see that not only the relative coordinate, but also the Lorentz boost of CM coordinate should be taken into account. This fact is not considered seriously in the usual treatment of RCQM. After boosting both relative and CM degrees of freedom, we calculate the electromagnetic matrix element and extract the magnetic moment of the system. We will find that the Lorentz boost of CM degrees of freedom is important to take non-relativistic limit of the matrix element correctly, and to get the magnetic moment of the system. The last section is devoted to the summary and discussions.

CP594, *Hadrons and Nuclei: First International Symposium,* edited by Il-T. Cheon et al.

TABLE 1. Proton and neutron magnetic moment in both non-relativistic and relativistic models are summarized. The unit of magnetic moment is $\mu_N = \frac{e}{2M_N}$. The former two column with star, loosely binding limit is applied.

	NRCQM*	RCQM* [1]	RCQM [2]	exp.[4]
μ_p	3	2	1.568	2.79
μ_n	-2	$-\frac{4}{3}$	-1.045	-1.91

RELATIVISTIC THREE-BODY SYSTEM

Let us consider the simple three body system where each particle has the same mass m and the charge e. First, we write down the wave function of this system in center of mass (CM) frame ($\mathbf{P}_T = 0$), which is given by

$$\varphi_{\mathcal{M}} = \left[\begin{pmatrix} 1 \\ \frac{\sigma_1 \cdot \mathbf{p}_1}{2m+k} \end{pmatrix} \otimes \begin{pmatrix} 1 \\ \frac{\sigma_2 \cdot \mathbf{p}_2}{2m+k} \end{pmatrix} \otimes \begin{pmatrix} 1 \\ \frac{\sigma_3 \cdot \mathbf{p}_3}{2m+k} \end{pmatrix} \right] \phi(\mathbf{p}_\rho, \mathbf{p}_\lambda)\chi_{SM} , \tag{1}$$

where $k \equiv \frac{1}{3}(U + \mathcal{M} - 3m)$, U is the potential energy and $\mathcal{M}$ is the total energy in CM frame. Physically, k corresponds to the kinetic energy of each particle. The subscript of Pauli matrix $i = 1,2,3$ are used to represent that σ_i acts only on the spinor of particle i. χ_{SM} is a constant spinor with total spin S and its z component M, which is composed by the combination of the three constant spinors $\chi_{\frac{1}{2}m_i}$ $(i = 1,2,3)$ with appropriate Clebsch-Gordan coefficients as

$$\chi_{SM} = \left[[\chi_{\frac{1}{2}m_1}\, \chi_{\frac{1}{2}m_2}]_{sm}\, \chi_{\frac{1}{2}m_3}\right]_{SM} . \tag{2}$$

The momentum of each particle is denoted as $\mathbf{p}_i$. In order to treat such a three-body system, we introduce Jacobi coordinate $(\mathbf{R}, \boldsymbol{\rho}, \boldsymbol{\lambda})$,

$$\begin{cases} \mathbf{R} = \frac{1}{\sqrt{3}}(\mathbf{r}_1 + \mathbf{r}_2 + \mathbf{r}_3) \\ \boldsymbol{\rho} = \frac{1}{\sqrt{2}}(\mathbf{r}_1 - \mathbf{r}_2) \\ \boldsymbol{\lambda} = \frac{1}{\sqrt{6}}(\mathbf{r}_1 + \mathbf{r}_2 - 2\mathbf{r}_3) \end{cases} , \tag{3}$$

and corresponding CM momentum $\mathbf{P}$ and relative momentum $(\mathbf{p}_\rho, \mathbf{p}_\lambda)$ as

$$\begin{cases} \mathbf{P} = \frac{1}{\sqrt{3}}\mathbf{P}_T = \frac{1}{\sqrt{3}}(\mathbf{p}_1 + \mathbf{p}_2 + \mathbf{p}_3) \\ \mathbf{p}_\rho = \frac{1}{\sqrt{2}}(\mathbf{p}_1 - \mathbf{p}_2) \\ \mathbf{p}_\lambda = \frac{1}{\sqrt{6}}(\mathbf{p}_1 + \mathbf{p}_2 - 2\mathbf{p}_3) \end{cases} , \tag{4}$$

where $\mathbf{r}_i$ $(i = 1,2,3)$ is the position of particle i. Though $\mathbf{P}$ is not physical CM momentum, this notation will turn out to be useful for the following calculation. To obtain the wave function in an arbitrary reference frame $(E, \mathbf{P}_T)$ from the CM frame $(\mathcal{M}, \mathbf{P}_T = \mathbf{0})$, we apply Lorentz transformation to each subspinor in (1)

$$\varphi_{E\mathbf{P}_T} = \hat{S}_1 \hat{S}_2 \hat{S}_3 \varphi_{\mathcal{M}} , \tag{5}$$

where

$$\hat{S}_i = \sqrt{\frac{E+\mathcal{M}}{2\mathcal{M}}} \left(1 + \frac{\boldsymbol{\alpha}_i \cdot \mathbf{P}_T}{\mathcal{M}+E}\right) , \qquad \boldsymbol{\alpha}_i = \begin{pmatrix} 0 & \boldsymbol{\sigma}_i \\ \boldsymbol{\sigma}_i & 0 \end{pmatrix} \tag{6}$$

is the usual Lorentz transformation operator for the spinor. Final expression of (5) in coordinate space becomes

$$\begin{aligned}
\varphi_{E\mathbf{P}_T}(\mathbf{r}_1, \mathbf{r}_2, \mathbf{r}_3, R^0) &= \left(\frac{E+\mathcal{M}}{2\mathcal{M}}\right)^{3/2} \int \frac{d^3p_\rho}{(2\pi)^{3/2}} \frac{d^3p_\lambda}{(2\pi)^{3/2}} \mathcal{N}_{p_1,p_2,p_3} \\
\times &\left[\begin{pmatrix} 1 + \frac{\boldsymbol{\sigma}_1 \cdot \sqrt{3}\mathbf{P}}{\mathcal{M}+E} \frac{\boldsymbol{\sigma}_1 \cdot \left(\frac{1}{\sqrt{2}}\mathbf{p}_\rho + \frac{1}{\sqrt{6}}\mathbf{p}_\lambda\right)}{2m+k} \\ \boldsymbol{\sigma}_1 \cdot \left(\frac{\sqrt{3}\mathbf{P}}{\mathcal{M}+E} + \frac{\frac{1}{\sqrt{2}}\mathbf{p}_\rho + \frac{1}{\sqrt{6}}\mathbf{p}_\lambda}{2m+k}\right) \end{pmatrix} \otimes \begin{pmatrix} 1 + \frac{\boldsymbol{\sigma}_2 \cdot \sqrt{3}\mathbf{P}}{\mathcal{M}+E} \frac{\boldsymbol{\sigma}_2 \cdot \left(-\frac{1}{\sqrt{2}}\mathbf{p}_\rho + \frac{1}{\sqrt{6}}\mathbf{p}_\lambda\right)}{2m+k} \\ \boldsymbol{\sigma}_2 \cdot \left(\frac{\sqrt{3}\mathbf{P}}{\mathcal{M}+E} + \frac{-\frac{1}{\sqrt{2}}\mathbf{p}_\rho + \frac{1}{\sqrt{6}}\mathbf{p}_\lambda}{2m+k}\right) \end{pmatrix} \right. \\
\otimes &\left. \begin{pmatrix} 1 + \frac{\boldsymbol{\sigma}_3 \cdot \sqrt{3}\mathbf{P}}{\mathcal{M}+E} \frac{\boldsymbol{\sigma}_3 \cdot \left(-\frac{2}{\sqrt{6}}\mathbf{p}_\lambda\right)}{2m+k} \\ \boldsymbol{\sigma}_3 \cdot \left(\frac{\sqrt{3}\mathbf{P}}{\mathcal{M}+E} + \frac{-\frac{2}{\sqrt{6}}\mathbf{p}_\lambda}{2m+k}\right) \end{pmatrix} \right] \phi(\mathbf{p}_\rho, \mathbf{p}_\lambda) \chi_{SM} \\
&\times \exp(i\tilde{\boldsymbol{\rho}} \cdot \mathbf{p}_\rho) \exp(i\tilde{\boldsymbol{\lambda}} \cdot \mathbf{p}_\lambda) \exp(i\mathbf{R} \cdot \mathbf{P}) \exp(-iR^0 E) ,
\end{aligned} \tag{7}$$

where $(R^0, \mathbf{R})$ is the CM coordinate and $\mathcal{N}$ is a normalization constant. In (7), $\tilde{\boldsymbol{\rho}}, \tilde{\boldsymbol{\lambda}}$ include the Lorentz-Fitzgerald contraction of the wave function. Practically, we investigate only the small $\mathbf{P}_T$ case, where the contraction effect is negligible.

As a physical state, we can consider a wave packet, constructed by superposition of an arbitrary small range of momentum eigenstates $\Phi(\mathbf{P})$ and introduce following notation,

$$|\boldsymbol{\rho}, \boldsymbol{\lambda}, \mathbf{R}\rangle \equiv \int \frac{d^3P}{(2\pi)^{3/2}} \sqrt{\frac{\mathcal{M}}{E}} \Phi(\mathbf{P}) \varphi_{E\mathbf{P}_T}(\mathbf{r}_1, \mathbf{r}_2, \mathbf{r}_3, R^0) \tag{8}$$

$$\begin{aligned}
\equiv &\left[\begin{pmatrix} 1 + \frac{\boldsymbol{\sigma}_1 \cdot \sqrt{3}\mathbf{P}}{\mathcal{M}+E} \frac{\boldsymbol{\sigma}_1 \cdot \left(\frac{1}{\sqrt{2}}\mathbf{p}_\rho + \frac{1}{\sqrt{6}}\mathbf{p}_\lambda\right)}{2m+k} \\ \boldsymbol{\sigma}_1 \cdot \left(\frac{\sqrt{3}\mathbf{P}}{\mathcal{M}+E} + \frac{\frac{1}{\sqrt{2}}\mathbf{p}_\rho + \frac{1}{\sqrt{6}}\mathbf{p}_\lambda}{2m+k}\right) \end{pmatrix} \otimes \begin{pmatrix} 1 + \frac{\boldsymbol{\sigma}_2 \cdot \sqrt{3}\mathbf{P}}{\mathcal{M}+E} \frac{\boldsymbol{\sigma}_2 \cdot \left(-\frac{1}{\sqrt{2}}\mathbf{p}_\rho + \frac{1}{\sqrt{6}}\mathbf{p}_\lambda\right)}{2m+k} \\ \boldsymbol{\sigma}_2 \cdot \left(\frac{\sqrt{3}\mathbf{P}}{\mathcal{M}+E} + \frac{-\frac{1}{\sqrt{2}}\mathbf{p}_\rho + \frac{1}{\sqrt{6}}\mathbf{p}_\lambda}{2m+k}\right) \end{pmatrix} \right. \\
\otimes &\left. \begin{pmatrix} 1 + \frac{\boldsymbol{\sigma}_3 \cdot \sqrt{3}\mathbf{P}}{\mathcal{M}+E} \frac{\boldsymbol{\sigma}_3 \cdot \left(-\frac{2}{\sqrt{6}}\mathbf{p}_\lambda\right)}{2m+k} \\ \boldsymbol{\sigma}_3 \cdot \left(\frac{\sqrt{3}\mathbf{P}}{\mathcal{M}+E} + \frac{-\frac{2}{\sqrt{6}}\mathbf{p}_\lambda}{2m+k}\right) \end{pmatrix} \right] |\boldsymbol{\rho}, \boldsymbol{\lambda}, \mathbf{R})
\end{aligned} \tag{9}$$

where bracket denotes the wave function including boosted spinor structure and parenthesis the wave function which has only constant spinor.

MAGNETIC MOMENT OF THREE-BODY SYSTEM

The magnetic moment of the three-body system is extracted from the expectation value of the electromagnetic interaction Hamiltonian,

$$H_{\rm int} = e(\boldsymbol{\alpha}_1 \cdot \boldsymbol{A}(\mathbf{r}_1) + \boldsymbol{\alpha}_2 \cdot \boldsymbol{A}(\mathbf{r}_2) + \boldsymbol{\alpha}_3 \cdot \boldsymbol{A}(\mathbf{r}_3)) \,, \tag{10}$$

where the photon field is defined as $\boldsymbol{A}(\mathbf{r}) = \frac{\boldsymbol{\varepsilon}}{\sqrt{2\omega}} \exp[i\mathbf{k} \cdot \mathbf{r}]$ with a polarization vector $\boldsymbol{\varepsilon}$, the momentum transfer $\mathbf{k}$ and the energy transfer ω between the initial and the final states.

First term of the matrix element $\langle f\,|H_{int}|\,i\rangle$ becomes

$$\begin{aligned}
&\langle f\,|\boldsymbol{\alpha}_1 \cdot \boldsymbol{A}(\mathbf{r}_1)|\,i\rangle \\
&= (f\,|\frac{1}{\sqrt{2\omega}}\left[\left(1 + \frac{\boldsymbol{\sigma}_1 \cdot \sqrt{3}\mathbf{P}'}{\mathcal{M}+E}\frac{\boldsymbol{\sigma}_1 \cdot \left(\frac{1}{\sqrt{2}}\mathbf{p}'_\rho + \frac{1}{\sqrt{6}}\mathbf{p}'_\lambda\right)}{2m+k} \quad \boldsymbol{\sigma}_1 \cdot \left(\frac{\sqrt{3}\mathbf{P}'}{\mathcal{M}+E} + \frac{\frac{1}{\sqrt{2}}\mathbf{p}'_\rho + \frac{1}{\sqrt{6}}\mathbf{p}'_\lambda}{2m+k}\right)\right)\right. \\
&\quad \times \begin{pmatrix} 0 & \boldsymbol{\sigma}_1 \cdot \boldsymbol{\varepsilon} \\ \boldsymbol{\sigma}_1 \cdot \boldsymbol{\varepsilon} & 0 \end{pmatrix} \left.\begin{pmatrix} 1 + \frac{\boldsymbol{\sigma}_1 \cdot \sqrt{3}\mathbf{P}}{\mathcal{M}+E}\frac{\boldsymbol{\sigma}_1 \cdot \left(\frac{1}{\sqrt{2}}\mathbf{p}_\rho + \frac{1}{\sqrt{6}}\mathbf{p}_\lambda\right)}{2m+k} \\ \boldsymbol{\sigma}_1 \cdot \left(\frac{\sqrt{3}\mathbf{P}}{\mathcal{M}+E} + \frac{\frac{1}{\sqrt{2}}\mathbf{p}_\rho + \frac{1}{\sqrt{6}}\mathbf{p}_\lambda}{2m+k}\right) \end{pmatrix}\right] \exp[i\mathbf{k} \cdot \mathbf{r}_1]|\,i) \\
&\equiv (f\,|(\text{electric part}) + \ (\text{magnetic part}) + O(\mathbf{p}^2, \mathbf{P}^2)|\,i)
\end{aligned} \tag{11}$$

where

$$\begin{aligned}
&(\text{electric part}) \\
&= \frac{1}{\sqrt{2\omega}}\left(\frac{\sqrt{3}(\mathbf{P}' + \mathbf{P})}{\mathcal{M}+E} + \frac{\frac{1}{\sqrt{2}}(\mathbf{p}'_\rho + \mathbf{p}_\rho) + \frac{1}{\sqrt{6}}(\mathbf{p}'_\lambda + \mathbf{p}_\lambda)}{2m+k}\right) \cdot \boldsymbol{\varepsilon} \exp[i\mathbf{k} \cdot \mathbf{r}_1],
\end{aligned} \tag{12}$$

$$\begin{aligned}
&(\text{magnetic part}) \\
&= \frac{i}{\sqrt{2\omega}}\boldsymbol{\sigma}_1 \cdot \left(\frac{\sqrt{3}(\mathbf{P}' - \mathbf{P})}{\mathcal{M}+E} + \frac{\frac{1}{\sqrt{2}}(\mathbf{p}'_\rho - \mathbf{p}_\rho) + \frac{1}{\sqrt{6}}(\mathbf{p}'_\lambda - \mathbf{p}_\lambda)}{2m+k}\right) \times \boldsymbol{\varepsilon} \exp[i\mathbf{k} \cdot \mathbf{r}_1].
\end{aligned} \tag{13}$$

To evaluate the magnetic moment of the system, it is enough to calculate only the magnetic part in (11). Integrating by parts and taking loosely binding limit

$$\frac{1}{2m+k} \sim \frac{1}{2m}\,, \qquad\qquad \frac{1}{\mathcal{M}+E} \sim \frac{1}{6m}\,, \tag{14}$$

we get

$$(\text{magnetic part}) = (f\,|\left[\left(\frac{1}{2m}\right)\boldsymbol{\sigma}_1 \cdot \boldsymbol{B}(\mathbf{r}_1)\right]|\,i)\,. \tag{15}$$

We can deal with the second and the third terms of the interaction Hamiltonian (10) in the same way. Finally, $\langle f\,|H_{int}|\,i\rangle$ can be written as

$$\langle f\,|H_{int}|\,i\rangle = (f\,|\left\{(\text{electric part}) + \frac{e}{2m}(\boldsymbol{\sigma}_1 \cdot \boldsymbol{B}(\mathbf{r}_1) + \boldsymbol{\sigma}_2 \cdot \boldsymbol{B}(\mathbf{r}_2) + \boldsymbol{\sigma}_3 \cdot \boldsymbol{B}(\mathbf{r}_3))\right\}|\,i)\,, \tag{16}$$

where $\boldsymbol{B}(\mathbf{r}_i) = \nabla_i \times \boldsymbol{A}(\mathbf{r}_i)$ is the magnetic field. The coefficient of $\boldsymbol{B}$, $\frac{e}{2m}\boldsymbol{\sigma}$, is nothing but the magnetic moment $\boldsymbol{\mu}$ of the Dirac particle with the mass m and the charge e. To summarize, with boosted wave function, magnetic moment is just the sum of the magnetic moment of each particle as in NRCQM,

$$\boldsymbol{\mu} = \boldsymbol{\mu}_1 + \boldsymbol{\mu}_2 + \boldsymbol{\mu}_3. \tag{17}$$

It should be noted that when we do not apply the Lorentz boost on the spinors as ordinary RCQM, the CM momentum $\mathbf{P}$ does not appear in the calculation, which leads to small magnetic moment as

$$\boldsymbol{\mu} = \frac{2}{3}(\boldsymbol{\mu}_1 + \boldsymbol{\mu}_2 + \boldsymbol{\mu}_3). \tag{18}$$

This means that lack of the Lorentz boost reduces the magnetic moment of nucleon factor $2/3$, which is considered as the origin of under estimation in Refs [1, 2].

SUMMARY AND DISCUSSIONS

We have studied the magnetic moment of relativistic composite system and found that the lack of Lorentz boost of CM motion reduces magnetic moment $2/3$.

What we did in this paper is the $\frac{1}{m}$ expansion of the matrix element of three-body system with electromagnetic interaction. On the other hand, as in the text book, the operator of the magnetic moment in NRCQM is obtained from the $\frac{1}{m}$ expansion of the current operator, for instance, by using Foldy-Wouthuysen (FW) transformation. Then the consistency between our result and usual NRCQM treatment is natural, rather trivial.[1] In other words, the effect of the Lorentz boost is necessary to get correct non-relativistic limit of the matrix elements.

During the calculation, we have also found that the Peierls-Yoccoz (PY) momentum projection method, to remove the CM motion, can mimic the Lorentz transformation effect on the spinor structure within the first order of $\mathbf{P}$. Such the PY projection method was already used in Refs. [1, 2], however, owing to the lack of consideration about the change of the spinor structure during the projection, they have missed the contribution from the CM motion on the magnetic moment.

Since the CM motion effect appear in the small component of the spinor, we expect that the same effect exists in other quantities which are directly related to the small component of the spinor such as $g_{\pi NN}$.

ACKNOWLEDGMENTS

The authors acknowledge fruitful discussions with Y. Nemoto, Y. Koma and A. Hosaka.

---

[1] At higher order of $\frac{1}{m}$, there are some differences between FW transformation and direct expansion of the matrix element [3].

REFERENCES

1. Tegen, R., Brockmann, R., and Weise, W., *Z. Phys*, **A307**, 339–350 (1982).
2. Dong, Y., Faessler, A., and Shimizu, K., *Eur. Phys. J.*, **A6**, 203 (1999).
3. Brodsky, S., and Primack, J., *Ann. Phys.*, **52**, 315–365 (1969).
4. Groom, D.E. *et al.* (Particle Data Group), *Eur. Phys. J.*, **C15**, 1 (2000).

STRANGENESS NUCLEAR PHYSICS

Strange vector form factors of the nucleon

Antonio Silva*, Hyun-Chul Kim† and Klaus Goeke*

*Institut für Theoretische Physik II, Ruhr-Universität Bochum, D–44780 Bochum, Germany
†Department of Physics, Pusan National University,609-735 Pusan, Republic of Korea

Abstract. In this talk, we revisit the strange vector form factors of the nucleon within the framework of the SU(3) chiral quark-soliton model. We employ in this new work the prescription of the SU(3) quantization in order to remedy the charge problem arising from the nonlocality in time. We compute the neutral weak and strange vector form factors. We also incorporate two different asymptotics of the soliton field: pion and kaon tails. The results turn out to be compatible with the recent experimental data from the SAMPLE and HAPPEX collaborations. We also present the prediction for the A4 experiment.

INTRODUCTION

Strangeness in the nucleon has been one of the most important issues for decades, since it paves the way to understanding the internal quark structure of the nucleon. In particular, the strangeness in the vector channel is of particular interest [1], as the strange vector form factors were measured very recently [2, 3, 4, 5] by the SAMPLE and HAPPEX collaborations. The most recent measurement by the SAMPLE collaboration [4] is as follows:

$$G_M^s(Q^2 = 0.1(\mathrm{GeV}/c)^2) = +0.14 \pm 0.29 \pm 0.31 \quad (\mathrm{SAMPLE}). \tag{1}$$

The HAPPEX collaboration also announced the measurement of the strange vector form factors [5]:

$$\frac{(G_E^s + 0.392 G_M^s)}{(G_M^p/\mu_p)}(Q^2 = 0.477(\mathrm{GeV}/c)^2) = 0.091 \pm 0.054 \pm 0.039 \quad (\mathrm{HAPPEX}). \tag{2}$$

Theoretically, there is already a great deal of work [6, 7, 8]. The strange vector form factors were also studied within the framework of the chiral quark-soliton model(χQSM) [9]. However, it is recently found that the collective SU(3) quantization in the χQSM disobeys the Gell-Mann–Nishijima relation, *i.e.* it breaks the baryon charge [10, 11]. It is due to the inherent time-nonlocality of the collective operators in the χQSM and the asymmetric transitions between the solitonic states and vacuum ones. On account of this discrepancy, redundant terms exist in the theoretical expressions for various observables. While a firm theoretical reason of the charge dilemma is not known to date, Ref. [11] suggests that one can pick out those terms and discard them, using the fact that the χQSM reproduces the results of the nonrelativistic quark model in the zero-size limit of the soliton ($R \to 0$). In this talk, we will present the recent investigation

CP594, *Hadrons and Nuclei: First International Symposium,* edited by Il-T. Cheon et al.

of the strange vector form factors of the nucleon and related strange observables in the χQSM, incorporating this new scheme of the SU(3) quantization.

FORMALISM

The low-energy partition function in Euclidean space characterizes the χQSM [12], which is expressed by the functional integral over pseudoscalar meson (π^a) and quark fields(ψ):

$$\begin{aligned} \mathcal{Z} &= \int \mathcal{D}\psi\mathcal{D}\psi^\dagger\mathcal{D}\pi^a \exp\left[\int d^4x\psi_f^{\dagger\alpha}\left(i\partial\!\!\!/ + iMe^{i\gamma_5\lambda^a\pi^a}\right)_{fg}\psi_g^\alpha\right], \\ &= \int \mathcal{D}\pi^a \exp(-S_{\mathrm{eff}}[\pi]), \end{aligned} \tag{3}$$

where S_{eff} is the effective action

$$S_{\mathrm{eff}}[\pi] = -\mathrm{Tr}\ln D. \tag{4}$$

Tr denotes the functional trace. D represents the Dirac differential operator

$$D = i\partial\!\!\!/ + i\hat{m} + iMU^{\gamma_5} \tag{5}$$

with the pseudoscalar chiral field

$$U^{\gamma_5} = \exp(i\pi^a\lambda^a\gamma_5) = \frac{1+\gamma_5}{2}U + \frac{1-\gamma_5}{2}U^\dagger. \tag{6}$$

$\hat{m}$ is the matrix of the current quark mass given by $\hat{m} = \mathrm{diag}(m_{\mathrm{u}}, m_{\mathrm{d}}, m_{\mathrm{s}})$. λ^a designate the usual Gell-Mann matrices normalized as $\mathrm{tr}(\lambda^a\lambda^b) = 2\delta^{ab}$. Here, we have assumed isospin symmetry ($m_{\mathrm{u}} = m_{\mathrm{d}}$). M stands for the dynamical quark mass arising from the spontaneous chiral symmetry breaking, which is in general momentum-dependent. We regard M as a constant and introduce the proper-time regularization for convenience. The Dirac operator can be expressed in terms of the Euclidean time derivative and the Dirac one-particle Hamiltonian $H(U^{\gamma_5})$:

$$D = \partial_\tau + H(U^{\gamma_5}) + \gamma_4\hat{m} - \gamma_4\bar{m} \tag{7}$$

with

$$H(U^{\gamma_5}) = \frac{\boldsymbol{\alpha}\cdot\nabla}{i} + \gamma_4 MU^{\gamma_5} + \gamma_4\bar{m}\mathbf{1}. \tag{8}$$

The parameter $\bar{m}$ is introduced in such a way that it produces a correct Yukawa-type asymptotic behavior of the profile function.

The U is assumed to have a structure corresponding to the so-called trivial embedding of the SU(2)-hedgehog into SU(3) [13]:

$$U = \begin{pmatrix} U_0 & 0 \\ 0 & 1 \end{pmatrix}, \tag{9}$$

with

$$U_0 = \exp[i\mathbf{n}\cdot\boldsymbol{\tau}P(r)]. \tag{10}$$

The partition function $\mathcal{Z}$ of Eq.(3) is simplified by the stationary phase approximation, which is justified in the large N_c limit of $\mathcal{Z}$. One ends up with one stationary profile function $P(r)$ which is determined numerically by solving the Euler-Lagrange equation corresponding to $\frac{\delta S_{\text{eff}}}{\delta P(r)} = 0$. This yields a selfconsistent classical field U_0 and a set of single quark energies and corresponding states E_n and Ψ_n. Note that the E_n and Ψ_n do not constitute the nucleon $|N\rangle$ yet because the collective spin and and isospin quantum numbers are missing. Those are obtained by the semiclassical quantization procedure.

Here, we need to explain the role of the parameter $\bar{m}$ somewhat in detail. Because of the trivial embedding shown above, the profile function $P(r)$ has a specific tail determined by $\bar{m}$. It means that one profile function represents all Goldstone boson fields. Because of that, one is not able to distinguish the pion from the kaon as far as the tails are concerned. While ordinary observables of the proton such as the electromagnetic form factors prefer the pion tail [14], the kaon plays a pivotal role in the strange vector form factors. Hence, we want to incorporate both pion and kaon tails to see their effects. From the theoretical point of view, it may of course seem to be inconsistent to incorporate the kaon tails instead of the pion ones, since we treat the strange quark mass perturbatively. However, here, our aim is not to be strict within the model but rather to follow general wisdoms found by other authors. Thus, we will utilize both of kaon and pion tails to investigate the strange vector form factors and look upon the difference between results with pion and kaon tails as the uncertainy of the model.

The strangeness content of the nucleon in the vector channel is probed by the Z^0 boson through parity-violating elastic electron scattering from the nucleon. The neutral weak electromagnetic form factors to lowest order represent the response of the nucleon to the Z^0:

$$G^Z_{E,M}(q^2) = \frac{1}{4}\left(G^p_{E,M}(q^2) - G^n_{E,M}(q^2)\right) - \sin^2\theta_W G^p_{E,M}(q^2) - \frac{1}{4}G^s_{E,M}(q^2), \tag{11}$$

where $G^p_{E,M}(G^n_{E,M})$ denote the electromagnetic form factors of the proton (neutron). $G^s_{E,M}$ stand for the strange vector form factors. $\sin^2\theta_W$ is the weak mixing angle determined by experiment [15]: $\sin^2\theta_W = 0.23147$. q^2 is the square of the four momentum transfer $q^2 = -Q^2$ with $Q^2 > 0$. The strange vector form factors can be expressed in the quark matrix elements as follows:

$$\langle N(p')|J^s_\mu|N(p)\rangle = \bar{u}_N(p')\left[\gamma_\mu F^s_1(q^2) + i\sigma_{\mu\nu}\frac{q^\nu}{2M_N}F^s_2(q^2)\right]u_N(p), \tag{12}$$

where M_N and $u_N(p)$ stand for the nucleon mass and its spinor, respectively. The strange quark current J^s_μ can be expressed in terms of the baryon current and the hypercharge one:

$$J^s_\mu = \bar{q}\gamma_\mu\hat{Q}_s q = \bar{s}\gamma_\mu s, \tag{13}$$

which can be written

$$\bar{q}\gamma_\mu\hat{Q}_s q = \frac{N_c}{3}J^B_\mu - J^Y_\mu. \tag{14}$$

J_μ^B and J_μ^Y are the baryon and hypercharge currents, respectively:

$$\begin{aligned} J_\mu^B &= \frac{1}{N_c}\bar{q}\gamma_\mu q \\ J_\mu^Y &= \frac{1}{\sqrt{3}}\bar{q}\gamma_\mu\lambda_8 q. \end{aligned} \tag{15}$$

$N_c = 3$ denotes the number of colors of the quark, which is correctly introduced to make it sure that the baryon number must be equal to unity. $\hat{Q}_s = \mathrm{diag}(0,0,1)$ is called *strangeness operator*. The baryon and hypercharge currents are equal to the singlet and octet currents, respectively.

The strange Dirac form factors F_1^s and F_2^s can be written in terms of the strange Sachs form factors, $G_E^s(Q^2)$ and $G_M^s(Q^2)$:

$$\begin{aligned} G_E^s(Q^2) &= F_1^s(Q^2) - \frac{Q^2}{4M_N^2}F_2^s(Q^2) \\ G_M^s(Q^2) &= F_1^s(Q^2) + F_2^s(Q^2). \end{aligned} \tag{16}$$

In the non–relativistic limit($Q^2 \ll M_N^2$), the Sachs-type form factors $G_E^s(Q^2)$ and $G_M^s(Q^2)$ are related to the time and space components of the strange current, respectively:

$$\begin{aligned} G_E^s(Q^2) &= \langle N'(p')|J_0^s(0)|N(p)\rangle \\ G_M^s(Q^2) &= iM_N\varepsilon_{ilk}\frac{q_l}{6q^2}\mathrm{tr}\left(\langle p',\lambda'|J_i^s|p,\lambda\rangle\sigma_k\right). \end{aligned} \tag{17}$$

where σ_j stand for Pauli spin matrices. The $|\lambda\rangle$ is the corresponding spin state of the baryon.

Since the hedgehog solution given in Eq.(9) is not invariant under the rotational and translational symmetries of the effective action, we need to restore these continuos symmetries, *i.e.* to consider quantum fluctuations around the classical solution, so that the quantum numbers of the baryon are well determined. The quantum fluctuations in the direction of the zero modes play a particular role, since they are not at all small. There are three translational zero modes and seven rotational ones for the SU(3) soliton.

Explicitly, the zero modes are taken into account by adiabatically rotating and translating the classical hedgehog configuration. The rotational zero modes can be considered as follows:

$$U(\boldsymbol{x},t) = R(t)U_c(\boldsymbol{x})R^\dagger(t), \tag{18}$$

where $R(t)$ denotes the unitary time-dependent SU(3) orientation matrix of the soliton and $U_c(\boldsymbol{x})$ is the stationary meson configuration. The translational zero modes can be taken into account as follows:

$$U(\boldsymbol{x},t) = U_c(\boldsymbol{x}-\boldsymbol{Z}(t)) = T_{\boldsymbol{Z}(t)}U_c(\boldsymbol{x})T_{\boldsymbol{Z}(t)}^\dagger, \tag{19}$$

where $T_{\boldsymbol{Z}(t)}$ is the unitary translational operator which causes a translation $\boldsymbol{Z}$.

It can be understood that the functional integral over the collective operators will be identified as the time-ordering of the collective operators:

$$\int_{R(T_1)}^{R(T_2)} DRO(R(t_1))\cdots O(R(t_n))e^{-S_{\rm eff}}$$
$$= \langle R(T_2),T_2|\mathcal{T}[\hat{O}(R(t_1))\cdots\hat{O}(R(t_n))]|R(t_1),T_1\rangle. \quad (20)$$

While this time-ordering makes the χQSM describe the observables quantitatively in SU(2), it has a flaw in SU(3): It violates the Gell-Mann–Nishijima relation [10]. It is deeply related to the non-locality in time inherited in the χQSM. Therefore, we need to modify the quantization scheme in such a way that it satisfies the Gell-Mann–Nishijima relation [11].

Having carried out a lengthy calculation, we obtain the expression for the strange vector form factors for the octet baryons:

$$G_{E,N}^{(s)}(Q^2) = G_{E,N}^{(s),m_s^0}(Q^2)+G_{E,N}^{(s),m_s^1,{\rm op}}(Q^2)+G_{E,N}^{(s),m_s^1,{\rm wf}}(Q^2),$$
$$G_{M,N}^{(s)}(Q^2) = G_{M,N}^{(s),m_s^0}(Q^2)+G_{M,N}^{(s),m_s^1,{\rm op}}(Q^2)+G_{M,N}^{(s),m_s^1,{\rm wf}}(Q^2), \quad (21)$$

where $G_{E,N}^{(s),m_s^0}(Q^2)(G_{M,N}^{(s),m_s^0}(Q^2))$ stands for the SU(3) symmetric part of the strange electric (magnetic) form factors, whereas the symmetry breaking parts $G_{E,N}^{(s),m_s^1,{\rm op}}(Q^2)(G_{M,N}^{(s),m_s^1,{\rm op}}(Q^2))$ and $G_{E,N}^{(s),m_s^1,{\rm wf}}(Q^2)(G_{M,N}^{(s),m_s^1,{\rm wf}}(Q^2))$ correspond to the symmetry breaking in the operator and in the baryon wave functions, respectively. Explicit expressions for Eq.(21) will appear elsewhere [16].

RESULTS

Figure 1 draws the neutral weak electric form factors with $\mu = m_\pi \simeq 140$ MeV and $\mu = m_K \simeq 490$ MeV, respectively. At $Q^2 = 0$, the G_E^Z is $1/4 - \sin^2\theta_W \simeq 0.019$ as shown in Eq.(11). There is an appreciable difference between the G_E^Z with the pion tail and that with the kaon one. It can be easily understood. The neutral weak electric form factor in Eq.(11) contains the neutron electric and strange electric form factors, both of which are very sensitive to the tails.

In Fig. 2, we show the neutral weak magnetic form factors with $\mu = 140$ MeV (with the pion tail) and $\mu = 490$ MeV (with the kaon tail), respectively. Remarkably, the G_M^Z with the kaon tail is almost the same as that with the pion one. The magnetic form factors of the nucleon are rather insensitive to the tails and the strange magnetic form factor is much smaller that those of the proton and neutron, so that the different tails hardly affect the G_M^Z.

Fig. 3 draws the strange electric form factors with $\mu = 140$ MeV and $\mu = 490$ MeV, respectively. The result is basically the same as in Ref. [9].

In Fig. 4, we show the strange magnetic form factors with $\mu = 140$ MeV and $\mu = 490$ MeV, respectively. it is found that the effect of the kaon tail is appreciable. Compared to the result with the pion tail, the $G_M^s(Q^2)$ with the kaon tail is enhanced more than 50%.

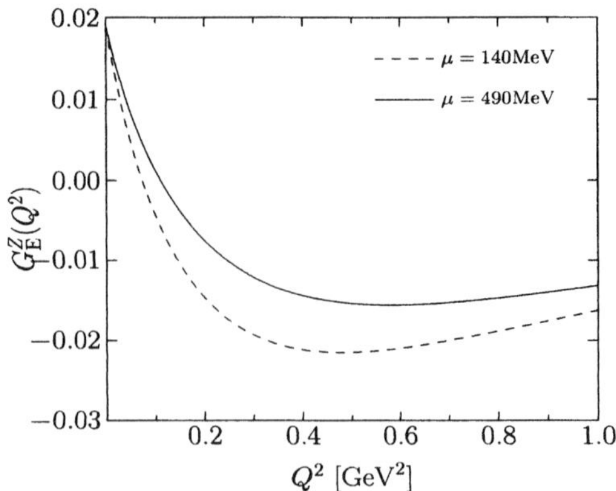

FIGURE 1. The neutral weak electric form factor G_E^Z as a function of Q^2. The solid curve represents the one with the kaon asymptotic tail ($\mu = 490$ MeV), whereas the dashed curve denotes the one with the pion tail ($\mu = 140$ MeV).

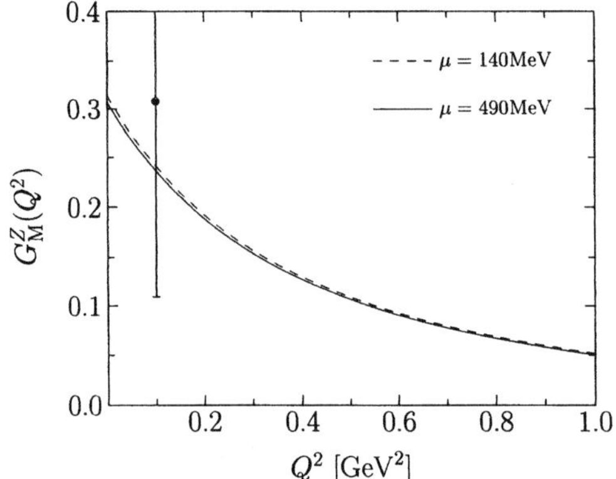

FIGURE 2. The neutral weak magnetic form factor G_M^Z as a function of Q^2. The solid curve represents the one with the kaon asymptotic tail ($\mu = 490$ MeV), whereas the dashed curve denotes the one with the pion tail ($\mu = 140$ MeV). The error bar corresponds to the recent measurement [2].

Since the SAMPLE collaboration measured the value of the $G_M^s(Q^2 = 0.1(\mathrm{GeV}/c)^2)$, it is of great interest to confront our theoretical result with the SAMPLE data. The error bar represents the data from the SAMPLE collaboration [4]. The present result enters the range of the experimental error.

The HAPPEX data presents the ratio $(G_E^s + 0.392G_M^s)/(G_M^p/\mu_p)$, not the absolute value for each form factor. In fact, the HAPPEX collaboration addresses basically two different results depending on each data for the elctromagnetic form factors. They preferably announce the experimental value: $(G_E^s + 0.392G_M^s)/(G_M^p/\mu_p) = 0.091 \pm 0.054 \pm 0.039$. However, they also give the result $(G_E^s + 0.392G_M^s)/(G_M^p/\mu_p) = 0.143 \pm 0.054 \pm 0.047$, employing the data for the G_M^n by Bruins *et al.* [17]. Because of the fact that the HAPPEX collaboration measured the ratio $(G_E^s + 0.392G_M^s)/(G_M^p/\mu_p)$,

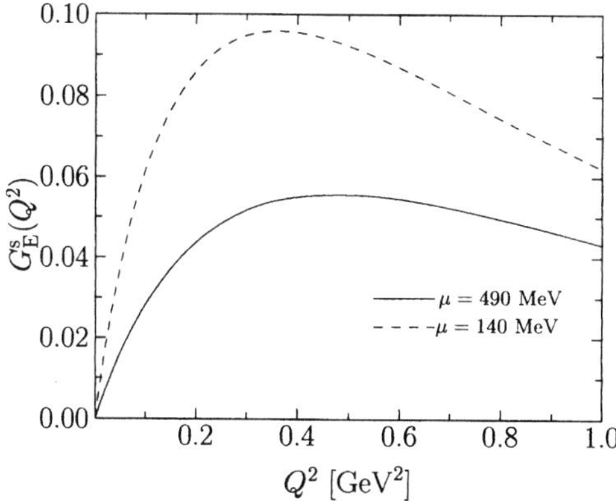

FIGURE 3. The strange electric form factor G^s_E as a function of Q^2. The solid curve represents the one with the kaon asymptotic tail ($\mu = 490$ MeV), whereas the dashed curve denotes the one with the pion tail ($\mu = 140$ MeV).

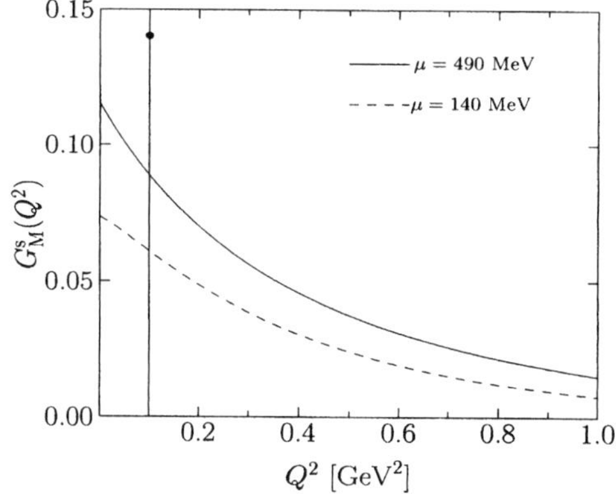

FIGURE 4. The strange magnetic form factor G^s_M as a function of Q^2. The solid curve represents the one with the kaon asymptotic tail ($\mu = 490$ MeV), whereas the dashed curve denotes the one with the pion tail ($\mu = 140$ MeV). The error bar corresponds to the recent measurement [4].

it requires the consistent treatment of the strange vector and electromagnetic form factors. In the present work, we also recalculated the electromagnetic form factors with the spurious terms causing the charge problem discarded. The value of the kinematic factor 0.392 comes from the assumption that the Q^2 dependence of the G^p_E is the same as that of G^p_M. However, these two quantities are independent each other in the present model. Thus, we derive the above quantity at the same Q^2 and θ as in the Happex measurement ($Q^2 = 0.48$ (Gev/c)^2 and $\theta = 12.5°$). Furthermore, the A4 collaboration will measure the same quantity at different Q^2 and θ. In Table 1 and 2 we list the corresponding results of the $(G^s_E + \alpha(Q^2,\theta)G^s_M)/(G^p_M/\mu_p)$ with the kinematic factor α to the SAMPLE, HAPPEX, and A4 measurements. The result at $Q^2 = 0.48$ (GeV/c)^2 and $\theta = 12.5°$ with

TABLE 1. The values of $(G_E^s + \alpha G_M^s)/(G_M^p/\mu_p)$ at various Q^2 and θ with the pion tail.

Q^2(GeV^2)	θ(°)	M =400 MeV	420 MeV	450 MeV
0.10	35	0.090	0.084	0.079
0.11	6	0.097	0.091	0.085
0.23	35	0.179	0.166	0.155
0.23	145	0.375	0.373	0.373
0.48	12.5	0.296	0.276	0.257
0.48	145	0.679	0.717	0.763

TABLE 2. The values of $(G_E^s + \alpha G_M^s)/(G_M^p/\mu_p)$ at various Q^2 and θ with the kaon tail.

Q^2(GeV^2)	θ(°)	M =400 MeV	420 MeV	450 MeV
0.10	35	0.053	0.043	0.035
0.11	6	0.056	0.046	0.038
0.23	35	0.114	0.097	0.080
0.23	145	0.387	0.386	0.385
0.48	12.5	0.211	0.182	0.154
0.48	145	0.783	0.834	0.883

the kaon tail is qualitatively in agreement with the HAPPEX measurement.

In Table 3 we list the model results for the strange electric and magnetic radii and the strange magnetic moments, varying the consistuent quark mass M with the pion and kaon tails, respectively.

SUMMARY

In this talk, our aim has been to show the recent investigation of the strange vector form factors of the nucleon within the framework of the SU(3) chiral quark-soliton model. We have taken into account the linear m_s and rotational $1/N_c$ corrections, employing the new scheme of the SU(3) quantization. We also have considered two different tails, that is, the pion and kaon tails. The difference between the results with the pion and kaon tails is regarded as the uncertainty which is inherent in the present model. The results are in a good agreement with the SAMPLE and HAPPEX results.

This work has partly been supported by the BMBF, the DFG, the COSY–Project (Jülich), and Sapiens/32845/99. The work of HCK is supported by Korean Research Foundation Grant (KRF-2000-015-DP0069). The work of AS is supported by Praxis XXI/BD/15681/98. HCK is grateful to M. Praszałowicz for useful discussions and critical comments. The authors wish to thank F. Maas (MAMI A4) for clarifying the related experimental points.

TABLE 3. The strange magnetic moments and mean-square strange radius varying with the constituent quark mass. Our final values in this table are with $M = 420$ MeV.

M [MeV]	400		420		450	
μ [MeV]	139	490	139	490	139	490
$\langle r^2 \rangle_E^s$ [fm^2]	−0.231	−0.078	−0.220	−0.095	−0.211	−0.069
μ_s [μ_N]	0.076	0.114	0.074	0.115	0.072	0.118
$\langle r^2 \rangle_M^s$ [fm^2]	0.55	0.78	0.30	0.63	0.46	0.49

REFERENCES

1. D. B. Kaplan and A. Manohar, Nucl. Phys. **B310** (1988) 527.
2. B. Mueller *et al.* [SAMPLE Collaboration], Phys. Rev. Lett. **78** (1997) 3824 [nucl-ex/9702004].
3. D. T. Spayde *et al.* [SAMPLE Collaboration], form-factor," Phys. Rev. Lett. **84** (2000) 1106.
4. R. Hasty *et al.* [SAMPLE Collaboration], Science**290** (2000) 2117 [nucl-ex/0102001].
5. K. A. Aniol *et al.* [HAPPEX Collaboration], nucl-ex/0006002.
6. R. L. Jaffe, Phys. Lett. **B229** (1989) 275; W. Koepf, E.M. Henley and S.J. Pollock, Phys. Lett. **288B** (1992) 11, T.D. Cohen, H.Forkel and M. Nielsen, Phys. Lett. **B316** (1993) 1; M.J. Musolf and M. Burkardt, Zeit. Phys. **C61** (1994) 433; H. Forkel, M. Nielsen, X. Jin and T.D. Cohen, Phys. Rev. **C50** (1994) 3108; M.J. Musolf *et al.*, Phys. Rep. **239** (1994) 1; H. Forkel, Phys. Rev.**C56** (1997) 510; U.-G. Meißner, V. Mull, J. Speth, and J.W. Van Orden, Phys. Lett. **B408** (1997) 510; T. R. Hemmert, U. Meissner and S. Steininger, Phys. Lett. B **437** (1998) 184.
7. D.B. Leinweber, Phys. Rev. **D53** (1996) 5115; . B. Leinweber and A. W. Thomas, Nucl. Phys. A **680** (2000) 117.
8. N.W. Park, J. Schechter, and H. Weigel, Phys. Rev. **D43** (1991) 869; N.W. Park and H. Weigel, Nucl. Phys. **A541** (1992) 453; H. Weigel, A. Abada, R. Alkofer, and H. Reinhardt, Phys. Lett.**B353** 20; S.T. Hong, B.Y. Park and D.P. Min, Phys. Lett. **B414** (1997) 229; L. Hannelius and D. O. Riska, Phys. Rev. C **62** (2000) 045204.
9. H.-Ch. Kim, T. Watabe and K. Goeke, Nucl. Phys. **A616** (1997) 606 [hep-ph/9606440].
10. T. Watabe and K. Goeke, Nucl. Phys. **A629** (1998) 152C [hep-ph/9706461].
11. M. Praszałowicz, T. Watabe and K. Goeke, Nucl. Phys. **A647** (1999) 49 [hep-ph/9806431].
12. D. Diakonov, V. Y. Petrov and P. V. Pobylitsa, Nucl. Phys. **B306** (1988) 809.
13. E. Witten, Nucl. Phys. B **223** (1983) 422.
14. C. V. Christov *et al.*, Prog. Part. Nucl. Phys. **37** (1996) 91 [hep-ph/9604441].
15. Particle Data Group, Euro. Phys. J. **C15** (2000) 1.
16. A. Silva, H.-Ch. Kim, and K. Goeke, in preparation.
17. E. E. Bruins *et al.*, Phys. Rev. Lett. **75** (1995) 21.

Prospects of Hypernuclear Structure

Toshio Motoba

Laboratory of Physics, Osaka Electro-Communication University, Nayagawa, Osaka 572-8530, Japan

Abstract. Recent topics in hypernuclear physics are discussed with emphasis on the experimental progress. Correspondingly, the role of structure calculations with ΛN meson theoretical potentials are also discussed by showing typical examples in close relation with spin-dependent interactions.

INTRODUCTION

Hypernuclear physics has several novel features in view of the baryon many-body problems. Participation of hyperon(s) gives rise to more important role than what is meant by the term "impurity": As the hyperon is free from the nucleon Pauli principle, the coupling to the nuclear motion is rather simplified and it can probe deeply inside the nucleus which is hardly known only through the nucleon itself. As there are several hyperons with different properties, we have many kinds of exotic systems which should provide us with opportunities for the study of many-body dynamics.

One of the goals of hypernuclear physics is to study properties of baryon-baryon interactions in a wider perspective. However the elementary hyperon-nucleon and hyperon-hyperon scattering data are very scarce and, in addition, we cannot expect such experiments in near future due to the practical difficulty. Thus the hypernuclear structural analyses are quite important to get an insight to interaction properties: There might be several bridges with both directions between hypernuclear phenomena and basic properties of the meson theoretical potentials.

After the Brookhaven (π^+, K^+) experiment [1], the systematic reaction spectroscopy of light to heavy nuclei is an important achievement of the superconducting Kaon spectrometer (KEK-E140a, E336, E369)[2, 3, 4]. In addition to the expected large peaks, they also disclosed interesting side peaks which are quite meaningful in view of the core-excitation. It is truly remarkable that higher-quality experimental data through the γ-ray measurements appeared quite recently, giving high precision level energies of some hypernuclei [5, 6]. Furthermore people are planning to perform another experiments to measure more γ-rays systematically. We also expect new data will come soon from another spectroscopy using $(e, e'K^+)$ reaction at JLab.

This situation suggests that hypernuclear study is now entering into a new stage as far as the Λ-hypernuclei are concerned. It is therefore timely that Rijken *et al.*[7] proposed new meson theoretical potentials in which the unknown parameters are allowed to change systematically so as to see the effect on hypernuclear structure outputs. Thus detailed comparison of hypernuclear structures and basic properties of realistic ΛN potentials becomes more and more meaningful in understanding strangeness many-body

CP594, *Hadrons and Nuclei: First International Symposium,* edited by Il-T. Cheon et al.

systems. In this report we will focus our attention to spin-dependent interactions and we will also discuss typical topics in Λ-hypernuclear study.

NEW SYMMETRIC STATES CONFIRMED IN $^9_\Lambda$BE

The addition of hyperon(s) to a nucleon many-body system leads to new symmetric states with respect to the radial and/or spin-isospin space. Even in the hypernuclear low-lying states, the s-state hyperon plays a role of adding spin-1/2 to the nuclear core and hence generating the "core analog states" with different spins. As one of the most interesting structures, possibility of a supersymmetric state in a shell model sense has been theoretically predicted many years ago[8, 9] in which the Λ particle in the $^9_\Lambda$Be excited states can occupy the same p-orbit as of filled nucleons and hence the $[f](\lambda\mu) = [54](50)$ symmetry should appear. In contrast to this, a nucleon added to the ^{8}Be nucleus cannot occupy the filled [44] space due to the Pauli principle, hence such symmetric state never appear in ordinary nuclei. In the microscopic $\alpha + x + \Lambda$ cluster model for light p-shell hypernuclei (here x stands for p, n, d, ^{3}H, ^{3}He, and α) [10], we have pointed out the appearance of "genuine hypernuclear" state ($x = \alpha$), which turned out to be a cluster-model name of the 'super symmetric' state. What we found as a consequence of the calculation is that the $^9_\Lambda$Be states can be classified into the following five groups:

(A) $[(\alpha\alpha)\otimes s^\Lambda]_{K=0^+}$ $E_\Lambda(L=0^+) \simeq -6.7$ MeV (^{8}Be-analog): $L = 0^+, 2^+, 4^+$.
(B) $[(\alpha\alpha)\otimes p^\Lambda_\parallel]_{K=0^-}$ $E_\Lambda(L=1^-) \simeq 0$ (genuine hypernuclear): $L = 1^-, 3^-, 5^-$.
(C) $[(\alpha\alpha)\otimes p^\Lambda_\perp]_{K=1^-}$ $E_\Lambda(L=1^-) \simeq 6$ MeV (^{9}Be-analog): $L = 1^-, 2^-, 3^-, 4^-$.
(D) $[(\alpha\alpha^*)\otimes s^\Lambda]$ $E_\Lambda \simeq 13$ MeV (core excited, T=0, 1)
(E) $[(\alpha\alpha^*)\otimes p^\Lambda]$ $E_\Lambda \simeq 20$ MeV (core excited, T=0, 1)

Here $p^\Lambda_\parallel$ and $p^\Lambda_\perp$ denote the Λ p-states parallel and perpendicular to the $\alpha-\alpha$ deformation axis, respectively. E_Λ is the energy relative to the ^{8}Be$(\alpha+\alpha)+\Lambda$ threshold. The rotational band (B) does not have the corresponding states in ordinary nuclei, while the band (C) manifests the ^{9}Be-analog except the Λ-spin coupling. The D- and E-groups, where α^* denotes the α-breaking degrees of freedom, are based on many core-excited states of ^{8}Be observed around the ^{7}Li+p and ^{7}Be+n thresholds. As another example we list the $^8_\Lambda$Li case: The system includes the α+$^4_\Lambda$H(0^+) structure as the normal ^{8}Be-analog state on one hand, but on the other hand it can also have the α+$^4_\Lambda$H(1^+) structure as the "spin=1 ^{8}Be-analog state".

The essential problem has been how to excite the genuinely hypernuclear states with such new symmetry, since other A-, C- and E-groups have been already excited by the (K^-, π^-) reaction with 'recoilless condition' [11]. Since the structure with such symmetry is much different from the target nuclear one, the excitation needs a reaction with a sizable momentum transfer. As a promising method, the (π^+, K^+) reaction done at BNL[1] seemed to show the first excitation, but the energy resolution was not enough to confirm them. Finally these states have been clearly confirmed by the recent (π^+, K^+) reaction experiment (E336) done at KEK by using the SKS spectrometer with much

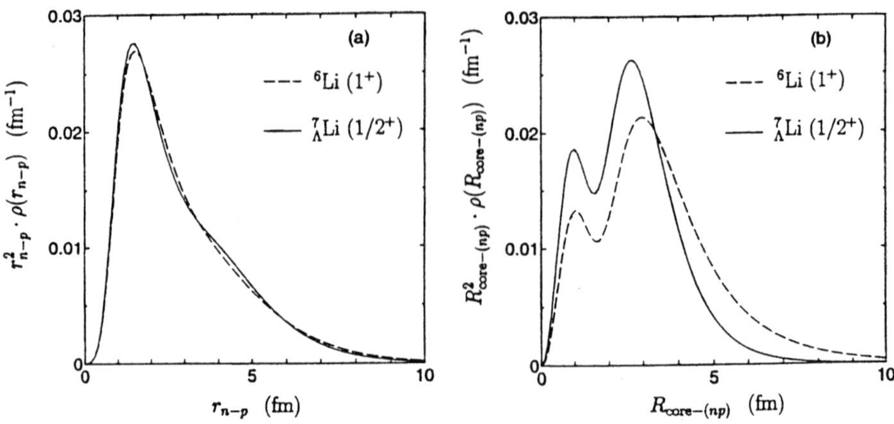

FIGURE 1. Comparison of the density distributions for the p-n relative coordinate r_{p-n} (*left*) and the relative coordinate $R_{\alpha-pn}$ between (pn) pair CM and α (*right*) [13].

better energy resolution [3]. An extensive calculation for $^9_\Lambda$Be was carried out including the α-core excited configurations [12]. For the A-, B-, and C-bands, one may also refer to Ref.[10]. The predicted (π^+, K^+) cross section turned out to be in good agreement with this measurement in which the $L = 1^-$ and 3^- members of the genuinely hypernuclear group (B) were clearly excited.

NUCLEAR RESPONSE TO THE Λ PARTICIPATION

One of the interesting predictions found in the extensive microscopic $\alpha + x + \Lambda$ cluster model ($x = N$, d, ^{3}H, ^{3}He, α) [10] was that the shrinkage of nuclear core ($\alpha + x$) due to the Λ participation should be sizable: The Λ-addition gives rise to 10–18% contraction of the α-x distance in a hypernuclus, so that hypernuclear B(E2) values are reduced to be about half of the corresponding nuclear B(E2) values. This remarkable effect is called the glue-like role of Λ. As one of the feasible experiments to see this remarkable effect, the measurement of γ-transition *rates* in $^7_\Lambda$Li which can be compared with the existing core-transition rate from the Coulomb excitation ^{6}Li(E2;$1^+ \to 3^+$). In fact this measurement has been successfully performed by H. Tamura *et al.* at KEK through the ^{7}Li$(\pi^+, K^+\gamma)^7_\Lambda$Li coincidence experiment[5].

Recently the theoretical predictions for $^7_\Lambda$Li have been updated by applying the $^5_\Lambda$He$+p+n$ model [13] in which proton and neutron are allowed to move freely rather than the "deuteron" model assumed previously [10]. All the rearrangement channels are taken into account together with the Gaussian basis, making the results more reliable in view of the numerical accuracy. The calculation shows such essentially new results that only the relative density distribution $\rho(R_{core} - (np)_{CM})$ between the core(α) and the center-of-mass of the n-p pair is changed remarkably when the Λ particle participates,

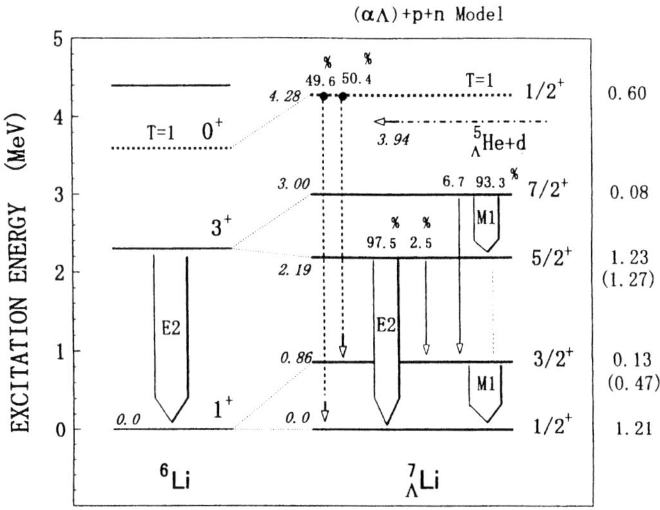

FIGURE 2. Theoretical γ-decay scheme of $^7_\Lambda$Li. On the extreme right are shown the population rates of the states calculated in combination with the calculated (π^+,K^+) cross sections.

while the $n-p$ relative density distribution $\rho(r_{n-p})$, i.e. the internal motion of the n-p pair, remains almost unchanged. See Fig. 1. Thus the angle part of the E2 operator is not affected by the contraction of core-(np) distance, so that the B(E2) value depends on the distance only.

In order to see the amount of shrinkage due to the glue-like role of Λ, we propose to check the ratio defined as

$$\Gamma_B \equiv \frac{B(E2;^7_\Lambda\mathrm{Li},5/2^+\to 1/2^+)}{(7/9)B(E2;^6\mathrm{Li},3^+\to 1^+)} = \frac{B(E2;3^+_c\to 1^+_c)_H}{B(E2;^6\mathrm{Li},3^+\to 1^+)} \qquad (1)$$

Here the factor of 7/9 in the denominator takes account of the branching relation between the hypernuclear B(E2) and the "core transition", and the subscript c in the right hand part denotes the transition occurred between core states in $^7_\Lambda$Li as deduced from the result mentioned above.

Before deducing the size contraction, we show the theoretical γ decay scheme of $^7_\Lambda$Li in Fig. 2. The level energies and branching ratios were all calculated before the experiment. The most important results concerned here are the theoretical branching ratios: the $5/2^+$ decays 97.5% to $1/2_{gs}$(E2) and 2.5% to $3/2^+$(M1), while the $7/2^+$ state decays 93.3% to $5/2^+$(M1) and 6.7% to $3/2^+$, as clearly shown in Fig. 2 Thus the lifetime of the $5/2^+$ state is determined by the E2 transition to the ground state, and that of the $3/2^+$ state by the M1 transition. In fact the experimental γ-ray yields are related to these transitions.

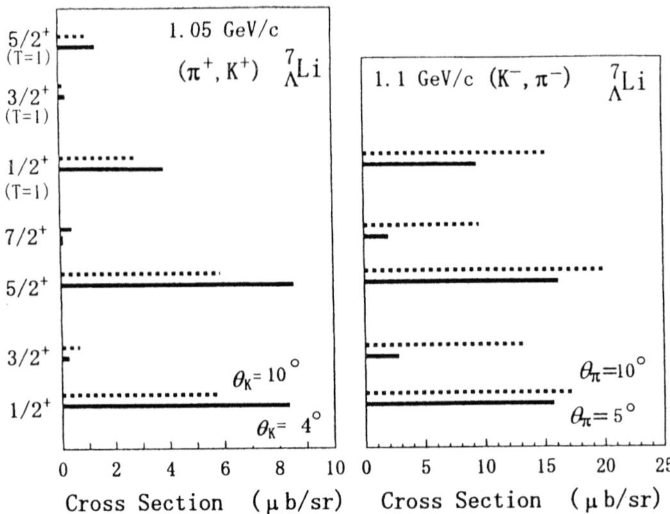

FIGURE 3. Calculated cross sections for the (π^+,K^+) and (K^-,π^-) reactions leading to ${}^7_\Lambda$Li. The former values (*left*) were used to estimate the level populations. The latter (*right*) demonstrates to show the ability of spin-flip excitation of $3/2^+$ and $7/2^+$.

In order to make a realistic comparison with the γ-ray yields and lifetime data of the ${}^7\mathrm{Li}(\pi^+,K^+\gamma){}^7_\Lambda\mathrm{Li}$ experiment, we combine γ-ray transition probabilities with the (π^+,K^+) cross sections (cf. Fig. 3) integrated over $\theta_K = 0° - 15°$: $\bar{\sigma}(J) = 1.21\mu\mathrm{b}(1/2^+_{gs})$, $0.13\mu\mathrm{b}(3/2^+)$, $1.23\mu\mathrm{b}(5/2^+)$, and $0.60\mu\mathrm{b}(1/2^+_{T=1})$. It is quite interesting that we can reproduce the experimental yield ratio between the E2$(5/2^+ \to 1/2^+_{gs})$ and M1$(3/2^+ \to 1/2^+_{gs})$ transitions (161±23 : 148±31)[5] only when we take γ-cascades from the $1/2_{T=1}$ state down to $3/2^+$ into account. The theoretical γ-ray yields for the interesting three transitions are obtained as (in arbitrary units)
$Y(\mathrm{E2};5/2^+ \to 1/2^+_{gs}) : Y(\mathrm{M1};3/2^+ \to 1/2^+_{gs}) : Y(\mathrm{M1};7/2^+ \to 5/2^+) = 1.27 : 0.47 : 0.07$

The major results are summarized in Table 1. The new cluster model results[13] are not much different from the original estimates[10]. In the E419 experiment[5], the observed lifetime of the $5/2^+$ state was deduced to be $\tau(5/2^+) = 5.2 \pm 1.4$ ps, which corresponds to $B(\mathrm{E2};5/2^+ \to 1/2^+) = 4.1 \pm 1.1\ e^2\mathrm{fm}^4$ [5]. This experimental B(E2) value is substantially smaller than the shell-model prediction ($8.6\ e^2\mathrm{fm}^4$)[14]. Therefore the dynamical shrinkage effect embodied in the cluster models is essential to explain the transition rate observed for the first time in the hypernuclear system. Thus prediction of the glue-like role of the Λ hyperon has been verified in this E419 experiment, although the cluster models seem to give a slight overestimation for the shrinkage effect (see Table 1). This is related with possible fine tuning of the ΛN interaction to be more appropriate to reproduce the known level energies.

TABLE 1. Predictions of electromagnetic properties of ${}^{7}_{\Lambda}$Li and the experiment.

	Dalitz-Gal (1978) [14] Shell model	Motoba *et al.* (1983) [10] ^{4}He$+d+\Lambda$	Hiyama *et al.* (1998) [13] ${}^{5}_{\Lambda}$He$+p+n$	Tamura *et al.* (1998) [5] Exp. E419
$B(\mathrm{M1};3/2^+ \to 1/2^+)$ [μ_N^2]	0.364	0.352	0.322	–
$B(\mathrm{E2};5/2^+ \to 1/2^+)$ [$e^2\mathrm{fm}^4$]	8.6	2.46	2.42	4.1 ±1.1
$B(\mathrm{E2};5/2^+ \to 3/2^+)$ [$e^2\mathrm{fm}^4$]	3.1	0.40	0.74	–
$B(\mathrm{E2};5/2^+ \to \mathrm{sum})/B_c$ *	1 †	0.44	0.33	–
Γ_B as of Eq.(1)	–	0.49	0.32	–
$R_{c-d}({}^{7}_{\Lambda}\mathrm{Li})/R_{\alpha-d}({}^{6}\mathrm{Li})$	–	0.83	0.75	0.87
Lifetime $\tau(5/2^+)$ [ps]	–	6.56	6.67	5.2±1.4

* B_c denotes $B(\mathrm{E2};{}^{6}\mathrm{Li},3^+ \to 1^+)$ in ^{6}Li.

† In Ref.[14] this ratio is assumed to be 1.

ΛN SPIN-SPIN AND SPIN-ORBIT INTERACTIONS

Traditionally the meson theoretical hyperon-nucleon potentials have been supplied by the effort of the Nijmegen group as model-D, model-F and the soft-core models[15, 16] and also by the Jülich group as the version-A and -B [17, 18]. Here we denote these potentials as ND, NF, NSC89, JA and JB, respectively. Starting with these potentials, the G-matrices have been derived in nuclear matter, and then they are expressed by Yamamoto[20] in terms of three-range gaussians with a parameter of the nuclear Fermi momentum k_F: $v_\alpha(r) = \sum_{i=1}^{3}(a_i + b_i k_F + c_i k_F^2)\exp[-(r/\beta_i)^2]$. Here α distinguishes the central, tensor, LS, and antisymmetric LS interactions. Their basic properties have been discussed in Refs. [19, 20]. As shown in the upper half of Table 2 where the divided contributions to the Λ one-body potential are listed, these potential models give very different nature: The spin-triplet attraction is very strong in JA and JB, while in NSC89 the spin-singlet interaction becomes stronger. One also notices the difference in the even/odd shares. The differences should affect hypernuclear structures, since such character remains unaveraged especially in light systems.

There are only about 35 data points for the YN scattering with mostly low energies, while for NN there are thousands scattering data points. Therefore it is important to feed back the results of hypernuclear structure analyses to the meson-theoretical YN potentials. In order to make such comparison useful, the potential models should have also some allowance of changing unknown parameters. To meet this requirement, Rijken *et al.* [7] presented the new soft-core model which consists of 6 versions (NSC97*a*, *b*, *c*, *d*, *e*, *f*) depending on the different $F/(F+D)$ ratios and meson mixing angles. It is remarked that all of these 6 versions can reproduce the YN scattering data equally well. Miyagawa *et al.*[21] have tested these potentials in the 3-body calculation of the simplest hypernucleus ${}^{3}_{\Lambda}$H and shown that only the NSC97f is acceptable in getting the the Λ binding energy. It should be noted, however, that the ${}^{3}_{\Lambda}$H ground state provides us with only $S = 1/2$ information, so that the calculation of ${}^{4}_{\Lambda}$H is quite important [22].

The Λ one-body potential has been also calculated with these potentials. From the

TABLE 2. ΛN-state contributions to the Λ one-body potential U_Λ.

	$^1S_0$	$^3S_1$	P	SUM(U_Λ)	$\lvert^1E\rvert : \lvert^3E\rvert$
JA	−3.6	−27.2	1.8	−29.0	1 : 7.6
JB	−0.5	−34.4	3.6	−31.3	1 :68.8
ND	−7.4	−25.2	−8.0	−40.5	1 : 3.4
NF	−10.0	−20.7	−0.9	−31.6	1 : 2.0
NSC89	−15.3	−8.7	0.2	−23.8	1 : 0.6
NSC97a	−3.8	−30.7	0.6	−33.9	1 : 8.1
b	−5.5	−30.0	1.4	−34.1	1 : 5.5
c	−7.8	−29.7	2.1	−35.3	1 : 3.8
d	−11.0	−27.7	3.5	−35.1	1 : 2.5
e	−12.8	−26.0	4.5	−34.3	1 : 2.0
f	−14.4	−22.9	6.2	−31.1	1 : 1.6

lower half of Table 2, the singlet-even contribution increases gradually as going from NSC97*a* to *f*, while the triplet-even one decreases and the repulsive *p*-state contribution increases. These changes should be reflected in the change of the spin-spin and spin-orbit components and hence the change of hypernuclear energy levels.

First of all, by employing the sample hypernucleus consisting ^{16}O+Λ, we show in Fig. 4 the basic values of symmetric and antisymmetric spin-orbit (ALS) components. In each version of YNG interactions, the k_F parameter is chosen to reproduce the *p*-orbit binding energy in $^{16}_{\Lambda}$O. The LS component increases gradually as going from NSC97*a* to *f* version, while the ALS component remains almost constant in negative sign and the absolute value amounts to about 30% of the LS component. As a result the total spin-orbit splitting increases from NSC97*a* to reach 0.98 MeV at NSC97*f*. Note that the NSC97*f* version is acceptable from the test of $^{3}_{\Lambda}$H. In the same figure we insert a preliminary values calculated from the extended soft-core model (ESC)[23].

Now we can make a realistic comparison in the case of $^{13}_{\Lambda}$C, since Ajimura *et al.*[6]

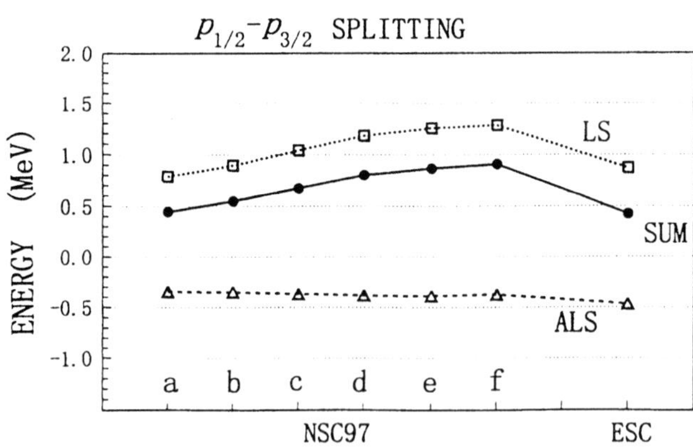

FIGURE 4. Spin-orbit splittings calculated for in $^{17}_{\Lambda}$O and the LS and ALS contributions.

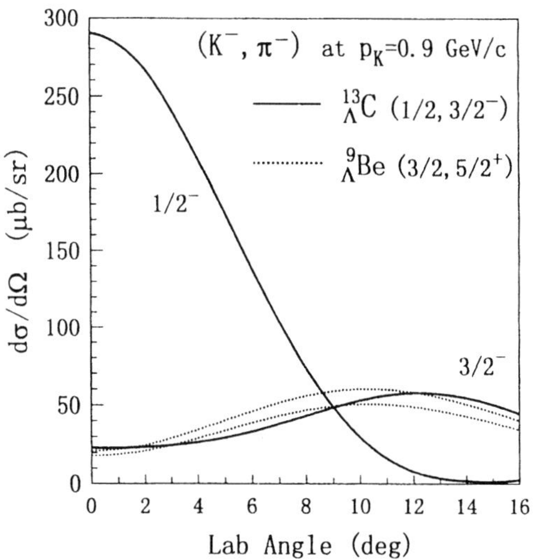

FIGURE 5. Calculated angular distribution of the (K^-,π^-) reaction cross sections for producing ${}^{13}_{\Lambda}$C and ${}^{9}_{\Lambda}$Be.

succeeded in measuring γ rays from the Λ single-particle states ($p_{3/2}$ and $p_{1/2}$) in this hypernucleus. In the analysis of the ^{13}C$(K^-,\pi^-\gamma){}^{13}_{\Lambda}$C reaction, they made use of the characteristic angular distribution as shown in Fig. 5 which was calculated in DWIA. In the forward angle the $1/2^-$ states is predominantly excited due to the recoilless condition, while at larger angle the share of the $3/2^-$ states increases. They deduced the Λ p-state spin-orbit splitting of $\Delta E = 0.154 \pm 0.054$ MeV[6]. How about the theoretical

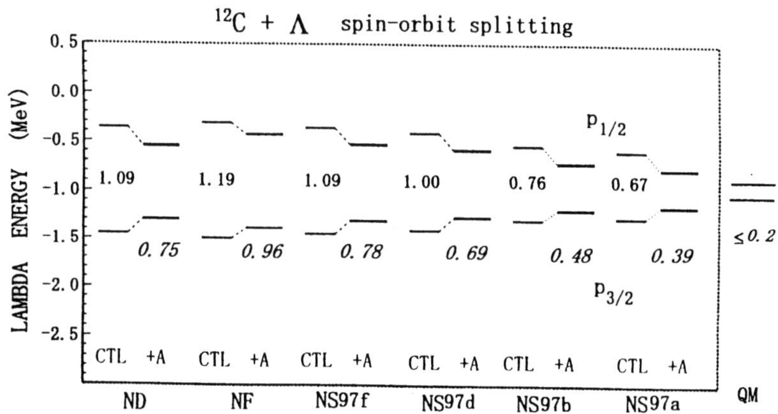

FIGURE 6. Spin-orbit splittings calculated for the Λ p-state in ${}^{13}_{\Lambda}$C. CTL denotes the central+ tensor+LS interaction and "+A" means the addition of the ALS interaction. On the extreme right the quark-model prediction[24] is inserted for reference.

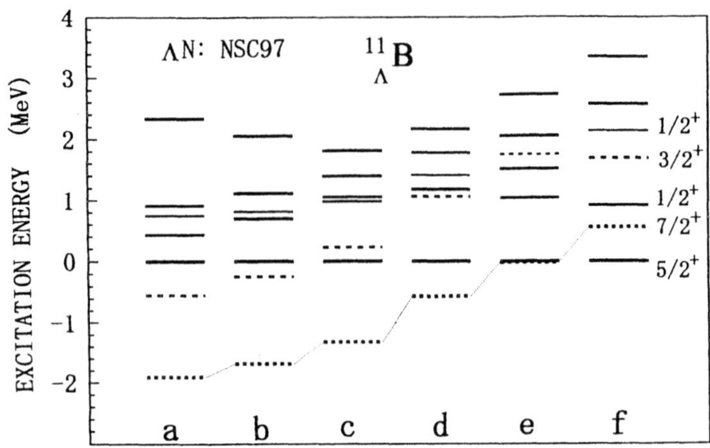

FIGURE 7. Comparison of low-lying energy levels of $^{11}_{\Lambda}$B calculated with NSC97 potentials. Relative excitations are shown with respect to the $5/2^+$ state.

values when we adopt the NSC97 potential models, which are summarized in Fig. 6. The general trend is similar to that of $^{17}_{\Lambda}$O shown in Fig. 4.

In general the energy level structure of p-shell hypernuclei is sensitive to the spin-spin and spin-orbit interactions because they are both active between p-shell nucleons and s-shell Λ. Among others, here we adopt the $^{11}_{\Lambda}$B case, because the ground state of the core nucleus ^{10}B has $J_c = 3^+$ consisting of 'large' orbital angular momentum ($L = 2$) and 'large' spin ($S = 1$), so that we can see the effect of different spin-dependent interaction easily in the behavior of $J_H = J_c \pm 1/2 = 5/2^+, 7/2^+$ states. In fact one sees in Fig. 7 that the relative positions of the doublet change quite sensitively as going from NSC97a to f. From the pionic decay data, the ground state of $^{11}_{\Lambda}$B is known to have $J_H = 5/2^+$, but we have no information on $7/2^+$.

Next Fig. 8 shows the individual roles of the spin-spin, LS, and antisymmetric LS (ALS) interactions, respectively. As seen from Table 2, the $^1S_0$ interaction is negative and largest in NSC97f, so that this favors the energy gain of $J = 5/2^+$. On the other hand , the $^3S_1$ interaction becomes minimum to unfavor $J = 7/2^+$. In both cases the LS interaction affects the position of $J = 7/2^+$ quite sensitively.

CONCLUDING REMARKS

Two of the most interesting predictions in hypernuclear structure have been finally confirmed by the great progress of experiments: One is the existence of "genuinely hypernuclear state" in $^{9}_{\Lambda}$Be having a supersymmetry. The evidence for the predicted "glue like role" of the Λ particle has been quantitatively established through the measurement of the absolute γ-decay rate for the first time. Then we have also demonstrated, in the

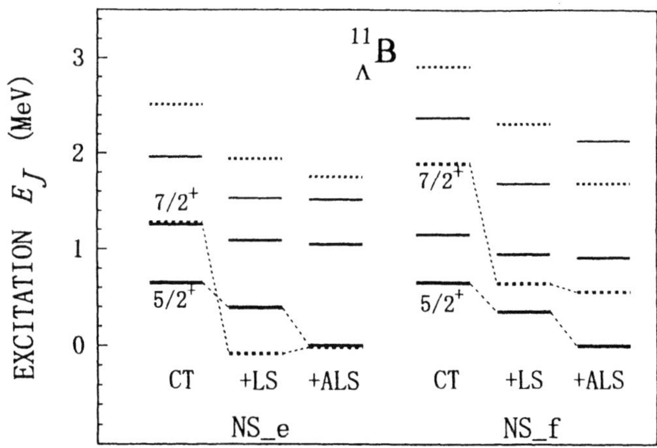

FIGURE 8. Change of energy levels of $^{11}_{\Lambda}$B when the LS and ALS interactions are added to the central+ tensor result(CT).

case of $^{11}_{\Lambda}$B, how the energy level structure changes under the individual effect of the spin-spin, LS, and ALS interactions. Through these examples we have emphasized the importance of knowing detailed level structures of hypernuclei in getting certain information on the particular components of YN interactions.

The author acknowledges Y. Yamamoto for providing the YNG effective interactions based on the G-matrix calculations. He is also grateful to E. Hiyama, M. Kamimura, Th. Rijken, and T. Yamada for collaboration and discussion. This work has been done as a part of the JSPS international collaboration projects (2000-2001).

REFERENCES

1. Pile, P.H. *et al.*, *Phys. Rev. Lett.* **66**, 2585 (1991).
2. Hasegawa,T. *et al.*, *Phys. Rev. Lett.* **74**, 224(1995); *Phys. Rev. C* **53**,1210 (1996).
3. Hashimoto, O.,*Proc. Strangeness Nuclear Physics*, edited by Il-T. Cheon, S.W. Hong and T. Motoba, World Scientific (1999) p.98 and references therein.
4. Nagae, T., ibid, p.110.
5. Tamura, H., *Nucl. Phys.* **A639**, 83c (1998); Tamura, H. *et al.*, *Phys. Rev. Lett.* **84**, 5963 (2000); Tanida, K. *et al.*, *Phys. Rev. Lett.* **86**, 1982 (2001).
6. Ajimura, S. *et al.*, *Phys. Rev. Lett.* **86**, 4255 (2001).
7. Rijken, Th.A., Yamamoto,Y. and Stoks, V.G.J., *Phys. Rev. C* **59**, 21 (1999); Rijken, Th.A. *et al.*, *Strangeness Nuclear Physics*, edited by Il-T. Cheon, S.W. Hong and T. Motoba, World Scientific, Singapore (1999), pp.5-12.
8. Dalitz, R.H., and Gal, A., *Phys. Rev. Lett.* **36**, 362 (1976).
9. Zhang, Z.Y., Li,G.L., and Yu, Y.W., *Phys. Lett.* **B108**, 261 (1982).
10. Motoba, T., Bando, H., and Ikeda, K. *Prog. Theor. Phys.* **70**, 189 (1983); Motoba, T. *et al.*, ibid. Suppl. **81**, 42 (1985).

11. Brückner, W. *et al.*,*Phys. Lett.* **B55**, 107 (1975); **B79**, 157 (1978).
12. Yamada, T., Ikeda, K., Bando, H. and Motoba, T., *Phys. Rev.* **C38**, 854 (1988).
13. Hiyama, E., Kamimura, M., Miyazaki, K. and Motoba, T., *Phys. Rev. C***59**, 2351 (1999).
14. Dalitz, R.H. and Gal, A. *Ann. Phys.(N.Y.)* **116**, 167 (1978); *J. Phys. G***4**, 889 (1978).
15. Nagels, M.M., *et al.*, *Phys. Rev.* **D12**, 744 (1975); **D15**, 2547 (1977); **D20**, 1633(1979) 1633.
16. Maessen, P.M.M.,*et al.*, *Phys. Rev.* **C40**, 2226 (1989).
17. Holzenkamp, B., *et al.*, *Nucl. Phys.* **A500**, 485 (1989).
18. Reuber, A., *et al.*,*Czech. J. Phys.* **42**, 1115 (1992) and references therein.
19. Yamamoto, Y. et al., *Czech J. Phys.* **42**, 1249 (1992);
Motoba, T. and Yamamoto, Y. *Nucl. Phys.***A585**, 29c (1995).
20. Yamamoto, Y., Motoba, T., Himeno, H., Ikeda, K. and Nagata, S., *Prog. Theor. Phys. Suppl.* No.**117** (1994) 361, and references therein.
21. Miyagawa, K., *et al.*, *Phys. Rev. C* **51**, 2905 (1995) and private communication.
22. Hiyama, E., *Proc. Int. Conf. Hypernuclear and Strange Prticle Physics*, Torio (2000), in press.
23. Rijken, Th.A. and Yamamoto, Y., private communication (2000).
24. Fujiwara, Y., Nakamoto, C., Suzuki, Y. *Phys. Rev. Lett.* **76**, 2242 (1996).

Hypernuclear γ spectroscopy and ΛN interactions

H. Tamura

Department of Physics, Tohoku University, Sendai 980-8578, Japan

Abstract. Our experimental project to study ΛN interaction using high-resolution hypernuclear γ spectroscopy is described. Employing a germanium detector array (Hyperball), we investigated level structure of ${}^{7}_{\Lambda}$Li and ${}^{9}_{\Lambda}$Be in high precision and obtained quantitative information on the strengths of the spin-spin force and the Λ-spin-dependent spin-orbit force between Λ and N. More experiments are in progress to investigate the tensor force and the nucleon-spin-dependent spin-orbit force.

INTRODUCTION

ΛN spin-dependent interactions

Investigation of hyperon-nucleon (YN) and hyperon-hyperon (YY) interactions is of particular importance in hadron physics. Since YN scattering experiments are difficult, study of hypernuclear structure plays an essential role to obtain information on the YN interactions. A significant progress in experimental studies of Λ hypernuclear structure has been made, which has revealed properties of the ΛN interaction.

The potential of the two body effective interaction between a Λ and a nucleon may be written as:

$$V_{\Lambda N}^{eff}(r) = V_0(r) + V_\sigma(r)\,\sigma_\Lambda\sigma_N + V_\Lambda(r)\,\mathbf{l}_{\Lambda N}\sigma_\Lambda + V_N(r)\,\mathbf{l}_{\Lambda N}\sigma_N + V_T(r)\left[3\frac{(\sigma_\Lambda\mathbf{r})(\sigma_N\mathbf{r})}{\mathbf{r}^2} - \sigma_\Lambda\sigma_N\right] \quad (1)$$

The strength of the spin-averaged central force, V_0, is well known from the depth of the nuclear potential for a Λ in hypernuclei, which is about 2/3 of that of the potential for a nucleon. On the other hand, the other terms with spin dependence, namely, the spin-spin term, the Λ-spin-dependent spin-orbit term, the nucleon-spin-dependent spin-orbit term, and the tensor term, have not been well know. Although the strengths of the spin-spin term and the Λ-spin-dependent spin-orbit term were suggested from several experimental data, they were not quantitatively established because of theoretical difficulties in interpretation, insufficient data quality, or inconsistency between different data, as described later.

In shell-model description, low-lying level energies of p-shell hypernuclei with a Λ in the 0s orbit can be described with five radial integrals for $s_\Lambda p_N$ interaction on the

CP594, *Hadrons and Nuclei: First International Symposium,* edited by Il-T. Cheon et al.

five terms in Eq. 1. The four integrals of the spin-dependent terms are denoted as Δ, S_Λ, S_N, and T, respectively, for the spin-spin, the Λ-spin-dependent spin-orbit, the nucleon-spin-dependent spin-orbit, and the tensor terms [1, 2]. These integrals (parameters) can be experimentally determined from various hypernuclear level energies.

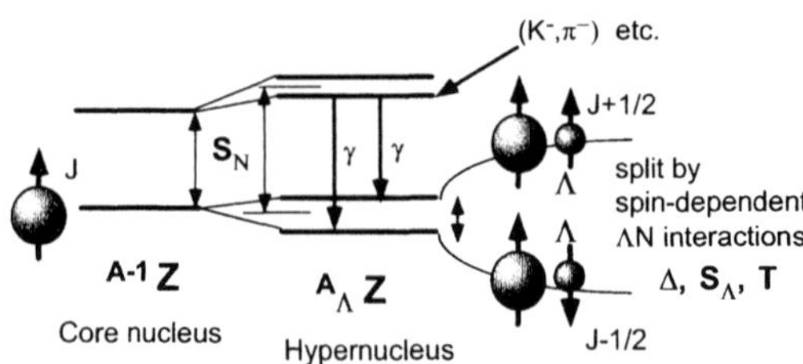

FIGURE 1. Hypernuclear fine structure and γ spectroscopy. High resolution of Ge detector is essential to resolve two levels (fine structure) split by ΛN spin-dependent interactions. See text for details.

When a Λ in the $0s$ orbit is coupled to a core nucleus (spin $J{\neq}0$), a doublet ($J-\frac{1}{2}$, $J+\frac{1}{2}$) appears as shown in Fig. 1. The energy spacing of such a spin doublet is determined by the spin-dependent terms of Δ, S_Λ and T. We call such spin doublets "hypernuclear fine structure", because their spacings are expected to be small, typically less than 100 keV. The other spin-dependent term S_N appears as the difference of the energy spacing of two levels in the core nucleus and that in the hypernucleus, where the spin-weighted average of each doublet is taken for hypernuclei, as shown in Fig. 1. Such a difference is also of the order of 100 keV. Therefore, energy resolution of the conventional method of the (K^-,π^-) and (π^+,K^+) spectroscopy using magnetic spectrometers, being currently 1.5 MeV (FWHM) at best, is far from required resolution for study of the ΛN spin-dependent forces.

It is necessary to determine each parameter from specific hypernuclear levels which have a large contribution of the parameter and smallest contributions of the other parameters. Table 1 and Fig. 2 show important hypernuclear levels to determine each of the spin-dependent terms. Our experimental plan described below is based on this table.

Hyperball Project

High-resolution γ-ray spectroscopy with germanium (Ge) detectors having a few keV (FWHM) resolution is almost the only method to investigate the ΛN spin-dependent forces from Λ hypernuclear levels. γ spectroscopy has additional merits; γ-ray cascade enables us to investigate various excited states including those which cannot be populated by the direct reactions, such as spin-flip states of Λ. Electromagnetic moments and spin-parities can be also determined in γ spectroscopy through transition probabilities and angular correlations.

Hypernuclear γ spectroscopy with Ge detectors was almost impossible before because of technical difficulties due to huge background of beam halo and scattered beam particles. After some technical studies, we solved the problem and constructed a large-acceptance Ge detector system dedicated to hypernuclear γ spectroscopy, Hyperball [4].

TABLE 1. Level spacings of p-shell Λ hypernuclei relevant to determine the ΛN spin-dependent interaction strengths and their level energies described by shell model calculations. The third column shows the term(s) which dominate(s) in the spacing. $\overline{E}$ denotes spin-averaged energy of the doublet.

Hypernuclear levels	calculation	term
${}^{7}_{\Lambda}$Li: $E(\frac{3}{2}^+) - E(\frac{1}{2}^+)$	$1.44\Delta + 0.05S_\Lambda + 0.02S_N - 0.27T$	Δ
${}^{9}_{\Lambda}$Be: $E(\frac{3}{2}^+) - E(\frac{5}{2}^+)$	$-0.04\Delta + 2.47S_\Lambda + 0.94T$	S_Λ
${}^{16}_{\Lambda}$O: $E(0^-) - E(1^-)$	$-0.38\Delta + 1.38S_\Lambda - 0.03S_N + 7.85T$ *	T
${}^{15}_{\Lambda}$N: $E(\frac{1}{2}^+) - E(\frac{3}{2}^+)$	$-0.73\Delta + 2.23S_\Lambda - 0.04S_N + 10.66T$ *	T
${}^{7}_{\Lambda}$Li: $E(\frac{7}{2}^+) - E(\frac{5}{2}^+)$	$1.31\Delta + 2.14S_\Lambda + 0.02S_N - 2.32T$	Δ, S_Λ
${}^{7}_{\Lambda}$Li: $\overline{E}(\frac{7}{2}^+, \frac{5}{2}^+) - \overline{E}(\frac{3}{2}^+, \frac{1}{2}^+)$	$-0.06\Delta + 0.08S_\Lambda + 0.68S_N - 0.05T$	S_N
${}^{12}_{\Lambda}$C: $E(2^-) - E(1^-)$	$0.48\Delta + 1.50S_\Lambda + 0.03S_N - 2.08T$	Δ, S_Λ

* Millener et al.[2], others: Millener [3]

Hyperball consists of fourteen sets of coaxial N-type Ge detectors (relative efficiency of 60% for each) equipped with fast electronics and BGO scintillation counters around each Ge crystal. The photo-peak efficiency for all the Ge detectors is 2.5% at 1 MeV. The BGO counters were used not only for Compton suppression but for rejection of high-energy photons from π^0 and high-energy charged particles which cause a serious background. More descriptions on Hyperball are found in Ref. [4, 5, 6, 7].

The first experiment with Hyperball was performed in 1998 at KEK. In this experiment (KEK E419) we succeeded in observing well-identified hypernuclear γ transitions using Ge detectors for the first time. The Hyperball project has thus started to investigate the hypernuclear levels shown in Table 1 and Fig. 2 for the purpose of studying all the ΛN spin-dependent forces.

SPIN-SPIN FORCE: ${}^{7}_{\Lambda}$LI

It is naively expected that the ΛN spin-spin force can be determined from ground-state doublet spacing of s-shell hypernuclei. The $M1(1^+ \rightarrow 0^+)$ transitions between the ground-state doublet of ${}^{4}_{\Lambda}$H and ${}^{4}_{\Lambda}$He were observed at 1.1 MeV with NaI counters [8]. This result suggests a spin-spin strength of $\Delta \sim 0.5$ MeV [2]. However, the three body ΛNN force is expected to give a large contribution to this splitting [9], which does not allow us to extract the strength without this effect quantitatively. In addition, another experiment for ${}^{10}_{\Lambda}$B suggests a very small strength ($\Delta < 0.22$ MeV) which seems contradictory to the $A = 4$ data [10]. Therefore, more data for level splittings of other p-shell hypernuclei have been awaited. In particular, the ground-state doublet spacing of ${}^{7}_{\Lambda}$Li is the best candidate, because the ^{6}Li(1^+) ground state has almost pure ${}^{3}S_1$ configuration and the hypernuclear doublet spacing is determined almost only by the spin-spin force, as also demonstrated by a shell model calculation [3] shown in Table 1. In this case, the effect of the three body force is expected to be small because the isospin of the ^{6}Li ground state is zero.

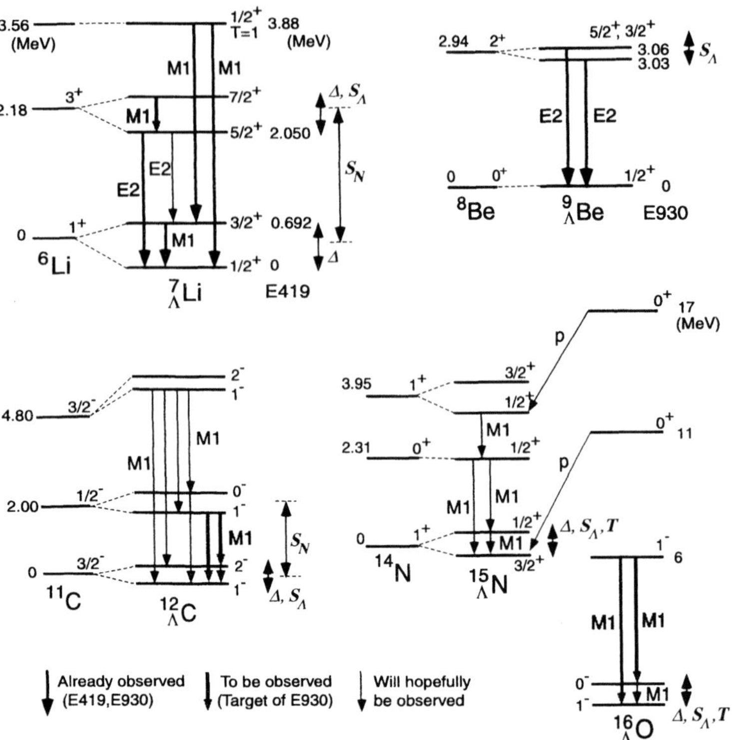

FIGURE 2. Level scheme and γ transitions of several p-shell hypernuclei which are investigated in our Hyperball project in order to determine all the spin-dependent interaction terms (Δ. S_Λ, S_N, T). Also see Table 1.

As the first experiment with Hyperball (KEK E419), we investigated ${}^{7}_{\Lambda}$Li. We produced ${}^{7}_{\Lambda}$Li bound states using the ^{7}Li(π^+,K^+) reaction employing the K6 beam line and the SKS spectrometer, and measured γ rays with Hyperball in coincidence. More descriptions are found in Refs. [5, 6, 7].

Figure 3 is the γ-ray spectrum when the bound-state region of ${}^{7}_{\Lambda}$Li is selected. We observed four γ-ray peaks at 691.7±0.6±1.0 keV, 2050.4±0.4±0.7 keV, 3186±4±6 keV, and 3877±5±7 keV. The 692 keV peak is uniquely assigned as the spin-flip $M1(\frac{3}{2}^+ \to \frac{1}{2}^+)$ transition, and the 2050 keV peak as the $E2(\frac{5}{2}^+ \to \frac{1}{2}^+)$ transition. The shapes of these peaks are consistent with Doppler broadening estimated from expected lifetimes of those states. The 692 keV peak becomes sharp after event-by-event Doppler-shift correction as shown in the left inset of Fig. 3 (left). The peaks at 3186 keV and 3877 keV, which were observed in the Doppler-shift corrected spectrum Fig. 3 (top-right), are assigned as the $M1$ transitions from the $\frac{1}{2}^+(T{=}1)$ state to the ground state

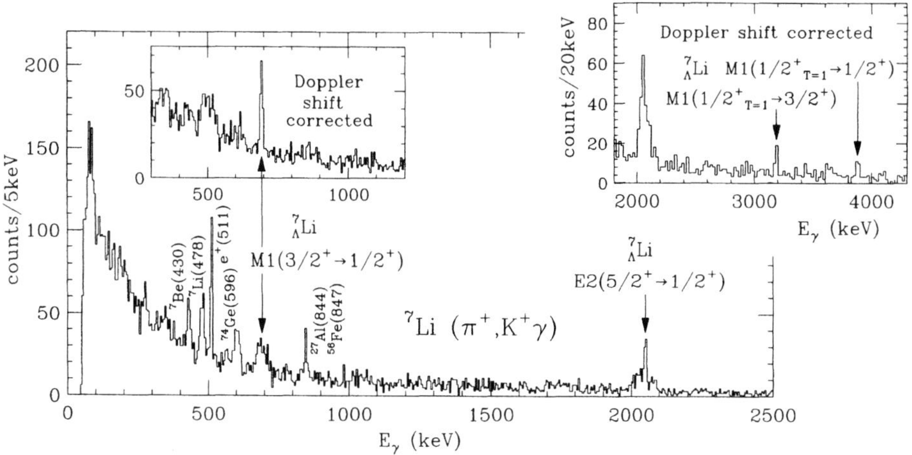

FIGURE 3. γ-ray spectrum of $^{7}_{\Lambda}$Li measured with Hyperball. Four hypernuclear transitions, $M1(\frac{3}{2}^{+}\rightarrow\frac{1}{2}^{+})$, $E2(\frac{5}{2}^{+}\rightarrow\frac{1}{2}^{+})$, $M1(\frac{1}{2}^{+}(T{=}1)\rightarrow\frac{3}{2}^{+})$, and $M1(\frac{1}{2}^{+}(T{=}1)\rightarrow\frac{1}{2}^{+})$, were observed.

doublet $(\frac{1}{2}^{+},\frac{3}{2}^{+})$. We have thus established the level scheme of $^{7}_{\Lambda}$Li as shown in Fig. 2 (top-left), where our observed transitions are shown in thick black arrows.

By comparing the experimental data of 692 keV with the shell-model calculation by Millener [3], the observed $M1$ energy unambiguously gives the strength of the spin-spin force, $\Delta = 0.48$ MeV. This result gives a strong restriction on baryon-baryon interaction models, because different versions of models predict a variety of spin-spin force strengths.

Λ-SPIN-DEPENDENT SPIN-ORBIT FORCE: $^{9}_{\Lambda}$BE

It has been long accepted that the Λ*N* spin-orbit interaction is very small compared with that of *NN*, based on the (K^{-},π^{-}) spectra of $^{16}_{\Lambda}$O and $^{13}_{\Lambda}$C [11, 12], although a finite size of the spin-orbit splitting has not been measured yet. Recently, it was pointed out that some new experimental data imply much larger spin-orbit force of about 1/3-1/5 of the nucleon case [13, 14]. This puzzling situation can be solved by γ spectroscopy of specific hypernuclei such as $^{9}_{\Lambda}$Be and $^{13}_{\Lambda}$C. Since quark models predict a very small spin-orbit force of Λ due to cancellation between the symmetric and antisymmetric *LS* forces [15], measurement of a finite size of the spin-orbit force is of particular interest.

As the second experiment with Hyperball, we studied $^{9}_{\Lambda}$Be employing the ^{9}Be(K^{-},π^{-}) reaction at 0.9 GeV/c utilizing a high intensity K^{-} beam at BNL AGS. This is one of the series of experiments (E930) to study all the spin-dependent Λ*N* interactions by γ spectroscopy of various *p*-shell hypernuclei. See Ref. [16] for details of the $^{9}_{\Lambda}$Be experiment.

Figure 4 shows the γ-ray spectrum when the bound-state region of $^{9}_{\Lambda}$Be in the (K^{-},π^{-})

spectrum is gated. We successfully observed a fine structure, twin peaks at 3029 keV and 3060 keV. They are assigned as the $E2(\frac{5}{2}^+ \to \frac{1}{2}^+, \frac{3}{2}^+ \to \frac{1}{2}^+)$ transitions shown in Fig. 2 (top-right). These transitions were observed at 3.08±0.04 MeV in the past experiment, where the two peaks were not separated due to a limited resolution of NaI counters [17].

The observed structure was well fitted by two peaks having partly Doppler-broadened shape expected from a simulation. The observed yields and the almost equal relative intensities of the two peaks, as well as the absolute yields (about 100 counts for each peak), are consistent with calculated cross sections of $\frac{3}{2}^+$ and $\frac{5}{2}^+$ [18]. It is to be noted that we cannot assign which peak is $\frac{5}{2}^+$ or $\frac{3}{2}^+$.

Since the $(\frac{5}{2}^+, \frac{3}{2}^+)$ doublet corresponds to the ^{8}Be(2^+) state with a Λ in 0s orbit, their energy spacing is determined almost purely by the Λ-spin-dependent ΛN spin-orbit interaction (S_Λ). According to a shell-model calculation by Millener [3], the spacing is written as shown in Table 1. Although T is unknown, it is expected to be small ($T \sim 0 - 0.1$ MeV). So the observed 31 keV spacing suggests a very small spin-orbit term, of the order of $|S_\Lambda| \sim 0.01$ MeV. Recent cluster-model calculations by Hiyama *et al.* [19] using several versions of the meson-exchange baryon-baryon interaction models predicted a spacing of 80 – 200 keV. It indicates that the present meson-exchange models, which give the strength of the Λ-spin-dependent spin-orbit force of the order of $\sim$1/4 of the NN spin-orbit force, have to be modified.

Our observation thus confirmed a very small ΛN spin-orbit force which is much less than 1/10 of the NN spin-orbit force. Our result seems consistent with the recent observation of a small spin-orbit splitting ($p_{1/2} - p_{3/2}$) of the Λ single particle states in $^{13}_{\Lambda}$C [20].

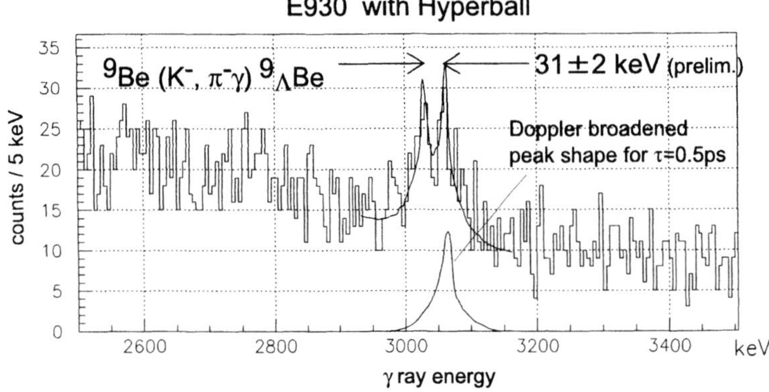

FIGURE 4. Measured γ-ray spectrum of $^9_\Lambda$Be. The twin peak structure was well fitted using a expected partly-Doppler-broadened peak shape. The two peaks are assigned as the $E2(\frac{5}{2}^+, \frac{3}{2}^+ \to \frac{1}{2}^+)$ transitions.

TENSOR FORCE: ${}^{16}_{\Lambda}$O

The next beam time for BNL E930 is scheduled in the summer in 2001, when we will study ${}^{16}_{\Lambda}$O(${}^{15}_{\Lambda}$N), ${}^{7}_{\Lambda}$Li, ${}^{12}_{\Lambda}$C, etc. with Hyperball. The purpose of the ${}^{16}_{\Lambda}$O experiment is to obtain information on the ΛN tensor force which is completely unknown at present. As shown in Table 1, the ground-state doublet (0^-,1^-) spacing has a large contribution of the tensor term.

One pion exchange, which is responsible for the NN tensor force, is forbidden for ΛN, but two pion exchange with ΛN-ΣN mixing is expected to give a significant contribution to the tensor force. Both of the $M1$ transitions from the 6 MeV 1^- state to both of the ground doublet states (0^-, 1^-) will be separately detected with a few weeks' beam time at BNL.

The 11 MeV-excited $[(p_{1/2})_n^{-1}(p_{1/2})_\Lambda]_{0^+}$ state of ${}^{16}_{\Lambda}$O is expected to decay to excited states of ${}^{15}_{\Lambda}$N with sizable branching ratios, and γ rays are emitted. The ground-state doublet spacing of ${}^{15}_{\Lambda}$N, which also has a large contribution of the ΛN tensor force, may be able to be observed simultaneously [4].

NUCLEON-SPIN-DEPENDENT SPIN-ORBIT FORCE: ${}^{7}_{\Lambda}$LI

The observed ${}^{7}_{\Lambda}$Li $E2(\frac{5}{2}^+ \to \frac{1}{2}^+)$ energy (2.05 MeV) suggests a large value of the nucleon-spin-dependent spin-orbit term, $S_N \sim -0.4$ MeV [3]. In order to obtain better information on S_N, we are planning to measure the $\frac{7}{2}^+ \to \frac{5}{2}^+$ spin-flip $M1$ transition, because the energy spacing between the weighted averages of the ($\frac{7}{2}^+$,$\frac{5}{2}^+$) doublet and the ($\frac{3}{2}^+$,$\frac{1}{2}^+$) doublet is determined only by S_N term, as shown in Table 1.

In E930 we will investigate ${}^{7}_{\Lambda}$Li again using 1.1 GeV/c (K^-,π^-) reaction which has a large cross section to produce the spin-flip state $\frac{7}{2}^+$. Both data of S_N and S_Λ enable us to separate strengths of the symmetric LS ($\propto \mathbf{l}_{\Lambda N}(\sigma_\Lambda + \sigma_N)$) and the antisymmetric LS ($\propto \mathbf{l}_{\Lambda N}(\sigma_\Lambda - \sigma_N)$) interactions. The separate information is necessary to discuss validity of baryon-baryon interaction models, not only various meson exchange models but quark models.

In addition, the spacing of the ($\frac{7}{2}^+$,$\frac{5}{2}^+$) doublet gives information on strengths of both of the spin-spin and the Λ-spin-dependent spin-orbit forces (Δ and S_Λ). Consistencies with the ${}^{7}_{\Lambda}$Li ground-doublet data for Δ and the ${}^{9}_{\Lambda}$Be doublet data for S_Λ can be checked.

SUMMARY

We started a project of hypernuclear γ spectroscopy. By using a Ge detector array, Hyperball, we precisely investigated level structure of ${}^{7}_{\Lambda}$Li and ${}^{9}_{\Lambda}$Be for the purpose of studying the ΛN spin-dependent interactions. We observed γ transitions in ${}^{7}_{\Lambda}$Li at KEK and obtained the strength of the ΛN spin-spin force from the spin-flip $M1$ transition between the ground-state doublet. We also observed two $E2$ transitions in ${}^{9}_{\Lambda}$Be at

BNL and confirmed a very small spin-orbit force of Λ, which cannot be explained by existing meson-exchange models of baryon-baryon interactions. The tensor force and the nucleon-spin-dependent spin-orbit force will be investigated with ${}^{16}_{\Lambda}O$ and ${}^{7}_{\Lambda}Li$ in the next beam time at BNL in 2001.

ACKNOWLEDGEMENTS

The author thanks to all the members of KEK E419 and BNL E930. He is also grateful to D. J. Millener, T. Motoba, and E. Hiyama. This work is supported by Grant-In-Aid from The Ministry of Education of Japan, No. 08239102 and No.11440070.

REFERENCES

1. R.H. Dalitz and A. Gal, Ann. Phys. 116 (1978) 167; J. Phys. G 6 (1978) 889.
2. D.J. Millener, A. Gal, C.B. Dover and R.H. Dalitz, Phys. Rev. C31 (1985) 499.
3. D.J. Millener, Proc. Workshop on "Hypernuclear Physics with Hadronic Probes" (HYPJLAB99), Hampton, December 1999, in press.
4. H. Tamura, Nucl. Phys. A639 (1998) 83c.
5. H. Tamura *et al.*, Phys. Rev. Lett. 84 (2000) 5963.
6. K. Tanida *et al.*, Phys. Rev. Lett. 86 (2001) 1982.
7. K. Tanida, Ph. D thesis, University of Tokyo, 2000.
8. M. Bedjidian *et al.*, Phys. Lett. 62B (1976) 467; M. Bedjidian *et al.*, Phys. Lett. 83B (1979) 252.
9. B.F. Gibson and D.R. Lehman, Phys. Rev. C **37** (1988) 679; Y. Akaishi *et al.*, Phys. Rev. Lett. **84** (2000) 3539.
10. R.E. Chrien *et al.*, Phys. Rev. C 41 (1990) 1062.
11. W. Brückner *et al.*, Phys. Lett. 79B (1978) 157.
12. M. May *et al.*, Phys. Rev. Lett. 47 (1981) 1106.
13. T. Nagae, Proc. 23rd INS symp. on Nuclear and Particle Physics with meson beams in the 1 GeV/c region, ed. S.Sugimoto and O. Hashimoto, (Universal Academic Press, Tokyo, 1995) 175; T. Motoba, ibid 187.
14. R.H. Dalitz *et al.*, Nucl. Phys. A 625 (1997) 71.
15. S. Takeuchi *et al.*, Prog. Theor. Phys. Suppl. 137 (2000) 83 and references therein.
16. H. Akikawa *et al.*, Proc. Int. Conf. on Hypernuclear and Strange particle physics (HYP2000), Torino, October 2000; to be published.
17. M. May *et al.*, Phys. Rev. Lett. 51 (1983) 2085.
18. T. Motoba, private communication (1996).
19. E. Hiyama *et al.*, Phys. Rev. Lett. 85 (2000) 270.
20. S. Ajimura *et al.*, Phys. Rev. Lett. 86 (2001) 4255.

Nuclear $\bar{K}$ bound states in light nuclei

Y. Akaishi[1] and T. Yamazaki[2]

[1] *Institute of Particle and Nuclear Studies, KEK, Tsukuba 305-0801, Japan*
[2] *RI Beam Science Laboratory, RIKEN, Wako 351-0198, Japan*

Abstract. Possible existence of deeply-bound nuclear states of $\bar{K}$ is investigated theoretically for few-body systems. The nuclear ground states of $\bar{K}$ in ^{3}He and ^{4}He are predicted to be discrete states with binding energies of 108 MeV and of 86 MeV and widths of 20 MeV and of 34 MeV, respectively. The formation of the $T = 0$ $K^- \otimes {}^3$He + $\bar{K}^0 \otimes {}^3$H state in the ^{4}He(stopped K^-, n) reaction is proposed with a calculated branching ratio of about 2 %. It is discussed that a kaon injected into ^{9}Be forms a nuclear object of very high density and low temperature.

INTRODUCTION

One of the most important but yet unsolved problems of hadrons is how the hadron masses change in nuclear medium. In the strangeness sector, possibility of dropping K^- mass in nuclei has been theoretically asserted [1, 2]. This problem is connected to an exciting issue of whether strangeness condensation in nuclear matter could occur or not. Despite its importance, no clear experimental method for deducing in-medium masses has yet been established. Recently, a new type of "in-medium hadron-mass spectroscopy" has been successfully carried out by observing deeply-bound states of pion [3, 4]. Here we investigate possible existence of discrete nuclear bound states of $\bar{K}$ and how to populate such states.

Recently a definite knowledge on the strong-interaction level-shift of kaonic hydrogen atom has been obtained from the KpX experiment at KEK [5, 6], which indicates a negative "repulsive-type" shift of the 1*s* orbit. For heavier nuclei Batty *et al.* [7, 8] re-analyzed all the existing data of K^- atoms, allowing a density dependent term for the K^-N scattering length, and deduced the optical potential with a strongly attractive real part V and a strongly absorptive imaginary part W as

$$V_0^{\text{atom}} = -183 \text{ MeV and } W_0^{\text{atom}} = -70 \text{ MeV}. \tag{1}$$

The reason for such an highly attractive potential despite the fact that the strong-interaction shifts are all negative (of "repulsive" type) comes from the assertion that the $\Lambda(1405)$ state is not an "elementary particle" but the nuclear ground state of the $\bar{K}$ + N system. For this potential one can expect deeply-bound nuclear states in heavier nuclei, but their widths are of 50 MeV or more if the parameters of Eq.(1) are rigorously applied. In this paper we examine the coupled-channel nature of $\bar{K}N$ interaction and show some property of deeply-bound kaonic states in few-body systems.

CP594, *Hadrons and Nuclei: First International Symposium*, edited by Il-T. Cheon et al.

FORMALISM

First we construct phenomenologically a $\bar{\mathrm{K}}\mathrm{N}$ interaction model as simple as possible by using free $\bar{\mathrm{K}}\mathrm{N}$ scattering data [9], the X-ray data of kaonic hydrogen atom [5, 6] and the binding energy and the width of $\Lambda(1405)$ which can be regarded as an isospin $I = 0$ bound state of $\bar{\mathrm{K}}$ + N. The $I = 0$ and $I = 1$ $\bar{\mathrm{K}}\mathrm{N}$ interactions are found to be

$$v^I_{\bar{\mathrm{K}}\mathrm{N}}(r) = v^I_{\mathrm{D}}\exp[-(r/0.66\,\mathrm{fm})^2], \tag{2}$$

$$v^I_{\bar{\mathrm{K}}\mathrm{N},\pi\Sigma}(r) = v^I_{\mathrm{C}_1}\exp[-(r/0.66\,\mathrm{fm})^2], \tag{3}$$

$$v^I_{\bar{\mathrm{K}}\mathrm{N},\pi\Lambda}(r) = v^I_{\mathrm{C}_2}\exp[-(r/0.66\,\mathrm{fm})^2] \tag{4}$$

with

$$v^{I=0}_{\mathrm{D}} = -436\,\mathrm{MeV},\ \ v^{I=0}_{\mathrm{C}_1} = -412\,\mathrm{MeV},\ \ v^{I=0}_{\mathrm{C}_2} = \mathrm{none}, \tag{5}$$

$$v^{I=1}_{\mathrm{D}} = -62\,\mathrm{MeV},\ \ v^{I=1}_{\mathrm{C}_1} = -285\,\mathrm{MeV},\ \ v^{I=1}_{\mathrm{C}_2} = -285\,\mathrm{MeV}, \tag{6}$$

where $v^I_{\pi\Sigma}(r) = v^I_{\pi\Lambda}(r) = 0$ is taken for reducing simply the number of parameters. The $I = 0$ interaction produces the unstable bound state of $\Lambda(1405)$ with $E_{\bar{\mathrm{K}}\mathrm{N}} = -29.5$ MeV (-27 MeV) from the $I = 0$ threshold (from the K^-+ p threshold) and $\Gamma = 40$ MeV. The interaction gives a scattering length of $a^{I=0} = -1.76 + i0.46$ fm, which can be compared to $a^{I=0} = (-1.70 \pm 0.07) + i(0.68 \pm 0.04)$ fm of Martin's analysis [9]. It is noted that the data of $\Lambda(1405)$ and the scattering length enable us to determine both the strength and the range of the potential. The $I = 1$ interaction, having no bound state, gives a scattering length $a^{I=1} = 0.37 + i0.60$ fm reproducing Martin's $I = 1$ value. The K^-p scattering length calculated from the $I = 0$ and $I = 1$ interactions,

$$a_{\mathrm{K}^-\mathrm{p}} = \frac{1}{2}(a^{I=0} + a^{I=1}) = -0.70 + i0.53\,\mathrm{fm}, \tag{7}$$

is in good agreement with the data obtained by the KpX experiment [5, 6]

$$a_{\mathrm{K}^-\mathrm{p}} = (-0.78 \pm 0.15 \pm 0.03) + i(0.49 \pm 0.25 \pm 0.12)\,\mathrm{fm}. \tag{8}$$

The binding of $\bar{\mathrm{K}}$ in nuclei is calculated in the framework of the Brueckner-Hartree-Fock theory. We employ the QTQ prescription for intermediate states in reaction-matrix equation [10], considering that the higher-order energy due to the insertion of intermediate-energy spectra would be largely canceled out with the rearrangement energy of the nuclear medium [11, 12]. The $\bar{\mathrm{K}}\mathrm{N}$ g-matrix in nuclear matter is defined by

$$g = v + v\frac{Q_{\mathrm{N}}}{E_{\mathrm{st}} - Q_{\mathrm{N}}\hat{T}Q_{\mathrm{N}}}g, \tag{9}$$

where E_{st} is a starting energy [13]. In this case, the binding effects of $\bar{\mathrm{K}}$ and N are properly taken into account through E_{st}. For atomic states we take $E_{\mathrm{st}} = E_{\mathrm{K}^-} + E_{\mathrm{N}} \approx E_{\mathrm{N}} \approx -20$ MeV and calculate g-matrices with $k_{\mathrm{F}} = 1.36\ \mathrm{fm}^{-1}$. It is convenient for practical use to approximate the obtained g-matrices with zero-range effective interactions having same volume-integral values:

$$g^I \approx g_0^I\,\delta(\vec{r}), \tag{10}$$

$$g_0^{I=0} = -2175 - i849\ \mathrm{MeV}\cdot\mathrm{fm}^3, \tag{11}$$

$$g_0^{I=1} = -323 - i225\ \mathrm{MeV}\cdot\mathrm{fm}^3. \tag{12}$$

The K^--nucleus optical potential for atomic states is now easily derived by folding these interactions with nuclear density distribution:

$$U_{\mathrm{opt}}^{\mathrm{atom}} = \frac{V_0^{\mathrm{atom}} + iW_0^{\mathrm{atom}}}{1+\exp[(r-R_0)/a_s]}, \qquad V_0^{\mathrm{atom}} + iW_0^{\mathrm{atom}} = \frac{1}{4}(g_0^{I=0} + 3g_0^{I=1})\,\rho_0. \tag{13}$$

For heavy nuclei of isospin $T=0$ we get

$$V_0^{\mathrm{atom}} = -134\ \mathrm{MeV}\ \text{and}\ W_0^{\mathrm{atom}} = -65\ \mathrm{MeV}. \tag{14}$$

This optical potential is considerably deeper than those obtained in previous theoretical works [14, 15, 16, 17, 18] and is compatible with that of Eq.(1) obtained phenomenologically by Batty *et al.* [7, 8]. It is noted that this potential can not be used to deeply-bound K^- nuclear states, because the imaginary part would become much smaller if the $\Sigma+\pi$ channel is closed. This situation has been assumed by Kishimoto to search for deeply-bound K^- nuclear state in medium-A nuclei by means of (in-flight K^-, N) reactions [19].

RESULTS AND DISCUSSION

Now let's move on to search for nuclear $\bar{K}$ bound states in ^{3}He and ^{4}He. The $\bar{K}N$ effective interactions, *i.e.* g-matrices, have following weights in the nuclear $\bar{K}$ states:

$$3\,(\frac{1}{2}g^{I=0} + \frac{1}{2}g^{I=1}) \quad \text{for}\ {}^{3}_{\bar{K}}\mathrm{H}(T=0), \tag{15}$$

$$3\,(\frac{1}{6}g^{I=0} + \frac{5}{6}g^{I=1}) \quad \text{for}\ {}^{3}_{\bar{K}}\mathrm{H}(T=1), \tag{16}$$

$$4\,(\frac{1}{4}g^{I=0} + \frac{3}{4}g^{I=1}) \quad \text{for}\ {}^{4}_{\bar{K}}\mathrm{H}(T=1/2), \tag{17}$$

where we employ the conventional nomenclature to express composite nuclei, namely, ${}^{3}_{\bar{K}}\mathrm{H}(T=0,1)$ are the states of $K^-\otimes{}^3$He + $\bar{K}^0\otimes{}^3$H $(T=0,1)$ and ${}^{4}_{\bar{K}}\mathrm{H}(T=1/2)$ is $K^-\otimes{}^4$He $(T=1/2)$. It is to be noted that the ${}^{3}_{\bar{K}}\mathrm{H}(T=0)$ state acquires the largest weight 3/2 of the more attractive interaction $g^{I=0}$ among the three states.

By using density distributions of the core nuclei ($A=3,4$) we calculate the bound state energies $E_{\bar{K}}$ in such a way that E_{st}, g^I and $E_{\bar{K}}$ become to be self-consistent. The obtained results are $E_{\bar{K}}$ = -76 MeV and Γ = 82 MeV for ${}^{3}_{\bar{K}}\mathrm{H}(T=0)$, and $E_{\bar{K}}$ = -69 MeV and Γ = 66 MeV for ${}^{4}_{\bar{K}}$H. The bindings are fairly large as expected, but the widths are also large, since the bound states are located above the $\Sigma+\pi$ threshold.

Here a question arises; do the sizes of the core nuclei remain unchanged when the $\bar{K}$ interacts with several tens MeV binding? Due to the strong $\bar{K}N$ attractive force the bound $\bar{K}$ would be expected to combine the nucleons closer. Suppose a zero-size limit of the core nucleus ^{3}He: the $\bar{K}\otimes^3$He system is considered as a "quasi-Λ(1405)" with $M_N \to 3M_N$ and $v^{I=0} \to \frac{3}{2}v^{I=0}$. The change of $v^{I=0} \to \frac{3}{2}v^{I=0}$ increases the binding energy from 27 MeV to 142 MeV, and the change of $M_N \to 3M_N$ brings additional binding of 47 MeV. Then, the fictitious "quasi-Λ(1405)" becomes a bound state of $E_{\bar{K}\text{"}N\text{"}}$ = -189 MeV with no decay width. Of course, such a shrinkage effect induced by $\bar{K}$ must be counterbalanced by the incompressibility of the core nucleus. By taking into account the incompressibility with Hasegawa-Nagata's effective NN interaction [20], we minimize the sum of the $\bar{K}$ energy $E_{\bar{K}}$ and the core internal energy E_{core} with respect to a core-size parameter.

TABLE 1. Energies $E_{\bar{K}}$, widths Γ and r.m.s. radii of core nuclei $R^{\bar{K}}_{core}$ of the $\bar{K}$ nuclear states. The energies are measured from the respective K^-+ nucleus threshold.

State	$E_{\bar{K}}$(MeV)	Γ (MeV)	$R^{\bar{K}}_{core}$(fm)	$R^{\bar{K}}_{core}/R^{free}_{core}$
$^3_{\bar{K}}$H($T=0$)	−108	20	0.97	0.60
$^3_{\bar{K}}$H($T=1$)	−21	95	1.20	0.74
$^4_{\bar{K}}$H($T=1/2$)	−86	34	1.12	0.76

The final results obtained from this size optimization are presented in Table 1, which shows substantial lowering of the bound state energies and shrinkage of the core-nucleus radii. The $\bar{K}$ in the light nuclei creates by itself compressed surroundings and provides information about properties of the hadron in high-density nuclear medium, which is currently an important key issue [2]. Figure 1 depicts the optimized $\bar{K}$-nucleus potentials $U_{\bar{K}}$ of the K^- + ^{3}He and K^- + ^{4}He systems with the bound states of $^3_{\bar{K}}$H and of $^4_{\bar{K}}$H, which are extensions of the basic K^- + p system with Λ(1405) as its bound state. The predicted bound states may be called *strange tri-baryon* and *strange tetra-baryon*, respectively.

The nuclear bound state of $K^-\otimes^4$He has the binding energy of 86 MeV and the width of 34 MeV. Since the state is slightly below the Σ + π threshold and is open only for the Λ + π channel, the width becomes narrow. Although such a deeply-bound narrow-width K^- state in ^{4}He was predicted by Staronski and Wycech [21, 22], the mechanism of its appearance given in this paper is quite different from theirs.

More interesting is the system composed of $K^-\otimes^3$He + $\bar{K}^0\otimes^3$H shown in Fig. 2. A deeply-bound $\bar{K}$ nuclear state with a much narrower width appears in this system of the isospin $T=0$. Its energy level lies by 108 MeV below the ^{3}He + K^- threshold, *i.e.* by 13 MeV below the Σ + π threshold, and the level width is 20 MeV which is about 20 % of the binding energy. On the contrary, the other isospin $T=1$ state has a very large width

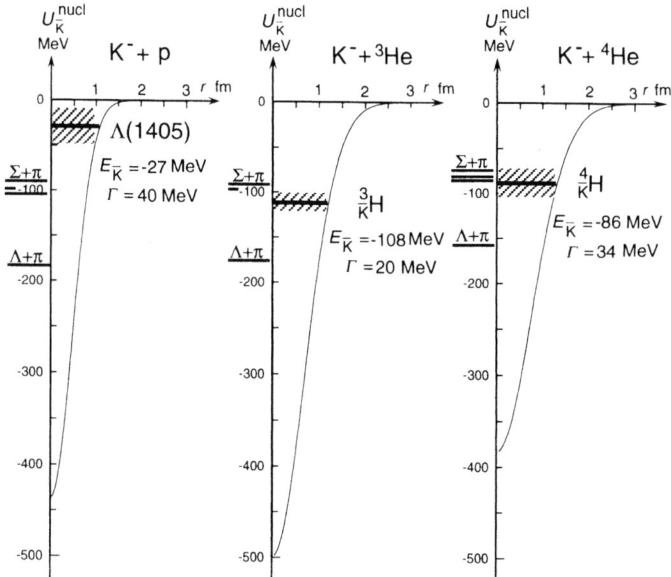

FIGURE 1. Calculated $\bar{K}$-N and $\bar{K}$-nucleus potentials and the bound levels, Λ(1405), $^3_{\bar{K}}$H and $^4_{\bar{K}}$H for the K⁻-p, -³He and -⁴He systems, respectively. The nuclear shrinkage effects are taken into account.

of 95 MeV with only 21 MeV binding energy. The different features of these two states can be well understood by counting the contribution of $I = 0$ and $I = 1$ $\bar{K}$N interactions to these states. In the $T = 0$ state the $I = 0$ interaction has a weight of 3/2 which is 3 times larger than that in the $T = 1$ state. This provides the $T = 0$ state with stronger attraction and deeper binding, and thus the $T = 0$ state lies below the $\Sigma + \pi$ threshold and cannot decay by the major $I = 0$ interaction to the open $\Lambda + \pi$ channel due to the isospin selection. The width originates exclusively from the $I = 1$ $\bar{K}$N interaction, of which the weight in the $T = 0$ state is only a half of that in the deep $K^- \otimes {}^4$He state. Thus, the deeply-bound $^3_{\bar{K}}$H$(T = 0)$ state has a markedly narrow width.

Now let us show that $^3_{\bar{K}}$H$(T = 0)$ may be formed and identified in K^- absorption at rest by ^{4}He. The K^- in an atomic orbit falls into the deeply-bound nuclear orbit of $^3_{\bar{K}}$H$(T = 0)$, emitting a neutron which helps to form the core nucleus with one neutron less and simultaneously serves as a spectator of the formed state. The partial decay rate for the formation of $^3_{\bar{K}}$H from the atomic $2p$ absorption is calculated to be

$$\Gamma = 1.2 \text{ eV}. \tag{18}$$

The branching ratio $\mathcal{B}$ for the formation of $^3_{\bar{K}}$H$(T = 0)$ is "modestly" estimated as

$$\mathcal{B} = \frac{\Gamma}{\Gamma_{\text{tot}}} = 2.2^{+3.5}_{-0.9}\ \%, \tag{19}$$

where we use experimental data of $\Gamma_{\text{tot}} = 55 \pm 34$ eV [23] which is known as a shift anomalously larger than theoretical estimations [24].

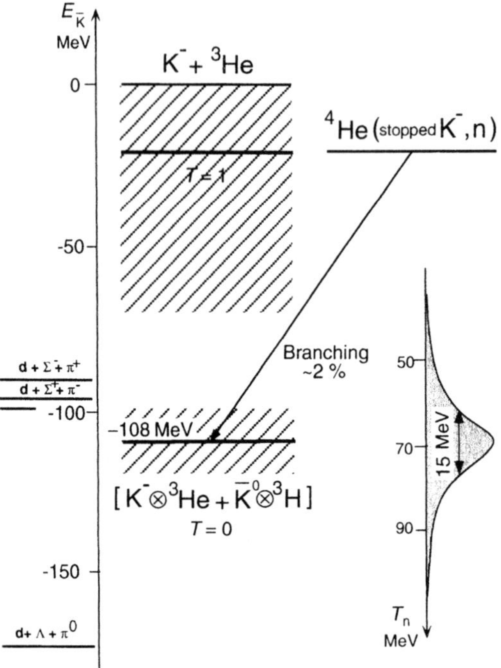

FIGURE 2. Energy level diagram of the $K^- + {}^3He$ system. The $T = 0$ state can be excited and signaled with discrete neutron emission from stopped K^- on ${}^4He$.

The ${}^9Be$ nucleus is known to have a well-developed α-cluster structure. Now let's consider the case when a kaon is injected into this ${}^9Be$ nucleus. The kaon plays a drastic role in this system as shown below. We solve the three-body system $\alpha\alpha\bar{K}$ variationally by the ATMS method [25], neglecting a p-shell neutron for simplicity. The $\alpha\alpha$ relative motion can be described with the orthogonality condition [26] as

$$\left[-\frac{\hbar^2}{2\mu_{\alpha\alpha}}\frac{d^2}{dr^2} + V_{\alpha\alpha}(r) + U^{\mathrm{K}}_{\alpha\alpha}(r)\right] u_{\alpha\alpha}(r) + \Lambda \sum \langle r|\phi_{\mathrm{F}}\rangle\langle\phi_{\mathrm{F}}|u_{\alpha\alpha}\rangle = E u_{\alpha\alpha}(r), \tag{20}$$

where $V_{\alpha\alpha}$ and $U^{\mathrm{K}}_{\alpha\alpha}$ are an $\alpha\alpha$ folding potential [27] and an $\bar{K}$-mediated potential, respectively, and ϕ_{F} is the Pauli-forbidden state to be excluded with a large positive constant Λ from the relative wave function $u_{\alpha\alpha}$.

The energy minimum of the $\alpha\alpha\bar{K}$ system is obtained at α-cluster with $\hbar\omega = 30$ MeV which is much shrunk from the alpha particle with $\hbar\omega = 21.6$ MeV. The behavior of the relative motion is shown in Fig. 3. When a kaon is added to the Be nucleus, it increases strikingly the binding energy to be 113 MeV which is well below the $\Sigma + \pi$ threshold. The width is 38 MeV decaying to the $\Lambda + \pi$ channel. Most interesting is the structure change of the system: The kaon attracts all nucleons within its interaction range and build a high-density and low-temperature system of $(0s)^4(0p)^4$ nucleon configuration, changing drastically the structure from loose "cluster" to compact "shell" as is indicated in Fig. 3. The central nucleon-density of the system attains to almost three times the one of the alpha particle, *i.e.* ~ 4 times the normal density ρ_0 [25]. The formation of deeply-bound $\bar{K}$ nuclear states in Be and other nuclei would provide a means to obtain high-density nuclear matter at "ultra-low temperatures" and a new way to investigate properties of hadrons in cold high-density nuclear medium. It would be also an interesting problem to consider how such high-density matter behaves after the kaon has died as $\bar{K}N \to \Lambda\pi$ or $\bar{K}NN \to \Lambda N$.

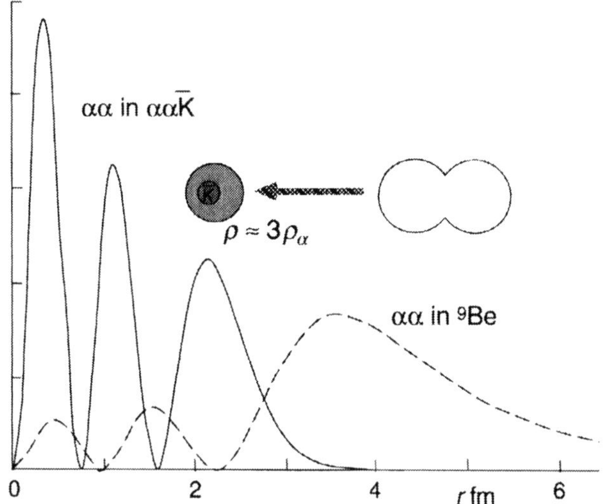

FIGURE 3. Density distribution $r^2\rho(r)$ of αα relative motion in the αα$\bar{K}$ system, which is compared to that in ^{9}Be. The kaon behaves as a "contractor" and makes the central nucleon-density about three times as high as that of the alpha particle.

SUMMARY

Nuclear $\bar{K}$ bound states with narrow widths are predicted to appear in ^{3}He and ^{4}He. They are the ground states of $K^-+{}^3$He and $K^-+{}^4$He with binding energies of 108 MeV and 86 MeV and widths of 20 MeV and 34 MeV, respectively. In the state the core nucleus is largely compressed due to the strong binding of $\bar{K}$ which accommodates a nucleus of higher density.

An experimental method is proposed to populate the ${}^3_{\bar{K}}H(T=0)$ state, by measuring a discrete neutron component of about 70 MeV from the ^{4}He(stopped K^-, n) reaction. The branching ratio is estimated to be about 2 % of the total absorption width, which makes the experiment feasible. An experimental proposal to search for the narrow peak in neutron spectra has been made by Iwasaki *et al.* [28], which takes into account possible neutron background from Σ-, Λ-decays and other processes. Observation of such deeply-bound $\bar{K}$ nuclear states would provide information of fundamental importance about the properties of hadrons in nuclear medium in relation to strangeness condensation.

ACKNOWLEDGMENTS

It is my (Y.A.) pleasure to thank the organizers for providing a very pleasant and memorable symposium dedicated to Professor Il-Tong Cheon. We (Y.A. and T.Y.) would like to take this opportunity to express our sincere gratitude to Il-Tong san for his warm friendship and partnership on many occasions of Korea-Japan physics collaborations.

REFERENCES

1. G.E. Brown, C.H. Lee, M. Rho and V. Thorsson, Nucl. Phys. **A567** (1994) 937: G.E. Brown, Nucl. Phys. **A574** (1994) 217c.
2. T. Waas, N. Kaiser and W. Weise, Phys. Lett. **B365** (1996) 12; **B379** (1996) 34.
3. T. Yamazaki *et al.*, Z. Phys. **A355** (1996) 219.
4. T. Yamazaki *et al.*, Phys. Lett. **B418** (1998) 246.
5. M. Iwasaki *et al.*, Phys. Rev. Lett. **78** (1997) 3067.
6. T.M. Itoh *et al.*, Phys. Rev. **C58** (1998) 2366.
7. E. Friedman, A. Gal and C.J. Batty, Phys. Lett. **B308** (1993) 6; Nucl. Phys. **A579** (1994) 518.
8. C.J. Batty, E. Friedman and A. Gal, Phys. Rep. **287** (1997) 385.
9. A.D. Martin, Nucl. Phys. **B179** (1981) 33.
10. Y. Akaishi, H. Bando, A. Kuriyama and S. Nagata, Prog. Theor. Phys. **40** (1968) 288.
11. J. Dabrowski and H.S. Köhler, Phys. Rev. **136** (1964) B162.
12. J. Dabrowski and J. Rozynek, to appear in Prog. Theor. Phys.
13. Y. Akaishi and S. Nagata, Prog. Theor. Phys. **48** (1972) 133.
14. M. Alberg, E.M. Henley and L. Wilets, Ann. Phys. **96** (1976) 43.
15. R. Brockmann, W. Weise and L. Tauscher, Nucl. Phys. **A308** (1978) 365.
16. M. Mizoguchi, S. Hirenzaki and H. Toki, Nucl. Phys. **A567** (1994) 893.
17. V. Koch, Phys. Lett. **B337** (1994) 7.
18. A. Ohnishi, Y. Nara and V. Koch, Phys. Rev. **C56** (1997) 2767.
19. T. Kishimoto, Phys. Rev. Lett. **83** (1999) 4701.
20. A. Hasegawa and S. Nagata, Prog. Theor. Phys. **45** (1971) 1786.
21. R. Staronski and S. Wycech, Czech. J. Phys. **B36** (1986) 903.
22. S. Wycech, Nucl. Phys. **A450** (1986) 399c.
23. S. Baird *et al.*, Nucl. Phys. **A392** (1983) 297.
24. A. Gal, Friedman and C.J. Batty, Nucl. Phys. **A606** (1996) 283.
25. Y. Akaishi, Int. Rev. Nucl. Phys. **4** (1986) 259.
26. S. Saito, Prog. Theor. Phys. **40** (1968) 893.
27. H. Furutani *et al.*, Prog. Theor. Phys. Suppl. **68** (1980) 193.
28. M. Iwasaki, K. Itahashi, A. Miyajima, H. Outa, Y. Akaishi and T. Yamazaki, Nucl. Inst. Methods **A** (2001).

Roles of Σ in Weak and Electromagnetic Interactions of Hypernuclei

M. Oka*, K. Saito*, K. Sasaki* and T. Inoue†

*Department of Physics, Tokyo Institute of Technology, Meguro, Tokyo 152-8551 Japan
†Dept de Física Teòrica, IFIC, Universitat de València-CSIC, 46100 Burjassot (València) Spain

Abstract. Roles of virtual mixings of the Σ hyperon in hypernuclear structure and interactions are studied in the magnetic moments and transitions as well as nonmesonic weak decays of hypernuclei. It was found that effects of the mixing are significantly large so that their information is valuable in the study of the hyperon nucleon interaction.

INTRODUCTION

The Σ is an isospin excited state of the Λ, and has a mass 75–80 MeV heavier than the Λ. The $\pi\Lambda\Sigma$ and $K\Lambda N$ couplings are both strong so that the virtual mixing of Σ to hypernuclear states is expected to be important. This is similar to the roles of the Δ resonance in nuclei, which is a spin-isospin excited state of the nucleon. The comparison of these two excitations may be instructive. The Σ has advantages: the small excitation energy, about one-fourth of $N-\Delta$, and the smaller width. (Although the free Σ decays only through the weak decay, the Σ in nuclei decays via $\Sigma N \to \Lambda N$ process. Therefore the Σ width is expected to be of order 10 MeV. The Δ emits the pion in the decay and its width is about 120 MeV.) While the Δ may affect the spin and isospin properties of nuclei, the Σ mixing will be sensitive to the isospin structure without spin flip. For example, the exchange currents for the weak and electromagnetic (EM) vertices are known to involve baryon excitations. Thus the Σ mixing may give significant effects on the isovector properties of hypernuclei in general, such as the isovector electric and magnetic transitions, charge symmetry breaking, isospin changing weak decays and so on. It should also be pointed out that the virtual mixing of Σ is the main source of three-body interactions of $\Lambda-N-N$, as the Δ is responsible for the nuclear three-body force.

Recent experimental efforts in hypernuclear structure and decays revealed that few baryon systems are suitable in studying the details of hyperon-nucleon interactions and mechanisms and properties of EM and weak transitions. For the lightest hypernucleus, ${}^{3}_{\Lambda}$H, it was pointed out by a full three-body calculation[1] that the three-body force induced by the mixing of the virtual Σ is critical in reproducing the tiny binding energy of the system.

The Σ mixing may also be strong in $A=4$ hypernuclear system, ${}^{4}_{\Lambda}\mathrm{He}-{}^{4}_{\Lambda}\mathrm{H}$. The importance of the mixing was stressed in explaining the large charge symmetry breaking effects in $A=4$[2]. Recently Akaishi *et al.* pointed out that the overbinding problem of the s-shell hypernuclei may be solved by considering a large Σ mixing in $A=4$

CP594, *Hadrons and Nuclei: First International Symposium*, edited by Il-T. Cheon et al.

hypernuclei[3]. A coupled channel few-body calculation by Hiyama *et al.* has indeed shown a significant mixing of Σ in ${}^4_\Lambda$He[4].

Microscopic dynamics of the Σ mixing is not completely fixed. The hyperon-nucleon interactions are not very well determined from the direct scattering experiment, whose data are very scarce. Various theoretical models of the hyperon-nucleon interactions have been proposed. The pion and kaon exchanges are main candidates of the long range interactions, while the quark exchange between the baryons is expected to dominate the short-distance interactions. The latter predicts strong spin-isospin channel dependence of the $L = 0$ $\Sigma - N$ interactions. For instance, the $I = 3/2$, ${}^3S_1$ $\Sigma - N$ state has strong repulsion at short distances, the origin of which is the Pauli exclusion principle among the valence quarks[5]. It was also pointed out that the $L = 1$ $\Lambda N - \Sigma N$ $(I = 1/2)$ system feels a strong antisymmetric spin-orbit force (ALS)[6]. Such a strong ALS force is able to explain the observed small spin-orbit force for the Λ in nuclei. A recent paper by Takeuchi has demonstrated the importance of the Σ mixing in explaining the cancellation of the symmetric and antisymmetric spin-orbit forces in the YN two-body system[7].

So far the study of the Σ mixing is focused mainly on the binding energies and the level structures of hypernuclei. The mixing of Σ, however, will be most explicit not in the energy spectrum but in the electromagnetic and weak properties of hypernuclei. The transitions are also sensitive to the phase of the mixing, which will in turn help to pin down the mixing mechanism. Several attempts to reveal the roles of the Σ mixings have been recently carried out.

(1) The nonmesonic weak decay of Λ in hypernuclei. The Σ intermediate state is considered in the nonmesonic weak decay of ${}^4_\Lambda$He, and ${}^4_\Lambda$H. Especially, the coherent mixing of Σ^+ will cause a special $\Sigma^+ p \to pp$ decay mode, that may be a good signal of the Σ mixing[8].

(2) The weak decay emitting π^+ was considered as a signal of Σ^+ mixing in ${}^4_\Lambda$He. We examine the role of the Σ in the π^+ decay[9].

(3) The magnetic properties of the Λ hypernuclei are considered in the context of the Σ mixing. We relate the Σ mixing with the general isovector exchange current for the magnetic moment and transitions. We compare the exchange current calculation with the virtual Σ contribution[10].

The second subject gives a negative result that the Σ is not a main provider of π^+ as the S-wave $\Sigma \to n\pi^+$ decay is strongly hindered by chiral symmetry. We skip this subject in this note and refer the readers to another publication[9]. We start with (3) and then discuss (1) in this note.

EXCHANGE CURRENTS FOR MAGNETIC MOMENTS AND TRANSITIONS

The magnetic moments of hypernuclei have not been measured yet, while the recent hyperball experiments succeeded in achieving gamma ray spectroscopy of hypernuclei and observed M1 transitions though the M1 transition rates are not determined yet[11]. The magnetic properties of hypernuclei are quite suitable for the study of possible medium effects on the hyperon. Some time ago, it was pointed out that the Pauli exclusion principle on the valence quarks may deform Λ in the nucleus, and thus the magnetic moments of Λ will be affected[12]. The quark cluster model calculation, however, shows that the effect seems not very large[13]. None is confirmed. Yet the idea of modified Λ in nuclear medium is extremely interesting to pursue further.

Dover *et al.* pointed out that the magnetic moments of the hypernuclei will be a good probe of the virtual Σ^0 mixing[15]. The virtual Σ^0 emits an isovector M1 photon in $\Sigma^0 \to \Lambda\gamma$, and thus contributes to the isovector M1 transition (Fig. 1). They estimated the effects for the magnetic moments of ${}^4_\Lambda$He (1^+) state, and ${}^{15}_\Lambda$C, both of which have nonzero isospin and therefore the Σ can be mixed coherently (see below).

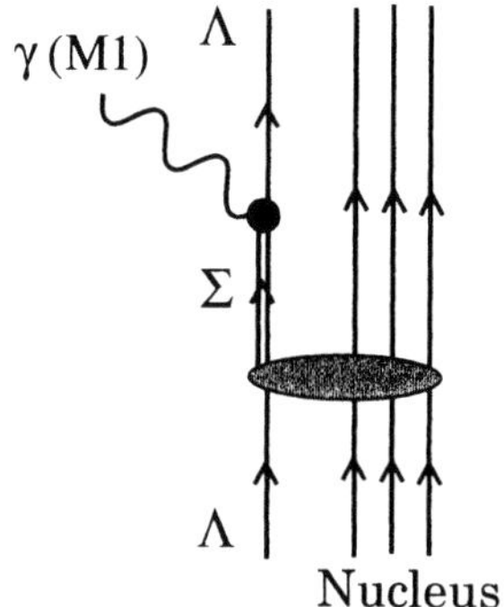

FIGURE 1. The Σ virtual mixing for the magnetic moments and transitions in hypernuclei.

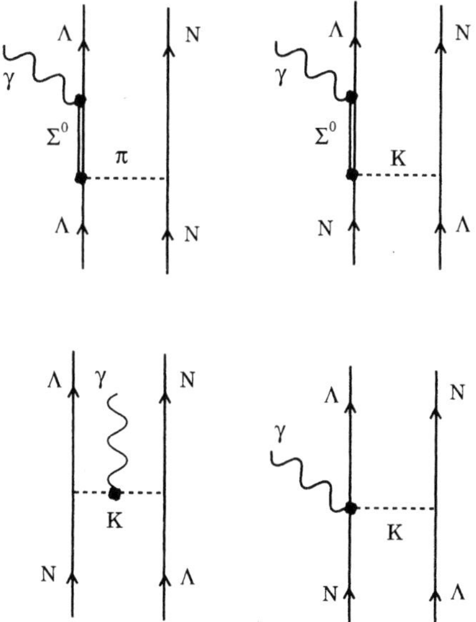

FIGURE 2. The exchange current diagrams for magnetic moments and transitions of hypernuclei.

In order to confirm their results and further to study other possibilities, we have calculated the Σ mixing effects as a part of the exchange current contribution to the magnetic operator[10]. We find that the pion and kaon exchange currents contain the Σ excitation contribution as well as other diagrams, which include both isoscalar and isovector components. Note that the Σ excitation is purely isovector.

We have considered the pion and kaon exchange currents diagrams shown in Fig. 2. We employ the effective Lagrangian approach for the derivation of the exchange current operators, using the phenomenological meson-baryon coupling constants with the minimal coupling to the magnetic field. The $\Sigma\Lambda\gamma$ vertex is given also by the tensor coupling with its strength obtained from the Σ^0 decay rate. We also include the isovector *NN* exchange current contribution, according to Hyuga et al.[14]

TABLE 1. Magnetic moments and $M1$ transitions of $s-p$ shell hypernuclei

		s.p.	Born (K)	ΛN Σ^0 (K)	Σ^0 (π)	NN
${}^4_\Lambda\mathrm{He}(1^+, \frac{1}{2})$	μ(I.S.)	-0.173	0.031	—	—	—
	μ(I.V.)	-2.353	0.031	0.053	-0.035	-0.286
	μ(sum)	-2.526	0.063	0.053	-0.035	-0.286
$(1^+,\frac{1}{2}) \to (0^+,\frac{1}{2})$	B(M1)	0.635	0.031	0.013	0.052	0.000
${}^5_\Lambda\mathrm{He}(\frac{1}{2}^+, 0)$	μ(I.S.)	-0.613	0.063	—	—	—
	μ(sum)	-0.613	0.063	—	—	—
${}^7_\Lambda\mathrm{Li}(\frac{1}{2}^+, 0)$	μ(I.S.)	0.778	-0.031	—	—	—
	μ(sum)	0.778	-0.031	—	—	—
$(3/2^+,0) \to (1/2^+,0)$	B(M1)	-0.584	0.043	—	—	—
$(1/2^+,1) \to (3/2^+,0)$	B(M1)	-3.235	0.006	0.009	0.000	0.185
$(1/2^+,1) \to (1/2^+,0)$	B(M1)	2.332	-0.005	-0.006	0.083	-0.131
${}^{15}_\Lambda\mathrm{C}(\frac{1}{2}^+, 1)$	μ(I.S.)	-0.619	0.071	—	—	—
	μ(I.V.)	-0.084	-0.019	-0.022	-0.001	-0.001
	μ(sum)	-0.703	0.052	-0.022	-0.001	-0.001

TABLE 2. Comparison between the virtual Σ mixing and the Σ^0-excitation(K and π) currents

	$R_\mu : {}^4_\Lambda\mathrm{He}(1^+)$	$R_B : {}^4_\Lambda\mathrm{He}(1^+ \to 0^+)$	$R_\mu : {}^6_\Lambda\mathrm{He}(1^-)$	$R_\mu : {}^{15}_\Lambda\mathrm{C}(1/2^+)$
Virtual Σ mixing	$0.74\beta_1$	$\{1-0.61(\beta_0+\beta_1)\}^2$	1.6β	-3.2β
Σ^0 exchange currents	-0.007	1.216	-0.019	0.033
β	$\beta_1 = -0.01$	$\beta_0 = -0.16, \beta_1 = -0.01$	$\beta = -0.01$	$\beta = -0.01$

We have calculated both the magnetic moments and the $M1$ transitions for $A \leq 17$ hypernuclei. The exchange currents and the single particle magnetic moment operators are evaluated in the s, p shell model of the harmonic oscillator basis. We use the YNG and Cohen-Kurath effective interactions. Details of the calculation is given in [10].

The results are summarized in Table 1. For the isoscalar magnetic moments and transitions, there exists no contribution from the pion exchange. The kaon Born terms, the kaonic and seagull diagrams, suppresses the single Λ magnetic moment by about -10%, while the $M1$ transition of $^{7}_{\Lambda}$Li is reduced about $7-8\%$.

The exchange current contributions to the magnetic moment of $I=1$ hypernuclei are found to be small. It is so especially for $^{4}_{\Lambda}\mathrm{He}(1^{+})$ state because of the cancellation between the Σ^{0} excitations by K and π exchanges. It is, however, found that the $M1$ transition rate, $M1(1^{+}\to 0^{+})$, has large Σ^{0} contribution. In Table 2, we compare our results with those of Dover et al.[15] The comparison shows that they are consistent, if we assume that the Σ mixing is strong in 0^{+}, but it is negligible in the 1^{+} state of $^{4}_{\Lambda}$He. Similarly, the Σ contribution to the magnetic moment of $^{15}_{\Lambda}$C is suppressed.

Thus we conclude that the $M1$ transition rate of $^{4}_{\Lambda}$He must be a good indicator of the virtual Σ mixing.

WEAK DECAYS OF HYPERNUCLEI

Nonmesonic weak decay of hypernuclei is a unique weak interaction process, which involves only baryons. The main decay channel is a two-body transition, $\Lambda N\to NN$, and does not emit pions. It is considered as a weak baryon-baryon interaction, which is analogous to the parity violating part of the nuclear force.

The microscopic picture of non-mesonic decay of hypernuclei is a long standing problem, as the traditional calculation clearly disagrees with experiment. The typical problem is the Γ_n/Γ_p puzzle, i.e., disagreement of the ratio of $\Gamma_n=\Gamma(\Lambda n\to nn)$ and $\Gamma_p=\Gamma(\Lambda p\to pn)$ transitions. A traditional calculation, which assumes one-pion-exchange (OPE) mechanism, is unable to explain the ratio at all. The weak OPE has a similar structure as the strong one, in both of which the tensor transition is strong. As it enhances Γ_p, the Γ_n/Γ_p ratio is suppressed. Typical prediction of OPE is $\Gamma_n/\Gamma_p\sim 0.1$, while most experimental data indicate $\Gamma_n/\Gamma_p\sim 1$ or larger.

It was then realized, however, the momentum transfer in this process is large due to the $\Lambda-N$ mass difference, and therefore the short range interaction must be important. The lifetime measurements of heavy hypernuclei observed saturation at large A and therefore suggest importance of short range interactions[16].

Recently we proposed to treat the short range part using the valence quark picture of the baryon and the effective four quark weak hamiltonian[17, 18]. We found that our model, called the direct quark (DQ) process, gives significantly large contribution and shows qualitatively different features from the meson exchange mechanism, especially in its isospin structure.

We employ a quark-meson hybrid model, which consists of the DQ transition potential supplemented by the long-range part that comes from one-pion (OPE) and one-kaon (OKE) exchanges. We showed that the model enhances Γ_n significantly, giving fairly large Γ_n/Γ_p[20, 19, 8]. We apply the model to the weak decays of $^{4}_{\Lambda}$He, $^{4}_{\Lambda}$H, and $^{5}_{\Lambda}$He.

Recently, it is advocated that the Σ-mixing is crucial in solving the overbinding problem of the s-shell hypernuclei[3]. Namely, the coherent Σ mixing, which is important in $A=4$ hypernuclei, gives enough attraction for the $A=4$ binding energy even if we take weaker central attraction that is preferable for the smaller binding of $^{5}_{\Lambda}$He. A sophisticated four-body calculation of $A=4$ hypernuclear structure also indicates significant mixing of virtual Σ of 1-2% level and thus supports the above idea[4].

If the mixing probability of the virtual Σ is 1%, the mixing amplitude β is $|\beta| \sim 0.1$. Although its effects on the binding energy are of order $|\beta|^2$ (perturbatively), those on the transition amplitude is proportional to $|\beta|$. The latter is also sensitive to the phase of the mixing and therefore to the mixing mechanism. Thus we here consider the coherent Σ mixing in non-mesonic decays of the $A = 4$ hypernuclei.

The diagrams shown in fig.3 are two types of nonmesonic weak decays of the virtual Σ in nuclei. A previous study [21] considered two-pion exchange process between Λ and N, one of which induces weak transition, (fig.3(a)). The intermediate Σ-N state is restricted to $I = 1/2$. It is, however, possible that the virtual Σ decays with the assistance of a second nucleon, (fig.3(b)). This "three body" type process is taken into account by considering the coherent Σ mixing. They are important in two reasons.

(1) It involves the weak interaction of the Σ-N ($I = 3/2$) states, which does not contribute in fig.3(a).
(2) For a hypernucleus with $I = 0$, such as ${}^5_\Lambda$He, the fig.3(a) and 3(b) interfere destructively as the Σ excitation by the strong interaction is suppressed by the isospin conservation. Indeed the ground state of ${}^5_\Sigma$He has $I = 1$ and can not mix with ${}^5_\Lambda$He.

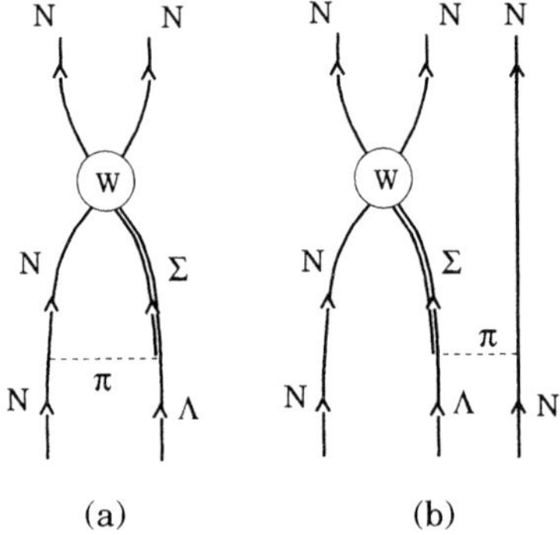

FIGURE 3. Diagrammatic representation of two and three body type Σ mixing for the $\Lambda N \to NN$ transition.

A perturbative estimate of the mixing amplitude β is given by

$$\beta(0^+) = -\frac{\langle \Sigma + {}^3\mathrm{He}|V|\Lambda + {}^3\mathrm{He}\rangle}{M_\Sigma - M_\Lambda}. \tag{1}$$

Using the transition potential given as the D2 potential of the paper [3], we obtain

$$\beta(0^+) \sim 0.08 \tag{2}$$

This is consistent with recent more sophisticated calculations[4].

In Table 3, we show the decay rates of A=4 hypernuclei for several values of β. One sees that the mixing changes the total decay rate by about 20% for $\beta^2 \sim 1\%$. The sign of β determined above is positive and the positive β is found to reduce the proton induced decay rate, which makes the agreement better. On the other hand, we find that the mixing does not change the qualitative behaviors of the results, such as Γ_n/Γ_p ratio and the proton asymmetry. It is noted that the sign of β taken here is different from that given in Table 2. The β in Table 2 is determined so that the exchange current contribution agrees with that from the Σ mixing. Although the exchange currents contain not only coherent mixing but also all higher excited states, this discrepancy is a little puzzling.

An advantage of the $A = 4$ and 5 hypernuclei regarding the nonmesonic weak decay is that their selectiveness of the two-body channels. A new interesting decay channel is available when we consider the Σ^+ mixing in ${}^4_\Lambda$He. The system consists of a virtual Σ^+, p and two n. When Σ^+ meets the proton, it decays into two p, *i.e.*, $\Sigma^+ p \to pp$ decay. The calculated decay rate for $\Sigma^+ p \to pp$ is

$$\Gamma_{pp} = 0.009\Gamma_\Lambda \tag{3}$$

at $\beta = 0.1$, Although the branching ratio is tiny, it gives a clean signal as a back to back $p - p$ in the final state. This will be a direct evidence of virtual Σ mixing in Λ hypernuclei.

TABLE 3. Nonmesonic decay observables of ${}^4_\Lambda$He. $\Gamma_{\rm tot}$, Γ_p and Γ_n are the total, proton-induced and neutron-induced decay rates, respectively, in units of Γ_Λ. PV/PC is the ratio of the parity violating decay rate to the parity conserving one. α is the proton asymmetry parameter.

${}^4_\Lambda$He	$\Gamma_{\rm tot}$	Γ_p	Γ_n	Γ_n/Γ_p	PV/PC	α
π	0.272	0.250	0.022	0.089	0.353	−0.417
π+K	0.155	0.145	0.009	0.064	0.146	−0.357
DQ	0.032	0.021	0.011	0.516	2.093	−0.373
π+K+DQ	0.218	0.214	0.004	0.019	2.321	−0.679
$\beta = -0.1$	0.261	0.256	0.005	0.021	—	−0.679
$\beta = +0.1$	0.178	0.175	0.003	0.017	—	−0.656
EXP [22]	0.19±0.04	0.15±0.02	0.04±0.02	0.27±0.14	—	—

CONCLUSION

The virtual mixing of Σ hyperon in hypernuclei is found to be significant in both the magnetic properties and weak decays of hypernuclei. One may question whether the virtual Σ has the same interaction as the on-shell Σ. It is, however, noted that the mass difference of Σ and Λ is only 80 MeV and the virtuality is not too large compared with the binding energy. It is certainly much closer to the on-mass-shell than the Δ isobar in the nucleus.

We pointed out that the coherent mixing of Σ of order 1% is significant to change the magnetic moments and transitions, as well as the nonmesonic decay rates of hypernuclei. Both of which are being measured to the accuracy level of 10% in near future. It is also stressed that the transition data are sensitive to the phase of the mixed Σ state and therefore have more information on the mixing dynamics, such as the transition potential.

We note that the $\Sigma^+ p \to pp$ decay of hypernuclei, which is special for Σ mixed state and therefore a good signature of Σ participation. This makes a good future experimental plan.

REFERENCES

1. K. Miyagawa, H. Kamada, W. Glöckle, V. Stoks, Phys. Rev. **C51** (1995) 2905.
2. B.F. Gibson, E.V. Hungerford, Phys. Rep. **C257** (1995) 349.
3. Y. Akaishi, T. Harada, S. Shinmura, and K.S. Myint, Phys. Rev. Lett. **84** (2000) 3539.
4. E. Hiyama, M. Kamimura, T. Motoba, T. Yamada, Y. Yamamoto, Nucl. Phys. **A670** (2000) 273.
5. M. Oka, K. Shimizu. K. Yazaki, Nucl. Phys. **A464** (1987) 700; K. Shimizu, S. takeuchi, A.J. Buchmann, Prog. Theor. Phys. Supple. **S137** (2000) 43.
6. S. Takeuchi, O. Morimatsu, Y. Tani, M. Oka, Prog. Theor. Phys. Supple. **S137** (2000) 83.
7. S. Takeuchi, preprint, hep-ph/0008185.
8. K. Sasaki, T. Inoue, and M. Oka, in preparation.
9. M. Oka, Nucl. Phys. **A647** (1999) 97.
10. K. Saito, M. Oka, T. Suzuki, Nucl. Phys. **A625** (1997) 95.
11. H. Tamura, et al., Few Body Systems Supple. **12** (2000) 342.
12. Hungerford, Biedenharn (1984), T. Yamazaki, Nucl. Phys. **A446** (1985) 467c.
13. S. Takeuchi, K. Shimizu, Phys. Lett, **B179** (1986) 197, and unpublished.
14. H. Hyuga, A. Arima, K. Shimizu, Nucl. Phys.**A336** (1980) 363.
15. C. Dover, H. Feshbach, A. Gal, Phys. Rev. **C51** (1995) 541.
16. H. Bhang, et al., Phys. Rev. Lett. **81** (1998) 4321.
17. T. Inoue, S. Takeuchi, and M. Oka, Nucl. Phys. **A577** (1994) 281c; *ibid* **A597** (1996) 563.
18. K. Maltman and Shmatikov, Phys. Lett. **B331** (1994) 1.
19. K. Sasaki, T. Inoue, and M. Oka, Nucl. Phys. **A669** (2000) 331; *ibid* **A678** (2000) 455 (E).
20. T. Inoue, M. Oka, T. Motoba, and K. Itonaga, Nucl. Phys. **A633** (1998) 312.
21. H. Bando, Y. Shono, H. Takaki, Int. J. Mod. Phys. A3 (1988) 1581.
22. V.J. Zeps, E788 Collaboration, Nucl. Phys. **A639** (1998) 261-268.

Non-Mesonic Weak Decay of Λ-Hypernuclei and the Final State Interaction

H. Bhang, M.J. Kim and J.H. Kim

School of Physics, Seoul National University, Seoul 151-747, Korea

Abstract. In spite of the recent progresses achieved at BNL and KEK, a consistent understanding of the non-mesonic weak decay is yet to be achieved. The rapid saturation of total decay rate in its mass dependence, Γ_n/Γ_p ratio puzzle, opposite sign of decay asymmetry parameter between ${}^{5}_{\Lambda}$He and ${}^{12}_{\Lambda}$C, and ΔI=1/2 rule of non-mesonic weak decay are the current concerns, but not understood yet. The difficulties are largely in the treatment of the final state interaction. In the paper, we discuss on the role of final state interaction in the interpretation of the newly measured decay particle spectra in the non-mesonic weak decay of Λ hypernuclei.

INTRODUCTION

The Λ hyperon inside a nucleus eventually decays through either a mesonic ($\Lambda \rightarrow p\pi^-$; Γ_{π^-}, $n\pi^0$; Γ_{π^0}) or a non-mesonic weak decay (Λ+p$\rightarrow$n+p; Γ_p, Λ+n$\rightarrow$n+n; Γ_n). Both modes are of comparable strength in light Λ hypernuclei, but the non-mesonic weak decay(NMWD) dominates in heavy($A \geq 12$) Λ hypernuclei. In NMWD, a Λ decays mostly through interaction with a neighbor nucleon by emitting two energetic nucleons of about 400MeV/c via a strangeness changing weak two-baryonic process. This non-mesonic weak decay process provides a unique opportunity for the study of the strangeness changing baryon-baryon weak interaction, since the process is very difficult to realize in the scattering state.

In this paper we discuss on the recent results of the experimental studies on NMWD, though some of them are preliminary yet, and on the role of the final state interaction (FSI) in their interpretation. The data discussed are mainly from the experiments, KEK-PS E307 and E369.

In the experiment E307, the lifetimes, namely the total decay widths, and the decay particle spectra of medium heavy Λ hypernuclei, ${}^{12}_{\Lambda}$C, ${}^{28}_{\Lambda}$Si and ${}^{56}_{\Lambda}$Fe were measured. Their total decay rates, essentially Γ_{nm}, were determined with uncertainties of 6$\sim$8 percent and found to saturate to a value about 25 percent increased from that of the free Λ hyperon [1].

In the experiment E369 the decay neutron spectra have beem measured from the weak decay of ${}^{12}_{\Lambda}$C and ${}^{89}_{\Lambda}$Y. The threshold energy was $\sim$10 MeV much lower than that of proton measurements. With such a low threshold energy, the spectra contained essentially all the effects of the final state interaction and would provide valuable information on the final state interactions on the emitted nucleons in the NMWD process.

So far the partial decay rates of the NMWD, the neutron induced- and the proton induced-decay, have not been determined accurately yet. Especially the relative ratio of

CP594, *Hadrons and Nuclei: First International Symposium,* edited by Il-T. Cheon et al.

the two has been a long standing puzzle in that the experimental ratios including the most recent KEK data [2] indicate the dominance of the neutron induced channel while theoretical predictions, so far mostly based on the boson exchange model, show the predominance of the proton induced channel.

However, the uncertainties of the current experimental data for the partial decay rates are still so large that it fails to give a clear guide for the theoretical models. One main source of the large uncertainties in the experimental data is due to the difficulties in the treatment of the final state interaction (FSI).

Although the recent intranuclear cascade (INC) calculation for the FSI has shown some success to reproduce the emitted proton spectra from NMWD of ${}^{12}_{\Lambda}C$ with $\Gamma_n/\Gamma_p \sim 1.24$, it seems to have a difficulty for its neutron counterpart [3, 4]. In order to study the FSI effect on NMWD systematically, we formulated an INC calculation adopting experimental nucleon-nucleon cross sections and more explicit energy balance through the INC process. The present INC calculation has reproduced the spectral shape of the emitted nucleons better and indicated a consistent Γ_n/Γ_p value somewhat reduced than that of the previous INC calculation.

RECENT EXPERIMENTAL PROGRESS ON NMWD

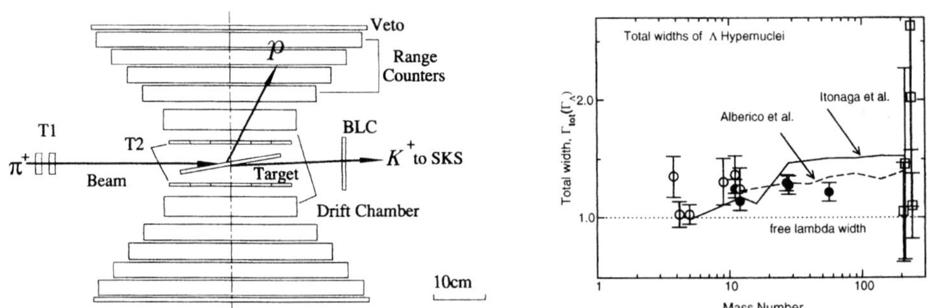

FIGURE 1. The schematic view of the decay particle counter system in KEK-PS E307 experiment is shown in the left figure. The energy of the charged particles are measured via ΔE(T2)-E telescope method. In the right figure, mass dependence of total decay width of Λ hypernuclei is shown. Open symbols are the previous data and closed circles those from KEK E307 experiment.

In the E307 experiment the outgoing decay particles, protons and π^-, in the decay of ${}^{12}_{\Lambda}C$, ${}^{28}_{\Lambda}Si$ and ${}^{56}_{\Lambda}Fe$ were measured with two coincidence decay counters placed symmetrically above and below the target as shown in the Fig. 1. Each counter was made up of a timing counter, a drift chamber, and a range counter. The lifetimes, i.e. total widths, were obtained by the direct measurement of the production and decay time of the Λ hypernucleus with fast plastic scintillators. Details of the experiment can be found in the references [5, 6].

Fig. 1 shows the recent results (dark circles) on the total widths of medium heavy Λ hypernuclei along with previous determinations [1, 6]. The results of ${}^{27}_{\Lambda}Al$, ${}^{28}_{\Lambda}Si$ and

${}^{56}_{\Lambda}$Fe show that the total widths of Λ hypernuclei do not vary much in the mass region beyond carbon, saturating at a value $\sim 25\%$ increased from that of free Λ. Along with the data, the calculations based on the boson exchange models are shown [7, 8].

The partial decay widths, Γ_n and Γ_p, have been a central concern in the study of the non-mesonic weak decay of Λ hypernuclei. Especially the large discrepancies in the ratio, Γ_n/Γ_p, between the experimental data and the theoretical calculations have been a longstanding puzzle, even if the extensive studies of NMWD have been made both experimentally and theoretically. The experimental ratios, from those of the old emulsion data to those of the recent counter experiments [2, 9], have been greater than or close to unity while the theoretical calculations based on boson exchange models imposing $\Delta I=1/2$ rule predict much smaller $\Gamma_n/\Gamma_p(\ll 1)$ ratios [7, 10]. There have also been efforts to construct models of hybrid types explicitly combining a short-range model of the ΛN weak interaction with a meson exchange process for the longer range interaction [11, 12]. None of these theoretical efforts has succeeded in explaining the experimental total width and Γ_n/Γ_p ratio together. However, the uncertainties of the experimental ratios are large and we need to take the situation with a grain of salt.

In the experiment E307, the spectral shapes and absolute yields of protons emitted in the non-mesonic weak decay of Λ hypernuclei were measured after gating the inclusive hypernuclear mass spectrum of the bound regions. The energy spectra of protons for the ${}^{12}_{\Lambda}$C and ${}^{28}_{\Lambda}$Si are shown in Fig. 2 and compared to the calculations of Ramos et al. for three Γ_n/Γ_p ratios, 0.1, 1.0 and 2.0 [3, 13]. The calculated energy spectra were obtained applying Monte-Carlo INC calculation for their final state interaction with the residual nucleus and taking into account the energy loss in the target and the detector materials (T2 in the Fig. 1) on the detected protons. The solid lines represent those for the fit values of Γ_n/Γ_p whose preliminary values are 1.24 ± 0.5 and 1.5 ± 0.6 for ${}^{12}_{\Lambda}$C and ${}^{28}_{\Lambda}$Si, respectively. Though the spectra above 50-60 MeV are reasonably reproduced, the INC failed to predict large yield at the energies below 50-60 MeV. The INC spectra show broad bumps around 50-60 MeV whose position exactly correspond to the energy of the produced nucleons right after the weak $\Lambda N \to NN$ interaction, but before the propagation inside the residual nucleus. The figures show that the INC spectra maintain the original peak position as broad bumps without energy shifts while the experimental ones do not show such bumps with monotonically increasing yield toward the low energy side.

One of the main difficulties in the investigation of the NMWD is the large uncertainties in the experimental data. They are mainly due to two facts: The first is that the neutron decay widths, Γ_n, were determined indirectly in most of the cases. The second is that the low energy section of the proton spectrum, which is not easily detectable due to the large detection threshold energy, constitutes a significant portion of the total proton yield and has to be estimated by relying on some model calculations of the final state interaction on the emitted protons. In a direct neutron measurement one can observe a much lower threshold energy so that one can detect almost the entire spectrum.

The neutron energy spectra in the NMWD of ${}^{12}_{\Lambda}$C and ${}^{89}_{\Lambda}$Y were measured in the KEK-PS E369. The schematics of the setup is shown in the Fig. 3 and more details can be found elsewhere [4]. The energies of the decay product particles were measured by arrays of TOF fast scintillator counters, T0, T1 and T2. The signal to background ratio was ~ 0.02, greatly improved from that of the previous experiment in which the ratio was ~ 1. The neutron energy spectra were obtained down to a kinetic energy of 10 MeV

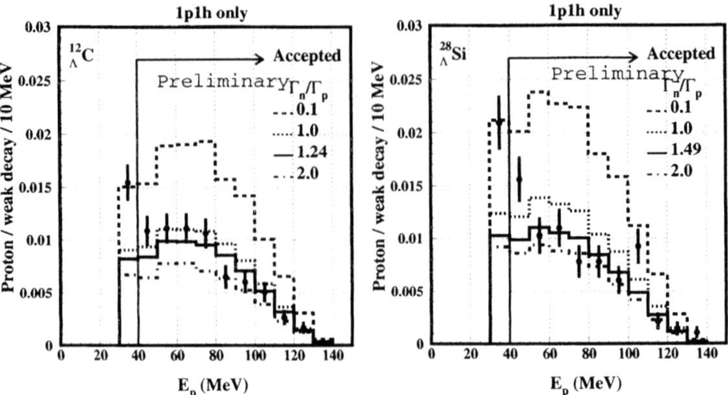

FIGURE 2. The energy spectra of protons in the NMWD of $^{12}_{\Lambda}C$ and $^{28}_{\Lambda}Si$ measured in KEK-PS E307 are shown and compared with the calculated spectra (histograms) of Ramos et al. for four Γ_n/Γ_p ratios.

as shown in Fig. 3. The spectra were compared with the previous Monte-Carlo INC predictions (indicated as 1N- and 2N-induced), and those of SINC (histogram) on which we will explain more in this paper. The previous INC predicted a broad peak around E_n=70-80 MeV while the measured spectrum of carbon did not show such a broad peak, instead showing a rather flat region between 30-90 MeV. The neutrons below 60 MeV are largely underestimated in the previous INC calculation.

The spectrum of the decay neutrons from the NMWD of $^{89}_{\Lambda}Y$ in the Fig. 3b was obtained in the experiment E369 for the first time [4]. A stronger effect of the final state interaction is clearly observed in $^{89}_{\Lambda}Y$ than in $^{12}_{\Lambda}C$. The Γ_n approaches to Γ_{nm} in the NMWD of $^{89}_{\Lambda}Y$. This situation would be relaxed if we included the 2N-induced component of NMWD whose validity has so far not been proven experimentally. An explicit measurement of this NMWD mode and further study on the NMWD of $^{5}_{\Lambda}He$ are now in progress [14].

THE INC CODE (SINC) AND ITS APPLICATIONS

In order to improve the situation and to investigate further the dependence on the treatment of the final state interaction, we have formulated an INC calculation (SINC) which we have tested for the various differential cross-sections of the experimental inelastic and charge exchange reactions. The INC calculations were made for a wide range of target masses and beam energies to study the systematics of the INC calculation with regard to the energy and mass dependences. It turned out that the present INC calculation calibrated with the scattering data reproduced the shape of the neutron spectrum well except for the low energy end and predicted a smaller Γ_n/Γ_p ratio than that of the previous INC calculation.

The angle integrated (total) reaction spectra for the proton inelastic scattering are very well reproduced over a wide dynamic range of emitted particles for all three

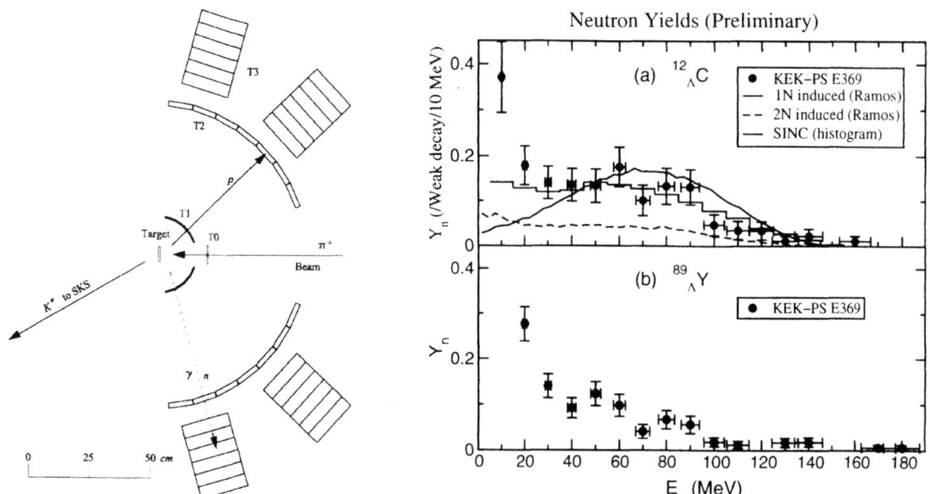

FIGURE 3. The schematic view of the decay particle counter system in KEK-PS E369 experiment is shown in the left side. The energy of the particles are measured via time of flight method. The neutron spectra from (a) ${}^{12}_{\Lambda}$C and (b) ${}^{89}_{\Lambda}$Y are shown in the right side. That of ${}^{12}_{\Lambda}$C is compared with those of INC calculations of Ramos et al and Mijung et al.(SINC).

targets, C, Al and Fe. The INC calculation we have formulated reproduces the mass dependence very well without adjusting parameters. The energy dependence of the reaction cross section has been checked with the charge exchange reactions and has also been reasonably reproduced. Overall the INC calculation reproduces the total reaction cross sections of (p,p') and (p,n) reactions, well over a wide range of beam energy and target mass with fixed parameters, providing a nice ground of its application for various situations in which nuclear reactions are involved.

The main ingredients of the developed INC code are explained in the following.

The nuclear potential: The nuclear potential, $V_\alpha(r)$, is taken as

$$V_\alpha(r) = E_{F,\alpha}(r) + E_s.$$

where $E_{F,\alpha}(r)$ is the Fermi surface energy of a type α nucleon of the at r and E_s the seperation energy.

The nucleon-nucleon cross sections: For the identical nucleon channels, i.e. p-p and n-n collisions, the nuclear scattering cross section was considered identical Differential cross section data for the intranuclear nucleon-nucleon collisions were directly adopted from those of the free nucleon-nucleon scattering experiment. The differential and the total cross sections of the nucleon-nucleon scatterings used for our INC calculation are adopted from the reference [16].

Event generation: A straight line approximation is used for the trajectory of a particle. The momentum distribution of bound nucleons that we adopted is that of a degenerate Fermi gas of nucleons in the spherical potential well $V_\alpha(r)$.

Selection of a collision partner and an Interaction point: An interaction point is generated according to the collision probability, $1 - exp(-s/\lambda_\alpha)$, where the λ_α is a mean free path of the nucleon, α, determined by $\lambda_\alpha = 1/(\rho_N \sigma_{t,\alpha})$. ρ_N is the nucleon density and $\sigma_{t,\alpha}$ either a $\sigma_{t,p}$ or a $\sigma_{t,n}$ meaning the total cross section of the moving nucleon, p or n, in nuclear medium. They are given by

$$\sigma_{t,p} = \frac{1}{A}(Z\sigma_{pp} + N\sigma_{pn}), \quad \sigma_{t,n} = \frac{1}{A}(Z\sigma_{pn} + N\sigma_{pp}),$$

where σ_{pp} and σ_{pn} are the p-p and p-n nucleon-nucleon scattering cross sections. A collision partner at an interaction point is selected according to the probability, $P_{\alpha\beta}$, which is determined by the two factors, the number densities of the proton and neutron and the nucleon-nucleon cross sections, σ_{pp} and σ_{pn}, as followings:

$$P_{pp} = \frac{Z\sigma_{pp}}{Z\sigma_{pp}+N\sigma_{pn}}, \quad P_{pn} = \frac{N\sigma_{pn}}{Z\sigma_{pp}+N\sigma_{pn}}, \quad P_{np} = \frac{Z\sigma_{pn}}{Z\sigma_{pn}+N\sigma_{pp}}, \quad P_{nn} = \frac{N\sigma_{pp}}{Z\sigma_{pn}+N\sigma_{pp}},$$

where A is the mass number, Z proton number and N neutron number.

Fig. 4 (left) shows the dσ/dE spectra for proton inelastic scattering from the ^{12}C, ^{27}Al and ^{56}Fe at the incident proton energy 62 MeV. The histograms represent the INC results and the symbols the experimental data [15]. We have fixed the radius and thickness parameters of the nuclear shape to the values of the global optical model potential parameter set [17], r_0=1.16 fm and a=0.75 fm for all the target nuclei, except for carbon. For C, we used somewhat reduced values, r_0=1.10 fm and a=0.50 fm. The intermediate energy region of the spectra, from 10 MeV to 50 MeV, have been excellently reproduced for all three targets with the INC processes without adjusting parameters.

Fig. 4 (right) show the dσ/dE spectra for (p,n) charge exchange reaction from the ^{90}Zr target at the incident proton energies, 45, 80 and 160 MeV [18, 19, 20]. We have used the same radius and thickness parameters of the nuclear shape as those of the (p,p) scattering, namely r_0=1.16 fm and a=0.75 fm for all the energies. The INC calculation works well for 80 MeV and excellently for 160 MeV, while at 45 MeV it begin to show some deviations. This better performance of INC at higher nucleon energy confirm our expectation considering that the treatment of INC is strictly classical one. This means that the current simple model reproduce the angle integrated cross sections very nicely over the full dynamic range of the reaction regions for the nucleon energies from ~50 MeV almost up to 200 MeV both for the inelastic and charge exchange reactions.

The main scheme of the above INC is applied for the energy spectrum of the nucleons emitted in the NMWD of Λ hypernuclei. Since the energies of the emitted nucleons are in the range from tens of MeV to above 100 MeV, the above INC is well suited for the application in this situation. We report here the current status of the application whose results are encouraging.

In the application, we have adopted the harmonic oscillator wave function for Λ in a hypernucleus with $\hbar\omega = 45A^{-1/3} - 25A^{-2/3}$ [21]. Then the Λ interacts with a nucleon with a Fermi momentum at the site selected according to $|\psi_\Lambda(1s)|^2$. After the interaction, two energetic nucleons are produced and propagate outward the nucleus via the INC process. In the current calculation we have not included the contribution from two

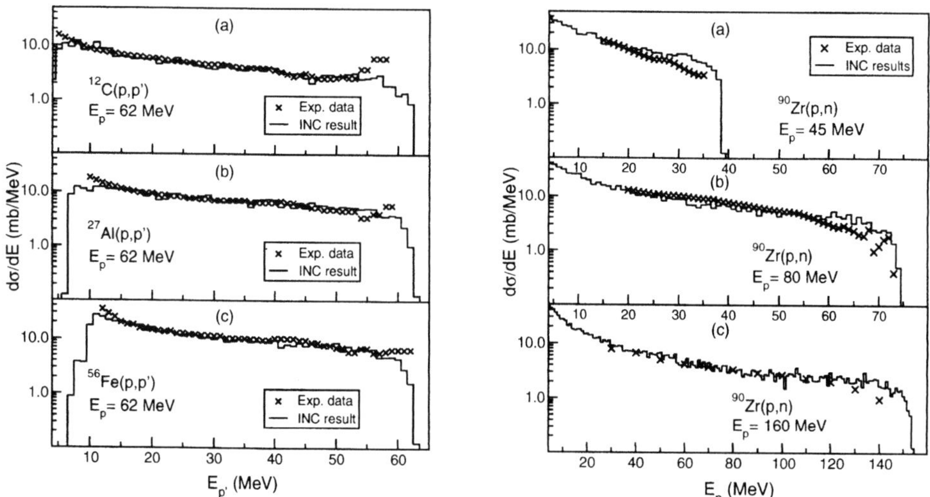

FIGURE 4. In the left figure, the experimental data [15] and the present INC calculations are compared for the angle integrated cross sections as a function of the outgoing proton energy for ^{12}C(p,p')^{12}C, ^{27}Al(p,p')^{27}Al and ^{56}Fe(p,p')^{56}Fe reaction at 62 MeV incident energy. In the right figure, angle integrated cross sections as a function of the outgoing neutron energy for ^{90}Zr(p,n')^{90}Nb reaction at 45, 80 and 160 MeV incident energy [18, 19, 20] are compared with the present INC calculations.

nucleon induced NMWD yet, namely, Λ+NN $\rightarrow$ NNN process, which was conjectured as a could be important component of NMWD. In the two nucleon induced process, the mass difference of Λ and neutron is shared among three nucleons instead of two nucleons and each one carries much less kinetic energy. Furthermore the low energy nucleons are more subjected to the final state interaction and the seperation energy costs them significant shares of their energies so that its contribution would be mainly localized in the low energy end of the energy spectrum. Therefore when we use the proton energy spectrum in the higher energy end above the threshold energy, 40 MeV, its influence may not be significant.

In the Fig. 5 (left), the energy spectrum of the emitted proton from $^{12}_{\Lambda}$C is compared to the present INC (SINC) results for 5 different $\Gamma_n/\Gamma_p(\equiv \gamma)$ values, namely, 0.1, 0.5, 1.0, 1.5 and 2.0. The INC energy spectra of protons were obtained taking into account the energy loss in the target and the detector materials. The spectral shape was well reproduced. We can see this result more clearly in the Fig. 5 (right) in which the integrated yield above the energy on the x-axis are compared and reproduced well with the γ value about 0.75. In the Fig. 6 the energy spectrum of the emitted neutron from $^{12}_{\Lambda}$C is compared to the present INC similarly. The uncertainties of the neutron data are quite large and from two major sources, the statistical and instrumental one, namely that of the efficiency of the neutron counters. At the moment the error bars are too large to show a clear distinguishability on the Γ_n/Γ_p ratio. However, with the ratio ($\sim$0.75) obtained from the proton spectrum, both experimental spectra of the emitted proton and neutron could be well reproduced.

In summary, recent experimental studies on NMWD of Λ hypernuclei have produced

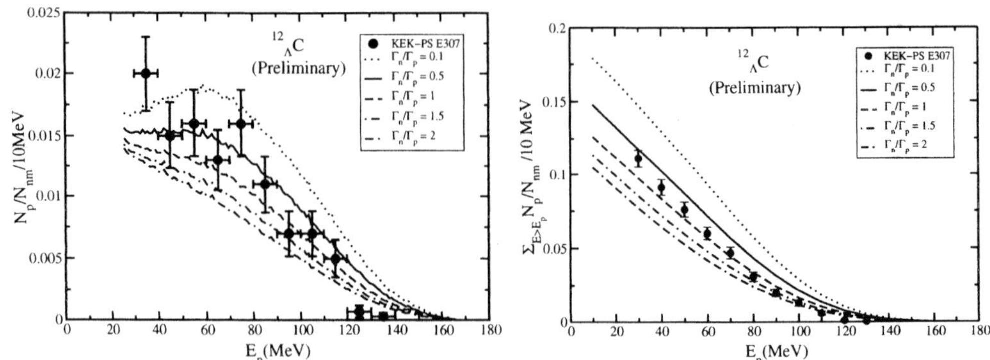

FIGURE 5. In the left figure, the energy spectrum of the emitted proton in the NMWD of $^{12}_{\Lambda}$C is compared with the results of the INC calculation (SINC) for 5 different γ values, 0.1, 0.5, 1.0, 1.5 and 2.0. In the rightt figure, the energy integrated proton yield in the NMWD of $^{12}_{\Lambda}$C is compared with the results of the INC calculation (SINC) for 5 different γ values, 0.1, 0.5, 1.0, 1.5 and 2.0.

many important results. The total widths (lifetimes) of Λ hypernuclei over a broad mass range were accurately determined and were found to saturate at a value about 25 percent increased from that of the free Λ. The decay particle (p and n) spectra of high quality were measured over the mass range from $^{12}_{\Lambda}$C to $^{89}_{\Lambda}$Y. The proton spectra again confirmed the dominance of neutron channel in NMWD. The present newly deveoped INC code (SINC) for the final state interaction showed an encouraging indication that it reproduced the spectral shapes of the emitted proton and neutron spectra well with a consistent Γ_n/Γ_p value. Our preliminary results again confirm the longstanding puzzle in that the experimental Γ_n/Γ_p ratios are of the order of unity, which contradicts all current theoretical predictions.

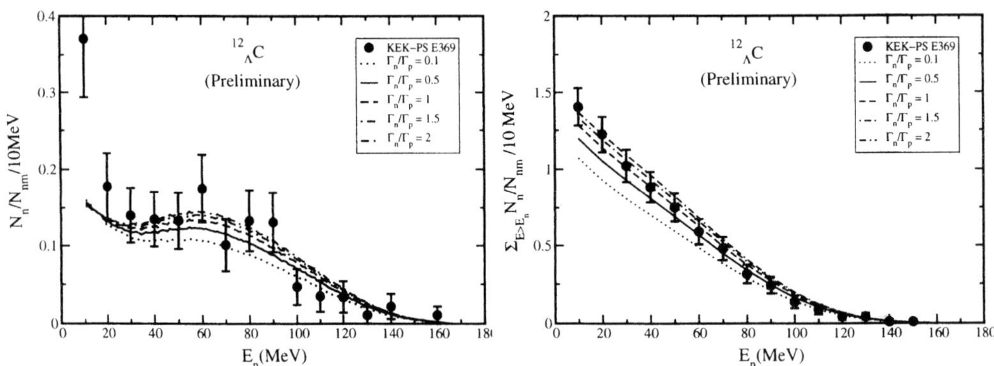

FIGURE 6. In the left figure, the energy spectrum of the emitted proton in the NMWD of $^{12}_{\Lambda}$C is compared with the results of the INC calculation (SINC) for 5 different γ values, 0.1, 0.5, 1.0, 1.5 and 2.0. In the right figure, the energy integrated proton yield in the NMWD of $^{12}_{\Lambda}$C is compared with the results of the INC calculation (SINC) for 5 different γ values, 0.1, 0.5, 1.0, 1.5 and 2.0.

We would like to thank our colleagues, O. Hashimoto, Y.D. Kim, H. Nagae, H. Noumi, H. Outa, H. Park and H. Sato, who collaborated with us in the experiment E307 and E369 and provided us valuable discussions on the development on the new INC code. H.B. acknowledges the support from KOSEF(2000-2-11100-004-4), KRF(2000-015-DP0084), and the BK21 program at Seoul National University.

REFERENCES

1. H. Bhang *et al*, Phys. Rev. Lett. 81 (1998) 4321.
2. H. Noumi *et al*, Phys. Rev. C52 (1995) 2936.
3. A. Ramos, M.J. Vincente-Vacas and E. Oset, Phys. Rev. C55 (1997) 735.
4. J. Kim *et al*, Frontiers in Nuclear Physics, Sungkyunkwan Univ. Press, (2000) 79.
5. Y.D. Kim *et al*., Nuc. Inst. Meth. A372 (1996) 431.
6. H. Park *et al*, Phys. Rev. C61 (2000) 054004.
7. K. Itonaga, Proc. Int. symp. on the Spectroscopy of Hypernuclei, Sendai, (1998) 207.
8. W.M. Alberico *et al*, Phys. Rev. C61 (2000) 044314.
9. J.J. Szymanski *et al*, Phys. Rev. C43 (1991) 849.
10. E. Oset and A. Ramos, Prog. Par. Nucl. Phys. 41 (1998) 191.
11. M. Oka *et al*, Nucl. Phys. A597 (1996) 563.
12. J. Jun *et al*, Nuovo Cimento 112A (1999) 649.
13. Y. Sato, Ph. D. Dissertation, Tohoku University (1999).
14. H. Outa *et al*, Proposal of KEK-PS E462 (2000).
15. F. Bertrand *et al*., Phys. Rev. C8 (1973) 1045.
16. K. Chen *et al*., Phys. Rev. 166 (1967) 949.
17. J. Menet *et al*., Phys. Rev. C4 (1971) 1114.
18. A. Galonsky, *et al*., Phys. Rev. C14 (1976) 748.
19. M. Trabandt, *et al*., Phys. Rev. C39 (1989) 452.
20. W. Scobel, *et al*., Phys. Rev. C41 (1990) 2010.
21. E. Oset and L.L. Salcedo, Nucl. Phys. A443 (1985) 704.

Doubly Strange Nuclei by a Hybrid-Emulsion Experiment E373 at KEK

J. K. Ahn[a1], Y. Akaishi[b], H. Akikawa[a], S. Aoki[c], K. Arai[d], S. Y. Bahk[e], K. M. Baik[f], B. Bassalleck[g], J. H. Chung[h], M. S. Chung[f], D. H. Davis[i], G. B. Franklin[j], T. Fukuda[b2], K. Hoshino[k], A. Ichikawa[a], M. Ieiri[b], K. Imai[a], Y. H. Iwata[h], Y. S. Iwata[h], H. Kanda[a3], M. Kaneko[h], T. Kawai[k], C. O. Kim[f], J. Y. Kim[l], S. J. Kim[l], S. H. Kim[m], Y. Kondo[a4], T. Kouketsu[h], Y. L. Lee[n], J. W. C. McNabb[j], M. Mitsuhara[h], T. Motoba[o], Y. Nagase[h], C. Nagoshi[p], K. Nakazawa[h5], H. Noumi[b], S. Ogawa[d], H. Okabe[q], K. Oyama[d], H. M. Park[h6], I. G. Park[m], J. Parker[j], Y. S. Ra[h7], J. T. Rhee[n], A. Rusek[r], H. Shibuya[d], K. S. Sim[f], P. K. Saha[b], D. Seki[a], M. Sekimoto[b], J. S. Song[m], H. Takahashi[a], T. Takahashi[s], F. Takeutchi[t], H. Tanaka[u], K. Tanida[v], J. Tojo[a], H. Torii[a], S. Torikai[h], D. N. Tovee[i], N. Ushida[u], K. Yamamoto[a8], Y. Yamamoto[w], N. Yasuda[x], J. T. Yang[f], C. J. Yoon[a], C. S. Yoon[m], M. Yosoi[a], T. Yoshida[y] and L. Zhu[a]

[a]*Dept. of Phys., Kyoto Univ., Kyoto 606-8502, Japan*
[b]*KEK, High Energy Accelerator Research Organization, Tsukuba 305-0801, Japan*
[c]*Faculty of Human Development, Kobe Univ., Kobe 657-8501, Japan*
[d]*Dept. of Phys., Toho Univ., Funabashi 274-8510, Japan*
[e]*Wonkwang University, Iri 570-749, Korea*
[f]*Dept. of Phys., Korea Univ., Seoul 136-701, Korea*
[g]*Dept. of Phys. and Astronomy, Univ. of New Mexico, Albuquerque, NM 87131, USA*
[h]*Phys. Dept., Gifu Univ., Gifu 501-1193, Japan*
[i]*Dept. of Phys. and Astronomy, Univ. College London, London WC1E 6BT U.K.*
[j]*Dept. of Phys., Carnegie Mellon Univ., Pittsburgh, PA 15213, USA*
[k]*Dept. of Phys., Nagoya University, Nagoya, 464-8601, Japan*
[l]*Dept. of Phys., Chonnam National Univ., Kwangju 500-757, Korea*
[m]*Dept. of Phys., Gyeongsang National Univ., Jinju 660-701, Korea*
[n]*Institute for Advanced Phys., Konkuk Univ., Seoul 143-701, Korea*
[o]*Laboratory of Physics, Osaka Electro-Communication Univ., Neyagawa 572-8530, Japan*
[p]*Higashi Nippon International Univ., Iwaki 970-8023, Japan*
[q]*Osaka Pref. Education Center, Osaka 558-0011, Japan*
[r]*Brookhaven National Laboratory, NY 11973, USA*
[s]*Dept. of Phys., Tohoku Univ., Sendai 980-8578, Japan*
[t]*Faculty of Science, Kyoto Sangyo Univ., Kyoto 603-8555, Japan*
[u]*Aichi Univ. of Edu., Kariya 448-8542, Japan*
[v]*Dept. of Phys., Univ. of Tokyo, Tokyo 113-0033, Japan*
[w]*Phys. Section of Tsuru Univ., Tsuru 402-8542, Japan*
[x]*NIRS, National Institute of Radiological Science, Chiba 263-0024, Japan*
[y]*Dept. of Phys., Osaka City Univ., Osaka 558-8585, Japan*

CP594, *Hadrons and Nuclei: First International Symposium,* edited by Il-T. Cheon et al.

Abstract. A hybrid emulsion experiment E373 at KEK has been carried out to study doubly strange nuclear system. By the 8% data analysis of all, one twin single-Λ hypernuclei event and two events of double-Λ hypernucleus have been successfully detected. The twin single-Λ event is uniquely interpreted as $\Xi^- + {}^{14}N \rightarrow {}^{5}_{\Lambda}He + {}^{5}_{\Lambda}He + {}^{4}He + n$ for the first time. In the first double-Λ hypernucleus event, "*Demachi-Yanagi*", it is interpreted as $\Xi^- + {}^{12}C \rightarrow {}^{10}_{\Lambda\Lambda}Be$(or ${}^{10}_{\Lambda\Lambda}Be^*$)$+t$. The second double-Λ hypernucleus event, "*NAGARA*", shows a clearly recognized topology and the interpretation is unique as $\Xi^- + {}^{12}C \rightarrow {}^{6}_{\Lambda\Lambda}He + {}^{4}He + t$, ${}^{6}_{\Lambda\Lambda}He \rightarrow {}^{5}_{\Lambda}He + \pi^- + p$, ${}^{5}_{\Lambda}He \rightarrow p + d + 2n$ *etc*. By the preliminary result of the ${}^{6}_{\Lambda\Lambda}He$, the attractive ΛΛ interaction has been confirmed with the $\Delta B_{\Lambda\Lambda} \sim 1$ MeV.

INTRODUCTION

The baryon interaction between nucleons is well understood in $SU(2)$ isospin symmetry, and that is considered as a part of the $SU(3)_f$ symmetry. For the $SU(3)_f$ two baryon systems, the flavor singlet 1_s sector of a doubly strange system is only studied through $S = -2$ nuclear system. R.L.Jaffe proposed the H-dibaryon as a confined 1_s six quark state based on an attractive force between two baryons using the MIT bag model with a color-magnetic interaction[1]. The nuclear system with $S = -2$ is also interesting because it gives us information about hyperon mixing of $\Lambda\Lambda - \Xi N - \Sigma\Sigma$. Since the H state may appear in the above mixing, it will be expected for various nuclear states, for example a H-nucleus. However, at present, experimental information is very limited and the theoretical prediction about ΛΛ interaction energy is in the range from -6 MeV to 10 MeV[2], for example.

Three different experiments have presented a double-Λ hypernucleus in each. One was a ${}^{6}_{\Lambda\Lambda}He$ event by Prowse[3], where only a sketch of the event is left and it has not been studied independently by any other scientists. A ${}^{10}_{\Lambda\Lambda}Be$ event has been presented by Danysz et al. and they have reported that the number of Ξ^- capture event might be one or two in the experiment[4]. In the hybrid-emulsion experiment of E176 at KEK, we have successfully detected a sequential weak decay of double-Λ hypernucleus in $\sim$ 80 events of Ξ^- capture at rest[5]. The nuclide of the event has been presented as ${}^{10}_{\Lambda\Lambda}Be$ or ${}^{13}_{\Lambda\Lambda}B$ with the ΛΛ interaction energy ($\Delta B_{\Lambda\Lambda}$) of $-4.8^{+0.7}_{-0.8}$ MeV (repulsive interaction) or $+4.9\pm0.8$ MeV (attractive), respectively, where the $\Delta B_{\Lambda\Lambda}$ is defined as;
$\Delta B_{\Lambda\Lambda}({}^{A}_{\Lambda\Lambda}Z) = B_{\Lambda\Lambda}({}^{A}_{\Lambda\Lambda}Z) - 2B_{\Lambda}({}^{A-1}_{\Lambda}Z)$.
For the ${}^{13}_{\Lambda\Lambda}B$ case of E176, the $\Delta B_{\Lambda\Lambda}$ is in good agreement with former two cases of $\Delta B_{\Lambda\Lambda}$= 4$\sim$5 MeV. However, all of the past $\Delta B_{\Lambda\Lambda}$ by those three events could say only

[1] Present Address: RCNP, Osaka Univ.
[2] Present Address: Lab. of Phys., Osaka Elec.-Comm. Univ.
[3] Present Address: Dept. of Phys., Tohoku Univ.
[4] Present Address: Japan Atomic Energy Research Institute, Tokai 319-1195, Japan
[5] e-mail: nakazawa@cc.gifu-u.ac.jp, tel.: +81-58-293-2249, fax.: +81-58-293-2207.
[6] Present Address: Chonnam Nat. Univ.
[7] Present Address: Wonkwang Univ.
[8] Present Address: Dept. of Phys., Osaka City Univ.

upper limit because there remained possibility for the production and/or decay of the excited state of the nucleus. Then, the value of the $\Delta B_{\Lambda\Lambda}$ has not been decided, yet. The number of expected Ξ^- stopping events in the past experiments was too small to fix the value of the $\Delta B_{\Lambda\Lambda}$.

Therefore, the experiment E373 at KEK has been carried out to obtain Ξ^- stopping events with ten times more statistics than that of the previous E176, where the number of the events becomes 10^3. Until now, the analysis of about 8% for all emulsion data has been finished. Among them, we have detected one event of twin single-Λ hypernuclei and two events of double-Λ hypernucleus. In this report, these events are discussed and the $\Delta B_{\Lambda\Lambda}$ is presented as the preliminary result.

EXPERIMENT

There are two ways to form $S = -2$ nuclear system. One is a direct formation through (K^-,K^+) reaction at the reaction point, and another method is the formation using Ξ^- capture reaction at rest. The Ξ^- particles are produced via quasi-free (K^-,K^+) reactions. Since the formation rate of $S = -2$ nuclear system by the later method is known to be higher than by the former one[6], the E373 experiment at KEK was planned to obtain Ξ^- capture events with use of hybrid-emulsion technics.

The beam exposure was carried out at the K2 beam line of the KEK Proton Synchrotron. It is shown a schematic diagram of the experimental setup in Fig.1, and shown in Fig.2 is a drawing around the target area. The (K^-,K^+) reaction was tagged by beam-line detectors and a spectrometer system. 1.66GeV/c K^- particles with $K^-/\pi^- \sim 25\%$ in the beam were provided on the target through the heavy metal slit. In the experiment, a diamond block in the dimension of 20mm square and 30mm length was used as the target for the quasi-free (K^-,K^+) reaction taking advantage of the largest density among low mass number materials. Comparing the previous E176, the diamond target enabled us to minimize emulsion volume of only twice for ten times more statistics. The emulsion has been used as an absorber of kinetic energy of the Ξ^- and a detector of $S = -2$ system. The tracking detector of newly developed fiber-bundle[7] has reconstructed Ξ^- particles and predicted their positions and angles on the top emulsion plate, and fully automated scanning systems[8] scan the Ξ^- tracks in the emulsion plate. The Ξ^- tracks are followed until they come to end points in thick emulsion plates. This following method is successfully applied to save scanning time than the way done by E176. In E176, (K^-,K^+) reaction points were located on the trace of K^+ tracks, at first, and the emitted Ξ^- tracks were followed. At the end point of the Ξ^-, the formation of $S = -2$ system is checked for sequential weak decay topologies. When some energetic decay daughters are emitted, they are possible to escape from the emulsion stack. Those particles are able to be detected by two SCIFI-Blocks[9], which were located to sandwich the emulsion for measuring kinetic energies of such particles. These SCIFI-Blocks could also detected the decay of the H, such as $H \to \Sigma^- + p$, and the decay of the Λ with a characteristic energy caused by $\Lambda\Lambda$ weak interaction, $\Lambda\Lambda \to \Lambda N$.

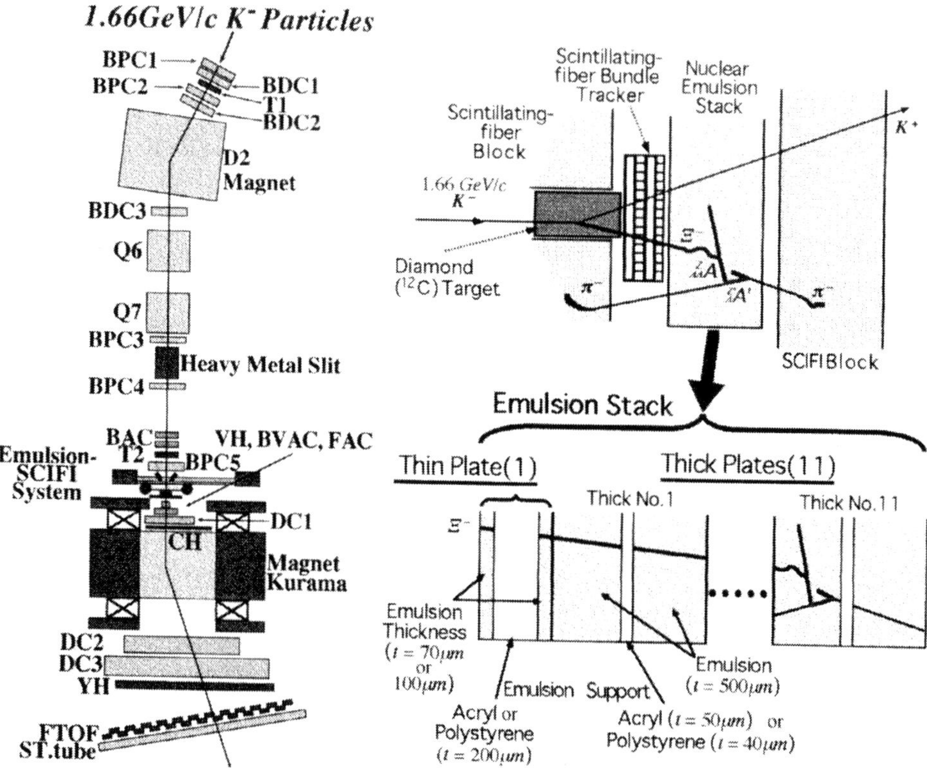

Fig.1 A schematic drawing of the experimental setup.

Fig.2 Around target area and the emulsion stack.

The beam has been successfully exposed to 100 emulsion stacks ($\sim 70 liters$ emulsion volume[10]) in '98, '99 and 2000. The number of accumulated (K^-,K^+) reactions is 9×10^4 after the spectrometer data analysis. The image data from the SCIFI detectors are scanned to pick up the Ξ^- tracks, and the analyses of the Ξ^- capture events in the emulsion have been done for $\sim$ 8% in all (K^-,K^+) reactions. The statistics of the Ξ^- stopping event have been almost the same as those of the previous E176 and we have succeeded to obtain the planned number of events.

EVENTS WITH DOUBLE STRANGENESS

Twin single - Λ hypernuclei

The twin single-Λ hypernuclei event is shown in Fig.3a) for the emulsion image, and in Fig.3b) with a schematic drawing. The analysis of the event is discussed in ref.[11] in detail. The Ξ^- has stopped at the point A. From this point, one stable nucleus and two single-Λ hypernuclei have been emitted. Two single-Λ hypernuclei have decayed at the points B and C, respectively. One particle from B is identified as a proton by its energy deposit in the emulsion. Another one from C is found to be a π^- by its topology at the end point. The formation and the decay of the event is interpreted as;

$$\Xi^- + {}^{14}\text{N} \rightarrow {}^{5}_{\Lambda}\text{He} + {}^{5}_{\Lambda}\text{He} + {}^{4}\text{He} + n.$$

This is the third event of this type and the first uniquely identified case reported until now. Combined the past two events of twin hypernuclei[12] and a double hypernucleus[5], statistical discussion for $S = -2$ system is started about the production rate of double-Λ to twin single-Λ hypernuclei , where its rate must strongly relate with the Λ-Λ interaction energy.

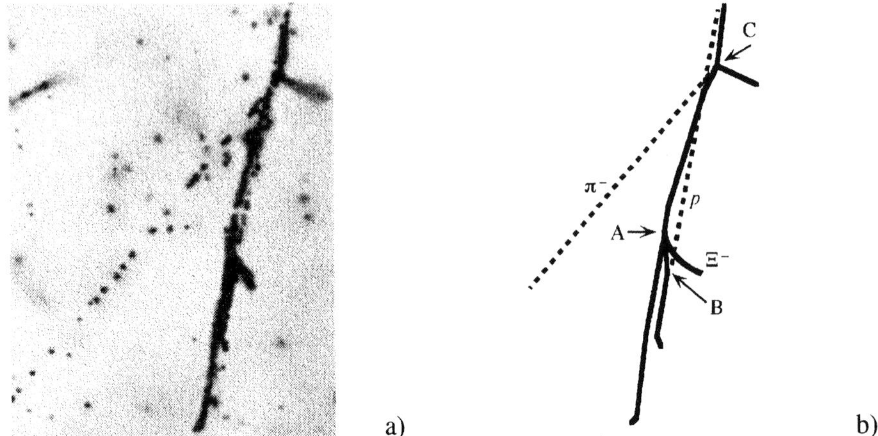

Figure 3. a): A projected image of twin single-Λ hypernuclei event in the emulsion obtained by E373. b): A schematic drawing of the event.

Double-Λ hypernucleus; "*Demachi-Yanagi*" event

The first double-Λ hypernucleus found in E373 is shown in Fig.4a) with a schematic drawing in Fig.4b), and called as "*Demachi-Yanagi*" in Kyoto. The double-Λ hypernucleus has been produced at the point A and decayed at the B. In the projected image of Fig.4a), it is not so easy to separate the B$-$C track from the A$-$B track. However, the point C is located in the different depth from the A$-$B track. Therefore, the decay of the single-Λ hypernucleus has been identified. An energetic particle from the point C is known as a proton by its energy loss measurement. The main background for a double-Λ hypernucleus must be backward-scattering and decay of a single-Λ hypernucleus at the point B and C, respectively. Its probability was, however, less than 1×10^{-4}, which is negligibly small in consideration of the number, 6 events, of detected single-Λ hypernucleus events in E373. At the Ξ^- capture point A, the momentum and energy conservation were checked by use of the production topology. If no neutron was escaping from the point A, following interpretations have been accepted;

$$\Xi^- + {}^{12}\text{C} \rightarrow {}^{10}_{\Lambda\Lambda}\text{Be} + t,$$

or an excited state production as ${}^{10}_{\Lambda\Lambda}\text{Be}^*$. Taking $0.15^{+0.3}_{-0.1}$ MeV into account as the level energy of the captured Ξ^- in the ${}^{12}\text{C}$, $B(\Xi^-)$, the $B_{\Lambda\Lambda}$ and $\Delta B_{\Lambda\Lambda}$ are obtained to be

$12.33^{+0.35}_{-0.21}$ MeV and $-1.09^{+0.35}_{-0.21}$ MeV, respectively, in the production case of the ground state. Assuming the ${}^{10}_{\Lambda\Lambda}$Be* with the excitation energy of 3 MeV, the $B_{\Lambda\Lambda}$ and $\Delta B_{\Lambda\Lambda}$ are $15.37^{+0.35}_{-0.21}$ MeV and $+1.95^{+0.35}_{-0.21}$ MeV, respectively. Although the directions of two tracks at the point A are collinear within measurement error (5°), the three body reaction as $\Xi^- + {}^{12}C \rightarrow {}^{11}_{\Lambda\Lambda}Be + p + n$ or $\Xi^- + {}^{12}C \rightarrow {}^{10}_{\Lambda\Lambda}Be + d + n$ is not denied. Such possibility by topological calculation is $\sim$ 4% or 6%, respectively, which is not negligible but very small. In the case of three body reaction with neutron(s) emission, the lowest value of the $\Delta B_{\Lambda\Lambda}$ has been obtained as $1.49^{+2.4}_{-0.66}$ MeV for the reaction, $\Xi^- + {}^{14}N \rightarrow {}^{13}_{\Lambda\Lambda}B + p + n$.

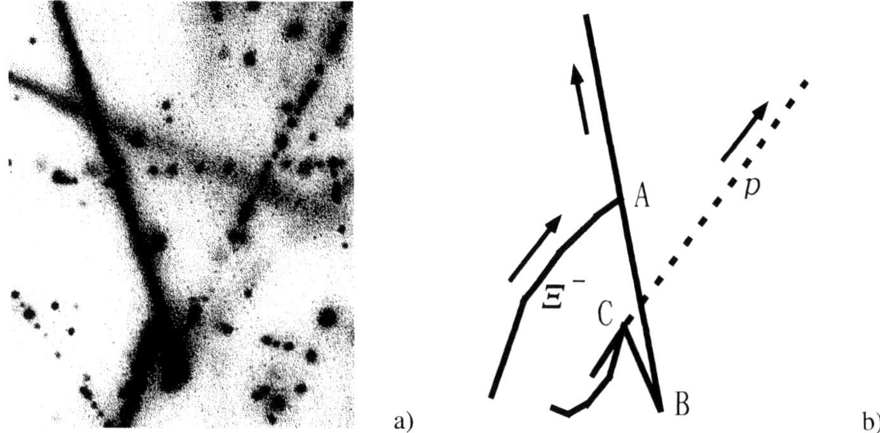

Figure 4. a): A projected emulsion image of the first double-Λ hypernucleus event, "*Demachi-Yanagi*", obtained by E373. b): A schematic drawing of the event.

Double-Λ hypernucleus; "*NAGARA*" event

The second event of double-Λ hypernucleus is very clearly recognized as shown in Fig.5a) and b), where the nucleus is produced at point A and decay at point B. The event is named "*NAGARA*", which is a river originating in Gifu, Japan. At the decay point B of the double-Λ hypernucleus, three charged particles have been emitted. One is ended in base film, the other one is identified as a π^- by its topology at the end point, and a single-Λ hypernucleus decays at the point C into two stable nuclei. The ranges of the double- and the single-Λ hypernucleus are 8.3 and 9.2 μm, respectively. The event is analyzed as follows;

- *Decay of Single-Λ hypernucleus* ; Point C
  Two tracks from the decay of single-Λ hypernucleus have long ranges. Assuming one of them to be multi-charged particle, energy-release from the decay becomes easily larger than 176 MeV, which is Q-value for the non-mesonic decay of single-Λ hypernucleus. On the other hand, mesonic decay can't be represented by the energy-release from the decay. Therefore, the single-Λ hypernucleus is the He nuclide, but its mass number can't be identified, here.

- *Decay of double-Λ hypernucleus* ; Point B
  Three tracks from the point B are quite well coplanar. However, we have reconstructed the vertex B even in the case of neutron(s) emission, where the neutron(s) have the momentum sum, p_{sum}, of three charged particles. After the reconstruction, about 20 decay modes are remained by the request of the $p_{sum} < 3$ standard deviations, 3σ, of measurement error for the case without neutron emission and $\Delta B_{\Lambda\Lambda} > -20$ MeV for all cases. All of the modes are the decay of He double-Λ hypernuclei.
- *Production of double-Λ hypernucleus* ; Point A
  Possible production modes are checked in all the cases of the Ξ^- capture by light nucleus, *i.e.* ^{12}C, ^{14}N or ^{16}O. The coplanarity of three tracks from the point A has also no problem within measurement error. By the reconstruction, about ten production modes are accepted, where we also set the criteria as the $p_{sum} < 3\sigma$ of measurement error for the case without neutron emission and $\Delta B_{\Lambda\Lambda} < 20$ MeV for all cases.
- *The double-Λ hypernucleus* ; $^{6}_{\Lambda\Lambda}$He
  By the analyses of the formation and the decay of this double-Λ hypernucleus, the interpretation of the nucleus was uniquely established as the following mode.

$$\begin{aligned} \Xi^- + {}^{12}\mathrm{C} &\rightarrow {}^{6}_{\Lambda\Lambda}\mathrm{He} + {}^{4}\mathrm{He} + t, \\ {}^{6}_{\Lambda\Lambda}\mathrm{He} &\rightarrow {}^{5}_{\Lambda}\mathrm{He} + \pi^- + p, \\ {}^{5}_{\Lambda}\mathrm{He} &\rightarrow p + d + 2n \; etc. \end{aligned}$$

The possibility for the production and/or decay in the excited state are completely rejected in this event. The values of the $B_{\Lambda\Lambda}$ and the $\Delta B_{\Lambda\Lambda}$ have been decided to be ~7 MeV and ~1 MeV with errors less than 0.5 MeV by the preliminary result[13].

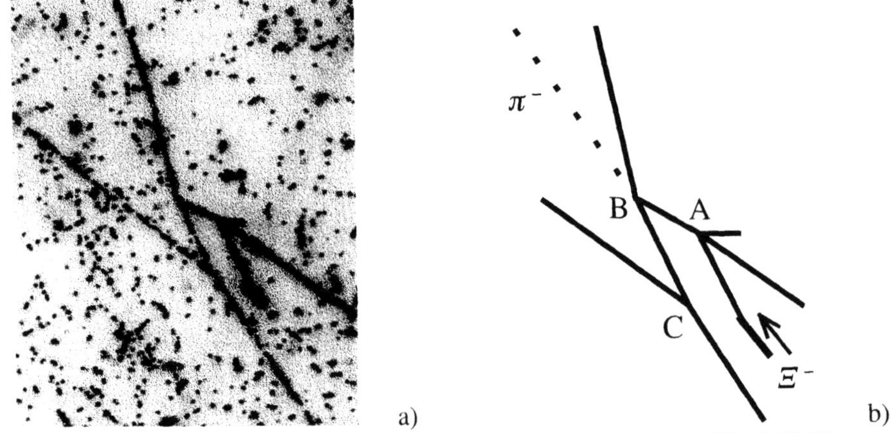

Figure 5. a): An emulsion image of the second double-Λ hypernucleus "*NAGARA*" event, $^{6}_{\Lambda\Lambda}$He, by E373. b): A schematic drawing of the event.

$\Lambda\Lambda$ interaction ; comparison with the past

Since the ${}^{6}_{\Lambda\Lambda}$He event has presented the unique $\Delta B_{\Lambda\Lambda}$, it is very interesting that the $\Delta B_{\Lambda\Lambda}$ from the past events will be compared. In Fig.6, possible values in the production case of the daughter single-Λ hypernucleus in the excited states are listed as ${}^{10}_{\Lambda\Lambda}$Be(${}^{9}_{\Lambda\Lambda}$Be*) and ${}^{13}_{\Lambda\Lambda}$B(${}^{13}_{\Lambda\Lambda}$C*). The plotted ${}^{10}_{\Lambda\Lambda}$Be in the right of the E176 result was calculated from the production topology, while ${}^{10}_{\Lambda\Lambda}$Be plotted in the left is by the decay topology in the original paper[5]. By the ${}^{6}_{\Lambda\Lambda}$He event "*NAGARA*", the past events except for Prowse's one are consistent to the $\Delta B_{\Lambda\Lambda} \sim 1$ MeV within errors.

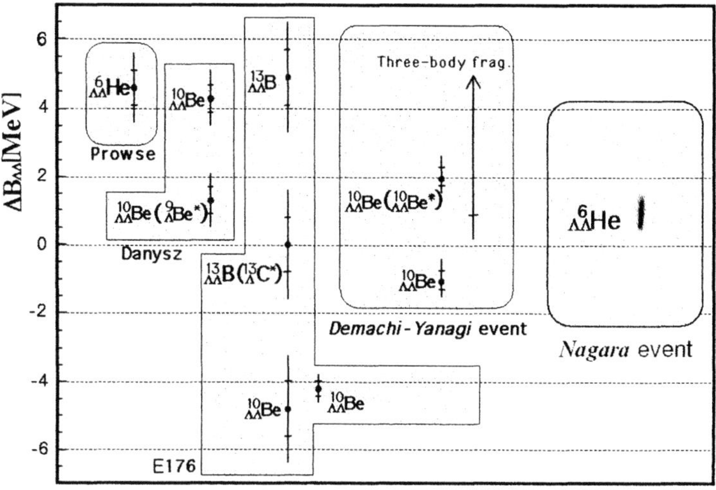

Figure 6. The $\Delta B_{\Lambda\Lambda}$ obtained from the past three events and the E373 events.

CONCLUSION

A hybrid emulsion experiment E373 at KEK is successfully carried out to study doubly strange nuclear system. Using 1.66 GeV/c K^- beam, totally 9×10^4 (K^-,K^+) events were obtained after the spectrometer analysis. The data analysis has been finished in 8% data of all and it is found that the planned one thousand events of the Ξ^- capture at rest has been recorded in the emulsion. Also, one twin single-Λ hypernuclei event and two events of double-Λ hypernucleus have been detected.

The twin single-Λ event is uniquely interpreted as $\Xi^- + {}^{14}\mathrm{N} \rightarrow {}^{5}_{\Lambda}\mathrm{He} + {}^{5}_{\Lambda}\mathrm{He} + {}^{4}\mathrm{He} + n$ for the first time. By the finding of the uniquely identified twin single-Λ event, the discussion about the production rate of double-Λ to Twin events has successfully started, and it will be very useful information about Λ-Λ interaction.

Two events of double-Λ hypernucleus give us a conclusive information about $\Lambda\Lambda$ interaction. In the first event "*Demachi-Yanagi*", the double-Λ hypernucleus is produced in back-to-back direction to one charged particle from Ξ^- stopping point. If a two-body decay of $\Xi^- - {}^{12}\mathrm{C}$ system is assumed, the event is interpreted as; $\Xi^- + {}^{12}\mathrm{C} \rightarrow {}^{10}_{\Lambda\Lambda}\mathrm{Be} + t$, or the production in excited state as ${}^{10}_{\Lambda\Lambda}$Be*. In each cases, the $B_{\Lambda\Lambda}$ and the $\Delta B_{\Lambda\Lambda}$

are obtained as $12.33^{+0.35}_{-0.21}$ or $15.37^{+0.35}_{-0.21}$ MeV and $-1.09^{+0.35}_{-0.21}$ or $+1.95^{+0.35}_{-0.21}$ MeV, respectively, where $0.15^{+0.3}_{-0.1}$ MeV as $B(\Xi^-)$ in the ^{12}C was taken into account. As for the case of three body decay, the lowest $\Delta B_{\Lambda\Lambda}$ is found to be $+1.49^{+2.4}_{-0.66}$ MeV.

The second event "*NAGARA*" has been found with a clearly recognized topology. We have succeeded to interpret the double-Λ hypernucleus without any ambiguity as; Ξ^- + $^{12}\mathrm{C} \rightarrow {}^{6}_{\Lambda\Lambda}\mathrm{He} + {}^{4}\mathrm{He} + t$, ${}^{6}_{\Lambda\Lambda}\mathrm{He} \rightarrow {}^{5}_{\Lambda}\mathrm{He} + \pi^- + p$, ${}^{5}_{\Lambda}\mathrm{He} \rightarrow p + d + 2n$ *etc*. By the preliminary result, the $\Delta B_{\Lambda\Lambda}$ is decided to be $\sim$1 MeV as an attractive interaction with an error of less than 0.5 MeV, and the $B_{\Lambda\Lambda}$ is $\sim$7 MeV for ${}^{6}_{\Lambda\Lambda}$He. The past experimental results except for the old ${}^{6}_{\Lambda\Lambda}$He seem to be consistent with the $\Delta B_{\Lambda\Lambda} \sim 1$ MeV. It has been cleared that the $\Delta B_{\Lambda\Lambda}$ has not been so large as 4$\sim$5 MeV.

A hybrid-emulsion experiment can detect real Ξ^- stopping events without ambiguity. By use of the information of Ξ^- stopping in the emulsion as an additional trigger signal to counter information, a very clean experiment against noise can be done. Therefore, when we carry out a new experiment to achieve 10^4 Ξ^- stopping events in reasonable time, very precious information for the doubly strange interaction via $S = -2$ nuclei will be obtained by a hybrid-emulsion experiment with some additional eyes, *e.g.* neutron detector, γ-ray detector and/or something.

REFERENCES

1. R.L.Jaffe, Phys. Rev. Lett. **38** (1977) 195.
2. H.Bando Prog. Theor. Phys. **67** (1977) 699; Y.Yamamoto *et al.*, Prog. Theor. Phys. **86** (1991) 867; Y.Yamamoto *et al.*, Prog. Theor. Phys. Suppl.**117** (1995) 361; H.Himeno *et al.*, Prog. Theor. Phys. **89** (1993) 109; S.B.Car *et al.*, Nucl. Phys. **A625** (1997) 143; T.Yamada and C.Nakamoto, Phys. Rev. **C62** (2000) 034319; J.Schaffner *et al.*, Ann. Phys. **235** (1994) 35.
3. D.J.Prowse, Phys. Rev. Lett. **17** (1966) 782.
4. M.Danysz *et al.*, Nucl. Phys. **49** (1963) 121; R.H.Dalitz *et al.*, Proc. R. Soc. Lond. **A426** (1963) 1.
5. S.Aoki *et al.*, Prog. Theor. Phys. **85** (1991) 1287.
6. K.Nakazawa *et al.*, Proc. of the 23rd INS symp. (edited by S.Sugimoto and O.Hashimoto) Universal Academy Press, Inc. and INS, Univ. of Tokyo (1995) 216; S.Aoki *et al.*, Nucl. Phys. **A644** (1998) 365.
7. A.Ichikawa *et al.*, Nucl. Instr. & Method **A417** (1998) 220.
8. A.Ichikawa *et al.*, will be submitted to Nucl. Instr. & Method A.
9. H.Takahashi *et al.*, will be submitted to Nucl. Instr. & Method A.
10. H.Akikawa *et al.*, submitted to Nucl. Instr. & Method A.
11. A.Ichikawa *et al.*, Phys. Lett. **B500** (2001) 37.
12. S.Aoki *et al.*, Prog. Theor. Phys. **89** (1993) 493; S.Aoki *et al.*, Phys. Lett. **B355** (1995) 45.
13. Final result will be submitted to Phys. Rev. Lett.

Four-body calclation of ${}^{4}_{\Lambda}$H and ${}^{4}_{\Lambda}$He with realistic YN and NN interactions

E. Hiyama*, M. Kamimura†, T. Motoba**, T. Yamada‡ and Y. Yamamoto§

*Institute of Particle and Nuclear Studies, High Energy Accelerator Research Organization (KEK), Tsukuba, 305-0801, Japan
†Department of Physics, Kyushu University, Fukuoka 812-8581, Japan
**Laboratory of Physics, Osaka Electro-Comm. University, Neyagawa 572-8530, Japan
‡Laboratory of Physics, Kanto Gakuin University, Yokohama 236-8501, Japan
§Physics Section, Tsuru University, Tsuru, Yamanashi 402-8555, Japan

Abstract. The four-body calculations for ${}^{4}_{\Lambda}$He and ${}^{4}_{\Lambda}$H with high accuracy have been performed in the framework of the variational method with Jacobian-coordinate Gaussian-basis functions. All the rearrangement channel of both $NNN\Lambda$ and $NNN\Sigma$ are explicitly taken into account for the first time using realistic NN and YN interactions. The important role of $\Lambda-\Sigma$ conversion and the amount of the virtual Σ-component in ${}^{4}_{\Lambda}$He and ${}^{4}_{\Lambda}$H are discussed.

INTRODUCTION

n hypernuclear physics, it is one of the most fundamental problems to extract novel information of YN interactions through precise calculations for a few-body systems such as ${}^{3}_{\Lambda}$H, ${}^{4}_{\Lambda}$He and ${}^{4}_{\Lambda}$H. Although a number of studies for ${}^{4}_{\Lambda}$He and ${}^{4}_{\Lambda}$H have been performed with the use of various models so far, the relations to the underlying YN interactions are dependent on the adopted models [1, 2, 3]. In order to explore the features of the YN interactions, it is highly required to perform four-body calculation without any restriction in the configuration space. An especially interesting issue in this relation is to make clear the role of $\Lambda-\Sigma$ conversion and to get a reliable estimate of the amount of the virtual Σ-component in Λ hypernuclei.

However, the $\Lambda N-\Sigma N$ coupling terms of the YN one-boson-exchange (OBE) models proposed so far have a lot of ambiguity due to little information from the YN scattering experiments. Recently, a series of the realistic YN interaction models called NSC97a$\sim$f [4] have been proposed in which the spin-spin strengths are varied within the acceptable range in view of the limited YN scattering data. Correspondingly, by including the $\Lambda N-\Sigma N$ coupling explicitly, Miyagawa *et al.* [5] performed the Faddeev calculations for ${}^{3}_{\Lambda}$H extensively to test these OBE potential models in this lightest system. They found that the $\Lambda-\Sigma$ coupling is crucial to get the bound-state of ${}^{3}_{\Lambda}$H and that only the version-f potential is acceptable among the NSC97 models. For the further examination on $\Lambda N-\Sigma N$ coupling, ${}^{4}_{\Lambda}$He and ${}^{4}_{\Lambda}$H are much more useful because both of the spin-doublet states have been observed. Regarding the representative calculations done so far, Gibson *et al.* employed the coupled two-body model [1] of ^{3}He(^{3}H)+Λ/Σ which was adopted originally by Dalitz and Downs [6], and then carried out the four-

CP594, *Hadrons and Nuclei: First International Symposium,* edited by Il-T. Cheon et al.

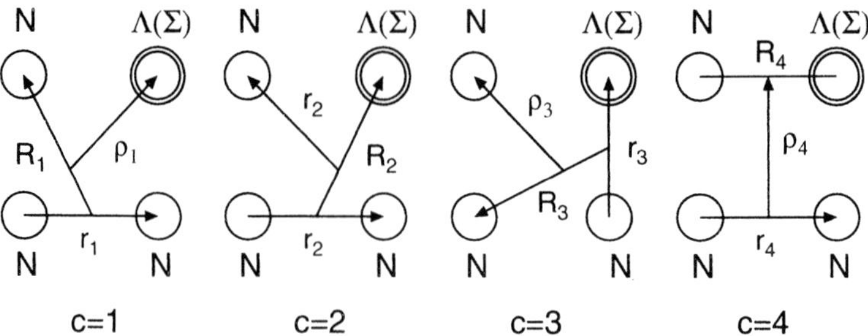

FIGURE 1. Jacobian coordinates for all the rearrangement channels of $3N+\Lambda(\Sigma)$ of ${}^4_\Lambda$H and ${}^4_\Lambda$He. The three nucleons have to be antisymmetrized.

body coupled-channel calculation with the separable potentials of central nature [2]. Akaishi *et al.* [3] recently analyzed the role of the $\Lambda N-\Sigma N$ coupling for the 0^+-1^+ splitting also in the framework of the coupled two-body model of ^{3}He+Λ/Σ. Now our concern is to perform precise four-body calculations of $NNN\Lambda$ and $NNN\Sigma$ irrespectively of any model assumptions. It should be noted that, when one goes from ${}^3_\Lambda$H to the $A=4$ strange systems by allowing for the $\Lambda-\Sigma$ conversion, the computational difficulty increases tremendously. Recently we have successfully performed the extensive four-body calculations without any restriction on the configuration space: Both the $NNN\Lambda$ and $NNN\Sigma$ channels have been incorporated explicitly and all the rearrangement channels of these baryons are taken into account. The variational method with the use of Jacobian-coordinate Gaussian-basis functions [7, 8] is adopted here; it has been proved to provide us with precise computational results for few-body systems [7, 8, 9, 10, 11].

The main purpose of this work is, first, to solve four-body problem of ${}^4_\Lambda$He and ${}^4_\Lambda$H by taking account of $NNN\Lambda(\Sigma)$ channels explicitly with the use of realistic NN and YN interactions and, secondly, to clarify the role of the $\Lambda N-\Sigma N$ coupling in the $A=4$ hypernuclei quantitatively. As the first step before going to the use of sophisticated OBE models, we employ here the $\Lambda N-\Sigma N$ coupled potential with central, spin-orbit and tensor terms [12] which simulates the scattering phase shifts given by NSC97f.

MODEL AND METHOD

Jacobian coordinates of ${}^4_\Lambda$H and ${}^4_\Lambda$He are illustrated in Fig. 1. The four-body wavefunctions are written as

$$\Psi_{JM}({}^4_\Lambda\mathrm{H},{}^4_\Lambda\mathrm{He}) = \sum_{c=1}^{4}\sum_{Y}\sum_{\alpha I}\sum_{s,s',S}\sum_{t_0,t_0} A^{(c)}_{Y\alpha Iss'St_0t} \times \mathcal{A}_{123}\left\{\left[\Phi^{(c)}_{\alpha I}(\mathbf{r}_c,\mathbf{R}_c,\rho_c)\right.\right.$$

$$\left.\times[[\eta_t(12)\eta_{\frac{1}{2}}(3)]_{t'}\eta_{t_Y}(Y)]_{T=\frac{1}{2}}\right\}, \qquad (1)$$

In Eq.(1), c denotes the channels of Fig. 1, $\mathcal{A}_{123}$ is the three-nucleon antisymmetrization operator and χ's and η's are the spin and isospin functions. The isospin $t_Y=0$ for $Y=\Lambda$

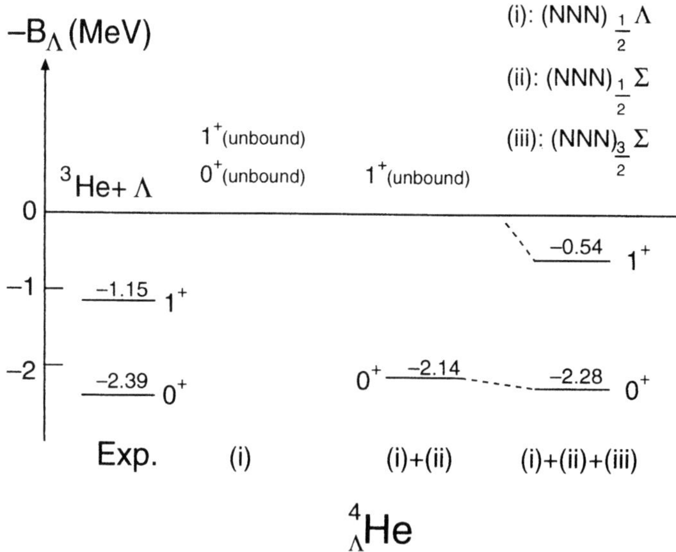

FIGURE 2. Calculated energy levels of $^4_\Lambda$He. The channels successively included are (i) $(NNN)_{\frac{1}{2}}\Lambda$, (ii) $(NNN)_{\frac{1}{2}}\Sigma$ and (iii) $(NNN)_{\frac{3}{2}}\Sigma$ where the isospin of the three nucleons is coupled to $t = 1/2$ or 3/2. Energy is measured from the ^{3}He+Λ threshold.

and $t_Y = 1$ for $Y = \Sigma$. Here, n, N and ν denote the size of the Gaussian basis functions for four-body systems. The eigenenegies of Hamiltonian and coefficients C are determined by Rayleigh-Ritz variational method.

We employ the AV8 potential as the NN interaction.

RESULTS

All the calculations have been performed both for $^4_\Lambda$He and $^4_\Lambda$H. Calculated B_Λ of the 0^+ ground state and the 1^+ excited state of $^4_\Lambda$He are illustrated in Fig. 2 together with the observed values. In the case of taking only the $NNN\Lambda$ channel, both of the two states are unbound. Here, the $NNN\Sigma$ sector is divided into the $(NNN)_{\frac{1}{2}}\Sigma$ and $(NNN)_{\frac{3}{2}}\Sigma$ channels, in which the three nucleons are coupled to isospin $t = 1/2$ and $3/2$, respectively. When the $(NNN)_{\frac{1}{2}}\Sigma$ channel is included, the 0^+ state becomes bound, but the 1^+ state is still unbound. Then we found that the 1^+ Then we found tht the 1^+ state becomes bound only when the $(NNN)_{\frac{3}{2}}\Sigma$ channel is switchted on. It is noted that the binding energy of the 0^+ state increases only slightly with this $t = 3/2$ channel. The Σ-channel components turn out to play an essential role in the binding mechanism of the $A = 4$ hypernuclei, the $(NNN)_{\frac{3}{2}}\Sigma$ channel being specially important in the 1^+ state. The calculated binding energy of the 0^+ state almost reproduces the observed binding

TABLE 1. Calculated energies of the 0^+ and 1^+ states of ${}^4_\Lambda$He and ${}^4_\Lambda$H. The energies E are measured from the $NNN\Lambda$ four-body breakup threshold. As for the ^{3}He (^{3}H) nucleus, the calculated binding energy is $-7.12(-7.77)$ MeV.

	${}^4_\Lambda$He		${}^4_\Lambda$H	
J	0^+	1^+	0^+	1^+
E (MeV)	−9.40	−7.66	−10.10	−8.33
$E^{\rm exp}$ (MeV)	−10.11	−8.87	−10.52	−9.53
B_Λ (MeV)	2.28	0.54	2.33	0.59
$B_\Lambda^{\rm exp}$ (MeV)	2.39	1.15	2.04	1.05

TABLE 2. The S-, P- and D-state and total probabilities of the $(NNN)_{\frac{1}{2}}\Lambda$, $(NNN)_{\frac{1}{2}}\Sigma$ and $(NNN)_{\frac{3}{2}}\Sigma$ channels in the 0^+ and 1^+ states of ${}^4_\Lambda$He. $(NNN)_t$ denotes three nucleons whose isospins are coupled to t.

		$(NNN)_{\frac{1}{2}}\Lambda$	$(NNN)_{\frac{1}{2}}\Sigma$	$(NNN)_{\frac{3}{2}}\Sigma$
0^+		97.92 %	2.04 %	0.04 %
	S-state	89.32	0.84	0.01
	P-state	0.08	0.04	0.01
	D-state	8.52	1.17	0.02
1^+		98.97 %	0.51 %	0.52 %
	S-state	90.38	0.10	0.09
	P-state	0.07	0.01	0.00
	D-state	8.52	0.40	0.43

energy, while the 1^+ state is less bound by 0.6 MeV and hence the $0^+ - 1^+$ splitting is larger than the observed splitting. The calculated B_Λ of ${}^4_\Lambda$H is similar to that of ${}^4_\Lambda$He as shown in Table I. The calculated value of $B_\Lambda({}^4_\Lambda\mathrm{He}(0^+)) - B_\Lambda({}^4_\Lambda\mathrm{H}(0^+)) = -0.05$ MeV is different from the experimental one, $+0.35$ MeV, although yje Coulomb potentials between charged particles $(p, \Sigma^\pm)$ are included. This difference should be attributed to the charge-symmetry-breaking component which is not included in our adopted YN interaction.

As listed in Table II, the calculated probabilies of the $NNN\Sigma$-channel admixture are 2.08% and 1.03% for the 0^+ and 1^+ states, respectively, in ${}^4_\Lambda$He. In the 0^+ state, the probability of the $(NNN)_{\frac{1}{2}}\Sigma$ channel is much larger than that of the $(NNN)_{\frac{3}{2}}\Sigma$ channel, while, in the 1^+ state they are nearly the same. We therefore confirm that $(NNN)_{\frac{3}{2}}\Sigma$ channel is especially important in the 1^+ state. The S-, P- and D-state probabilities of the channels are also listed in Table III. It is remarkable that, in the $NNN\Sigma$ channel, the D-state component is dominant both in the 0^+ and 1^+ states, since the $\Lambda N - \Sigma N$ coupling part of the present interaction is dominated by the tensor component. These properties are almost persistent in the case of ${}^4_\Lambda$H.

In order to show the physical effect of Σ-mixing in more detail, let us separate the

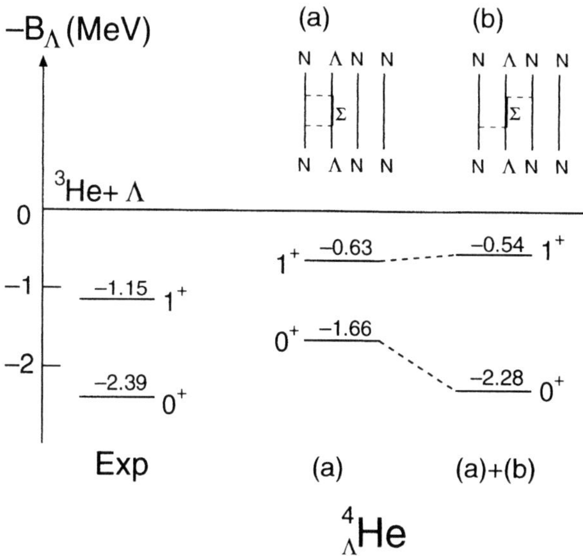

FIGURE 3. Calculated energy levels of $^4_\Lambda$He for case (a) and case (a)+(b); here, (a) denotes the two-body process and (b) denotes the three-body process.

whole contribution of the $\Lambda N - \Sigma N$ coupling interaction into the following two processes illustrated in Fig. 3: The first one is the process (a) which can be renormalized into the effective ΛN two-body force and the second one is the process (b) which can be represented by the effective ΛNN three-body force acting in the $NNN\Lambda$ space. We solve the Schrödinger equation by excluding the three-body process (b) so as to evaluate the contribution of the process (a) only. As shown in Fig.3, the process (a) is large enough to make both the 0^+ and 1^+ states bound. We found that the three-body process(b) is substantial: This process results in attraction by 0.6 MeV in the 0^+ state ,while repulsion by 0.1 MeV in the 1^+ state. Characteristic feature of our present result of the process (b) in the 0^+ state is similar to that given by Akaishi *et al* [3]. On the other hand, in the 1^+ state the effect of the process (b) is found to be different from the result of Ref.[3]: The process (b) works repulsively in our treatment, while the contribution of the process (b) is negligible in Ref. [3]. This difference may come from the fact that in the 1^+ state of $^4_\Lambda$He the probability of $(NNN)_{\frac{3}{2}}\Sigma$ is as large as that of $(NNN)_{\frac{1}{2}}\Sigma$ in our result, but $(NNN)_{\frac{3}{2}}\Sigma$ channel in the coupled two-body model of $^3\text{He} + \Lambda/\Sigma$ of Ref.[3] cannot be explicitly included.

SUMMARY

We have developed the calculational method of four-body bound-state problems so that it becomes possible to make precise four-body calculations of $^4_\Lambda$H and $^4_\Lambda$He taking both the $NNN\Lambda$ and $NNN\Sigma$ channels explicitly with the use of realistic NN and YN interactions. As a result, we succeeded in making clear the role of $\Lambda - \Sigma$ conversion and deriving the amount of the Σ-mixing in $A = 4$ hypernuclei quantitatively. However, our YN interaction empoyed here is not sufficient to reproduce the binding energy of the excited state of 1^+, although those of ground states of ${}^3_\Lambda$H, ${}^4_\Lambda$H and ${}^4_\Lambda$He are in good agreement with the experimental values. It is future problem to explore the feature of $\Lambda - \Sigma$ conversion in Λ hypernuclei with the use of more reasonable YN interactions through systematic study of structure for the few-body hypernuclear systems.

ACKNOWLEDGMENTS

The authors thank Prpfessor Y. Akaishi, B. F. Gibson and T. A. Rijken for helpful discussion.

REFERENCES

1. B. F. Gibson, A. Goldberg, and M. S. Weiss, Phys. Rev. **C 6**, 741 (1972).
2. B. F. Gibson, and D. R. Lehman, Phys. Rev. **C37**, 679 (1988).
3. Y. Akaishi, T. Harada, S. Shinmura, and Khin Swe Myint, Phys. Rev. Lett. **84**, 3539 (2000).
4. T. A. Rijken, V. G. J. Stoks, and Y. Yamamoto, Phys. Rev. **C59**, 21 (1999).
5. K. Miyagawa, H. Kamada, W. Glöckle, H. Yamamura, T. Mart, and C. Bennhold, Few-Body Systems Suppl. **12**, 324 (2000).
6. R. H. Dalitz and B. W. Downs, Phys. Rev. **111**, 967 (1958).
7. M. Kamimura, Phys. Rev. **A38**, 621 (1988).
8. H. Kameyama, M. Kamimura, and Y.Fukushima, Phys. Rev. **C40**, 974 (1989).
9. E. Hiyama, and M. Kamimura, Nucl. Phys. **A588**, 35c (1995).
10. E. Hiyama, M. Kamimura, T. Motoba, T. Yamada, and Y. Yamamoto, Phys. Rev. **C53**, 2075 (1996).
11. E. Hiyama, M. Kamimura, T. Motoba, T. Yamada, and Y. Yamamoto, Phys. Rev. Letters **85**, 270 (2000).
12. S. Shinmura (private communications).

Quark Pauli Principle in Λ-Hypernuclear Systems

H. Nemura* and Y. Suzuki†

*Institute of Particle and Nuclear Studies, KEK, Ibaraki 305-0801, Japan
†Department of Physics, Niigata University, Niigata 950-2181, Japan

Abstract. We explore a question of whether quark Pauli principle plays an important role or not in the binding mechanism of Λ hypernuclei. We introduce a special five-baryon configuration which is forbidden by the Pauli principle at the quark level. By excluding the forbidden state, we show that the Λ separation energy of ${}^{5}_{\Lambda}$He or ${}^{6}_{\Lambda\Lambda}$He is significantly reduced for a baryon size of 0.86 fm, while it is unchanged for a smaller size of 0.6 fm. We also show a qualitative estimation of the energy reduction for more heavier systems, ${}^{9}_{\Lambda}$Be and ${}^{10}_{\Lambda\Lambda}$Be, by an α cluster model.

The Pauli principle plays a vital role in the binding mechanism for systems of identical particles. Its importance is well known in atoms and atomic nuclei. However, in Λ hypernuclei where the Λ-particle is embedded in nuclei the role of the Pauli principle has not yet been explored from the point of view of the binding mechanism. Though Λ can be distinguished from the nucleon N at the baryon level, they are subjected to the Pauli principle at the quark level. We estimate how much the Λ separation energy of $A = 5,6,9,10$ Λ hypernuclei changes by the quark Pauli effect. The calculation is twofold: One is a complete A body treatment ($A \leq 6$) and another is an α cluster model approximation based on a resonating group method(RGM) formalism (for ${}^{9}_{\Lambda}$Be and ${}^{10}_{\Lambda\Lambda}$Be). The latter is rather a qualitative estimation of energy change for larger mass system than $A = 6$.

In a naive baryon picture of ${}^{5}_{\Lambda}$He, for example, Λ is distinguishable from the nucleon so that ${}^{5}_{\Lambda}$He is considered a spatially compact system as ^{4}He is. In such a compact system, quark effects might appear due to the strong overlapping of each baryon. If we assume that a baryon is described as a three quark wave function of $(0s)^3$ and all five baryons completely overlap, such a state must be forbidden due to quark Pauli principle. (This system contains seven u and d quarks.) This situation derived from the quark picture must be reflected in ${}^{5}_{\Lambda}$He based on the baryon model. Namely, there should exist some forbidden states in a model space describing ${}^{5}_{\Lambda}$He in a baryon picture. Even in case of the system consisting of less than five baryons, there exists the effect of the quark Pauli principle, as was discussed by Takeuchi and Shimizu[1]. Such a light system, however, does not produce a quark Pauli forbidden state mentioned above. It is important to note, however, that quark effects might be included into an appropriate effective baryon-baryon potential. The quark Pauli effect including five baryons cannot be represented by two- or three-body effective force.

Taking into account of the quark Pauli effect into a framework of baryon few body calculation amounts to constraining the model space of the variational calculation. In

CP594, *Hadrons and Nuclei: First International Symposium,* edited by Il-T. Cheon et al.

order to perform such a calculation, we must define a five-baryon configuration which is forbidden by the Pauli principle at the quark level. As explained in Ref.[2, 3, 4] in detail, we give briefly our idea below. We assume that a baryon is a $3q$ system, and the orbital motion of each quark is described by a $(0s)$ harmonic-oscillator function, $\exp\{-\rho^2/2b^2\}$, with a size parameter b. The relative motion function between the baryons of a 15-quark, five-baryon wave function becomes Pauli-forbidden when the function takes the form

$$\psi_{\rm F} = \left(\frac{\pi b^2}{3}\right)^{-\frac{3}{4}} \exp\left\{-\frac{3}{2b^2}\sum_{i=1}^{5}(\mathbf{r}_i - \mathbf{R})^2\right\}, \tag{1}$$

where $\mathbf{R} = \frac{1}{5}\sum_{i=1}^{5}\mathbf{r}_i = \frac{1}{15}\sum_{i=1}^{15}\rho_i$ is the center-of-mass coordinate of the five baryons. The ground state wave function Ψ must be subjected to remain in the subspace which is orthogonal to the forbidden state, that is, $\langle\psi_F|\Psi\rangle = 0$. The constraint applies to the hypernuclei with $A \geq 5$.

In order to evaluate quantitatively the binding energy reduction due to the quark Pauli effect, the variational trial function has to be chosen flexibly enough. We assumed that the trial function with the total orbital angular momentum $L = 0$ is given by

$$\Psi_{JMIM_I} = \sum_{k=1}^{K} c_k\Big(\varphi_k - \sum_{a}\Gamma_a\langle\Gamma_a|\varphi_k\rangle\Big),$$

$$\varphi_k = \mathcal{A}_{\rm B}\{G(\mathbf{x},A_k)\chi_{kJM}\eta_{kIM_I}\},\ \ G(\mathbf{x},A_k) \equiv \exp\left\{-\frac{1}{2}\mathbf{x}A_k\mathbf{x}\right\}, \tag{2}$$

where Γ_a is an orthonormal set of functions constructed from the Pauli forbidden space. In the case of $A = 5$, Γ_a can simply be represented by $\mathcal{A}\{\psi_{\rm F}\chi\eta\}$. For $A \geq 6$, however, the construction of Γ_a is never trivial. A method proposed in Ref. [3] enables us to practically eliminate the Pauli forbidden components. $\mathcal{A}_{\rm B}$ is the operator antisymmetrizing identical baryons, $\mathbf{x} = (\mathbf{x}_1, ..., \mathbf{x}_{A-1})$ stands for a set of relative (e.g., Jacobi) coordinates, A_k is positive definite matrix characterizing the k-th basis, and χ_{kSM_S} (η_{kIM_I}) is the spin (isospin) function of the system.

In the α cluster model, we replace the above basis function φ_k with

$$\varphi_k = \mathcal{A}_{\rm B}\{\phi_{\alpha_1}\phi_{\alpha_2}G(\mathbf{x},A_k)\chi_{kJM}\eta_{kIM_I}\},\quad \text{for } {}^{9}_{\Lambda}\text{Be},\ {}^{10}_{\Lambda\Lambda}\text{Be}. \tag{3}$$

The definition of $\mathcal{A}_{\rm B}$ is same as above, but $\mathbf{x} = (\mathbf{x}_1, ..., \mathbf{x}_{N-1})$ only stands for a set of *intercluster* relative coordinates (e.g., $N = 3$ for ${}^{9}_{\Lambda}$Be and $N = 4$ for ${}^{10}_{\Lambda\Lambda}$Be). Since the spin and the isospin are both zero for ^{4}He, and the spin of 2Λ couples to singlet, both the configurations of total spin and isospin function are trivial. ϕ_α is the intrinsic wave function of α cluster and is constructed from a harmonic oscillator $\prod_{i=1}^{4}\exp\left\{-\frac{1}{2\alpha^2}(\mathbf{r}_i - \mathbf{r}_{\alpha_{CM}})^2\right\}$ and the Slater determinant, where the size parameter is $\alpha = 1.39$ fm. In the present calculation of the α-cluster approximation the Pauli forbidden space spanned by Γ_a does not contain deformed component of the α cluster.

The basis element is increased one by one in the stochastic variational method[5, 6, 7] until the solution reaches a practical convergence. Here the basis set φ_k is chosen so as to minimize the energy, whereas the set Γ_a is chosen to maximize the energy. The accuracy of solutions is vital in the present calculation because we have to take account of very small components produced by the quark Pauli principle.

To consider the effect of the quark structure of baryons on the binding energy of Λ hypernuclei, it is desirable to use a ΛN potential that reproduces the Λ separation energies of $A = 3,4$ Λ hypernuclei. The NN and ΛN potentials were the same as employed in Ref. [4]. The NN potential reproduces the binding energies and sizes of ^{2}H, ^{3}H, ^{3}He, and ^{4}He very well[6], and likewise the ΛN potential is set to well reproduce the energies of $A = 3,4$ hypernuclei. The $\Lambda\Lambda$ potential was taken from Ref. [8].

Table 1 compares the Λ separation energies calculated for $A = 5$ systems ($^5_\Lambda$He, $^5_{\Lambda\Lambda}$H, $^5_{\Lambda\Lambda}$He). Spin, parity and isospin (J^π, I) are $(\frac{1}{2}^+, 0)$, $(\frac{1}{2}^+, \frac{1}{2})$, and $(\frac{1}{2}^+, \frac{1}{2})$, respectively. Table 1 also shows the result of $A = 6$ system ($^6_{\Lambda\Lambda}$He), which will be discussed later.

As shown in Table 1, the quark Pauli effect leads to a significant reduction in the binding energy $^5_\Lambda$He at $b = 0.86$ fm. Although the value of b is chosen referring to the experimental proton's charge radius, a reasonable value of b could be slightly smaller than 0.86 fm because we know that a baryon is made from not only three quarks but also other contributions, e.g. meson clouds. In the case of $b = 0.6$ fm, the energy reduction becomes small. The quark Pauli effect becomes much less in $^5_{\Lambda\Lambda}$H and $^5_{\Lambda\Lambda}$He, which reflects the fact that they are spatially extended more than $^5_\Lambda$He. In fact, the root mean square (rms) radius of $^5_{\Lambda\Lambda}$H (and also $^5_{\Lambda\Lambda}$He) turns out to be 2.0 fm, while that of $^5_\Lambda$He is 1.6 fm.

These results can be qualitatively understood as follows. Let Ψ_0 and E_0 denote the wave function and the energy of the ground state which is obtained without the quark Pauli effect. When the Pauli effect is taken into account, the wave function may be written according to the perturbation theory as

$$\Psi = \frac{1}{\sqrt{1-\varepsilon^2}}(\Psi_0 - \varepsilon\psi_F) \quad \text{with} \quad \varepsilon = \langle\psi_F|\Psi_0\rangle. \tag{4}$$

Assuming that ε is small, the energy difference is given by

$$E - E_0 = \langle\Psi|H|\Psi\rangle - E_0 \approx \varepsilon^2\langle\psi_F|H|\psi_F\rangle. \tag{5}$$

TABLE 1. The Λ separation energies, in MeV, of $A = 5$ and 6 systems as a function of the baryon size b (in fm). The case of $b = 0$ indicates that no quark Pauli effect is imposed in the calculation. Set A of ΛN potential parameters are used. (See Ref.[4].)

	Theory				Expt.
	$b = 0$	0.6	0.7	0.86	
$B_\Lambda(^5_\Lambda\text{He})$	4.98	4.90	4.70	3.63	3.12±0.02
$B_{\Lambda\Lambda}(^5_{\Lambda\Lambda}\text{H})$	5.63	5.59	5.52	5.12	
$B_{\Lambda\Lambda}(^5_{\Lambda\Lambda}\text{He})$	5.53	5.52	5.45	5.05	
$B_{\Lambda\Lambda}(^6_{\Lambda\Lambda}\text{He})$	14.3	13.9	12.9	9.4	10.9±0.6

The energy difference is determined by the product of ε^2, the probability of finding the forbidden component in Ψ_0, and the energy expectation value of the Pauli forbidden state. Let us use the above equation to estimate how much the energy difference is produced by the quark Pauli principle. In the case of ${}^{5}_{\Lambda}$He, $\varepsilon^2 \approx 0.004$, $\langle\psi_F|H|\psi_F\rangle \approx 500$ MeV for $b = 0.86$ fm, and thus $E - E_0 \approx 2$ MeV, while in the case of ${}^{5}_{\Lambda\Lambda}$H $\varepsilon^2 \approx 0.001$, $\langle\psi_F|H|\psi_F\rangle \approx 800$ MeV, and $E - E_0 \approx 1$ MeV. This estimation is in qualitatively good agreement with the result of Table 1.

We have calculated the binding energy of ${}^{6}_{\Lambda\Lambda}$He in a full six-body treatment. This system has three types of forbidden states (1), containing $ppnn\Lambda$, $ppn\Lambda\Lambda$, or $pnn\Lambda\Lambda$ five baryons. The energy obtained without the quark Pauli effect ($b = 0$) is apparently overbound. The elimination of the forbidden states leads to a surprisingly large reduction in the binding energy. The rms radius of ${}^{6}_{\Lambda\Lambda}$He is calculated to be 1.6 fm, considerably smaller than that (2.0 fm) of ${}^{5}_{\Lambda\Lambda}$H, which leads to a larger quark Pauli effect in ${}^{6}_{\Lambda\Lambda}$He.

In the calculations for ${}^{9}_{\Lambda}$Be and ${}^{10}_{\Lambda\Lambda}$Be we used the RGM-type basis function, that is, the basis functions have $2\alpha + \Lambda$ three body configuration for ${}^{9}_{\Lambda}$Be and $2\alpha + 2\Lambda$ four body configuration for ${}^{10}_{\Lambda\Lambda}$Be, respectively (See Eq. (3)). Table 2 shows the Λ separation energies for ${}^{9}_{\Lambda}$Be and ${}^{10}_{\Lambda\Lambda}$Be. Both B_Λ and $B_{\Lambda\Lambda}$ are values from the ground state level of ^{8}Be. In these calculations, the exchange mixture parameter u of the NN potential is set to $u = 0.95$ so as to reproduce the $\alpha + \alpha$ scattering. And the parameter u of the ΛN potential is set to $u = 1.3$ which is tuned to reproduce the energy of ${}^{9}_{\Lambda}$Be at the baryon size $b = 0.86$ fm. Then the energy of ${}^{10}_{\Lambda\Lambda}$Be is manifestly overbound at $b = 0$ fm, whereas the energy at $b = 0.86$ fm is surprisingly reduced to near the experimental value.

We studied the consequence of the quark structure of the baryons on the binding energies of Λ- and $\Lambda\Lambda$ hypernuclei. We showed that, though the Λ particle can be distinguished from the nucleon, the quark model predicts a special forbidden state involving five baryons. By excluding the forbidden states we found that the binding energies started to decrease at the baryon size of $b = 0.6$ fm and reduced significantly at the size of free baryons, $b = 0.86$ fm. Since the forbidden states considered here are many-body configurations involving five baryons, the Pauli effect cannot be revealed through a study on the baryon-baryon interaction alone. In other words, a full understanding of the binding mechanism of the hypernuclei may not be attainable only through the knowledge of the two- and three-body interactions of the baryons.

TABLE 2. The Λ separation energies, in MeV, of $A = 9$ and 10 systems calculated by the cluster model as a function of the baryon size b (in fm). The case of $b = 0$ indicates that no quark Pauli effect is imposed in the calculation. Set A of ΛN potential parameters are used. (See Ref.[4].)

	Theory		Expt.
	$b = 0$	0.86	
$B_\Lambda({}^{9}_{\Lambda}\text{Be})$	8.43	6.75	6.71 ± 0.04
$B_{\Lambda\Lambda}({}^{10}_{\Lambda\Lambda}\text{Be})$	22.3	18.2	17.7 ± 0.4

ACKNOWLEDGMENTS

We are thankful to Y. Fujiwara and C. Nakamoto for useful discussions. One of the authors (H.N.) would like to thank Y. Akaishi for helpful discussions and JSPS Research Fellowships for Young Scientists.

REFERENCES

1. Takeuchi, S., Shimizu, K., *Phys. Lett.* **B179**, 197-200 (1986).
2. Nemura, H., Suzuki, Y., Fujiwara, Y., Nakamoto,C., *Prog. Theor. Phys.* **101**, 981-986 (1999).
3. Suzuki, Y., Nemura, H., *Prog. Theor. Phys.* **102**, 203-208 (1999).
4. Nemura, H., Suzuki, Y., Fujiwara, Y., Nakamoto, C., *Prog. Theor. Phys.* **103**, 929-958 (2000).
5. Kukulin, V. I., Krasnopol'sky, V. M., *J. Phys.* **G3**, 795-811 (1977).
6. Varga, K., Suzuki, Y., *Phys. Rev.* **C52**, 2885-2905 (1995).
7. Suzuki, Y., Varga, K., *Stochastic Variational Approach to Quantum-Mechanical Few-Body Problems* (Lecture Notes in Physics, Vol. m54), Springer-Verlag, Berlin, 1998.
8. Hiyama, E., Kamimura, M., Motoba, T., Yamada, T., Yamamoto, Y., *Prog. Theor. Phys.* **97**, 881-899 (1997).

Shell-Model Calculations of ${}^{17}_{\Lambda}$O using Microscopic ΛN and ΣN Effective Interactions

Shinichiro Fujii*, Ryoji Okamoto† and Kenji Suzuki†

*RI Beam Factory Project Office, The Institute of Physical and Chemical Research (RIKEN), Wako 351-0198, Japan
†Department of Physics, Kyushu Institute of Technology, Kitakyushu 804-8550, Japan

Abstract. Shell-model calculations in a large model space are performed for the low-lying $1/2^+$, $3/2^-$ and $1/2^-$ states in ${}^{17}_{\Lambda}$O, introducing effective interactions among the Λ, Σ and nucleons. The effective interactions are derived from hyperon-nucleon and nucleon-nucleon interactions given in free space in the framework of the unitary-model-operator approach. The results are compared with those obtained in a perturbative method studied previously, and reliability of the present method is discussed.

INTRODUCTION

One of the challenging problems in theoretical study of Λ hypernuclei is to describe their structure starting from hyperon-nucleon (YN) and nucleon-nucleon (NN) interactions given in free space. For this purpose we need to introduce an effective interaction derived from a bare interaction, except for study of few-body systems.

In previous works [1, 2], we proposed a method for a microscopic description of Λ hypernuclei in the framework of the unitary-model-operator approach (UMOA) [3, 4]. The UMOA is a many-body theory that leads to an energy-independent and Hermitian effective interaction with the property of decoupling [5] between two states in a model space and excluded one. We applied this method to calculations of Λ single-particle energies in ${}^{17}_{\Lambda}$O and ${}^{41}_{\Lambda}$Ca, using various YN interactions given by the Nijmegen [6, 7] and the Jülich [8] groups. Some reasonable results were obtained, such as small spin-orbit splittings of Λ in comparison with those in nuclei. Recently some experiments on Λ spin-orbit splittings have been performed for several hypernuclei such as ${}^{89}_{\Lambda}$Y [9], ${}^{13}_{\Lambda}$C [10] and ${}^{9}_{\Lambda}$Be [11]. Considerably higher energy resolution has been attained to gain a better understanding of the Λ spin-orbit splitting and the underlying ΛN interaction. This marked progress in experiment realizes the importance of precise calculation of properties of Λ hypernuclei.

Shell-model calculation is one of the promising methods to make structure calculations of Λ hypernuclei [12, 13, 14]. If we want to understand hypernuclear structure in the shell-model calculation starting from a baryon-baryon interaction in free space, an effective interaction has to be introduced in a microscopic way. In the case of Λ hypernuclei, it is of great importance to take into account the coupling effect between the ΣN and ΛN channels because of the small mass difference between the Λ and Σ. In the shell-model calculations of this sort made so far, all the effects of the ΣN channel has been

CP594, *Hadrons and Nuclei: First International Symposium*, edited by Il-T. Cheon et al.

treated as renormalization into the ΛN effective interaction. The degrees of freedom of Σ have not been treated explicitly in the shell-model calculations. Therefore it is interesting to derive an effective YN interaction which includes not only the ΛN channel but also the ΣN one and to apply such an effective interaction to the shell-model calculation of Λ hypernuclei.

Lately we have been making a shell-model calculation of $^{17}_{\Lambda}$O in a large model space which contains the states of the Σ and nucleons in addition to those of the Λ and nucleons. The effective interaction including the ΛN and ΣN channels was introduced in the framework of the UMOA. In this report the present shell-model approach is outlined, and some preliminary results of the numerical calculation are shown. The results obtained are compared with a result of the previous calculation in a perturbative treatment, and reliability of the two methods is discussed.

CALCULATION PROCEDURE

Hamiltonian

The effective Hamiltonian for a system of nucleons and one hyperon is given by

$$H_{\rm SM} = \tilde{H} + H_{\beta}, \qquad (1)$$

where

$$\begin{aligned} \tilde{H} &= \sum_{\rm Y=\Lambda,\Sigma} t_{\rm Y}^{(1)} + \sum_{\rm Y=\Lambda,\Sigma} u_{\rm Y}^{(1)} + m_{\Sigma-\Lambda} + \sum_{i=1}^{A}(t_{\rm N}^{(i)} + u_{\rm N}^{(i)}) \\ &+ \sum_{\substack{i=1 \\ \rm Y,Y'=\Lambda,\Sigma}}^{A} \tilde{v}_{\rm YN-Y'N}^{(1,i)} + \sum_{i<j=1}^{A} \tilde{v}_{\rm NN-NN}^{(i,j)}, \qquad (2) \\ H_{\beta} &= \beta|1^{-}{}_{\rm CM}\rangle\langle 1^{-}{}_{\rm CM}|. \qquad (3) \end{aligned}$$

The terms t_k for $k = \Lambda, \Sigma$ and N are the kinetic energies of Λ, Σ and nucleons, respectively. The single-particle potential energies u_k for $k = \Lambda, \Sigma$ and N can be determined self-consistently with the two-body effective interactions $\tilde{v}_{\Lambda\rm N-\Lambda N}$, $\tilde{v}_{\rm NN-NN}$ and $\tilde{v}_{\Sigma\rm N-\Sigma N}$, respectively. The term $m_{\Sigma-\Lambda}$ denotes the mass difference between the Λ and Σ, and A stands for the mass number of a core nucleus.

It should be noted that, in the shell-model calculation, spurious states caused by the center-of-mass (CM) motion are mixed into physical states. In the present case of $^{17}_{\Lambda}$O, the one-particle one-hole spurious $1^-(T=0)$ state affects physical states, especially, the $3/2^-(T=0)$ and $1/2^-(T=0)$ states. Therefore, we need to remove the spurious 1^- state carefully. We follow the prescription as given in Ref. [15]. The value β in Eq. (3) should be sufficiently large. In the present calculation we take as $\beta = 3\hbar\omega$ with the harmonic-oscillator frequency which has been determined as $\hbar\omega = 14$MeV in the calculation of ^{16}O [3].

Model Space

The model space of two-particle states for the ΛN and ΣN channels is defined graphically in Fig. 1. The boundary number $\rho_{\rm Y}$ for Y = Λ or Σ is given by

$$\rho_{\rm Y} = 2n_{\rm Y} + l_{\rm Y} + 2n_{\rm N} + l_{\rm N} \tag{4}$$

with the harmonic-oscillator quantum numbers, and $\rho_{\rm X}$ is defined the same as $\rho_{\rm Y}$. The number $\rho_{\rm F}$ denotes the uppermost occupied state of the core nucleus. The ΛN and ΣN effective interactions $\tilde{v}_{\Lambda{\rm N}-\Lambda{\rm N}}$ and $\tilde{v}_{\Sigma{\rm N}-\Sigma{\rm N}}$ are determined self-consistently with the one-body potential u_Λ and u_Σ, by solving the decoupling equation [1] between the model space $\rm P_{\Lambda N} + P_{\Sigma N}$ and its complement $(\rm Q_{\Lambda N} - Q^{(X)}_{\Lambda N}) + (Q_{\Sigma N} - Q^{(X)}_{\Sigma N})$. It is noted here that the $\rm Q^{(X)}_{\Lambda N}$ and $\rm Q^{(X)}_{\Sigma N}$ spaces should be excluded due to the Pauli principle for nucleons. The effective interaction of the transition channel $\tilde{v}_{\Sigma{\rm N}-\Lambda{\rm N}}$ is also calculated simultaneously. The NN effective interaction $\tilde{v}_{\rm NN-NN}$ is derived self-consistently with the one-body potential $u_{\rm N}$ as in Ref. [4].

In principle, the values of ρ_Λ and ρ_Σ should be taken as large as possible so that the ρ_Λ and ρ_Σ dependence of the results in the numerical calculation becomes negligible. We here show only the result for $\rho_\Lambda = 8$ and $\rho_\Sigma = 5$.

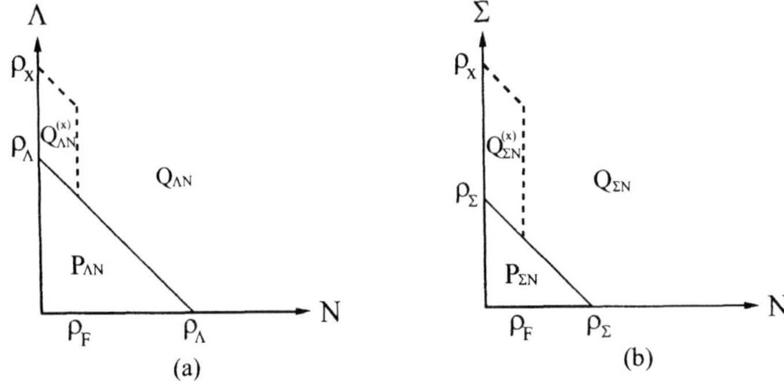

FIGURE 1. The model space of the ΛN and ΣN channels and its complement.

Shell-Model Diagonalization

The shell-model space we adopted is composed of the states, $d^\dagger_{\rm Y}|\Phi_0\rangle$ and $d^\dagger_{\rm Y}a^\dagger b^\dagger|\Phi_0\rangle$, where $d^\dagger_{\rm Y}$ is the creation operator of a hyperon Y, $a^\dagger(b^\dagger)$ the creation operator of a particle (hole). The state $|\Phi_0\rangle$ is the correlated ground state of the core nucleus which contains two-particle two-hole ($2p$-$2h$), $4p$-$4h$ and higher-order particle-hole components. The effective Hamiltonian in the UMOA is given through the unitary transformation as $H_{\rm eff} = e^{-S}He^{S}$. The essence of the UMOA is that the operator S is determined so that the transformed Hamiltonian does not contain interactions inducing $2p$-$2h$ excitations in the ground state of the core nucleus. The $2p$-$2h$ excitations and related higher-order many-particle many-hole excitations are incorporated into the correlated ground state

$|\Phi_0\rangle = e^S|\phi_0\rangle$, where $|\phi_0\rangle$ is the unperturbed ground state. In general the transformed Hamiltonian $H_{\rm eff}$ contains three-or-more-body effective interactions. However in the present calculation we neglect the many-body effective interactions, and take one- and two-body parts in $H_{\rm eff}$ as the effective Hamiltonian $\tilde{H}$ in Eq. (2). [1] In this approximation the direct coupling of $d_Y^\dagger|\Phi_0\rangle$ and $d_Y^\dagger a^\dagger a^\dagger b^\dagger b^\dagger|\Phi_0\rangle$ does not occur anymore. We may say that it is sufficient to take only the space of $d_Y^\dagger|\Phi_0\rangle$ and $d_Y^\dagger a^\dagger b^\dagger|\Phi_0\rangle$ as the shell-model space in which the effective Hamiltonian is diagonalized.

Unlinked-diagram contributions often emerge in shell-model calculations. In order to remove unlinked terms we calculate separately the correlation energy E_0 of the core nucleus in the space of $|\Phi_0\rangle$ and $a^\dagger b^\dagger|\Phi_0\rangle$. We subtract E_0 from the eigenvalue E of the Hamiltonian $H_{\rm SM}$ in Eq. (1). The value $E - E_0$ corresponds to the binding energy of Λ relative to the ground state energy of the core ^{16}O.

RESULTS AND DISCUSSION

We performed calculations employing the Nijmegen soft-core 97f (NSC97f) [7] potential for the YN interaction. As for the NN interaction, we used the Paris [16] potential. All the interaction matrix elements of the Hamiltonian in Eq. (2) can be derived from these bare interactions without any adjustable parameters in the framework of the UMOA.

In Fig. 2, calculated energy levels of $^{17}_{\Lambda}$O are shown for the boundary numbers, $\rho_\Lambda = 8$, $\rho_\Sigma = 5$ and $\rho_X = 12$. We also show a result obtained in the previous work [2], which we refer to as II, in a perturbative treatment. Note that the contributions of the effective three-body interactions are not included in these results.

The model space of the ΛN channel for $\rho_\Lambda = 8$ is considered to be sufficiently large for the convergence of the results, which was confirmed in II. In II the ΣN channel was treated as renormalization into the ΛN effective interaction. In other words, the ΣN states were not included in the model space, namely $\rho_\Sigma = 0$. The matrix elements of the ΛN effective interaction obtained in the present work are, in general, different from those in II even if the same YN interaction is adopted in the calculation. The results of the shell-model calculations are, however, quite similar with each other. This fact suggests that the renormalization of the effects of the ΣN channel into the ΛN effective interaction was done very well. Furthermore we see in Fig. 2 that the relative spacing of the relevant levels obtained in the present work agree fairly well with those in II, though the differences in the absolute energy of about 1MeV are observed. This feature means that two methods, the perturbative method and the shell-model diagonalization, would be reliable.

To obtain a definite conclusion, however, we should confirm the convergence of the calculated results by enlarging the size of the model space of the ΣN channel and investigate the dependence of the results on the YN interactions adopted. Calculated

[1] In the previous work [2], three-body cluster effects have been evaluated as one of the higher-order contributions.

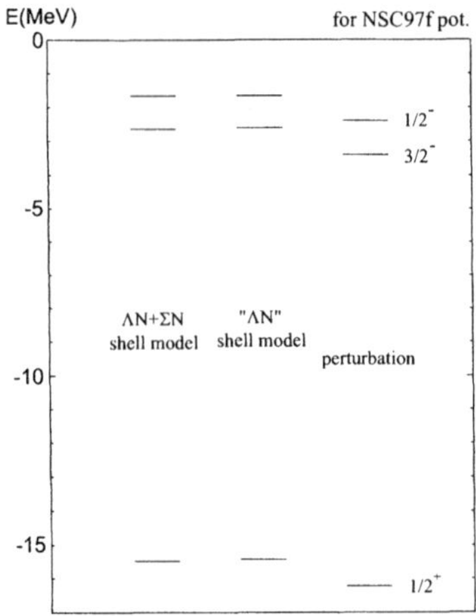

FIGURE 2. Calculated energy levels in $^{17}_{\Lambda}$O. The levels labeled as ΛN+ΣN are the results of the shell-model calculation in the model space including ΣN states. On the other hand, the levels with "ΛN" are the results in the model space of only ΛN states. The levels labeled as perturbation are the results obtained in the previous work.

results with sufficiently large model space of the ΣN channel and for various YN interactions will be reported in a forthcoming paper.

This work was supported in part by the Special Postdoctoral Researchers Program (S. F.) at RIKEN.

REFERENCES

1. Fujii, S., Okamoto, R., and Suzuki, K., *Nucl. Phys.* **A651**, 411-433 (1999); *ibid.* **A676**, 475-476 (2000).
2. Fujii, S., Okamoto, R., and Suzuki, K., *Prog. Theor. Phys.* **104**, 123-141 (2000).
3. Suzuki, K., Okamoto, R., and Kumagai, H., *Phys. Rev.* **C36**, 804-819 (1987).
4. Suzuki, K., Okamoto, R., *Prog. Theor. Phys.* **92**, 1045-1080 (1994).
5. Suzuki, K., and Lee, S. Y., *Prog. Theor. Phys.* **64**, 2091-2106 (1980).
6. Maessen, P. M. M., Rijken, Th. A., and de Swart, J. J., *Phys. Rev.* **C40**, 2226-2245 (1989).
7. Rijken, Th. A., Stoks, V. G. J., and Yamamoto, Y., *Phys. Rev.* **C59**, 21-40 (1999).
8. Reuber, A., Holinde, K., and Speth, J., *Nucl. Phys.* **A570**, 543-579 (1994).
9. Nagae, T., *Nucl. Phys.* **A670**, 269c-272c (2000).
10. Kishimoto, T., et al., in *Proceedings of the International Workshop on Hypernuclear Physics with Electromagnetic Probes*, 2-4 December 1999, Hampton, Virginia, to be published.
11. Tamura, H., et al., *Nucl. Phys.* **A663-664**, 481c-484c (2000).
12. Motoba, T., *Nucl. Phys.* **A639**, 135c-146c (1998).
13. Yiharn Tzeng, Tsay Tzeng, S. Y., Kuo, T. T. S., Lee, T. -S. H., and Stoks, V. G. D., *Phys. Rev.* **C61**, 031305(R) (2000).
14. Millener, D. J., *nucl-th*/0103017, to appear in *Nucl. Phys.* **A**.
15. Yamada, T., Motoba, T., Ikeda, K., and Bandō, H., *Prog. Theor. Phys.* Supplement 81, 104-146 (1985).
16. Lacombe, M., Loiseau, B., Richard, J. M., Vinh Mau, R., Côté, J., Pirès, P., and de Tourreil, R., *Phys. Rev.* **C21**, 861-873 (1980).

Pionic Weak-Decay of Lightest double-Λ Hypernucleus ${}^{4}_{\Lambda\Lambda}$H

Izumi Kumagai-Fuse* and Shigeto Okabe*

*Center for Information and Multimedia Studies, Hokkaido University, Sapporo 060-0811, JAPAN

Abstract. The weak π^--decay of ${}^{4}_{\Lambda\Lambda}$H are theoretically analyzed with the recent experimental data by BNL-E906. The two-body π^--decay width of ${}^{4}_{\Lambda\Lambda}$H is calculated to be 0.69 Γ_Λ, and its branching ratio to the total π^- decay is about 25%. The branching ratio of decay channels accompanied with a Λ particle to the total π^--decay is about 43%. The calculated π^- spectrum of ${}^{4}_{\Lambda\Lambda}$H has a broad peak around 100 MeV/c in nuclear continuum region, whose value is inconsistent with the experimental results of a peak structure around 103 MeV/c.

INTRODUCTION

Nuclear bound states with strangeness $S=-2$ can provide important information about a ΛΛ and a ΞN interaction, which is a basic knowledge to investigate structures of nuclear systems consisting of octet baryons, such as condensed matters at the center of neutron stars. Observations of the $S=-2$ nuclear systems are, however, very limited at present due to difficulties of an identification of $S=-2$ hypernuclear species. The identificated data are only for three cases of ${}^{10}_{\Lambda\Lambda}$Be [1], ${}^{6}_{\Lambda\Lambda}$He [2], and ${}^{13}_{\Lambda\Lambda}$B (or ${}^{10}_{\Lambda\Lambda}$Be) [3]. The first two cases have been observed in the early 1960's, and the last one has been observed in 1991.

Higher intense K^- beams produce more double-Λ hypernuclei by (K^-,K^+) and by stopped Ξ^- reactions. Characteristic weak-decay pions from the hypernuclei are the most useful tool to identify hypernuclear species. Recently, Fukuda *et al.* have reported that two π^- particles from ${}^{4}_{\Lambda\Lambda}$H sequential weak-decays might be measured by the BNL-E906 experiment concerning to the reaction of a stopped Ξ^- on ^{9}Be [4].

${}^{4}_{\Lambda\Lambda}$H is predicted to be the lightest double-Λ hypernucleus by a theoretical calculation as a few-body system [5]. The observation of ${}^{4}_{\Lambda\Lambda}$H solves the problem whether a ΛΛ interaction is attractive or not. Then the pionic decay of ${}^{4}_{\Lambda\Lambda}$H should be carefully analyzed not only experimentally but also theoretically. Yamamoto *et al.* have calculated the pionic-decay widths of ${}^{4}_{\Lambda\Lambda}$H but not decay spectra [6]. They have discribed ${}^{4}_{\Lambda\Lambda}$H as a system of ${}^{3}_{\Lambda}$H(core) and Λ. Then they have calculated decay widths only for the cases of ${}^{4}_{\Lambda\Lambda}\mathrm{H} \rightarrow {}^{4}_{\Lambda}\mathrm{He}(0^+)+\pi^-$ and ${}^{3}_{\Lambda}\mathrm{H}+\mathrm{p}+\pi^-$ for π^- decays, where the converted protons go to a bound state and continuum states. Not only ${}^{3}_{\Lambda}\mathrm{H}+\mathrm{p}+\pi^-$ channel but also $({}^{2}\mathrm{H}+\Lambda)_{3/2}+\mathrm{p}+\pi^-$ channel should be taking into account for the pionic decay of ${}^{4}_{\Lambda\Lambda}$H, since two spin configulations of ${}^{3}_{\Lambda}$H, 1/2 and 3/2, are mixed in ${}^{4}_{\Lambda\Lambda}$H, although there exist

CP594, *Hadrons and Nuclei: First International Symposium*, edited by Il-T. Cheon et al.

no bound states of $(^2\mathrm{H}+\Lambda)_{3/2}$. In this paper, we discuss the pionic decays of ${}^{4}_{\Lambda\Lambda}\mathrm{H}$, focusing on its π^--decay widths and sequential π^--decay spectra and compare the results to recent BNL-E906 experiment. We also take decay channels emitted with a Λ particle such as $^2\mathrm{H}+\Lambda+\mathrm{p}+\pi^-$ into consideration.

FRAMEWORK

All π^--decay modes of ${}^{4}_{\Lambda\Lambda}\mathrm{H}$ are given by:

$$
\begin{aligned}
{}^{4}_{\Lambda\Lambda}\mathrm{H} &\rightarrow {}^{4}_{\Lambda}\mathrm{He}(0^+)+\pi^- && (1)\\
&\rightarrow {}^{4}_{\Lambda}\mathrm{He}(1^+)+\pi^- && (2)\\
&\rightarrow {}^{3}_{\Lambda}\mathrm{H}+\mathrm{p}+\pi^- && (3)\\
&\rightarrow {}^{3}\mathrm{He}+\Lambda+\pi^- && (4)\\
&\rightarrow {}^{2}\mathrm{H}+\Lambda+\mathrm{p}+\pi^- && (5)\\
&\rightarrow \mathrm{p}+\mathrm{n}+\mathrm{p}+\Lambda+\pi^- && (6)
\end{aligned}
$$

A spin and a parity of ${}^{4}_{\Lambda\Lambda}\mathrm{H}$ have to be 1^+ because of the main configuration of $[^2\mathrm{H}(1^+)\otimes\Lambda\Lambda(0^+)]^{1^+}$. Therefore the process (2) dominates the two-body decay. The process (1) is largely suppressed since spin-non-flip processes dominate pionic decays. Yamamoto *et al.*, however, have not taken process (2) into consideration [6]. We neglect contributions from the process (6), since the branching ratio is expected to be small from the analysis by Kamada *et al.* about a decay of ${}^{3}_{\Lambda}\mathrm{H}\rightarrow \mathrm{p}+\mathrm{n}+\mathrm{p}+\pi^-$, whose branching ratio to the total π^- decay is about 1% [7].

We evaluate decay widths by an impulse approximation. For an N-body decay,

$$
\begin{aligned}
\Gamma^{NB}_{\pi^-} &= \frac{(\hbar c)^3}{(2\pi)^{3(N-2)+1}}\int d\vec{k}_{\pi^-}d\vec{k}_{c_1}\cdots d\vec{k}_{c_{N-1}}\frac{1}{E_\pi}\delta(E_f-E_i)\\
&\quad\times\delta(\vec{k}_{\pi^-}+\vec{k}_{c_1}+\cdots+\vec{k}_{c_{N-1}})\frac{1}{2J_i+1}\sum_{M_i,J_f,M_f}|\mathcal{M}_{fi}|^2, \qquad (7)
\end{aligned}
$$

$$
\mathcal{M}_{fi} \equiv \left\langle \Psi_f \left| \left[s_{\pi^-} + i p_{\pi^-}\frac{\left(\vec{\sigma}\cdot\vec{\nabla}_\pi\right)}{k^{(0)}_{\pi^-}}\right]\chi^{(-)*}_{\pi^-}O_{\Lambda\rightarrow p}\right| \Psi_{{}^{4}_{\Lambda\Lambda}\mathrm{H}}\right\rangle, \qquad (8)
$$

where $\int d\vec{k}_{c_i}$ is an integral with respect to a momentum of each emitted particle with a suffix c_i; e.g. $^2\mathrm{H}$ by c_1, Λ by c_2 and p by c_3 for the process (5). Coefficients s_{π^-} and p_{π^-} are interaction constants for a spin-non-flip and a spin-flip processes, respectively. The ratio $s^2_{\pi^-} : p^2_{\pi^-}$ is taken to be 0.88 : 0.12. A spin operator $\vec{\sigma}$ acts on a decaying hyperon and $O_{\Lambda\rightarrow p}$ is a one-body operator which converts a Λ particle into a proton. For a pion wave function $\chi^{(-)*}_{\pi^-}$, we take both cases of a plain wave (PW) and a distorted wave (DW) which is determined with an optical potential of the MSU group [8]. The modified parameter set 1 is used in the present calculation (See Table III and the figure

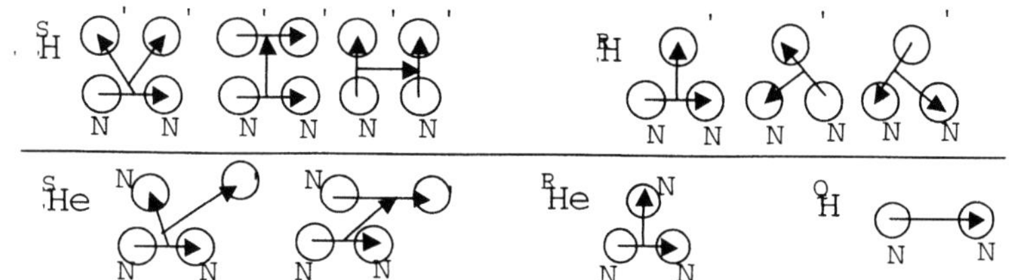

FIGURE 1. Spacial coordinates adopted in variational calculations of bound-states of ${}^{4}_{\Lambda\Lambda}$H,${}^{4}_{\Lambda}$He, ${}^{3}_{\Lambda}$H, ^{3}He and ^{2}H. Antisymmetrizated coordinates are also taken into account.

caption of Fig.5 in Ref.[8]). $\Psi_{{}^{4}_{\Lambda\Lambda}\mathrm{H}}$ and Ψ_f denote the initial- and the final-state nuclear wave functions, respectively.

The two-body pionic-decay width $\Gamma^{2B}_{\pi^-}$ is rewritten as

$$\Gamma^{2B}_{\pi^-} = \Gamma^{(0)}_{\pi^-} \cdot \frac{k_{\pi^-}}{k^{(0)}_{\pi^-}} \cdot \frac{1+\frac{E^{(0)}_{\pi^-}}{M_p c^2}}{1+\frac{E_{\pi^-}}{M_{{}^{4}_{\Lambda}\mathrm{He}} c^2}} \cdot \frac{1}{s^2_{\pi^-}+p^2_{\pi^-}} |\mathcal{M}|^2, \tag{9}$$

where $k^{(0)}_{\pi^-}$ $(E^{(0)}_{\pi^-})$ and $\Gamma^{(0)}_{\pi^-}$ are a π^- wave number (total energy) and a π^--decay width of a free Λ particle, respectively. The π^--decay width is empirically given by $\Gamma^{(0)}_{\pi^-} = 0.64\Gamma_\Lambda$, where Γ_Λ is the total decay width of a free Λ particle. The other three-body and four-body decay widths are also calculated similarly.

In order to get the nuclear bound states of the initial- and the final-states (${}^{4}_{\Lambda\Lambda}$H, ${}^{4}_{\Lambda}$He, ${}^{3}_{\Lambda}$H, ^{3}He, ^{2}H), we employ a variational method. The wave functions are expanded in terms of a Gaussian bases with rearrangement channels in the Jacobi coordinates [9], where we take only a zero-angular momentum for each coordinate. Spacial coordinates used here are shown in Figure 1

We use an interaction by Malfliet and Tjon [10] for NN and interactions by Shinmura and Akaishi [11] for NΛ and $\Lambda\Lambda$, which are determined to reproduce S-matrices of the Nijmegen D potential [12] at low-energies. Values of the strength are readjusted to reproduce binding energies. We treat the $\Lambda\Lambda$ interaction strength as a parameter to see dependences of the π^- decay widths on binding energy of ${}^{4}_{\Lambda\Lambda}$H. The values of parameters and calculated binding energies are listed in Table 1.

For a ${}^{3}_{\Lambda}\mathrm{H}-\mathrm{p}$ and a ${}^{3}\mathrm{He}-\Lambda$ channel, we obtain scattering wave functions by solving single-channel problems with folding potentails, where we use the calculated wave functions of the bound-states.

For the four-body decay of ${}^{2}\mathrm{H}+\Lambda+\mathrm{p}+\pi^-$, we assume that it proceeds as follows: First a Λ particle decays with keeping ${}^{3}_{\Lambda}$H being as a spectator. Next ${}^{3}_{\Lambda}$H breaks to scattering states of $({}^{2}\mathrm{H}+\Lambda)_{1/2,3/2}$.

TABLE 1. Interaction parameters and calculated binding energies (BE) of nuclei. For NN, $V(r) = V_R \frac{e^{-\mu_R r}}{r} - V_A \frac{e^{-\mu_A r}}{r}$. For NΛ and ΛΛ, $V(r) = V_R e^{-(r/\gamma_R)^2} - V_A e^{-(r/\gamma_A)^2}$. An NΛ interaction in triplet-odd state is the averaged one with weights of $(2J+1)$ over $^3P_0, ^3P_1$, and $^3P_2$.

	State	V_R[MeV]	μ_R[1/fm]	V_A[MeV]	μ_A[1/fm]
NN	$^3E(^1E)$	1458.247	3.110	635.393(520.943)	1.555
	State	V_R[MeV]	γ_R[fm]	V_A[MeV]	γ_A[fm]
$N\Lambda$	$^3E(^1E)$	763.11(1165.0)	0.5	83.938(105.12)	1.2
	$^3O(^1O)$	1862.756(272.02)	0.5	44.754(31.143)	1.2
ΛΛ	^1E	5000.0	0.355	332.97	0.8550

	BE_{exp}[MeV]	BE_{cal}[MeV]	f_A^{NN}	$f_A^{N\Lambda}$	$f_A^{\Lambda\Lambda}$	theshold
^{2}H	2.224	2.240	1.0	—	—	p+n
^{3}He	7.72	7.82	0.994	—	—	p+p+n
$^3_\Lambda$H	0.13	0.12	1.0	1.09	—	$d+\Lambda$
$^4_\Lambda$He(0^+)	2.39	2.31	0.994	1.1	—	$^3\mathrm{He}+\Lambda$
$^4_\Lambda$He(1^+)	1.24	1.22	0.994	1.05	—	$^3\mathrm{He}+\Lambda$
$^4_{\Lambda\Lambda}$H	—	0.30	1.0	1.09	0.8	$d+\Lambda+\Lambda$
	—	0.54	1.0	1.09	0.9	$d+\Lambda+\Lambda$
	—	1.24	1.0	1.09	1.0	$d+\Lambda+\Lambda$
	—	2.30	1.0	1.09	1.1	$d+\Lambda+\Lambda$

RESULTS AND DISCUSSIONS

Decay widths

The calculated decay widths of $^4_{\Lambda\Lambda}$H is summarized in Table 2 in the unit of Γ_Λ. In order to see dependences of the calculated two-body decay widths on the ΛΛ interaction, we change the strength of an attractive part of the interaction by a factor $f_A^{\Lambda\Lambda}$. The dependence of decay widths on the factor are shown in Table 2 in the cases of two-body decays.

A pionic-decay width depends essentially on overlappings between a Λ wave function in the initial-state and a proton wave function in the final-state, since both an annihilation of Λ and a creation of proton occurs at the same point due to a short-range nature of the weak interaction $\Lambda \to p\pi^-$. A Λ particle is weakly bound in comparison with nucleons in light Λ hypernuclei. Table 2 shows that the two-body decay width becomes larger with a larger value of $f_A^{\Lambda\Lambda}$, since the overlappings between wave functions of Λ and proton get better by more strongly binding of ΛΛ.

For the s-shell Λ hypernuclei, distortions of pion wave functions have the effect that two-body decay widths are reduced while three-body decay widths are enhanced. The transition matrix element of Eq.(8) is reduced (enhanced) in two-body (three-body) decays with $s(p)$-wave pions, since pion s-waves are pushed out by the distortion while pion p-waves are drawn in, which is caused by a characteristic feature of the π-N interaction through π-nucleus optical potentials in light nuclei. This feature holds in s-

TABLE 2. Calculated π^--decay widths from ${}^{4}_{\Lambda\Lambda}$H are shown in the cases of PW and DW (in parenthesis) in the unit of Γ_Λ. The $\Lambda\Lambda$ binding energy suggested by the BNL-E906 experiment is 0.6 MeV with experimental uncertainties of about 2 MeV.

BE($\Lambda\Lambda$) [*MeV*]	$f_A^{\Lambda\Lambda}$	${}^{4}_{\Lambda}$He(0^+) $\Gamma_{(1)}$	${}^{4}_{\Lambda}$He(1^+) $\Gamma_{(2)}$	${}^{3}_{\Lambda}$H+p $\Gamma_{(3)}$	^{3}He+Λ $\Gamma_{(4)}$	^{2}H+Λ+p $\Gamma_{(5)}$
0.30	0.8	0.01(0.01)	0.65(0.59)			
0.54	**0.9**	**0.01(0.01)**	**0.68(0.62)**	**0.80**	**0.10**	**1.18**
1.24	1.0	0.01(0.01)	0.73(0.66)			
2.30	1.1	0.01(0.01)	0.76(0.67)			

shell double-Λ hypernuclei. Unfortunately we have obtained no π-hypernucleus optical potentials. In order to see the effect of the pion distortion, we evaluate two-body decay widths with the optical potential of $\pi^- - {}^4$He instead of $\pi^- - {}^4_\Lambda$He. This effect is roughly 10 % as seen in Table 2. Since an optical potential of $\pi^- - {}^4$He would have stronger effect than that of $\pi^- - {}^4_\Lambda$He, distortion effects are expected to be less than 10%.

For the three-body and four-body decay widths, we take only a spin-non-flip term with $s^2_{\pi^-} = 1$ preliminarily. In three-body decays, main contributions to the widths come from scattering states of the relative p-waves in the ${}^3_\Lambda$H-p and the ^{3}He-Λ channels, which exhaust 65 % of the widths in each decay process. We have no resonance states of ${}^3_\Lambda$H $- p$ in the present calculation. Though the possibility of a resonance state is mentioned in ref.[4] concerning to the BNL-E906 experiment. Detailed are discussed in the following subsections.

The width of four-body decay of ^{2}H$+\Lambda+p+\pi^-$ has a large contribution ($\sim$43%) for the total π^--decay width, which are dominated by the configuration of $[{}^2\mathrm{H}(1^+)\otimes\Lambda]_{(3/2)}$. This is explained facts that nuclear states of three baryons with spin 3/2 are permitted in four-body decay modes, while they does not exist in three-body decays and that in ${}^{4}_{\Lambda\Lambda}$H, two spin configurations of ${}^3_\Lambda$H, 3/2 and 1/2, are mixed in the ratio of 3:1.

Decay spectrum

Figure 2 shows the calculated π^--decay spectrum from ${}^{4}_{\Lambda\Lambda}$H in the case when the binding energy is 0.53 MeV. In this case, the π^- spectum has two discrete peaks from two-body decays, where a large peak around $E_\pi = 181.8$ MeV (π^- momentum $P_\pi = 116.5$ MeV/c) from process (2) and a small peak around $E_\pi = 182.9$ MeV (118.2 MeV/c) from process (1). The peak of nuclear continuum region is located around $E_\pi = 171.5$ MeV (99 MeV/c). The position can be shifted if the binding energy of ${}^{4}_{\Lambda\Lambda}$H is changed. The shifted position of the peak is $E_\pi = 172.0$ MeV (100 MeV/c) at most. It is rather difficults to get the position shifted up to 173.8 MeV (103 MeV/c), which is a key value in the following subsection.

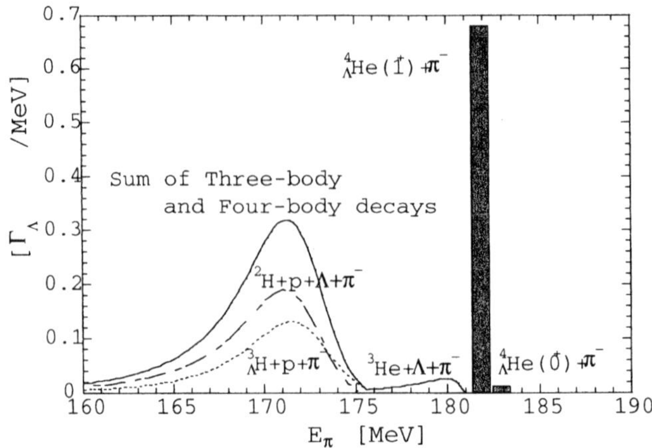

FIGURE 2. The π^--decay spectrum from ${}^{4}_{\Lambda\Lambda}$H. The abscissa is the total π^- energy and the ordinate is the decay strength. Each partial decay width in the unit of Γ_Λ is given by integration of each decay spectrum.

Comparison to the experimental data

In the BNL-E906 experiment, two weak-decay π^--momenta have been mesured in order to see decay sequences of double-Λ hypernuclei [4]. They have obtained several peak structures on two-dimensional scatter plot of the two π^- momenta, which is shown in Figure 3 with square marks. They have analyzed that the most prominent peak region of $P_\pi^> \sim 114$ MeV/c and $P_\pi^< \sim 103$ MeV/c, which is shown in Fig.3 as the region surrounded by ellips, is originated from sequential decays of ${}^{4}_{\Lambda\Lambda}$H. Here a larger (smaller) π^- momentum is labbeled by $P_\pi^>$ ($P_\pi^<$). Main sequences of π^- weak-decays from ${}^{4}_{\Lambda\Lambda}$H are summalized in Table 3. Branching ratios to the total width of ${}^{4}_{\Lambda\Lambda}$H are evaluated by using the calculated partial decay widths (Γ_c) of Table 2, $\mathrm{BR}_c^{1st} = \Gamma_c / \sum_c \Gamma_c \times 0.64$, where c denotes the each decay-process. We assume that nonmesonic decays of ${}^{4}_{\Lambda\Lambda}$H are negligible from the analysis on the ${}^{3}_{\Lambda}$H decay by Kamada *et al* [7], where branching ratio of nonmesonic decay is only 2%. We also assume that a π^--π^0 decay ratio of ${}^{3}_{\Lambda}$H (${}^{4}_{\Lambda\Lambda}$H) is the same to the one of a free Λ particle, since ${}^{3}_{\Lambda}$H (${}^{4}_{\Lambda\Lambda}$H) consists of a proton, a neutron and a (two) Λ particle(s) and the permitted decays are symmetric between π^-- and π^0-decays.

The calculated sequential π^--decay spectrum of ${}^{4}_{\Lambda\Lambda}$H is also displayed in Figure 3, binned in 2.5 MeV/c cells in order to be directly compared with the BNL-E906 experimental data. The cell size is proportional to each sequential decay widths.

The calculated spectrum has no remarkable peak in the region of $P_\pi^> \sim 114$ MeV/c and $P_\pi^< \sim 103$ MeV/c (see the black circles in the ellips region in Fig. 3). On the other hand, calculated widths are relatively larger than experimental π^- distributions in two-body decay region of $P_\pi^> \sim 116$ MeV/c and $P_\pi^< \sim 90-100$ MeV/c (see the white circles in Fig. 3) and four-body decay region of $P_\pi^> \sim 100$ MeV/c and $P_\pi^< \sim 90-100$ MeV/c (see the gray circles in Fig.3).

TABLE 3. Main sequences of π^- weak-decays from ${}^{4}_{\Lambda\Lambda}$H. Branching ratios BR to the total width of ${}^{4}_{\Lambda\Lambda}$H are evaluated by using the calculated decay widths (Γ_c) of Table 2. Sequential-decay branching ratio SBR is derived by multiplying BR^{1st} by BR^{2nd}.

First decays	BR^{1st}	Second decays	BR^{2nd}[*Ref.*]	SBR
(1) ${}^{4}_{\Lambda\Lambda}\mathrm{H} \rightarrow {}^{4}_{\Lambda}\mathrm{He}^{*} + \pi^-$ ($P^{>}_{\pi} \sim 116\mathrm{MeV}/c$)	0.16	${}^{4}_{\Lambda}\mathrm{He} \rightarrow {}^{3}\mathrm{He} + p + \pi^-$ ($P^{<}_{\pi} \sim 96\mathrm{MeV}/c$)	0.32 [13,14]	0.051
(2) ${}^{4}_{\Lambda\Lambda}\mathrm{H} \rightarrow {}^{3}_{\Lambda}\mathrm{H} + p + \pi^-$ ($P^{<}_{\pi} \sim 99\mathrm{MeV}/c$)	0.18	${}^{3}_{\Lambda}\mathrm{H} \rightarrow {}^{3}\mathrm{He} + \pi^-$ ($P^{>}_{\pi} = 114\mathrm{MeV}/c$)	0.24 [7]	0.043
(3) ${}^{4}_{\Lambda\Lambda}\mathrm{H} \rightarrow {}^{3}_{\Lambda}\mathrm{H} + p + \pi^-$ ($P_{\pi} \sim 99\mathrm{MeV}/c$)	0.18	${}^{3}_{\Lambda}\mathrm{H} \rightarrow {}^{2}\mathrm{H} + p + \pi^-$ ($P_{\pi} \sim 100\mathrm{MeV}/c$)	0.38 [7]	0.069
(4) ${}^{4}_{\Lambda\Lambda}\mathrm{H} \rightarrow {}^{3}\mathrm{He} + \Lambda + \pi^-$ ($P^{>}_{\pi} \sim 113\mathrm{MeV}/c$)	0.023	$\Lambda \rightarrow p + \pi^-$ ($P^{<}_{\pi} \sim 100\mathrm{MeV}/c$)	0.64	0.015
(5) ${}^{4}_{\Lambda\Lambda}\mathrm{H} \rightarrow {}^{2}\mathrm{H} + \Lambda + p + \pi^-$ ($P_{\pi} \sim 99\mathrm{MeV}/c$)	0.27	$\Lambda \rightarrow p + \pi^-$ ($P_{\pi} \sim 100\mathrm{MeV}/c$)	0.64	0.173

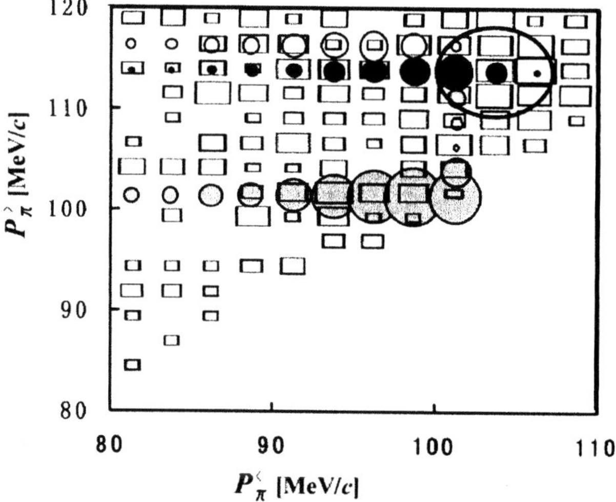

FIGURE 3. Sequential π^- decay spectrum of ${}^{4}_{\Lambda\Lambda}$H. Circles and squars show calculated partial widths and experimental event numbers, respectively. The white circle, the black circle, the gray circle with white center and the gray circle correspond to the process (1), (2), (4) and (5) in Table 3, respectively. The region surrounded by ellips indicates the prominent peak region in the BNL-E906 experiment.

Then not only decays of ${}^{4}_{\Lambda\Lambda}$H but also the other decays are required to give contributions in the ellips region, since various $S = -2$ hyperfragments (twin Λ and other double Λ hypernuclei) are produced by a reaction of a stopped Ξ^- on ^{9}Be. We are now investigating other possible decays around the region of $P^{>}_{\pi} \sim 114$ MeV/c and $P^{<}_{\pi} \sim 103$ MeV/c.

SUMMARY AND CONCLUSIONS

In the present paper, the pionic-decay width and spectra of ${}^{4}_{\Lambda\Lambda}$H are discussed and compared with the recent experiment by BNL-E906.

The calculated two-body decay width is $0.69\Gamma_{\Lambda}$ and the branching ratio of the decay to the total decay of ${}^{4}_{\Lambda\Lambda}$H is estimated to be about 25 %. Four-body decay modes exhauses 43 % of the π^{-}-decay width.

The calculated spectrum has a broad peak around 100 MeV/c, which disagree with a prominent peak of 103 MeV/c in experimental data. Not only decays of ${}^{4}_{\Lambda\Lambda}$H but also the other decays are required to give contributions in the region of 103 MeV/c peak, since various $S = -2$ hyperfragments are produced by a reaction of a stopped Ξ^{-} on ^{9}Be. Then, exclusive productions of double-Λ hypernuclei are necessary for making a clear conclusion about $\Lambda\Lambda$ interaction, for instance, ${}^{4}_{\Lambda\Lambda}$H production via stopped Ξ^{-} on ^{4}He [15] or ${}^{5}_{\Lambda\Lambda}$H production via a Ξ-nuclear state of ${}^{7}_{\Xi}$H [16]. Further experiments are desired.

ACKNOWLEDGMENTS

The authors would like to express their gratitude to Proffessor Y. Akaishi and Proffessor T. Fukuda for fruitful discussions.

REFERENCES

1. M. Danysz *et al.*, Nucl. Phys. **49** (1963) 121.
2. D. J. Prowse, Phys. Rev. Lett. **17** (1966) 782.
3. S. Aoki *et al.*, Prog. Theor. Phys. **85** (1991) 1287.
4. T. Fukuda et al., Proceedings of HYP2000 (Trino, 2000), publised to Nucl. Phys. **A**.
5. S. Nakaichi-Maeda and Y. Akaishi, Prog. Theor. Phys. **84** (1990) 1025.
6. Y. Yamamoto, M. Wakai, T. Motoba and T. Fukuda, Nucl. Phys. **A625** (1997) 107.
7. H. Kamada, J. Golak, K. Miyagawa, H. Witała, and W. Glöckle, Phys. Rev. **C57** (1998) 1595.
8. K. Stricker, H. McManus and J. A. Carr, Phys. Rev. **C19** (1979) 929; J. A. Carr, H. McManus and K. Stricker-Bauer, Phys. Rev. **C25** (1982) 952.
9. M. Kamimura, Phys. Rev. **A38** (1988) 621.
10. R. A. Malfliet and J. A. Tjon, Nucl. Phys. **A127** (1969) 161.
11. Y. Akaishi, T. Harada, S. Shinmura, and K. S. Myint, Phys. Rev. Lett. **84** (2000) 3539; S. Shinmura, private communication.
12. M. M. Nagels, T. A. Rijken and J. J. de Swart, Phys. Rev. **D12** (1975) 744; Phys. Rev. **D15** (1977) 2547; Phys. Rev. **D20** (1979) 1633.
13. H. Outa *et al.*, Nucl. Phys. **A585** (1995) 109c.
14. I. Kumagai-Fuse, S. Okabe and Y. Akaishi, Phys. Rev. **C54** (1996) 2843.
15. I. Kumagai-Fuse, S. Okabe and Y. Akaishi, in preparation.
16. I. Kumagai-Fuse and Y. Akaishi, Phys. Rev. **C54** (1996) R24.

Strange Dibaryon States in the Chiral Quark Soliton Model

N. Sawado[1]

Department of Physics, Faculty of Science and Technology, Frontier Research Center for Computational Science, Science University of Tokyo, Noda, Chiba 278-8510, Japan

Abstract. In this work we investigate an SU(3) extension of the axially symmetric $B = 2$ chiral quark soliton model. The classical soliton is extended to the SU(3) by trivial embedding. We expand the quark determinant in terms of the collective angular velocity up to the second order as well as in the quark mass difference of the first order. The diagonalization of the collective SU(3) hamiltonian is estimated by the method of Yabu-Ando. Finally the mass spectrum and the binding energy of the baryon-baryon channels down to strangeness $S = -6$ are obtained. Absolute mass are considerably larger than the values of experimental candidates because of the absence of all Casimir effects. The results suggest the possibility of several bound states especially for lower strangeness.

Since the first prediction of the H-particle in a MIT bag model calculation [1], there have been many efforts to study the spectrum of baryonic systems including strangeness. Most of their approaches can be classified into two classes: one is QCD-inspired potential models, and the other is the topological soliton models. The chiral soliton approach is of special interest because it provides a quite different point of view from the conventional potential model picture. One of the most general scheme of the chiral soliton is the Skyrme model. The Skyrme model is a nonlinear theory of interacting pions and describe well the ground state property of the baryonic systems. In this work we apply the chiral quark soliton model (CQSM) to the study of the various hexa-quark states. The CQSM is a simple quark model which provides spontaneous breaking of chiral symmetry. The model not only has the valence quarks but also the Dirac sea quark effects in a non-perturbative way. On the other hand, it is related to the Skyrme model in the large limit of the constituent quark mass.

The CQSM is characterized by the following vacuum functional form:

$$\mathcal{Z} = \int \mathcal{D}\pi\mathcal{D}\psi\mathcal{D}\psi^\dagger \exp\left[i\int d^4x\bar{\psi}(i\partial\!\!\!/ - MU^{\gamma_5})\psi\right], \tag{1}$$

with $U^{\gamma_5}(x) = e^{i\gamma_5\tau\cdot\pi(x)/f_\pi}$.

In the Skyrme model, the minimal energy pion configuration for baryon-number-two ($B = 2$) has axial symmetry [3]. We adopt the following ansatz for the pion fields with a

[1] sawado@ph.noda.sut.ac.jp

CP594, *Hadrons and Nuclei: First International Symposium*, edited by Il-T. Cheon et al.

winding number m :

$$U_0(x) = \cos F(\rho,z) + i\tau\cdot\hat{n}\sin F(\rho,z), \tag{2}$$

where

$$\hat{n} = (\sin\Theta(\rho,z)\cos m\varphi, \sin\Theta(\rho,z)\sin m\varphi, \cos\Theta(\rho,z)). \tag{3}$$

The profile functions $F(\rho,z)$, $\Theta(\rho,z)$ are determined by minimizing the total energy of the soliton solution. Under the constraint of axial symmetry, six valence quarks can occupy the lowest energy level [4]. The total energy of the soliton is

$$E_{static} = 2N_c E_{val}[U_0] + E_{field}[U_0] - E_{field}[U_0 = 1], \tag{4}$$

where

$$E_{field} = \frac{N_c}{2}\frac{1}{\sqrt{4\pi}}\sum_{\mu}|E_\mu|\Gamma\left(-\frac{1}{2},\left(\frac{E_\mu}{\Lambda}\right)^2\right) \tag{5}$$

within the proper-time regularization scheme. The E_{val} , E_{field} are the valence quark and the vacuum sea contribution of the total energy, which are calculated by the eigenvalues E_μ of the one-particle Dirac equation:

$$H(U_0^{\gamma_5})\phi_\mu(x) = E_\mu\phi_\mu(x), \quad H(U_0^{\gamma_5}) = -i\alpha\cdot\nabla + \beta M U_0^{\gamma_5}. \tag{6}$$

From the extremum conditions for the total energy E_{static} , the profile functions $F(\rho,z), \Theta(\rho,z)$ are uniquely determined. The new profile functions induce the new eigenequation. These procedures are continued until self-consistency is attained. In Figure 1 we present the results for the baryon number density, which is estimated in terms of the quark fields corresponding to a soliton. It is found that the baryon number density has a toroidal shape. This feature is consistent with other chiral invariant models with axial symmetry, such as the Skyrme model calculation [3].

In this work, the extension to SU(3) is performed by the trivial embedding [5]:

$$U(x,t) = A(t)\begin{pmatrix} U_0(\Lambda^i_j(t)x^j) & 0 \\ 0 & 1 \end{pmatrix} A^\dagger(t). \tag{7}$$

The $A(t)$ is the time-dependent SU(3) collective rotation matrix and the $\Lambda^i_j(t) = \frac{1}{2}Tr(\tau^i B\tau_j B^\dagger)$ is the spatial rotation matrix. By substituting (7) into the chiral quark lagrangian and going to the rotated frame of reference, the lagrangian becomes following form (in order to estimate the effects of the symmetry breaking of SU(3) explicitly, we add the quark mass matrix $\hat{m}$ =diag(m_u, m_u, m_s) to the lagrangian)

$$\begin{aligned} \mathcal{L}_{CQSM} &= \bar{\psi}(i\partial\!\!\!/ - MU^{\gamma_5}(x,t) - \hat{m})\psi \\ &\to \psi^\dagger_{AB}(i\partial_t - H(U_0^{\gamma_5}) - H_{SB} - \Omega_A + \Omega_B)\psi_{AB}. \end{aligned} \tag{8}$$

Here ψ_{AB} denotes the quark fields in the rotating system. In the rotating system, the quark fields feel the induced terms Ω_A, Ω_B and H_{SB}. The Ω_A, Ω_B are the angular velocity

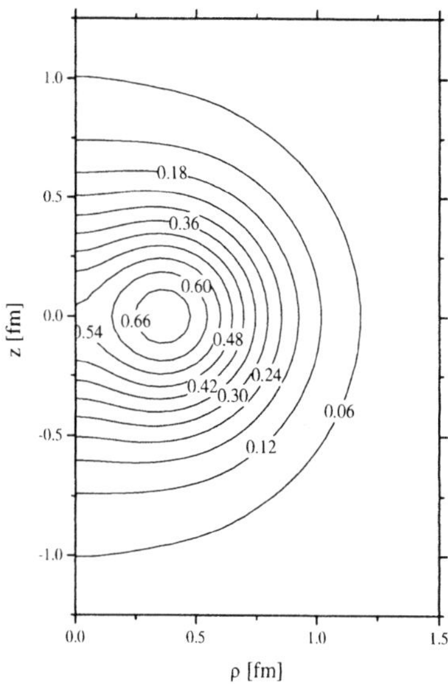

FIGURE 1. Contour plot of baryon number density.

operators for the right flavor rotation and the spatial rotation which are defined by

$$\Omega_A = -iA^\dagger \dot{A} = -i\dot{q}^p(t)A^\dagger(t)\partial_p A(t) = -i\frac{1}{2}\dot{q}^p(t)C_p^a\lambda_a = \frac{1}{2}\Omega_A^a\lambda_a\,, \tag{9}$$

$$\Omega_B = \Omega_B^a(\frac{1}{2}\varepsilon_{abc}\gamma^b\gamma^c - i(r\times\nabla)_a) = \Omega_B^a J_a\,, \tag{10}$$

where the q^p are the coordinates of SU(3) and the C_p^a are the vielbeins. The H_{SB} means the contribution to the hamiltonian due to the SU(3) symmetry breaking

$$H_{SB} = A^\dagger(t)\beta\Delta m\frac{1}{3}(1-\sqrt{3}\lambda_8)A(t)\,, \tag{11}$$

where $\Delta m = m_s - m_u$ denotes the difference of the strange- and up-current quark masses.

We assume that the rotational velocities and the mass difference are relatively small and the expansion in powers of Ω_A, Ω_B and Δm is rapidly convergent. We expand the quark determinant and the (valence quark) Green function in terms of the collective angular velocity up to the second order as well as in the quark mass difference of the first order. As a result, the effective lagrangian is of the form:

$$L = -E_0 - \frac{\sqrt{3}}{2}B[U_0]\Omega_A^8 - \frac{1}{2}\gamma(1-D_{88})$$

TABLE 1. Absolute mass, binding energy (in MeV) of the dibaryon states belonging to the $\{\overline{10}\}$ multiplets ($(p,q)=(0,3)$) for $J\leq 1$, except for $I=0, J=3$ which is the candidate of non-strange dibaryon $\Delta\Delta$.

		(I)		(II)		
$(S\ I\ J)$	Parity	Mass	E_B	Mass	E_B	Threshold
(0 0 1)	+	3385	103	3301	187	NN
(0 0 3)	+	3636	438	3678	395	$\Delta\Delta$
(-1 $\frac{1}{2}$ 1)	+	3533	90	3499	124	$N\Lambda$
(-2 1 1)	+	3685	109	3698	96	$N\Xi$
(-3 $\frac{3}{2}$ 1)	+	3844	158	3902	100	$\Sigma\Xi$

$$-K_{ab}^{A} D_{8a}\Omega_{A}^{b} - K_{ab}^{B} D_{8a}\Omega_{B}^{b} + \frac{1}{2} I_{ab}^{AA}\Omega_{A}^{a}\Omega_{A}^{b} + \frac{1}{2} I_{ab}^{BB}\Omega_{B}^{a}\Omega_{B}^{b}. \tag{12}$$

E_0 is the classical soliton energy and $D_{ab}(A)=\frac{1}{2}\mathrm{Tr}(\lambda_a A \lambda_b A^\dagger)$ is a SU(3) Wigner rotation matrix. The expansion yields various type of moments of inertia:

$$\gamma = -\Delta m \sum_{n\leq 0} \langle n|\gamma_0|n\rangle, \tag{13}$$

$$K_{ab}^{A} = \Delta m \frac{1}{\sqrt{3}} N_c \sum_{m>0,n\leq 0} \frac{\langle n|\beta\lambda_a|m\rangle\langle m|\lambda_b|n\rangle}{E_m - E_n}, \tag{14}$$

$$K_{ab}^{B} = \Delta m \frac{2}{\sqrt{3}} N_c \sum_{m>0,n\leq 0} \frac{\langle n|\beta\lambda_a|m\rangle\langle m|J_b|n\rangle}{E_m - E_n}, \tag{15}$$

$$I_{ab}^{AA} = \frac{N_c}{2} \sum_{m>0,n\leq 0} \frac{\langle n|\lambda_a|m\rangle\langle m|\lambda_b|n\rangle}{E_m - E_n}, \tag{16}$$

$$I_{ab}^{BB} = 2N_c \sum_{m>0,n\leq 0} \frac{\langle n|J_a|m\rangle\langle m|J_b|n\rangle}{E_m - E_n}. \tag{17}$$

We can estimate these terms by using the eigenvalues and the eigenfunctions of the Dirac equation (6). We would like to mention that above formalisms are rather superficial. In actual calculations we have to introduce appropriate cut-off for the summation of the one-particle spectrum.

The canonical quantization formulas for the collective coordinates [6,7]

$$R_a = -\pi_p C_a^p(q) = -\delta L/\delta \dot{q}^p C_a^p(q), \tag{18}$$

$$K_i = -\delta L/\delta \dot{\theta}^i \tag{19}$$

lead to the quantization prescriptions

$$R_a = \begin{cases} -\sum_j (I_{aj}^{AA}\Omega_A^j - I_{aj}^{AB}\Omega_B^j - K_{aj}^{A} D_{8j}), & a=1,2,3, \\ -\sum_b (I_{ab}^{AA}\Omega_A^b - K_{ab}^{A} D_{8b}), & a=4,5,6,7, \\ -\frac{\sqrt{3}}{2} B, & a=8 \end{cases} \tag{20}$$

TABLE 2. Also for the {27} multiplets ($(p,q) = (2,2)$).

$(S\,I\,J)$	Parity	(I) Mass	(I) E_B	(II) Mass	(II) E_B	Threshold
(0 1 0)	+	3426	62	3339	149	NN
(-1 $\frac{3}{2}$ 0)	+	3612	84	3565	131	$N\Sigma$
(-1 $\frac{1}{2}$ 0)	+	3529	94	3505	118	$N\Lambda$
(-2 2 0)	+	3812	92	3892	12	$\Sigma\Sigma$
(-2 1 0)	+	3665	129	3710	84	$N\Xi$
(-2 0 0)	+	3632	126	3670	88	$\Lambda\Lambda$
(-3 $\frac{3}{2}$ 0)	+	3857	145	3920	82	$\Sigma\Xi$
(-3 $\frac{1}{2}$ 0)	+	3763	166	3855	74	$\Lambda\Xi$
(-4 1 0)	+	3900	200	4043	57	$\Xi\Xi$

and

$$K_i = -\sum_j (I_{ij}^{BB}\Omega_B^j - K_{ij}^B D_{8j}) . \tag{21}$$

The contribution to the hamiltonian from the collective rotations as well as the mass difference are obtained such as

$$\begin{aligned} H_{rot} &= \frac{1}{2}\left(\frac{1}{I_{11}^{AA}} - \frac{1}{I_{44}^{AA}}\right)\sum_{i=1}^{3} R_i^2 + \frac{1}{2}\frac{1}{I_{11}^{BB}}\sum_{i=1}^{3} K_i^2 \\ &+ \frac{1}{2}\left(\frac{1}{I_{33}^{AA}} - \frac{1}{I_{11}^{AA}} - \frac{m^2}{I_{11}^{BB}}\right)R_3^2 - \frac{3}{8}\frac{1}{I_{44}^{AA}}B^2 + \mathcal{H}_{SB} , \end{aligned} \tag{22}$$

where $\mathcal{H}_{SB}$, which is induced by the SU(3) symmetry breaking effects, is difficult to handle due to the SU(3) Wigner rotation matrix in it. The leading terms are of the form

$$\mathcal{H}_{SB} = \frac{1}{2}\frac{1}{I_{44}^{AA}}\sum_{a=1}^{8} R_a^2 + \frac{1}{2}\gamma(1 - D_{88}) + \dots . \tag{23}$$

Within the Skyrme model, Yabu and Ando suggest to estimate the effects of mixing of the higher SU(3) representation by diagonalizing the collective hamiltonian in a basis [8]. By introducing an SU(3) "Euler angle", the eigenfunction of the collective hamiltonian is represented via [5]

$$\Psi_{YII_3,JJ_3}^{NL} = \sum_{M_L M_R} D_{I_3 M_L}^{I}(\alpha,\beta,\gamma) f_{M_L,M_R}^{IY,N2}(\nu) D_{M_R L}^{N}(\alpha',\beta',\gamma') e^{-2i\rho} D_{J_3,-mL}^{J}(B) . \tag{24}$$

The diagonalization including the leading terms of the $\mathcal{H}_{SB}$ reduce to the numerical problem of the coupled ordinarily differential equations of the angle ν. The next to leading terms are formed like $-\frac{K_{44}^A}{I_{44}^{AA}}(\frac{\sqrt{3}}{2} - \frac{\sqrt{3}}{2}BD_{88})$, and the calculation including these terms are also performed. The diagonalization of the full hamiltonian is rather tedious

TABLE 3. For the $\{35\}$ multiplets $((p,q) = (4,1))$.

$(S\ I\ J)$	Parity	(I) Mass	(I) E_B	(II) Mass	(II) E_B	Threshold
(0 2 1)	+	3701	80	3607	174	$N\Delta$
(-1 $\frac{5}{2}$ 1)	+	3933	56	3870	119	$\Delta\Sigma$
(-1 $\frac{3}{2}$ 1)	+	3789	108	3762	135	$N\Lambda^*$
(-2 2 1)	+	4007	98	4011	94	$\Lambda\Lambda^*$
(-2 1 1)	+	3950	137	3913	98	$N\Xi^*$
(-3 $\frac{3}{2}$ 1)	+	4064	139	4141	62	$\Sigma^*\Xi$
(-3 $\frac{1}{2}$ 1)	+	3950	165	4066	49	$N\Omega^-$
(-4 1 1)	+	4115	202	4269	48	$\Xi\Xi^*$
(-4 0 1)	+	4044	206	4214	36	$\Lambda\Omega^-$
(-5 $\frac{1}{2}$ 1)	+	4163	258	4395	26	$\Xi\Omega^-$

task. The remaining terms will be vanished when all values of the moments of inertia are equal, i.e. $K^A_{11} = K^A_{33} = K^A_{44}$ and $I^{AA}_{11} = I^{AA}_{33} = I^{AA}_{44}$. In actual, they are all different values. The contributions of the remaining terms will not be negligible for consistency of the calculation.

The classical soliton energy has already been calculated in [4]. In this work we investigate the quantized soliton energy for the case of (I) including the leading terms of the symmetry breaking hamiltonian (II) plus the next to leading terms. The resulting masses are displayed in Table 1-4 for the multiplets coupled to the $\{\overline{10}\}$, $\{27\}$, $\{35\}$, $\{28\}$-plets of dibaryons ($(p,q) = (0,3),(2,2),(4,1),(6,0)$, respectively). Also in the tables we present the binding energy of each dibaryons, which are evaluated by using the data from $B = 1$ hedgehog analysis in [7]. We would like to say that the model parameters and the formulation in [7] slightly differ from ours, then, for the binding energies we can only make a rough estimate.

In the whole soliton approach, the absolute mass always tend to be high values mainly due to the lack of Casimir effects – the loop corrections of the order of N_c^0 – in the calculations. In the $B = 1$ hedgehog case, the Casimir energies for the rotational and the translational zero modes were estimated. Their corrections for the total mass were about $(-0.7 \sim -0.8)$ GeV [9]. For the $B = 2$ the Casimir energies were not estimated yet [10]. The thorough analysis for the Casimir effects are needed in order to examine the state can be stable or not.

In the analysis of $B = 2$ torus-like Skyrmion, the authors suggest the possibility of bound states with higher strangeness [5]. In our calculation with (I), similar behavior is reproduced. (One should notice that these terms are treated in the same way within the Skyrme model calculation.) But for the case of (II) the situation is drastically changed. For growing absolute values of strangeness, the binding energies rapidly decrease. Therefore the possibility of existence of the bound states for higher strangeness seems to be hopeless. The full calculation of the symmetry breaking hamiltonian is quite desired for the completeness of the investigation.

TABLE 4. For the {28} multiplets $((p,q) = (6,0))$.

$(S\,I\,J)$	Parity	(I) Mass	(I) E_B	(II) Mass	(II) E_B	Threshold
(0 3 0)	+	3969	106	3867	207	$\Delta\Delta$
(-1 $\frac{5}{2}$ 0)	+	4043	147	3986	204	$\Delta\Lambda^*$
(-2 2 0)	+	4113	191	4151	153	$\Delta\Xi^*$
(-3 $\frac{3}{2}$ 0)	+	4190	218	4296	112	$\Delta\Omega^-$
(-4 1 0)	+	4258	266	4433	91	$\Sigma^*\Omega^-$
(-5 $\frac{1}{2}$ 0)	+	4325	313	4574	64	$\Xi^*\Omega^-$
(-6 0 0)	+	4386	356	4708	34	$\Omega^-\Omega^-$

REFERENCES

1. R. L. Jaffe, Phys. Rev. Lett. **38**, 195 (1977); **38**, 1617(E) (1977);

2. D. I. Diakonov and V. Yu. Petrov and P. V. Pobylitsa, Nucl. Phys. B **306**, 809 (1988).
M. Wakamatsu and H. Yoshiki, Nucl. Phys. A **524**, 561 (1991).
H. Reinhardt and R. Wünsch, Phys. Lett. B **215**, 577 (1988).

3. V. B. Kopeliovich and B. E. Stern, Pis'ma v ZhETF **45**, 165 (1987);
J. Verbaarschot, Phys. Lett. B **195**, 235 (1987);
H. Weigel, B. Schwesinger and G. Holzwarth, Phys. Lett. B **168**, 321 (1986);
N. S. Manton, Phys. Lett. B **192**, 177 (1987);
E. Braaten and L. Carson, Phys. Rev. D **38**, 3525 (1988);

4. N. Sawado and S. Oryu, Phys. Rev. C **58**, R3046 (1998);
N. Sawado, Phys. Rev. C **61**, 065206 (2000).

5. V. B. Kopeliovich, B. Schwesinger and B. E. Stern, Phys. Lett. **B242**, 145 (1990);
V. B. Kopeliovich, B. Schwesinger and B. E. Stern, Nucl. Phys. A **549**, 485 (1992).

6. N. Toyota, Prog. Theor. Phys. **77** (1987) 688.

7. A. Blotz, D. Diakonov, K. Goeke, N. W. Park, V. Petrov and P. V. Pobylitsa, Nucl. Phys. A Ą@**555**, 765 (1993).

8. H. Yabu, K. Ando, Nucl. Phys. B**301**, 601 (1988).

9. P. V. Pobylitsa, E. Ruiz-Arriola, Th. Meissner, F. Gruemmer, K. Goeke and W. Broniowski, Ą@J. of Phys. **G18**, 1455 (1992).

10. V. B. Kopeliovich, Nucl. Phys. A **639**, 75c (1998).

HADRONS IN NUCLEAR MATTER

Proton and Pion Distributions in Heavy-Ion Collisions at SIS Energies

Byungsik Hong* and the FOPI collaboration

*Department of Physics, Korea University, Seoul 136-701, South Korea

Abstract. We present the centrality dependence of proton and deuteron rapidity distributions in Ru + Ru collisions at 400*A* MeV. The ratio of baryon rapidity distributions in isospin asymmetric collision systems shows incomplete mixing and transparency of the projectile and target nucleons at this beam energy. In addition, we also present the results of thermal model calculations including both a modification of the $\Delta(1232)$ spectral function and the Coulomb interaction. The model calculations are compared with the published pion spectra for central Au + Au collisions at 1*A* GeV.

INTRODUCTION

Presently, relativistic heavy-ion collisions are a unique tool for studying excited nuclear matter in the laboratory [1]. Two to three times the normal nuclear matter density with a temperature below 100 MeV can be reached by nucleus-nucleus collisions at an incident beam energy around 1 GeV per nucleon [2, 3]. After a few tens of fm/*c* after the collision, the compressed nuclear matter starts to blow up and freeze-out. By analyzing hadron spectra in nuclear collisions at various beam energies, we can obtain *nearly* clean picture at freeze-out, which is important to constrain the initial compressed state of nuclear matter albeit model dependent way.

The degree of nuclear stopping in heavy-ion collisions is one of the essential parameters which are necessary to understand the reaction dynamics. It is a crucial variable to estimate the energy and particle densities of the compressed nuclear matter at early stage of the participant fireball. Experimental data shows that baryon rapidity distributions are wider than the isotropic source for a beam energy larger than 1*A* GeV [4, 5]. The origin of this observation has become one of the most disputable subjects, as very similar rapidity distributions can be produced by transparency, rebound, or the longitudinal expansion after thermal equilibrium of projectile and target nucleons. In order to shed some light in this subject, we will utilize isobar nuclei ($^{96}_{44}$Ru and $^{96}_{40}$Zr) at 400*A* MeV in the first part of this contribution [6].

In addition, we will develope a rather complete thermal model, which includes resonance decays, the modified spectral function of resonances, and the Coulomb interaction. Then we compare the results with the published pion spectra at 1*A* GeV by KaoS collabopration [7]. With this study we can test a modification of the spectral function for, especially, $\Delta(1232)$ resonances in multi-hadron system at freeze-out [8, 9].

In the following, all proton and deuteron spectra by the FOPI collaboration should be regarded as preliminary. We use the natural unit $\hbar = c = 1$.

CP594, *Hadrons and Nuclei: First International Symposium,* edited by Il-T. Cheon et al.

PROTON RAPIDITY DISTRIBUTION

The centrality of each event is determined by the ratio of total transverse ($E_\perp$) to longitudinal kinetic energies ($E_{||}$), which is so called E_{rat},

$$E_{rat} = \sum_i E_{\perp,i} / \sum_i E_{||,i}. \tag{1}$$

Previously, it has been demonstrated that E_{rat} is a suitable variable for event centrality, especially, in the most central collisions at beam energy under present investigation [10]. Table 1 summarizes the centrality criteria used in this analysis.

TABLE 1. Centrality criteria on the E_{rat} distribution in Ru + Ru collisions with the corresponding reaction cross section, σ_r, for the present analysis. The geometric impact parameter, b_{geom}, is determined by the sharp cut-off approximation. The impact parameters b_{IQMD} in the last column are determined from the IQMD(HM) model and the errors represent 1σ region with a Gaussian fit. Here H stands for the hard Equation-of-State (EoS) option with a compressibility coefficient $K = 380$ MeV and M stands for the momentum dependent interaction (MDI).

Centrality	E_{rat}	σ_r **(mb)**	b_{geom} **(fm)**	b_{IQMD} **(fm)**
$E_{rat}A$	$1.40 \leq$	≤ 89	≤ 1.7	≤ 1.5
$E_{rat}B$	(1.05, 1.40)	(89, 223)	(1.7, 2.7)	2.0 ± 0.5
$E_{rat}C$	(0.80, 1.05)	(223, 447)	(2.7, 3.8)	2.9 ± 0.4
$E_{rat}D$	(0.60, 0.80)	(447, 746)	(3.8, 4.9)	3.7 ± 0.4

The left panel of Fig. 1 shows the rapidity distributions of protons and deuterons in Ru + Ru collisions for four centrality conditions. We have to note that Fig. 1 may not contain full protons and deuterons, as the detector does not have an acceptance near zero p_t neither at projectile nor at target rapidities [4]. Bearing this in mind, we observe that protons and deutrons are concentrated more at midrapidity for more central collisions.

The comparison of the experimental $dN/dy^{(0)}$ spectrum for the most central event criterion ($E_{rat}A$) with the IQMD(HM) calculations ($b_{IQMD} \leq 1$ fm) is shown in the right panel of Fig. 1. Four assumptions on the nucleon-nucleon (NN) elastic cross-section (σ_{NN}) in the model are used. In the IQMD model, composite particles are produced by an afterburner coordinate space cluster algorithm, which implies that the yields can vary depending on the model parameter which allows to form composite particles. With default parameters, a coordinate space cluster algorithm starts to produce composite particles, including deuterons, after 200 fm in time, and it is well known effect that the model produces too few composites. Therefore, we emphasize only the spectral shapes of protons and deuterons, but not the absolute numbers in the model. All model calculations are normalized in such a way that the integration of $dN/dy^{(0)}$ is the same as data. In Fig. 1 we find that the experimental proton and deuteron $dN/dy^{(0)}$ distributions are in best agreement with $\sigma_{NN} = \sigma_{NN}^{free}$ assumption. But the choices of $0.5\sigma_{NN}^{free}$ and $5\sigma_{NN}^{free}$ can be clearly excluded from later considerations.

In order to study, experimentally, whether nucleons are transparent, rebound, or expanded longitudinally after equilibrium in nucleus-nucleus collisions, we analyzed the

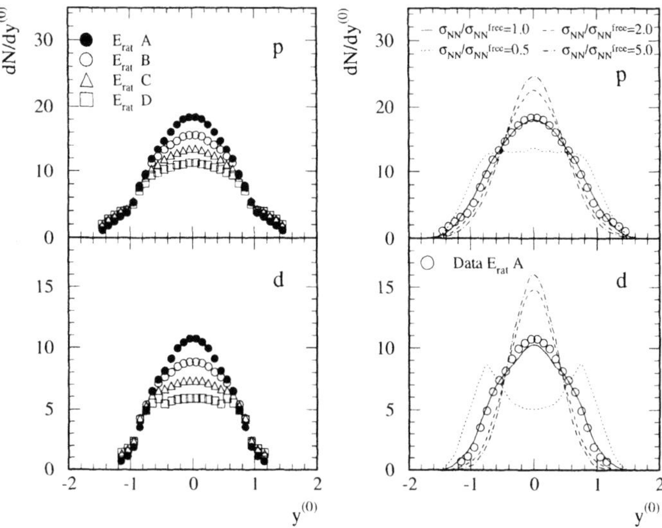

FIGURE 1. (Left) Centrality dependence of the proton and deuteron $dN/dy^{(0)}$ distributions in Ru + Ru collisions. (Right) Comparison of proton and deuteron $dN/dy^{(0)}$ distributions (open circles) with the results from the IQMD(HM) calculations (various lines). Data are for $E_{rat}A$ criterion, and the model calculations are for $b_{IQMD} \leq 1$ fm. In the model, σ_{NN} was changed from 0.5 to 5.0 of σ_{NN}^{free}.

ratio R_p for Ru + Zr and Zr + Ru collisions;

$$R_p = \frac{dN/dy^{(0)}(^{96}_{44}Ru + ^{96}_{40}Zr)}{dN/dy^{(0)}(^{96}_{40}Zr + ^{96}_{44}Ru)}. \tag{2}$$

The ratio of the proton rapidity distributions has advantage because the systematic uncertainties are eliminated to a large extent under the same experimental conditions, and moreover the signal of transparency or rebound will be enhanced. Note that R_p should be one if nucleons in the projectile and target nuclei reach isospin equilibrium. And if nucleons are transparent (rebound), R_p should converge to 0.91 near the target (projectile) rapidity and to 1.1 near the projectile (target) rapidity. Open circles in Fig. 2 show R_p measured in the central drift chamber (CDC) for $E_{rat}A$ condition. The fact that R_p crosses one at midrapidity confirms that the analysis is done right, because an equal number of projectile and target nucleons should be admixed in c.m. due to the mass symmetry of the collision systems. Solid triangles in Fig. 2 reflects the earlier results by the proton-like particle ratio in the backward hemisphere [11]. The current results agree nicely with the previous ones. The behavior of experimental R_p in Fig. 2 is clearly in good agreement with the transparency scenario for central collisions as R_p increases with $y^{(0)}$ at backward hemisphere.

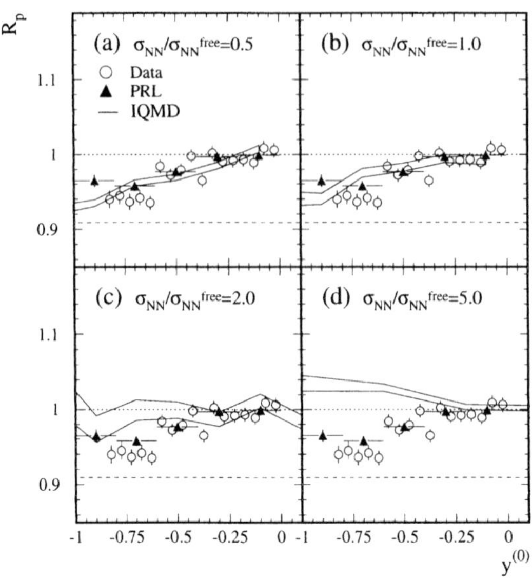

FIGURE 2. Comparison of the experimental R_p obtained by the CDC for $E_{rat}A$ criterion with the results from the IQMD(HM) calculations. Open circles are data, and thin solid bands are the limits estimated by model calculations. Cross section σ_{NN} in the model was assumed to be (a) one half of, (b) the same as, (c) twice of, and (d) five times of the σ_{NN}^{free}. Solid triangles reflect the earler results by the proton-like particle ratio in the backward hemisphere. Dashed line is the lower limit determined by the N/Z ratio of projectile and target nuclei.

The ratio R_p estimated by the IQMD(HM) model are shown by thin solid bands in Fig. 2 for various assumptions on σ_{NN}. The model calculations are for $b_{IQMD} \leq 1$ fm which is a little more central than data. The geometrical acceptance of the detector have been applied for the model calculations, too. However, the model calculations with and without filtering geometrical acceptance of the FOPI detector show very little difference. From recent theoretical investigations, the degree of nuclear stopping is expected to be rather sensitive to σ_{NN}. In order to investigate the effect of this parameter, we show experimental R_p with model calculations for σ_{NN} to be (a) half of, (b) the same as, (c) twice of, and (d) five times of σ_{NN}^{free}. The comparison in Fig. 2 shows that the ratio R_p is not such a sensitive parameter to distinguish various options in $\sigma_{NN}/\sigma_{NN}^{free}$ "quantitatively". However, it shows a clear transparency effect for $\sigma_{NN}/\sigma_{NN}^{free} < 2$. Data agree with model calculations best for $\sigma_{NN}/\sigma_{NN}^{free} = 1$, especially for $|y^{(0)}| \leq 0.6$ which is the most interesting region for nuclear stopping phenomenon (χ^2 per degrees of freedom for $\sigma_{NN}/\sigma_{NN}^{free} = 1$ is 1.04 which is superior to other options, *e.g.*, it is 5.06 for $\sigma_{NN}/\sigma_{NN}^{free} = 0.5$). Note that the options for model calculations with $\sigma_{NN}/\sigma_{NN}^{free} = 0.5$

and 5 were excluded by the comparison in proton and deuteron rapidity distributions in previous sections. The model calculation with $\sigma_{NN}/\sigma_{NN}^{free}$ = 5 can not reproduce data at all, showing even a negative slope which implys a rebound effect.

PION DISTRIBUTION

The thermal model adopted in this analysis has been described in Refs. [8, 9, 12]. For the phase-space distribution of the $\Delta(1232)$ resonance at freeze-out, we utilize the previous results obtained by the EOS collaboration: the freeze-out temperature $T_f = 81$ MeV, and the radial flow velocity $\beta_f = 0.32$ [13]. Then, the isotropically expanding $\Delta(1232)$ source at the center-of-mass (c.m.) becomes

$$\frac{d^3N}{dp^3} \propto \exp\left(-\frac{\gamma E}{T_f}\right)\,[(\gamma+\frac{T_f}{E})\frac{\sinh\alpha}{\alpha}-\frac{T_f}{E}\cosh\alpha], \tag{3}$$

where $\gamma = 1/\sqrt{1-\beta_f^2}$, $\alpha = (\gamma\cdot\beta_f\cdot p)/T_f$, and E and p are the total energy and momentum of $\Delta(1232)$ in the c.m. [14].

In addition, two different spectral functions of $\Delta(1232)$ were considered, one with and one without the πN interactions being included in the thermal fireball. Recently, Weinhold and collaborators exploited a detailed calculation for the thermodynamic potential of a system consisting of pions and nucleons [8, 9]. Two spectral functions of $\Delta(1232)$ are shown in the right panel of Fig. 3 where the dashed and the solid lines represent free and modified spectral functions, respectively. The free spectral function A_R is simply a normal Breit-Wigner type, whereas B_R is the modified spectral function;

$$B_R(E_{cm}) = 2\partial\delta_{33}(E_{cm})/\partial E_{cm}, \tag{4}$$

where the phase shift of $\Delta(1232)$, δ_{33}, can be deduced by

$$\tan\delta_{33}(E_{cm}) = -\Gamma(E_{cm})/2(E_{cm}-E_R). \tag{5}$$

The measured phase-shift factor δ_{33} is displayed in the left panel in Fig. 3, and the resultant $\Delta(1232)$ spectral function is displayed in the right panel. Figure 3 shows a clear difference between the two spectral functions A_R and B_R, especially, close to the threshold: the modified spectral function B_R is shifted to lower c.m. energy. Note that B_R can be uniquely determined in a model-independent way by using the measured phase-shift δ_{33} in Eq. 4.

Figure 4 shows a comparison of the published KaoS pion spectra [7] with the model calculations for various components. The KaoS data were taken at laboratory polar angles, θ_{lab}, from 40 to 48 degrees. The collision centrality was the most central 14 $\pm$ 4% of the total reaction cross-section and corresponded to the number of participating protons Z_{eff} = 110 $\pm$ 8. Figure 4 shows the double differential cross-section, $d^2\sigma/(dE_{cm}^{kin}d\Omega_{cm})$, of $\pi^{\pm}$ as a function of the kinetic energy E_{cm}^{kin} in the c.m. for Au + Au collisions. KaoS pion data have the advantage of covering the low E_{cm}^{kin} region down to 50 MeV, where the effects of the Δ spectral function and the Coulomb interaction are

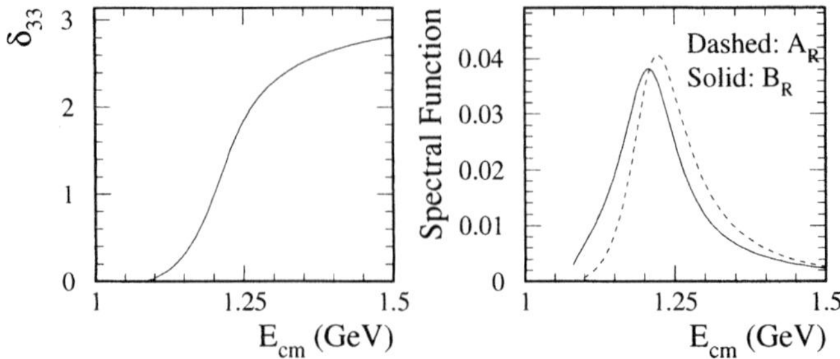

FIGURE 3. Phase shift δ_{33} (left) and spectral functions (right) of the $\Delta(1232)$ resonance. In the right panel, the dashed line represents a normal Breit-Wigner type spectral function while the solid line represents a thermodynamic spectral function including πN interactions.

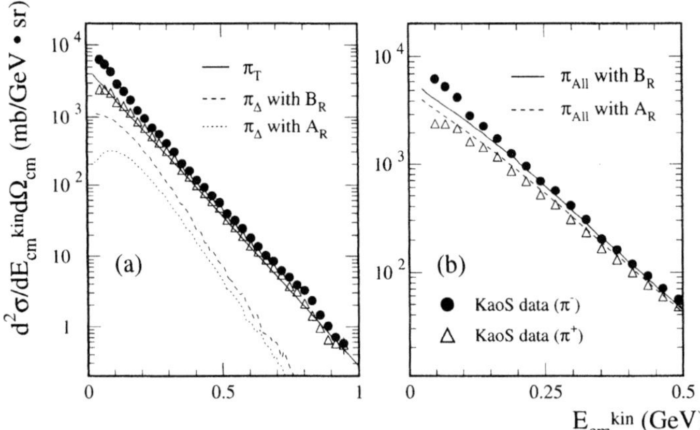

FIGURE 4. Comparison of the experimental data with the thermal model calculations. The solid circles and the open triangles show the π^- and the π^+ data, respectively, obtained by the KaoS collaboration. (a) The solid line represents thermal pions (π_T) while the dotted and the dashed lines represent the decay pions from the $\Delta(1232)$ resonances (π_Δ) with the A_R and the B_R spectral functions, respectively. (b) The solid line represents the sum of π_T and π_Δ with B_R while the dashed line represents the sum with A_R. Only the low E_{cm}^{kin} region is enlarged for a clearer display.

expected to be maximum. It is clear that the π^- yield exceeds the π^+ yield, especially in the low E_{cm}^{kin} region. In Fig. 4 (a), the data are shown along with the model calculations for thermal pions (π_T) and pions decayed from the $\Delta(1232)$ resonance (π_Δ) together. The solid line represents the π_T component, which is determined by the fit at a relatively high c.m. kinetic energy ($E_{cm}^{kin} > 0.5$ GeV). The dotted and the dashed lines show the π_Δ component for the spectral functions A_R and B_R, respectively. The yield ratio of π_Δ/π_T to best reproduce the KaoS data was 0.3 ± 0.1. We can see that the π_Δ spectra with the B_R function contribute more in the low E_{cm}^{kin} region. Here, the same geometrical acceptance in data was also applied to the model calculations. The total pion spectra (π_{All}), which is the sum of π_T and π_Δ, are shown by the dashed and the solid lines for the spectral functions A_R and B_R, respectively, in Fig. 4 (b) where the low E_{cm}^{kin} region is magnified for a clearer display. The spectra of π_{All} with the B_R function has a higher yield in the low E_{cm}^{kin} region and come nicely in between the π^- and the π^+ data. As we will show later, π_{All} spectra with the A_R function will underestimate the data after the Coulomb interaction is considered.

Finally, the effect of the Coulomb potential was considered in order to explain the difference between the π^- and the π^+ spectral shapes. With a static approximation, the Coulomb potential becomes $V_C = Z_{eff} \cdot e^2/R_f$, where R_f is the radius of the fireball at freeze-out and Z_{eff} is the effective charge contained in the fireball [7, 15]. The assumption of a static Coulomb field can be justified as we consider only E_{cm}^{kin} of pions larger than 50 MeV in the data. The dynamical consideration is necessary only when the velocity of a pion is less than the velocity of expansion ($\beta_\pi < \beta_f$) [16]. For $\beta_\pi = \beta_f$, the corresponding pion E_{cm}^{kin} is about 10 MeV, which is significantly lower than 50 MeV.

Then, the number of particles with momentum p, $N(p)$, can be related to $N(p_i)$ by the Jacobian:

$$N(p) = \left|\frac{\partial^3 p_i}{\partial^3 p}\right| N(p_i) = \frac{p_i E(p_i)}{pE(p)} N(p_i) = C^{\pm} N(p_i) \qquad (6)$$

where p_i is the initial momentum of the pions before the Coulomb interaction is considered and

$$C^{\pm} = \frac{p_i E(p_i)}{pE(p)} = \sqrt{p^2 \mp 2E(p)V_C + V_C^2} \cdot \frac{E(p) \mp V_C}{pE(p)} \qquad (7)$$

for positive (C^+) and negative (C^-) particles. In order to determine the strength of the Coulomb potential, V_C, we fit the experimental yield ratio of π^- to π^+ at a given momentum p, $N_{\pi^-}(p)/N_{\pi^+}(p)$, by

$$\frac{N_{\pi^-}(p)}{N_{\pi^+}(p)} = R_C \frac{\sqrt{p^2 + 2E(p)V_C + V_C^2}}{\sqrt{p^2 - 2E(p)V_C + V_C^2}} \left(\frac{E(p) + V_C}{E(p) - V_C}\right) \frac{N_{\pi^-}(p_i)}{N_{\pi^+}(p_i)}, \qquad (8)$$

where the normalization constant R_C and the Coulomb potential V_C are free fit parameters. The constant R_C is responsible for the height, and V_C is responsible for the slope of $N_{\pi^-}(p)/N_{\pi^+}(p)$ in the low E_{cm}^{kin} region. The values of $R_C = 1.05$ and $V_C = 25$ MeV are found to describe the data best.

Figure 5 shows a comparison of the KaoS pion spectra with the present thermal model calculations including the modification of $\Delta(1232)$ spectral function and Coulomb interactions. The agreement between the data and the model calculation is excellent over the entire E_{cm}^{kin} range. Therefore, Fig. 5 is evidence for the modification of the $\Delta(1232)$ spectral function due to the πN interactions and for the effect of Coulomb interactions

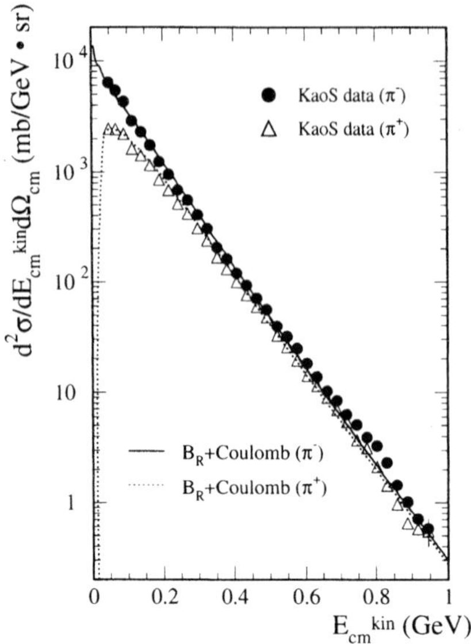

FIGURE 5. Comparison of the measured pion spectra with the thermal model calculations including the modification of the $\Delta(1232)$ spectral function and Coulomb interactions. Both data and model calculations are obtained at laboratory polar angles $\theta_{lab} = 44 \pm 4°$.

in pion spectra. In addition, the agreement in Fig. 5 demonstrates that the phase-space distribution of Δ at freeze-out is quite isotropic.

Once we have an estimate of the Coulomb potential V_C, it is possible to deduce the source size at freeze-out. With $Z_{eff} = 110 \pm 8$ [7], the freeze-out radius $R_f = Z_{eff} \cdot e^2/V_C = 6.3 \pm 0.5$ fm.

By using the given informations, we can further deduce the percentage of participating nucleons excited to the $\Delta(1232)$ resonance at freeze-out. As shown in Figs. 4 and 5, the experimental data can be described well by the thermal model with a yield ratio $\pi_\Delta/\pi_T = 0.3 \pm 0.1$, which implies that $23 \pm 10\%$ of all pions are from the decay of $\Delta(1232)$ resonances (π_Δ/π_{All}). The experimental value for total pion multiplicity per participating nucleon (π_{All}/A_{part}) is 0.086 ± 0.010 for Au + Au collisions at 1A GeV [17]. Then, the number of nucleons excited to the $\Delta(1232)$ resonance per participating nucleon (Δ/A_{part}) becomes

$$\frac{\Delta}{A_{part}} = \frac{\pi_{All}}{A_{part}} \times \frac{\pi_\Delta}{\pi_{All}} = 0.02 \pm 0.01. \tag{9}$$

Comparing this result with the result of a similar analysis performed for Ni + Ni collisions at 1A GeV [12], we find that the percentage of the number of participting nucleons excited to $\Delta(1232)$ is reduced by a factor of five for Au + Au collisions.

CONCLUSIONS

We used two experimental observables, which are truly complementary, to study nuclear stopping phenomenon at 400*A* MeV. The first obervable was the conventional method, *i.e.*, proton and deuteron rapidity distributions in Ru + Ru collisions. The second observable which was never tried before was the ratio of proton rapidity distributions in isospin asymmetric collisions by using isobaric nuclei, *i.e.*, Ru + Zr and Zr + Ru. We conclude that

- more stopping can be achieved in more central events,
- the *NN* elastic cross-section does not change in dense and hot nuclear matter within the framework of IQMD,
- the ratio of proton rapidity distributions in Ru + Zr and Zr + Ru collisions evidences a clear transparency effect of nucleons at 400*A* MeV.

In addition, we have developed a rather complete thermal model including the modification of the $\Delta(1232)$ spectral function due to πN scattering and the Coulomb interaction. Pions at low energy required the modification of the $\Delta(1232)$ spectral function. Finally, the Coulomb interaction was necessary to explain the difference between the π^- and the π^+ spectra.

ACKNOWLEDGMENTS

This article is dedicated to Prof. I.-T. Cheon on the occasion of his retirement.

REFERENCES

1. Stock, R., *Phys. Rep.*, **135**, 259 (1986).
2. Stöcker, H. and Greiner, W., *Phys. Rep.*, **137**, 277 (1986).
3. Hong, B., *J. Korean Phys. Soc.*, **36**, L131 (2000).
4. Hong, B. *et al.*, FOPI Collaboration, *Phys. Rev.*, **C57**, 244 (1998).
5. Barrette, J. *et al.*, E814 Collaboration, *Phys. Rev.*, **C50**, 3047 (1994).
6. Bass, S. A., Konopka, J., Bleicher, M., Stöcker, H. and Greiner, W., *GSI Scientific Report 94-1*, GSI, Germany, 1995, p. 66.
7. Wagner, A. *et al.*, KaoS Collaboration, *Phys. Lett.*, **B420**, 20 (1998).
8. Weinhold, W., Friman, B., and Nörenberg, W., *Phys. Lett.*, **B433**, 236 (1998).
9. Weinhold, W., Ph.D. thesis, Technische Universität Darmstadt, Germany, 1998.
10. Reisdorf, W. *et al.*, FOPI Collaboration, *Nucl. Phys.*, **A612**, 493 (1997).
11. Rami, F. *et al.*, FOPI Collaboration, *Phys. Rev. Lett.*, **84**, 1120 (2000).
12. Hong, B. *et al.*, FOPI Collaboration, *Phys. Lett.*, **B407**, 115 (1997).
13. Lisa, M. A. *et al.*, EOS Collaboration, *Phys. Rev. Lett.*, **75**, 2662 (1995).
14. Siemens, P. J. and Rasmussen, J. O., *Phys. Rev. Lett.*, **42**, 880 (1979).
15. Barrette, J. *et al.*, E877 Collaboration, *Phys. Rev.*, **C62**, 024901 (2000).
16. Ayala, A. and Kapusta, J., *Phys. Rev.*, **C56**, 407 (1997).
17. Senger, P. and Ströbele, H., *J. Phys.*, **G25**, R59 (1999).

Time Evolution of Quantum Many Body Systems in the Projection Operator Method

Masahiro Maruyama

Department of Physics, Tohoku University, 980-8578 Sendai, Japan

Abstract. The time evolution of quantum many body systems in the relativistic quatum field theory is studied. We use Shibata-Hashitsume's projection operator method which gives a systematic expansion of the Heisenberg equation of motion in terms of the interaction Hamiltonian. We calculate the time evolution of expectation values of creation and annihilation operators in a renormalizable model with unstable particles in a vacuum and in a medium. It is shown how to renormalize the ultraviolet divergences appearing in the equation of motion and calculate the medium effects on the time evolution. The mass and the wavefunction renormalizations are carried out in the lowest order. We find that the results coincide with the on-mass-shell values of the self-energy and its derivative which are calculated by the usual Feynman rule. We also discuss the initial divergence which remains at the initial time after mass renormalization. It is suggested that the projection operator method can be applied to the relativistic quantum field theory.

INTRODUCTION

In high energy heavy ion collisions it is expected that after multiparticle production thermalization is realized and if the energy density is high enough the system undergoes a phase transition. The process is a problem of relativistic quantum many body systems and we want to describe it by using the relativistic quantum field theory.

However, since the whole process of heavy ion collisions is very complicated, it is not easy to calculate the scattering matrix itself. Instead of doing so, we divide it into several parts such as equilibration process, phase transition and so on, and try to describe the time evolution of each part by solving the equation of motion as an initial value problem.

For this purpose we use the projection operator method which is the so-called damping theory[1]. Historically speaking, it is related to the linear response theory which had been developed by R. Kubo. In this theory, by adding a time dependent external source to a Hamiltonian the linear response of the system is calculated. Consequently, transport coefficients which are macroscopic quantities can be related to the Hamiltonian which contains microscopic information.

Transport equations with tranport coefficients can describe the time evolution of such systems. The projection operator method is a way to obtain the transport equations. It relates the Hamiltonian with the equations of motion for macroscopic variables. In the Shrödinger picture the projection operator method was originated from S. Nakajima and R. Zwanzig. In the Heisenberg picture there is a famous equation of motion called Mori's Langevin equation. A unified description of both pictures has been obtained by F. Shibata and N. Hashitsume. There are four types of equations. In the Schrödinger picture the equation of motion is called as the master equation, while in the Heisenberg picture it

CP594, *Hadrons and Nuclei: First International Symposium,* edited by Il-T. Cheon et al.

is called as the Langevin equation. In both pictures there exist the time-convolution type (TC) and the time-convolutionless type (TCL) equations. There are many applications such as spin relaxation processes, magnetic resonances, laser physics, and so on.

We want to apply the method to treat time dependent processes in the relativistic quantum field theory. Here we adopt TCL equation in the Heisenberg picture[2] among them, because we think that it is suitable for performing renormalization procedures. The method gives a systematic expansion of the Heisenberg equation of motion in terms of the interaction Hamiltonian, which corresponds to Wick's theorem in the usual S matrix calculation.

In the previous work[3] we carried out the mass renormalization in a model with unstable particles (or resonances) and studied the medium effects on the time evolution of expectation values of creation and annihilation operators, where they undergo damped oscillation toward an equilibrium value. We have shown that an divergence of the $\ln t$-type at the initial time $t = 0$ remains after mass renormalization. In this article we show how to carry out the wave function renormalization by introducing a composite operator and discuss the relation of it to the initial divergence. It is also shown that the on-mass-shell renormalization is realized in the Heisenberg equation of motion.

PROJECTION OPERATOR METHOD

Our starting point is the Heisenberg equation of motion,

$$\frac{d}{dt}A(t) = i[H,A(t)]. \tag{1}$$

Using the Liouville operator L:

$$LA = [H,A], \tag{2}$$

we can express the time evolution of the Hisenberg operator as

$$A(t) = \mathrm{e}^{iHt}A(0)\mathrm{e}^{-iHt} = \mathrm{e}^{iLt}A(0). \tag{3}$$

In order to solve the Heisenberg equation, it is necessary to make some approximations. For this purpose we introduce generic projection operators:

$$P^2 = P,\ Q = 1 - P,\ PQ = QP = 0, \tag{4}$$

and carry out coarse-grainings for some degrees of freedom. Since the time dependence of the operators is determined by e^{iLt}, the equation of motion is written as

$$\begin{aligned} \frac{d}{dt}\mathrm{e}^{iLt} &= \mathrm{e}^{iLt}iL = \mathrm{e}^{iLt}PiL + \mathrm{e}^{iLt}QiL, \\ \frac{d}{dt}\mathrm{e}^{iLt}Q &= \mathrm{e}^{iLt}PiLQ + \mathrm{e}^{iLt}QiLQ. \end{aligned} \tag{5}$$

Solving the second equation and substituting it into the first equation, we obtain[4]

$$\frac{d}{dt}A(t) = e^{iLt}PiLA(0)+e^{iLt}P\Sigma(t)\frac{1}{1-\Sigma(t)}iLA(0)+Qe^{iLQt}\frac{1}{1-\Sigma(t)}iLA(0), \tag{6}$$

where

$$\Sigma(t) = \int_0^t ds\, e^{-iL(t-s)}PiLQe^{iLQ(t-s)}. \tag{7}$$

Now we consider the case that the total system consists of the system and the environment. Then the total Hamiltonian can be divided into three parts, the system (S), the environment (E) and the interaction (I) between the system and the environment:

$$H = H_0+H_I = H_S+H_E+H_I. \tag{8}$$

The initial condition at $t=0$ is given by a density matrix which can describe either a pure state or a mixed state. It is noted that a state in thermal equlibrium is represented by a mixed state. Here we assume that the initial density matrix ρ is given by the direct product of the system density matrix ρ_S and the environment density matrix ρ_E:

$$\rho = \rho_S\otimes\rho_E. \tag{9}$$

Physically, it implies that there are no initial correlations between the system and the environment at $t=0$. We then define the projection operator as

$$PA = Tr_E(\rho_E A) \tag{10}$$

for any operator A. With this projection operator we replace an operator acting on the environment with a c-number. Using the definition of the projection operator, we have the relation:

$$QL_0Q = QL_0 \quad (L_0A\equiv[H_0,A]). \tag{11}$$

With this property we can derive a systematic expansion of the Heisenberg equation of motion:

$$\begin{aligned}\frac{d}{dt}A(t) &= e^{iLt}PiLA(0)+Qe^{iLQt}\frac{1}{1-\Sigma(t)}iLA(0)\\ &\quad -e^{iLt}Pe^{-iL_0t}C(t)Q\frac{1}{1+(C(t)-1)Q}e^{iL_0t}iLA(0),\end{aligned} \tag{12}$$

where $C(t)$ is a time ordered function of the interaction Liouville operator:

$$\begin{aligned}C(t) &= e^{iL_0t}e^{-iLt}\\ &= 1+\sum_{n=1}^{\infty}(-i)^n\int_0^t dt_1\int_0^{t_1}dt_2\cdots\int_0^{t_{n-1}}dt_n\check{L}_I(t_1)\check{L}_I(t_2)\cdots\check{L}_I(t_n),\end{aligned} \tag{13}$$

and

$$\breve{L}_I(t) = e^{iL_0t}L_Ie^{-iL_0t} \quad (L_IA \equiv [H_I,A]). \tag{14}$$

When we expand the last term in Eq.(12) up to first order in the interaction H_I, we have

$$\begin{aligned}\frac{d}{dt}A(t) &= e^{iLt}Pe^{-iL_0t}Pe^{iL_0t}iLA(0) + Qe^{iLQt}\frac{1}{1-\Sigma(t)}iLA(0) \\ &\quad + e^{iLt}Pe^{-iL_0t}P\int_0^t dse^{iL_0s}iL_Ie^{-iL_0s}Qe^{iL_0t}iLA(0).\end{aligned} \tag{15}$$

The details of the derivation are shown in ref.[4]. The equation of motion is called as the time-convolutionless equation, where the time variation of the operator $A(t)$ at time t is given by the values of operators only at t. The so-called memory terms which are usually represented by time-convolution integrals are expanded by using a time ordered function of the interaction Liouville operators[5]. This property plays an important role in performing renormalizations.

The transition probability for $|\phi_i> \rightarrow |\phi_f>$ can be written as

$$\begin{aligned}P(t) &= |<\phi_f|e^{-iHt}|\phi_i>|^2 \\ &= Tr\,\rho(e^{iLt}|\phi_f><\phi_f|) \quad (\rho = |\phi_i><\phi_i|).\end{aligned} \tag{16}$$

Therefore, when we regard $|\phi_f><\phi_f|$ as an operator $A(0)$ we can calulate the time evolution of the transition probability in a similar way.

RENORMALIZATION

We apply Eq.(15) to the σ-2π model defined by the Hamiltonian[3]

$$\begin{aligned}H &= \frac{1}{2}\int_V d^3x[\Pi_\sigma^2(x) + (\vec{\nabla}\phi_\sigma(x))^2 + m_\sigma^2\phi_\sigma^2(x)] \\ &\quad + \frac{1}{2}\int_V d^3x[\Pi_\pi^2(x) + (\vec{\nabla}\phi_\pi(x))^2 + m_\pi^2\phi_\pi^2(x)] \\ &\quad + g\int_V d^3x : \phi_\pi^2(x) : \phi_\sigma(x).\end{aligned} \tag{17}$$

Regarding the π meson field as the environment, we evaluate the time dependence of the σ meson and the π meson fields. Here we want to carry out the renormalization of ultraviolet divergences appearing in the σ meson self-energy. Then the interaction Hamiltonian is

$$\begin{aligned}H_I &= \frac{1}{2}(Z-1)\int_V d^3x\,[\Pi_\sigma^{(r)2}(x) + (\vec{\nabla}\phi_\sigma^{(r)}(x))^2 + m_\sigma^{(r)2}\phi_\sigma^{(r)2}(x)] \\ &\quad - \frac{1}{2}Z\delta m_\sigma^2\int_V d^3x\,\phi_\sigma^{(r)2}(x) + g\sqrt{Z}\int_V d^3x\, : \phi_\pi^2(x) : \phi_\sigma^{(r)}(x),\end{aligned} \tag{18}$$

where

$$\phi_\sigma(x) = \sqrt{Z}\phi_\sigma^{(r)}(x), \ \Pi_\sigma(x) = \sqrt{Z}\Pi_\sigma^{(r)}(x), \ m_\sigma^2 = m_\sigma^{(r)2} - \delta m_\sigma^2. \tag{19}$$

Here δm_σ^2 is the mass counterterm and Z is the wave function renormalization.

We consider the situation where at the initial time $t = 0$ the π meson field is in a thermal free gas phase of temperature T with H_E:

$$PA = \mathrm{Tr}_\mathrm{E}(\rho_\mathrm{E} A), \ \rho_\mathrm{E} = \mathrm{e}^{-H_\mathrm{E}/T}/\mathrm{Tr}_\mathrm{E}(\mathrm{e}^{-H_\mathrm{E}/T}). \tag{20}$$

The σ meson and the π meson fields and their conjugate fields are expanded as

$$\begin{aligned}
\phi_\sigma^{(r)}(\vec{x},t) &= \sum_{\vec{k}} \frac{1}{\sqrt{2\omega_\sigma(\vec{k})V}}(C_{\vec{k}}(t) + C^\dagger_{-\vec{k}}(t))\mathrm{e}^{i\vec{k}\vec{x}} \\
\Pi_\sigma^{(r)}(\vec{x},t) &= -i\sum_{\vec{k}} \sqrt{\frac{\omega_\sigma(\vec{k})}{2V}}(C_{\vec{k}}(t) - C^\dagger_{-\vec{k}}(t))\mathrm{e}^{i\vec{k}\vec{x}} \\
\phi_\pi(\vec{x},0) &= \sum_{\vec{p}} \frac{1}{\sqrt{2\omega_\pi(\vec{p})V}}(b_{\vec{p}}(0) + b^\dagger_{-\vec{p}}(0))\mathrm{e}^{i\vec{p}\vec{x}} \\
\Pi_\pi(\vec{x},0) &= -i\sum_{\vec{p}} \sqrt{\frac{\omega_\pi(\vec{p})}{2V}}(b_{\vec{p}}(0) - b^\dagger_{-\vec{p}}(0))\mathrm{e}^{i\vec{p}\vec{x}},
\end{aligned} \tag{21}$$

where $\omega_\sigma(\vec{k}) = \sqrt{\vec{k}^2 + m_\sigma^{(r)2}}$ and $\omega_\pi(\vec{p}) = \sqrt{\vec{p}^2 + m_\pi^2}$. We take the limit $V \to \infty$ at the end of the calculation. The Hamiltonian is expanded by the creation and annihilation operators at $t = 0$:

$$\begin{aligned}
H_S &= \sum_{\vec{k}} \omega_\sigma(\vec{k}) C^\dagger_{\vec{k}}(0) C_{\vec{k}}(0) \\
H_E &= \sum_{\vec{p}} \omega_\pi(\vec{p}) b^\dagger_{\vec{p}}(0) b_{\vec{p}}(0) \\
H_I &= (Z-1)\sum_{\vec{k}} \omega_\sigma(\vec{k}) C^\dagger_{\vec{k}}(0) C_{\vec{k}}(0) \\
&\quad - \frac{Z}{4}\delta m_\sigma^2 \sum_{\vec{k}} \frac{1}{\omega_\sigma(\vec{k})}(C_{\vec{k}}(0)C_{-\vec{k}}(0) + C^\dagger_{-\vec{k}}(0)C^\dagger_{\vec{k}}(0) \\
&\qquad\qquad + C^\dagger_{-\vec{k}}(0)C_{-\vec{k}}(0) + C_{\vec{k}}(0)C^\dagger_{\vec{k}}(0)) \\
&\quad + \frac{g}{2}\sqrt{\frac{Z}{2V}} \sum_{\vec{k},\vec{p}} \frac{1}{\sqrt{\omega_\sigma(\vec{k})\omega_\pi(\vec{p})\omega_\pi(\vec{k}+\vec{p})}}(C_{\vec{k}}(0) + C^\dagger_{-\vec{k}}(0)) \\
&\quad \times (b_{\vec{p}}(0)b_{-\vec{k}-\vec{p}}(0) + b^\dagger_{-\vec{p}}(0)b^\dagger_{\vec{k}+\vec{p}}(0) + b^\dagger_{\vec{k}+\vec{p}}(0)b_{\vec{p}}(0) + b^\dagger_{-\vec{p}}(0)b_{-\vec{k}-\vec{p}}(0)).
\end{aligned} \tag{22}$$

First of all, we consider the mass renormalization of the σ meson. We set $T = 0$. Using Eq.(15), we can derive an equation of motion for the expectation value of the creation oparator of the σ meson. In case of $m_\sigma^{(r)} < 2m_\pi$ we obtain

$$
\begin{aligned}
\frac{d}{dt} < C^\dagger_{\vec{k}}(t) > &= i\omega_\sigma(\vec{k}) Z < C^\dagger_{\vec{k}}(t) > \\
& - \frac{i}{2\omega_\sigma(\vec{k})}[\delta m_\sigma^2 - \frac{g^2}{V}\sum_{\vec{p}} \frac{1}{\omega_\pi(\vec{p})\omega_\pi(\vec{k}+\vec{p})} \\
& \times \frac{\omega_\pi(\vec{p}) + \omega_\pi(\vec{k}+\vec{p})}{\omega_\sigma^2(\vec{k}) - (\omega_\pi(\vec{p}) + \omega_\pi(\vec{k}+\vec{p}))^2}](< C_{-\vec{k}}(t) > + < C^\dagger_{\vec{k}}(t) >) \\
& + \frac{ig^2}{4V}\sum_{\vec{p}} \frac{1}{\omega_\sigma(\vec{k})\omega_\pi(\vec{p})\omega_\pi(\vec{k}+\vec{p})}[(\frac{e^{i(\omega_\sigma(\vec{k})+\omega_\pi(\vec{p})+\omega_\pi(\vec{k}+\vec{p}))t}}{\omega_\sigma(\vec{k}) + \omega_\pi(\vec{p}) + \omega_\pi(\vec{k}+\vec{p})} \\
& - \frac{e^{i(\omega_\sigma(\vec{k})-\omega_\pi(\vec{p})-\omega_\pi(\vec{k}+\vec{p}))t}}{\omega_\sigma(\vec{k}) - \omega_\pi(\vec{p}) - \omega_\pi(\vec{k}+\vec{p})}) < C_{-\vec{k}}(t) > \\
& + (\frac{e^{-i(\omega_\sigma(\vec{k})+\omega_\pi(\vec{p})+\omega_\pi(\vec{k}+\vec{p}))t}}{\omega_\sigma(\vec{k}) + \omega_\pi(\vec{p}) + \omega_\pi(\vec{k}+\vec{p})} \\
& - \frac{e^{-i(\omega_\sigma(\vec{k})-\omega_\pi(\vec{p})-\omega_\pi(\vec{k}+\vec{p}))t}}{\omega_\sigma(\vec{k}) - \omega_\pi(\vec{p}) - \omega_\pi(\vec{k}+\vec{p})}) < C^\dagger_{\vec{k}}(t) >].
\end{aligned}
\tag{23}
$$

We take the mass counterterm as

$$
\delta m_\sigma^2 = \frac{g^2}{V}\sum_{\vec{p}} \frac{1}{\omega_\pi(\vec{p})\omega_\pi(\vec{k}+\vec{p})} \frac{\omega_\pi(\vec{p}) + \omega_\pi(\vec{k}+\vec{p})}{\omega_\sigma^2(\vec{k}) - (\omega_\pi(\vec{p}) + \omega_\pi(\vec{k}+\vec{p}))^2}. \tag{24}
$$

Note that in the limit of $t \to \infty$ the third term on the r.h.s of Eq.(23) vanishes and then we have

$$
\frac{d}{dt} < C^\dagger_{\vec{k}}(t) > = i\omega_\sigma(\vec{k}) Z < C^\dagger_{\vec{k}}(t) > . \tag{25}
$$

It is found that at $V \to \infty$ the value of δm_σ^2 coincides with the on-mass-shell value ($k^2 = -m_\sigma^{(r)2}$) of the self-energy

$$
\Sigma_\sigma(k^2) = i\frac{2g^2}{(2\pi)^4}\int d^4p \frac{1}{p^2 + m_\pi^2 - i\varepsilon} \frac{1}{(k-p)^2 + m_\pi^2 - i\varepsilon} \tag{26}
$$

which is calculated by the usual Feynman rule. It is noted that since the self-energy is a function of the four-momentum squared k^2, the mass counterterm δm_σ^2 defined at the on-mass-shell is independent of the value of $\vec{k}$. The time evolution of the expectation value of the creation operator at $T = 0$ and $T \neq 0$ is shown in ref.[3].

In case of $m_\sigma^{(r)} > 2m_\pi$, at $t \to \infty$ the equation of motion becomes

$$\begin{aligned}\frac{d}{dt} < C^\dagger_{\vec{k}}(t) > &= i\omega_\sigma(\vec{k}) Z < C^\dagger_{\vec{k}}(t) > \\ &- \frac{\pi g^2 Z}{4\omega_\sigma(\vec{k})V} \sum_{\vec{p}} \frac{1}{\omega_\pi(\vec{p})\omega_\pi(\vec{k}+\vec{p})} \delta(\omega_\sigma(\vec{k}) - \omega_\pi(\vec{p}) - \omega_\pi(\vec{k}+\vec{p})) \\ &\times (< C^\dagger_{\vec{k}}(t) > - < C_{-\vec{k}}(t) >). \end{aligned} \quad (27)$$

The second term of the r.h.s. of Eq.(27) describes the decay of the σ meson. In this case δm_σ^2 should be modified as

$$\delta m_\sigma^2 = \frac{g^2}{V} \sum_{\vec{p}} \frac{1}{\omega_\pi(\vec{p})\omega_\pi(\vec{k}+\vec{p})} \mathcal{P} \frac{\omega_\pi(\vec{p}) + \omega_\pi(\vec{k}+\vec{p})}{\omega_\sigma^2(\vec{k}) - (\omega_\pi(\vec{p}) + \omega_\pi(\vec{k}+\vec{p}))^2}, \quad (28)$$

where $\mathcal{P}$ indicates the principal value.

As seen in ref.[3], after mass renormalization there appears a $\ln t$-type divergence at $t = 0$ which comes from the third term of the r.h.s. of Eq.(23). Note that before mass renormalization the third term and the term with the factor of the r.h.s. of Eq.(24) are cancelled each other at $t = 0$ in Eq.(23). This divergence is called as the initial divergence.

Let's consider the following composite operator

$$\begin{aligned} C^{(\mathrm{dr})\dagger}_{\vec{k}}(t) &= C^\dagger_{\vec{k}}(t) + \frac{g}{2}\sqrt{\frac{Z}{2V}} \sum_{\vec{p}} \frac{1}{\sqrt{\omega_\sigma(\vec{k})\omega_\pi(\vec{p})\omega_\pi(\vec{k}+\vec{p})}} \\ &\times [\frac{1}{\omega_\sigma(\vec{k}) + \omega_\pi(\vec{p}) + \omega_\pi(\vec{k}+\vec{p})} b_{\vec{p}}(t) b_{-\vec{k}-\vec{p}}(t) \\ &+ \frac{\mathcal{P}}{\omega_\sigma(\vec{k}) - \omega_\pi(\vec{p}) - \omega_\pi(\vec{k}+\vec{p}) + \varepsilon} b^\dagger_{-\vec{p}}(t) b^\dagger_{\vec{k}+\vec{p}}(t) \\ &+ \frac{1}{\omega_\sigma(\vec{k}) + \omega_\pi(\vec{p}) - \omega_\pi(\vec{k}+\vec{p})} b^\dagger_{\vec{k}+\vec{p}}(t) b_{\vec{p}}(t) \\ &+ \frac{1}{\omega_\sigma(\vec{k}) - \omega_\pi(\vec{p}) + \omega_\pi(\vec{k}+\vec{p})} b^\dagger_{-\vec{p}}(t) b_{-\vec{k}-\vec{p}}(t)], \end{aligned} \quad (29)$$

where $\varepsilon(= 0^+)$ is introduced to regularize the contribution from the pole[6]. As one can show, the initial divergence disappears in the equation of motion for $C^{(\mathrm{dr})\dagger}_{\vec{k}}(t)$. Actually, the coefficients of the π meson operators in Eq.(29) are determined in such a way that the initial divergence is cancelled out.

Then we obtain the wave function renormalization:

$$Z^{-1} = 1 + \frac{g^2}{4V\omega_\sigma(\vec{k})} \sum_{\vec{p}} \frac{1}{\omega_\pi(\vec{p})\omega_\pi(\vec{k}+\vec{p})}$$

$$\times [\mathcal{P}\frac{1}{\omega_\sigma(\vec{k})-\omega_\pi(\vec{p})-\omega_\pi(\vec{k}+\vec{p})+\varepsilon}\frac{1}{\omega_\sigma(\vec{k})-\omega_\pi(\vec{p})-\omega_\pi(\vec{k}+\vec{p})} - \frac{1}{(\omega_\sigma(\vec{k})+\omega_\pi(\vec{p})+\omega_\pi(\vec{k}+\vec{p}))^2}]. \tag{30}$$

It also coincides with the on-mass-shell value of the derivative of the self-energy in the case of $m_\sigma^{(r)} < 2m_\pi$. The value of Z is interpreted as the probability of finding a bare σ meson in a dressed σ meson.

TIME EVOLUTION OF TRANSITION AMPLITUDE

Renormalizations in the calculation of the time evolution of density matrices can be carried out in a similar way. For example, let's consider a density matrix

$$A(t) = e^{iLt}|1_{\vec{k}}>_{\sigma\sigma}<0|\otimes|0>_{\pi\pi}<0|, \tag{31}$$

where $|0>_\sigma, |0>_\pi$ are the vacua of the σ and π mesons, respectively, and $|1_{\vec{k}_0}>_\sigma$ is the one-particle state with momentum $\vec{k}_0$:

$$C_{\vec{k}}(0)|0>_\sigma = 0,\ b_{\vec{p}}(0)|0>_\pi = 0,\ |1_{\vec{k}_0}>_\sigma = C^\dagger_{\vec{k}_0}(0)|0>_\sigma . \tag{32}$$

The projection operator is chosen as

$$\begin{aligned} P = & (|0>_{\sigma\sigma}<0|\ Tr_S|0>_{\sigma\sigma}<0| + \sum_{\vec{k}}|1_{\vec{k}}>_{\sigma\sigma}<1_{\vec{k}}|\ Tr_S|1_{\vec{k}}>_{\sigma\sigma}<1_{\vec{k}}| \\ & + \sum_{\vec{k}}|1_{\vec{k}}>_{\sigma\sigma}<0|\ Tr_S|0>_{\sigma\sigma}<1_{\vec{k}}| + \sum_{\vec{k}}|0>_{\sigma\sigma}<1_{\vec{k}}|\ Tr_S|1_{\vec{k}}>_{\sigma\sigma}<0|) \\ & \otimes|0>_{\pi\pi}<0|\ Tr_E|0>_{\pi\pi}<0|. \end{aligned} \tag{33}$$

Then, we can calculate the time evolution of the transition amplitude between the vacuum and the one-particle state at $T=0$. The composite operator corresponding to Eq.(29) is given as

$$\begin{aligned} A^{(\mathrm{dr})}(t) = & e^{iLt}|1_{\vec{k}}>_{\sigma\sigma}<0|\otimes|0>_{\pi\pi}<0| \\ & +\frac{g}{2}\sqrt{\frac{Z}{2V}}\sum_{\vec{p}}\frac{1}{\omega_\sigma(\vec{k})\omega_\pi(\vec{p})\omega_\pi(\vec{k}+\vec{p})} \\ & \times[\frac{1}{\omega_\sigma(\vec{k})+\omega_\pi(\vec{p})+\omega_\pi(\vec{k}+\vec{p})}e^{iLt}|1_{\vec{k}}>_{\sigma\sigma}<1_{\vec{k}}|\otimes|0>_{\pi\pi}<1_{\vec{p}}1_{-\vec{k}-\vec{p}}| \\ & +\frac{\mathcal{P}}{\omega_\sigma(\vec{k})-\omega_\pi(\vec{p})-\omega_\pi(\vec{k}+\vec{p})+\varepsilon}e^{iLt}|0>_{\sigma\sigma}<0|\otimes|1_{-\vec{p}}1_{\vec{k}+\vec{p}}>_{\pi\pi}<0|] \end{aligned} \tag{34}$$

It is found that we can carry out the mass and the wave function renormalizations by using the same δm_σ^2 and Z.

SUMMARY AND DISCUSSIONS

Applying the projection operator method to the σ-2π model we have calculated the time evolution of the field operators in a vacuum and in a medium. The prescription for the ultraviolet divergence appearing in the Heisenberg equation of motion is given. First we carried out the mass renormalization in the lowest order, using a mass counterterm. Then, the initial divergence appears at the initial time $t = 0$. Considering the equation of motion for a composite operator with the σ meson and the π meson fields which corresponds to the dressed σ meson, we carried out the wave function renormalization where the initial divergence is cancelled. It is found that the values of the renormalization parameters coincide with the on-mass-shell values of the self-energy and its derivative calculated by the conventional way. The appearance of the initial divergence may be related to our assumption that there is no initial correlation in the initial density matrix. We need further study on this point.

The procedure to study medium effects on the time evolution of the field operators is the following. At $T = 0$ we take the narrowing limit, that is, use Eq.(25) or Eq.(27) with $Z = 1$ at all t, and then calculate the contribution from the environment at $T \neq 0$ to the equation of motion by applying the projection operator method.

It has been also shown that we can calculate the time evolution of the transition amplitude in a similar way by using the same renormalization parameters.

Recently we have applied the projection operator method to ϕ^4 theory and derived the quantum and the semiclassical Langevin equations[7]. Comparing the result with the one derived by using the influence functional method, we have discussed the correspondences and the differences between the two approaches. It is found that similar expressions are obtained for the self-energy part. Therefore, we believe that the projection operator method gives a basis for treating the time evolution of the relativistic quantum many body systems.

REFERENCES

1. Kubo, R., Toda, M., and Hashitsume, N., *Statistical Physics II*, Springer 1985.
2. Hashitsume, N., Shibata, F., and Shingu, M., *J. Stati.Phys.* **17**, 155-169 (1977).
3. Koide, T., Maruyama, M., and Takagi, F., *Prog.Theor.Phys.* **101**, 373-384 (1999).
4. Koide, T., and Maruyama, M., *Prog.Theor.Phys.* **104**, 575-594 (2000).
5. Shibata, F., and Arimitsu, T., *J.Phys.Soc.Jpn.* **49**, 891-897 (1980).
6. Araki, H., Munakata, Y., Kawaguchi, M., and Goto, T., *Prog.Theor.Phys.* **17**, 419-442 (1957).
7. Koide, T., Maruyama, M., and Takagi, F., hep-ph/0102272 (2001).

$\frac{\partial m}{\partial \mu}$ in the Nambu–Jona-Lasinio model

O. Miyamura and S. Choe

Dept. of Physics, Hiroshima University, Higashi-Hiroshima 739-8526, Japan

Abstract. Using the Nambu–Jona-Lasinio (NJL) model we study responses of the pion and kaon masses to changes in the chemical potential, $\frac{\partial m}{\partial \mu}$, at zero and finite chemical potential. We find that the behavior of $\frac{\partial m}{\partial \mu}$ for the pion is quite different from that for the kaon. Our results can give a clue for future studies of $\frac{\partial m}{\partial \mu}$ on the lattice.

1. INTRODUCTION

There are several methods to understand the behavior of matter under extreme conditions of temperature and/or density. One of them is the lattice QCD. While the structure of QCD at high temperature has been investigated in detail, little is known about matter at high baryon density due to the well-known "complex-action" problem [1]. One of possible ways on the lattice is to simulate the response of a hadron mass to changes in the chemical potential, $\frac{\partial m}{\partial \mu}$, at zero chemical potential ($\mu = 0$) [2, 3].

One the other hand, one can use effective models of QCD : e.g., the NJL model [4]. This model has been widely used for describing the phase transition as well as hadron properties in hot and/or dense matter [5].

In this work we present the NJL model calculations of $\frac{\partial m}{\partial \mu}$ for the pion and kaon. The primary goal of our study is to get the same quantities which are simulated on the lattice. Of course, the direct comparison between the lattice data and the NJL model calculations is not possible because $\frac{\partial m}{\partial \mu}$ on the lattice is for the screening mass while for the pole mass in the NJL model. Nevertheless, we can learn some ideas for future studies of $\frac{\partial m}{\partial \mu}$ on the lattice from this effective model calculations.

Following the notation of the lattice simulations we consider two kinds of the chemical potential. One is the isoscalar $\mu_S = \mu_u + \mu_s$ for the kaon (or $\mu_u + \mu_d$ for the pion). The other is the isovector $\mu_V = \mu_s - \mu_u$ (or $\mu_d - \mu_u$). In contrast to the lattice simulations we get $\frac{\partial m}{\partial \mu}$ at zero and finite chemical potential within the NJL model. Then, our study can give information about the role of the light quark chemical potential and/or the strange quark chemical potential in hot and/or dense matter.

The paper is organized as follows. In Sec. 2 we introduce some basic formulas to get $\frac{\partial m_K}{\partial \mu}$ for the kaon in the NJL model, and show results at zero and finite chemical potential. We present $\frac{\partial m_\pi}{\partial \mu}$ for the pion in Sec. 3. In Sec. 4 we summarize our results and discuss some uncertainties in our calculations.

CP594, *Hadrons and Nuclei: First International Symposium*, edited by Il-T. Cheon et al.

2. $\frac{\partial M_K}{\partial \mu}$ IN THE NJL MODEL

We use the generalized SU(3) NJL model with the anomaly term [5]:

$$\mathcal{L} = \bar{q}(i\gamma \cdot \partial - m)q + \frac{1}{2}g_S \sum_{a=0}^{8} \left[(\bar{q}\lambda_a q)^2 + \bar{q}(i\lambda_a \gamma_5 q)^2\right] + g_D \left[\det \bar{q}_i(1-\gamma_5)q_j + h.c.\right] , \quad (1)$$

where λ_a are the Gell-Mann matrices and m is a mass matrix for current quarks, m = diag(m_u, m_d, m_s). We take the following parameters in [5].

$$\begin{aligned} \Lambda = 631.4\ \text{MeV},\ g_S\Lambda^2 = 3.67,\ g_D\Lambda^5 = -9.29 \\ m_u = m_d = 5.5\ \text{MeV},\ m_s = 135.7\ \text{MeV}\ , \end{aligned} \quad (2)$$

where Λ is the momentum cut-off. The third term in Eq.(1) is a reflection of the axial anomaly, and causes a mixing in flavors. For example, the constituent quark masses are given as follows.

$$\begin{aligned} M_u &= m_u - 2g_S\alpha - 2g_D\beta\gamma\ , \\ M_d &= m_d - 2g_S\beta - 2g_D\alpha\gamma\ , \\ M_s &= m_s - 2g_S\gamma - 2g_D\alpha\beta\ , \end{aligned} \quad (3)$$

where $\alpha \equiv \langle\bar{u}u\rangle$, $\beta \equiv \langle\bar{d}d\rangle$, and $\gamma \equiv \langle\bar{s}s\rangle$. It means that a change of $\langle\bar{u}u\rangle$ results in a change of $\langle\bar{s}s\rangle$, and vice versa. Then, we can expect a change in the properties of the observables related with the strange quarks even in the nuclear matter.

In this work we concentrate mostly on the Case II in [5], where only g_D has a T-dependence

$$g_D(T) = g_D(T=0)\ \exp[-(T/T_0)^2] \quad (4)$$

while other coupling constants and the cut-off are independent of T and chemical potential (or density). Here, we set T_0 = 0.1 GeV taking into account the restoration of $U_A(1)$ symmetry as in [5]. It might be realistic to make the coupling constants and/or the cut-off dependent on temperature and chemical potential. However, at present, there is no such an estimate including all variations in the cut-off and the coupling constants except for a few estimates of the strength of the anomaly term g_D [6].

In the mean-field approximation the above Lagrangian leads to the following gap equation [5].

$$\langle\bar{q}_i q_i\rangle = 2N_c \sum_p \left(\frac{-M_i}{E_{ip}} f(E_{ip})\right)\ , \quad (5)$$

where $\langle\cdot\rangle$ means the statistical average and the index i denotes the u, d, and s quarks. N_c is the number of colors and M_i is the constituent quark mass, and $E_{ip} = \sqrt{M_i^2 + p^2}$. $f(E_{ip}) = 1 - n_{ip} - \bar{n}_{ip}$, where n_{ip} and $\bar{n}_{ip}$ are the distribution functions of the ith quark

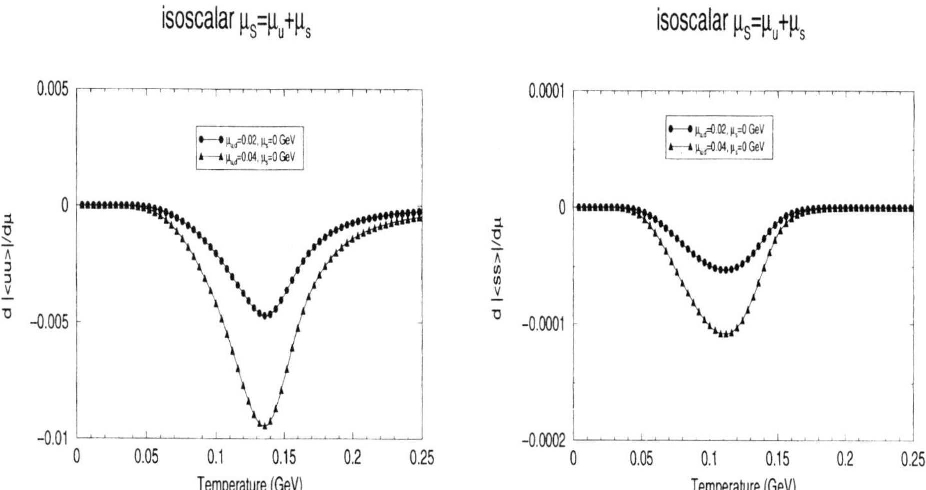

FIGURE 1. The responses of the u (left) and s (right) quark condensates.

and antiquark, respectively.

$$n_{ip} = \frac{1}{1+\exp\left((E_{ip}-\mu_i)/T\right)} ,$$
$$\bar{n}_{ip} = \frac{1}{1+\exp\left((E_{ip}+\mu_i)/T\right)} . \tag{6}$$

The right-hand side of Eq.(5) is a function of $F\ (\langle\bar{u}u\rangle, \langle\bar{d}d\rangle, \langle\bar{s}s\rangle, \mu_i, T)$. Then, we obtain responses of the quark condensates $\frac{\partial\langle\bar{u}u\rangle}{\partial\mu}$, $\frac{\partial\langle\bar{d}d\rangle}{\partial\mu}$, and $\frac{\partial\langle\bar{s}s\rangle}{\partial\mu}$ by differentiating both sides with respect to μ at a fixed T. These $\frac{\partial\langle\bar{q}q\rangle}{\partial\mu}$ will be used to get $\frac{\partial m_K}{\partial\mu}$ in the below.

Fig.1 shows $\frac{\partial\langle\bar{u}u\rangle}{\partial\mu_S}$ and $\frac{\partial\langle\bar{s}s\rangle}{\partial\mu_S}$ at finite chemical potential. At zero chemical potential both $\frac{\partial\langle\bar{q}q\rangle}{\partial\mu_S}$ and $\frac{\partial\langle\bar{q}q\rangle}{\partial\mu_V}$ are zero. We take two different values for the chemical potential, $\mu_u = \mu_d = 0.02$ and 0.04 GeV. In the figure we set the perpendicular axis as the absolute value of $\frac{\partial\langle\bar{q}q\rangle}{\partial\mu_S}$, i.e. $\frac{\partial|\langle\bar{q}q\rangle|}{\partial\mu_S}$, and thus the figure shows that the absolute value of the quark condensate decreases with increasing chemical potential. In addition, the figure shows that variations of the u quark condensate $\frac{\partial\langle\bar{u}u\rangle}{\partial\mu_S}$ are much larger than those of $\frac{\partial\langle\bar{s}s\rangle}{\partial\mu_S}$, and the variation of each quark condensate is proportional to the chemical potential.

Now, consider the dispersion equation for the kaon, e.g., the K^- [5].

$$D^R_{K^-}(\omega,\vec{q}=0)^{-1} \equiv -G_K^{-1}\left[1+2G_K I^p_{su}(\omega,\vec{q}=0)\right] = 0 , \tag{7}$$

where G_K is the coupling strength in this channel, $G_K \equiv g_S + g_D\beta$, and $I^p_{su}(\omega,\vec{q}=0)$ is the one-loop polarization due to u- and s-quarks. Differentiating both sides of the above equation with respect to μ_S (or μ_V) at the fixed T, and using $\frac{\partial\langle\bar{q}q\rangle}{\partial\mu_S}$ (or $\frac{\partial\langle\bar{q}q\rangle}{\partial\mu_V}$) we get $\frac{\partial m_K}{\partial\mu_S}$ (or $\frac{\partial m_K}{\partial\mu_V}$), i.e. the response of the kaon mass to changes in the isoscalar (or isovector) chemical potential μ_S (or μ_V).

First, we show $\frac{\partial m_K}{\partial \mu_S}$ for the K^- at zero chemical potential in Fig. 2. Below $T \sim 0.04$ GeV $\frac{\partial m_K}{\partial \mu_S}$ is almost zero, and this is because $\frac{\partial \langle \bar{q}q \rangle}{\partial \mu_S}$ is hardly changed in this region as shown in Fig. 1. Near the kaon Mott temperature T_{m_K} $\frac{\partial m_K}{\partial \mu_S}$ changes rapidly and becomes almost zero. Here, T_{m_K} is defined as a temperature at which the sum of the u and s constituent quark masses equals to the kaon mass, i.e. $M_u + M_s = m_K$. Above T_{m_K} the kaon becomes a resonance.

In the figure we do not show the points in the above T_{m_K} region because there may be a large uncertainty. We can not get a reliable kaon mass in this region, and hence $\frac{\partial m_K}{\partial \mu_S}$. In fact, the authors of [5] presented the kaon mass in this region using the imaginary part of the self-energy. However, the imaginary part is an artifact of the model and thus we need physical justifications before using this part. In this work we take only the real part and concentrate on the below T_{m_K} region.

Fig.3 shows $\frac{\partial m_K}{\partial \mu_S}$ and $\frac{\partial m_K}{\partial \mu_V}$ at zero and finite chemical potential. It shows that $\frac{\partial m_K}{\partial \mu_S}$ increases with increasing chemical potential, and there is a critical value between μ_u = μ_s = 0.06 and 0.08 GeV where the sign of $\frac{\partial m_K}{\partial \mu_S}$ is changed even at below T_{m_K}. This result is consistent with previous NJL model calculations [7]. As in the case of zero chemical potential $\frac{\partial m_K}{\partial \mu_S}$ changes rapidly near T_{m_K}. Now, consider the isovector case, where $\mu_V = \mu_s - \mu_u$. Then, one can expect that the sign of $\frac{\partial m_K}{\partial \mu_V}$ will be opposite to that of $\frac{\partial m_K}{\partial \mu_S}$ because the u quark plays a dominant role rather than the s quark does. $\frac{\partial m_K}{\partial \mu_V}$ decreases with increasing chemical potential as shown in the figure.

In Fig. 4 we present $\frac{\partial m_K}{\partial \mu_S}$ and $\frac{\partial m_K}{\partial \mu_V}$ for the K^+. They are obtained by replacing ω in Eq. (7) with $-\omega$. For comparison we also show $\frac{\partial m_K}{\partial \mu_S}$ and $\frac{\partial m_K}{\partial \mu_V}$ for the K^- at zero chemical potential.

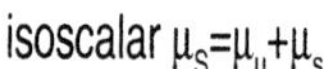

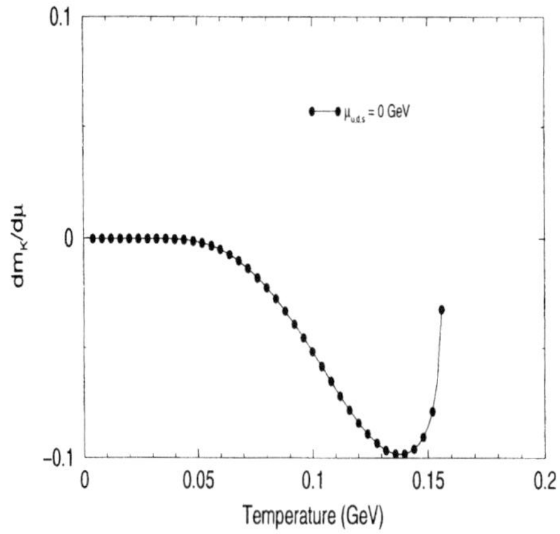

FIGURE 2. $\frac{\partial m_K}{\partial \mu_S}$ for the K^- at zero chemical potential.

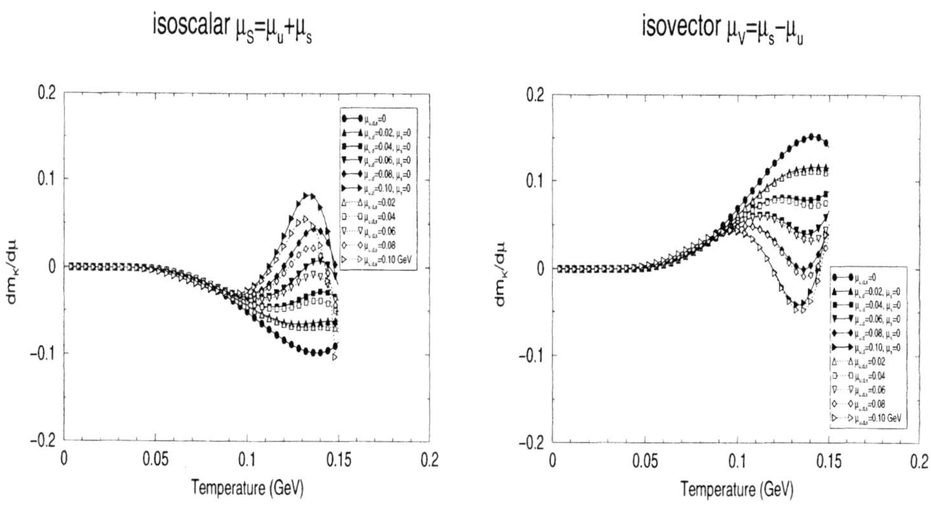

FIGURE 3. $\frac{\partial m_K}{\partial \mu}$ for the K^-.

3. $\frac{\partial M_\pi}{\partial \mu}$ IN THE NJL MODEL

In this section we show $\frac{\partial m_\pi}{\partial \mu}$ for the π^- and π^+. We use the same formulas in the previous section by replacing m_s, μ_s with m_d, μ_d, respectively. As for the dispersion equation a new coupling strength $G_\pi \equiv g_S + g_D \gamma$ is introduced [5].

First, consider the isoscalar $\mu_S = \mu_u + \mu_d$. In the case $m_u = m_d$, $\frac{\partial m_\pi}{\partial \mu_S} = 0$ at zero chemical potential. $\frac{\partial m_\pi}{\partial \mu_S}$ and $\frac{\partial m_\pi}{\partial \mu_V}$ for the π^- and π^+ at finite chemical potential are given in Fig.5.

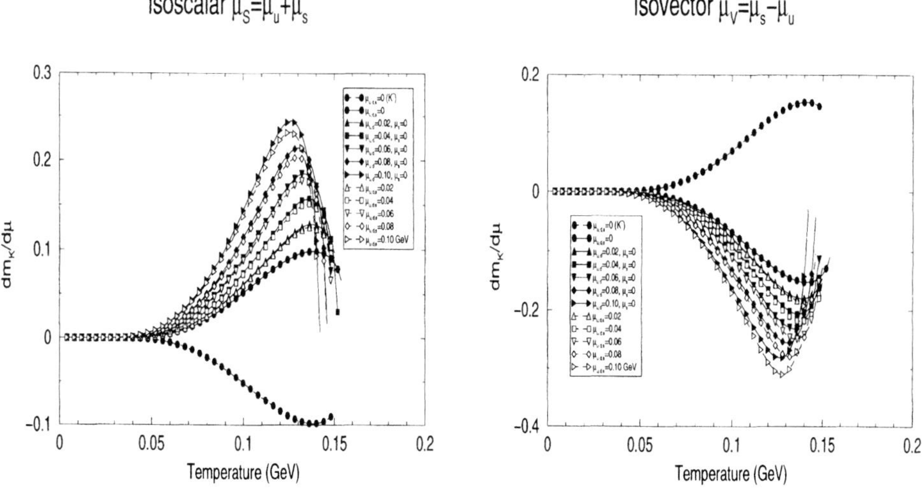

FIGURE 4. $\frac{\partial m_K}{\partial \mu}$ for the K^+.

Note that in the case of the isovector chemical potential we take $\mu_d = 2\,\mu_u$.

In the previous calculation we assumed $m_u = m_d = 5.5$ MeV. It will be interesting to consider different u and d quark masses, e.g., $m_u = 4$ MeV and $m_d = 7$ MeV. Although the cut-off and the coupling constants should be modified according to this change of the quark masses, we use the same parameters as before and study $\frac{\partial m_\pi}{\partial \mu_S}$ and $\frac{\partial m_\pi}{\partial \mu_V}$.

In the case of $\frac{\partial m_\pi}{\partial \mu_S}$ for the π^- a transition point appears between $\mu_u = \mu_d = 0.004$ GeV and 0.006 GeV as shown in Fig. 6. This transition point seems reasonable considering the mass ratios of m_s/m_u for the kaon and m_d/m_u for the pion. For comparison we show both $\frac{\partial m_\pi}{\partial \mu_S}$ for the π^- and the π^+. On the other hand, in the case of the isovector chemical potential $\frac{\partial m_\pi}{\partial \mu_V}$ for the π^- and π^+ are similar to the previous ones, i.e. the results for the pion with the degenerate u and d quark masses.

4. DISCUSSIONS

Using the NJL model we have calculated responses of the kaon and pion masses to changes in the chemical potential, $\frac{\partial m_K}{\partial \mu}$ and $\frac{\partial m_\pi}{\partial \mu}$, at zero and finite chemical potential, and found that $\frac{\partial m}{\partial \mu}$ is much dependent on the mass difference of two quarks, i.e. the mass difference between the u and s (or d) quarks.

Let us discuss some uncertainties in our calculations. First, we have considered the Lagrangian (Eq.(1)) without the vector and axial-vector terms. Although there are still arguments about the strength of the vector coupling g_V [9], a further analysis including these terms is required. In fact, one of the NJL model calculations showed that the K^- mass at finite density with $g_V \neq 0$ is quite different from that with $g_V = 0$ [8]. A

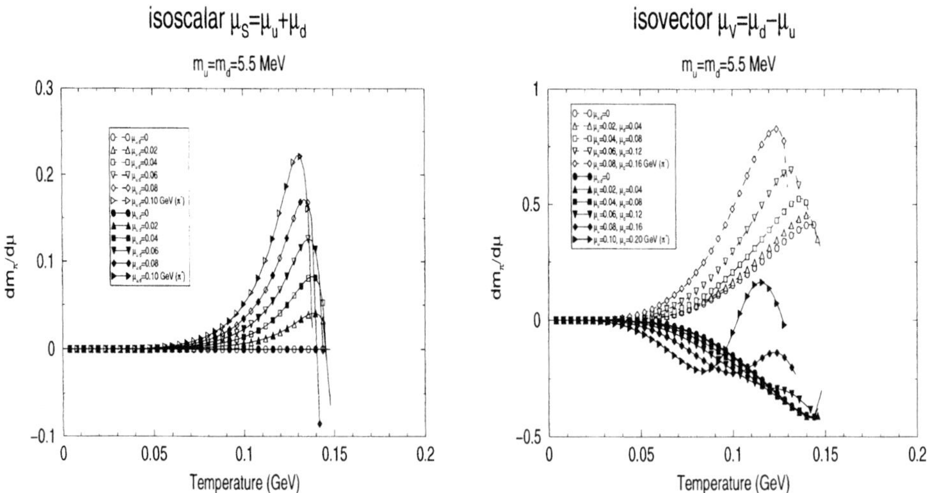

FIGURE 5. $\frac{\partial m_\pi}{\partial \mu}$ for the pion with $m_u = m_d = 5.5$ MeV.

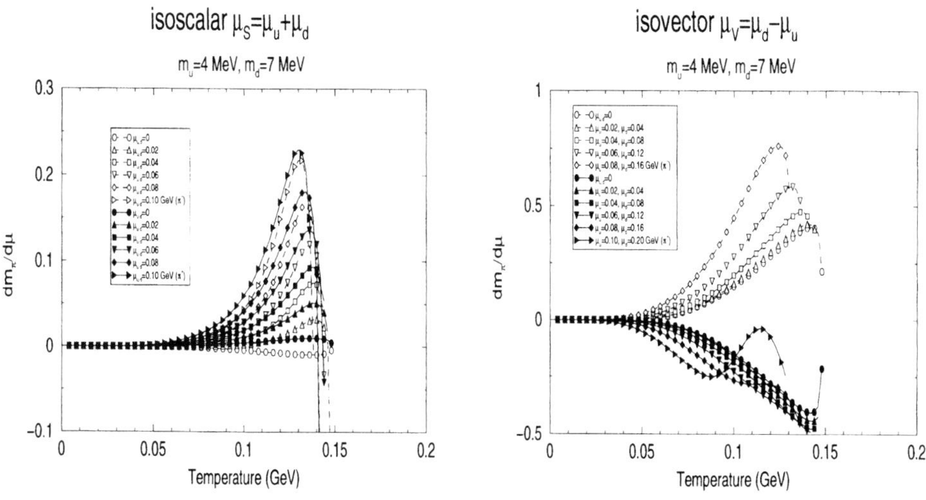

FIGURE 6. $\frac{\partial m_\pi}{\partial \mu}$ for the pion with m_u = 4 MeV, m_d = 7 MeV.

preliminary result of $\frac{\partial m_K}{\partial \mu_S}$ for the K^- with a non-zero g_V also confirms this [10].

Second, in the previous section we have also considered the different u, d quark masses for the pion (m_u = 4 MeV and m_d = 7 MeV) and assumed the other parameters are invariant under this change, and found that in the case of $\frac{\partial m_\pi}{\partial \mu_S}$ the result is slightly different from the previous one, i.e. the pion with the degenerate u and d quark masses ($m_u = m_d$ = 5.5 MeV). However, we have to take into account variations of the cut-off and coupling constants, although we expect that they would be very small. In the real world, SU(2) symmetry is slightly broken ($m_u \neq m_d$, $\langle \bar{u}u \rangle \neq \langle \bar{d}d \rangle$), thus a more careful analysis is needed in this case.

Third, in this work we have mainly considered the Case II in [5], where only g_D has the temperature dependence as shown in Eq.(4). It may be interesting to compare $\frac{\partial m_K}{\partial \mu}$ and $\frac{\partial m_\pi}{\partial \mu}$ for the Case II with those for the Case I, where all the coupling constants (g_S, g_D) and the cut-off Λ are independent of temperature and/or chemical potential. We have checked that the behaviors of $\frac{\partial m_K}{\partial \mu}$ and $\frac{\partial m_\pi}{\partial \mu}$ for the Case I are similar to those for the Case II except for the different Mott temperatures [10]. This is because g_D is rather irrelevant to the pion and kaon masses. However, further analyses including all variations of the cut-off and coupling constants at finite temperature and/or chemical potential are required before any firm conclusions may be drawn.

As a final remark, we find that the second order responses of the kaon and pion masses to the chemical potential, $\frac{\partial^2 m_K}{\partial \mu^2}$ and $\frac{\partial^2 m_\pi}{\partial \mu^2}$, are much larger than $\frac{\partial m_K}{\partial \mu}$ and $\frac{\partial m_\pi}{\partial \mu}$, respectively. Thus, one can see rather clearer signals than before.

ACKNOWLEDGEMENTS

We thank T. Kunihiro, K. Redlich, T. Hatsuda, and Su H. Lee for valuable comments. The work of O.M. was supported by Grant-in-Aide for Scientific Research by Monbusho, Japan (No. 11694085 and No. 11740159), and the work of S.C. was supported by the Japan Society for the Promotion of Science (JSPS).

REFERENCES

1. See, e.g., Proc. of Annual Lattice Conference.
2. de Forcrand, Ph., *et al.* (QCD-TARO Collaboration), *Nucl. Phys.* **B(Proc.Suppl.) 73**, 477 (1999); **83**, 408 (2000).
3. Miyamura, O., talk given at this symposium.
4. Nambu, Y. and Jona-Lasinio, G., *Phys. Rev.* **122**, 345 (1961); **124**, 246 (1961).
5. For a review, see Hatsuda, T. and Kunihiro, T., *Phys. Rep.* **247**, 221 (1994); and references therein.
6. Gross, D., Pisarski, R.D., and Yaffe, L.G., *Rev. Mod. Phys.* **53**, 43 (1981);
   Teper, M., *Phys. Lett.* **B202**, 553 (1987);
   Alkofer, R. and Reinhardt, H., *Z. Phys.* **C45**, 275 (1989).
7. Ruivo, M.C. and de Sousa, C.A., *Phys. Lett.* **B385**, 39 (1996);
   de Sousa, C.A. and Ruivo, M.C., *Nucl. Phys.* **A625**, 713 (1997).
8. Ruivo, M.C., de Sousa, C.A., and Providência, C., *Nucl. Phys.* **A651**, 59 (1999).
9. Yamawaki, K and Zakharov, V.I., hep-ph/9406373;
   Christov, Chr.V., Goeke, K., and Polyakov, M., hep-ph/9501383.
10. Miyamura, O. and Choe, S., work in preparation.

QCD sum rules for J/ψ in the nuclear medium

Sungsik Kim and Su Houng Lee

Department of Physics and Institute of Physics and Applied Physics, Yonsei University, Seoul 120-749, Korea

Abstract. We calculate the Wilson coefficients of all dimension 6 gluon operators with non zero spin in the correlation function between two heavy vector currents. We apply our results to investigate the mass of J/ψ in nuclear matter. Using an upper bound estimate on the matrix elements of the dimension 6 gluon operators to linear order in density, we find that the density dependent contribution from dimension 6 operators is less than 40% of the dimension 4 operators with opposite sign.

INTRODUCTION

In the correlation function of two heavy vector currents, only gluon operators contribute in the operator product expansion(OPE) because in the heavy quark system, all the heavy quark condensates are generated via gluonic condensates[1, 2, 3]. In dimension 6, there are scalar operators, twist-2 and twist-4 operators. For the scalar gluonic operators at dimension 6, there are two independent operators. In ref.[4], the two were identified and the corresponding Wilson coefficient were calculated . For twist-2 gluon operator, the calculation for the leading order(LO) Wilson coefficient is simple and its matrix element is just the second moment of the gluon structure function.

The twist-4 dimension 6 gluon operators are more involved. In this work, we have identified the three independent local gluon operators and calculated their corresponding LO Wilson coefficients in the correlation function between two heavy vector currents. This result is new and complimentary to a previous work[5] on gluon twist-4 operators, where they start from certain diagrams and identify the three independent twist-4 gluon structure functions. Together with the previous calculation of the Wilson coefficients for the dimension 6 scalar gluon operators by Nikolaev and Radyushkin[4], our result completes the list of all the Wilson coefficients of dimension 6 gluon operators in the correlation function between heavy vector currents.

As an application, we will use our result in QCD sum rule approach to investigate the property of J/ψ in nuclear matter. This is particularly interesting because the on-going discussion of J/ψ suppression in RHIC as a possible signal for quark gluon plasma[6], inevitably requires a detailed knowledge of the changes of J/ψ properties in "normal" nuclear matter[7]. Furthermore, the large charm quark mass $m_c \gg \Lambda_{QCD}$ provides a natural renormalization point for which a perturbative QCD expansion is partly possible.

CP594, *Hadrons and Nuclei: First International Symposium,* edited by Il-T. Cheon et al.

In fact, studies have shown that the multi-gluon exchange between a $c\bar{c}$ pair and nucleons might induce a bound $c\bar{c}$ state with even light nuclei[8, 9, 10, 11, 12, 13, 14]. In such analysis, the low energy multi-gluon potential was modeled either from the effective theory obtained in the infinitely large m_Q limit[10, 11, 12] or from extrapolating the high energy scattering via pomeron exchange to lower energy[8]. Although both approaches, gave similar binding for the J/ψ in nuclear matter, it is not clear how reliable these results are unless one systematically calculates the corrections.

In the next sections we calculate the Wilson coefficients for the independent set of gluon operators and show current conservation. We also calculate the moments and perform a moment sum rules analysis to calculate the J/ψ mass in nuclear matter. We conclude with some discussions.

POLARIZATION (OPE)

Three independent twist-4 operators at dimension 6 are here chosen as [18],

$$g^2 G^a_{\kappa\lambda} G^a_{\kappa\lambda;\mu\nu},\ g^2 G^a_{\mu\kappa} G^a_{\nu\lambda;\lambda\kappa},\ g^2 G^a_{\mu\kappa} G^a_{\kappa\lambda;\lambda\nu}. \tag{1}$$

Then we will calculate their LO Wilson coefficients in the correlation function between two vector currents made of heavy quarks, $j_\mu = \bar{h}\gamma_\mu h$.

$$\begin{aligned}\Pi_{\mu\nu}(q) &= i\int d^4x e^{iqx}\langle \mathrm{T}\{j_\mu(x) j_\nu(0)\}\rangle_\rho \\ &= i\int d^4x e^{iqx}\langle \mathrm{Tr}\left[\gamma_\mu S(x,0)\gamma_\nu S(0,x))\right]\rangle_\rho \\ &= i\int \frac{d^4k}{(2\pi)^4}\langle \mathrm{Tr}[\gamma_\mu S(k+q)\gamma_\nu \tilde{S}(k)]\rangle_\rho,\end{aligned} \tag{2}$$

where $\langle\cdot\rangle_\rho$ represents the expectation value at finite nuclear density ρ. The fourier transforms are defined by

$$\begin{aligned} iS(p) &= \int d^4x e^{ipx}\, iS(x,0) \\ i\tilde{S}(p) &= \int d^4x e^{-ipx}\, iS(0,x). \end{aligned} \tag{3}$$

$S(x,0)$ is the heavy quark propagator in the presence of external gluon field [15]. To calculate the Wilson coefficients, we obtain the quark propagator in the presence of external gauge fields in the Fock-Schwinger gauge [4, 15].

In general the polarization tensor in eq.(2) will have two invariant functions. They can be divided into the longitudinal and transverse part of the external three momentum, which are the following when the nuclear matter is at rest.

$$\Pi_L = \frac{1}{k^2}\Pi_{00},\quad \Pi_T = -\frac{1}{2}\left(\frac{1}{q^2}\Pi^\mu_\mu + \frac{1}{k^2}\Pi_{00}\right), \tag{4}$$

where $q = (\omega, k)$ is the external momentum.

In the vacuum or in the limit when $k \to 0$ they become the same so that there is only one invariant function

$$\Pi_L(\omega^2) = \Pi_T(\omega^2) = \frac{-1}{3\omega^2} g^{\mu\nu} \Pi_{\mu\nu} \equiv \Pi(\omega^2). \tag{5}$$

So in this work, we will construct the sum rule for $\Pi(\omega^2)$ at $k \to 0$ and nuclear matter at rest. However, in the calculation of the OPE, we will start from the general expression of eq.(2) at nonzero value of k and calculate each tensor structure separately. This way, current conservation will be a nontrivial check of our calculation and the generalization to $k \neq 0$ will be straightforward.

In the following we summarize the OPE for operators of dimension 4 and 6.

scalar contributions

$$\begin{aligned}\Pi^{\text{scalar}}_{\mu\nu}(q) &= (q_\mu q_\nu - g_{\mu\nu} q^2) \Big[C^{pert.} + C^0_{G^2} \langle g^2 G^a_{\alpha\beta} G^a_{\alpha\beta} \rangle \\ &\quad + C^0_{G^3} \langle g^3 f^{abc} G^a_{\alpha\beta} G^b_{\beta\lambda} G^c_{\lambda\alpha} \rangle + C^0_{j^2} \langle g^4 j^a_\kappa j^a_\kappa \rangle \Big],\end{aligned} \tag{6}$$

where [4]

$$\begin{aligned}C^0_{G^2} &= \frac{1}{48\pi^2 (Q^2)^2} [-1 + 3J_2 - 2J_3], \\ C^0_{G^3} &= \frac{1}{72\pi^2 (Q^2)^3} \left[\frac{2}{15} - \frac{1}{10} y + 4J_2 - \frac{31}{3} J_3 + \frac{43}{5} J_4 - \frac{12}{5} J_5 \right], \\ C^0_{j^2} &= \frac{1}{36\pi^2 (Q^2)^3} \left[\frac{41}{45} - \frac{3}{5} y + (\frac{2}{3} + \frac{1}{3} y) J_1 - J_2 - \frac{4}{9} J_3 - \frac{26}{15} J_4 + \frac{8}{5} J_5 \right],\end{aligned} \tag{7}$$

dimension 4 spin 2 operator

$$\begin{aligned}\Pi^{4,2}_{\mu\nu}(q) &= \frac{1}{Q^2} \Big[I^2_{\mu\nu} + \frac{1}{Q^2} (q_\rho q_\mu I^2_{\rho\nu} + q_\rho q_\nu I^2_{\rho\mu}) \\ &\quad + g_{\mu\nu} \frac{q_\rho q_\sigma}{Q^2} \mathcal{J}^2_{\rho\sigma} + \frac{q_\mu q_\nu q_\rho q_\sigma}{Q^4} (I^2_{\rho\sigma} + \mathcal{J}^2_{\rho\sigma}) \Big],\end{aligned} \tag{8}$$

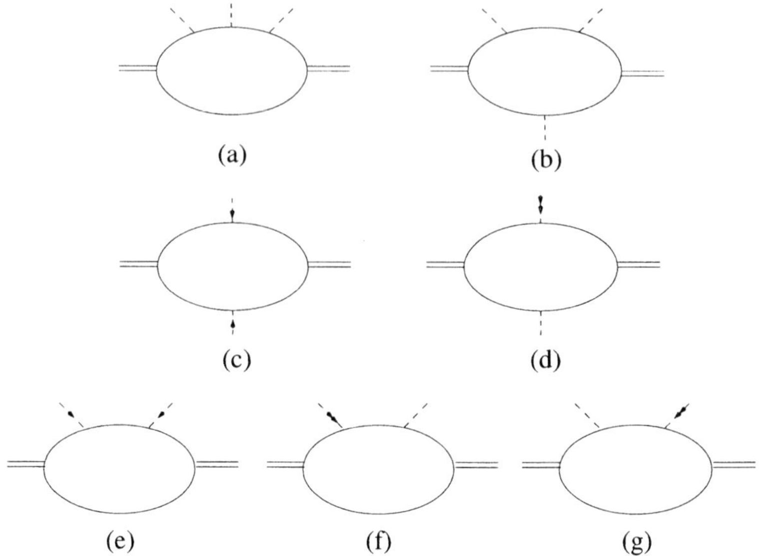

FIGURE 1. 1-loop diagrams describing interactions with the gluonic field giving dimension 6 operators: The dashed line represents an external gluon field G and the dashed line with n arrows represents an external gluon field with n derivatives D^nG.

where [16, 17]

$$\begin{aligned} I^2_{\mu\nu} &= \langle \frac{\alpha_s}{\pi} G^a_{\sigma\mu} G^a_{\sigma\nu} \rangle \left[\frac{1}{2} + (1 - \frac{1}{3}y) J_1 - \frac{3}{2} J_2 \right] \\ \mathcal{J}^2_{\mu\nu} &= \langle \frac{\alpha_s}{\pi} G^a_{\sigma\mu} G^a_{\sigma\nu} \rangle \left[-\frac{7}{6} + (1 + \frac{1}{3}y) J_1 - \frac{1}{2} J_2 + \frac{2}{3} J_3 \right] . \end{aligned} \tag{9}$$

contribution from dimension 6 and spin 4

$$\begin{aligned} \Pi^{6,4}_{\mu\nu}(q) &= \frac{q_\kappa q_\lambda}{(Q^2)^3} \Big[I^4_{\kappa\lambda\mu\nu} + \frac{1}{Q^2} (q_\rho q_\mu I^4_{\kappa\lambda\rho\nu} + q_\rho q_\nu I^4_{\kappa\lambda\rho\mu}) \\ &\quad + g_{\mu\nu} \frac{q_\rho q_\sigma}{Q^2} J^4_{\kappa\lambda\rho\sigma} + \frac{q_\mu q_\nu q_\rho q_\sigma}{Q^4} (I^4_{\kappa\lambda\rho\sigma} + J^4_{\kappa\lambda\rho\sigma}) \Big] , \end{aligned} \tag{10}$$

where [18]

$$I^4_{\mu\nu\rho\sigma} = \left[-\frac{266}{45} + \left(-\frac{20}{3} + \frac{22}{15} y \right) J_1 + \frac{138}{5} J_2 - \frac{916}{45} J_3 + \frac{16}{3} J_4 \right]$$

$$\times\langle\frac{\alpha_s}{\pi}G^a_{\rho\kappa}G^a_{\sigma\kappa;\mu\nu}\rangle$$

$$J^4_{\mu\nu\rho\sigma} = \left[\frac{362}{45}-\left(4+\frac{22}{15}y\right)J_1-\frac{94}{15}J_2-\frac{44}{45}J_3+\frac{16}{3}J_4-\frac{32}{15}J_5\right] \times\langle\frac{\alpha_s}{\pi}G^a_{\rho\kappa}G^a_{\sigma\kappa;\mu\nu}\rangle. \quad (11)$$

contributions from dimension 6 and spin 2

$$\Pi^{6,2}_{\mu\nu}(q) = \frac{1}{(Q^2)^2}\Big[I^2_{\mu\nu}+\frac{1}{Q^2}(q_\rho q_\mu I^2_{\rho\nu}+q_\rho q_\nu I^2_{\rho\mu}) +g_{\mu\nu}\frac{q_\rho q_\sigma}{Q^2}J^2_{\rho\sigma}+\frac{q_\mu q_\nu q_\rho q_\sigma}{Q^4}(I^2_{\rho\sigma}+J^2_{\rho\sigma})\Big], \quad (12)$$

where [18]

$$\begin{aligned}
I^2_{\mu\nu} &= \langle\frac{\alpha_s}{\pi}G^a_{\kappa\lambda}G^a_{\kappa\lambda;\mu\nu}\rangle\left[\frac{31}{240}-\frac{1}{60}y+\left(\frac{13}{24}+\frac{1}{48}y\right)J_1-\frac{115}{48}J_2+\frac{21}{8}J_3-\frac{9}{10}J_4\right]\\
&+\langle\frac{\alpha_s}{\pi}G^a_{\mu\kappa}G^a_{\nu\lambda;\lambda\kappa}\rangle\left[-\frac{739}{720}+\frac{2}{15}y+\left(-\frac{9}{8}+\frac{3}{16}y\right)J_1+\frac{133}{48}J_2+\frac{1}{72}J_3-\frac{19}{30}J_4\right]\\
&+\langle\frac{\alpha_s}{\pi}G^a_{\mu\kappa}G^a_{\kappa\lambda;\lambda\nu}\rangle\left[\frac{293}{240}-\frac{3}{10}y+\left(\frac{55}{24}+\frac{1}{16}y\right)J_1-\frac{131}{16}J_2+\frac{145}{24}J_3-\frac{41}{30}J_4\right]
\end{aligned}$$

$$\begin{aligned}
J^2_{\mu\nu} &= \langle\frac{\alpha_s}{\pi}G^a_{\kappa\lambda}G^a_{\kappa\lambda;\mu\nu}\rangle\left[\frac{103}{240}+\frac{1}{60}y+\left(\frac{5}{24}-\frac{7}{48}y\right)J_1-\frac{59}{48}J_2+\frac{31}{24}J_3-\frac{11}{10}J_4+\frac{2}{5}J_5\right]\\
&+\langle\frac{\alpha_s}{\pi}G^a_{\mu\kappa}G^a_{\nu\lambda;\lambda\kappa}\rangle\left[\frac{71}{240}-\frac{2}{15}y+\left(-\frac{1}{8}+\frac{1}{48}y\right)J_1+\frac{61}{48}J_2-\frac{61}{24}J_3+\frac{29}{30}J_4+\frac{2}{15}J_5\right]\\
&+\langle\frac{\alpha_s}{\pi}G^a_{\mu\kappa}G^a_{\kappa\lambda;\lambda\nu}\rangle\left[\frac{29}{240}+\frac{3}{10}y-\left(\frac{1}{24}+\frac{7}{16}y\right)J_1+\frac{31}{48}J_2-\frac{23}{24}J_3+\frac{11}{30}J_4-\frac{2}{15}J_5\right].
\end{aligned} \quad (13)$$

In the vacuum, only scalar operators will contribute. However, in medium, the tensor operators will also contribute. As an application of our result, we will apply our OPE to the QCD sum rule analysis for the J/ψ in nuclear matter.

Now, the OPE from dimension 4 and dimension 6 operators in eq.(8), eq.(10) and eq.(12) give the following contribution to Π.

$$\begin{aligned}
\Pi(\omega^2) &= \Pi_{4,2}(\omega^2)+\Pi_{6,2}(\omega^2)+\Pi_{6,4}(\omega^2)\\
&= \frac{\rho}{2m_N}\left[C_{G_2}G_2+(C_XX+C_YY+C_ZZ)+C_{G_4}G_4\right],
\end{aligned} \quad (14)$$

where

$$
\begin{aligned}
C_{G_2} &= \frac{m_N^2}{12(Q^2)^2}[9-(12+2y)J_1+9J_2-6J_3], \\
C_X &= \frac{m_N^2}{240(Q^2)^3}[-85-2y+(-70+25y)J_1+365J_2-390J_3+252J_4-72J_5], \\
C_Y &= \frac{m_N^2}{720(Q^2)^3}[25+48y+(270-45y)J_1-1185J_2+1370J_3-408J_4-72J_5], \\
C_Z &= \frac{m_N^2}{240(Q^2)^3}[-95-36y+(-130+75y)J_1+375J_2-190J_3+16J_4+24J_5], \\
C_{G_4} &= \frac{m_N^4}{108(Q^2)^3}[205-(210+33y)J_1+99J_2-262J_3+240J_4-72J_5], \qquad (15)
\end{aligned}
$$

where in our kinematical limit $Q^2 = -\omega^2$ and the scalar parts of the matrix elements in nuclear matter to leading density come from the following nucleon expectation values,

$$
\begin{aligned}
\langle N|\frac{\alpha_s}{\pi}G^a_{\sigma\mu}G^a_{\sigma\nu}|N\rangle &= G_2\,(p_\mu p_\nu - \frac{1}{4}m_N^2 g_{\mu\nu}), \\
\langle N|\frac{\alpha_s}{\pi}G^a_{\kappa\lambda}G^a_{\kappa\lambda;\mu\nu}|N\rangle &= X\,(p_\mu p_\nu - \frac{1}{4}m_N^2 g_{\mu\nu}), \\
\langle N|\frac{\alpha_s}{\pi}G^a_{\mu\kappa}G^a_{\nu\lambda;\lambda\kappa}|N\rangle &= Y\,(p_\mu p_\nu - \frac{1}{4}m_N^2 g_{\mu\nu}), \\
\langle N|\frac{\alpha_s}{\pi}G^a_{\mu\kappa}G^a_{\kappa\lambda;\lambda\nu}|N\rangle &= Z\,(p_\mu p_\nu - \frac{1}{4}m_N^2 g_{\mu\nu}), \\
\langle N|\frac{\alpha_s}{\pi}G^a_{\mu\kappa}G^a_{\nu\kappa;\alpha\beta}|N\rangle &= G_4\Big[p_\mu p_\nu p_\alpha p_\beta + \frac{m_N^4}{48}(g_{\mu\nu}g_{\alpha\beta}+g_{\mu\alpha}g_{\nu\beta}+g_{\mu\beta}g_{\nu\alpha}) \\
&\quad -\frac{1}{8}m_N^2(p_\mu p_\nu g_{\alpha\beta}+p_\mu p_\alpha g_{\nu\beta}+p_\mu p_\beta g_{\alpha\nu} \\
&\quad +p_\nu p_\alpha g_{\mu\beta}+p_\nu p_\beta g_{\mu\alpha}+p_\alpha p_\beta g_{\mu\nu})\Big].
\end{aligned}
\qquad (16)
$$

Note that here we chose the nucleon four momentum to be $p = (m_N, 0, 0, 0)$. We discuss the magnitudes of these nucleon matrix elements in section .

NUMERICAL ANALYSIS

As an application we use the moment sum rule. The moments of the polarization function is defined as,

$$
M_n(Q_0^2) = \frac{1}{n!}\left(-\frac{d}{dQ^2}\right)^n \Pi(Q^2)|_{Q^2=Q_0^2}. \qquad (17)
$$

where in our kinematics, $Q^2 = -\omega^2$.

Direct evaluation of these moments using the OPE gives, up to dimension 4,

$$M_n(\xi) = A_n^V(\xi)\left[1 + a_n(\xi)\alpha_s + b_n(\xi)\phi_b^4 + c_n(\xi)\phi_c^4\right] \text{ [16, 19]}. \tag{18}$$

The dimension 6 operators contribute to the moment as follows,

$$\begin{aligned}\Delta M_n^6(\xi) &= A_n^V(\xi)\Big[s_n(\xi)\phi_s^6 + t_n(\xi)\phi_t^6 + x_n(\xi)\phi_x^6 \\ &\quad + y_n(\xi)\phi_y^6 + z_n(\xi)\phi_z^6 + g_{4n}(\xi)\phi_{g4}^6\Big] \text{ [18]}.\end{aligned} \tag{19}$$

The inelastic channels opening due to the scattering of the $J/\psi + N \to \Lambda_c(2.28) + \bar{D}(1.87)$ is forbidden by kinematics. All other processes are OZI rule violating. Hence, the delta function approximation for the J/ψ is also good in the nuclear medium. For $n > 5$, the mass [18] is determined thus from

$$m_{J/\psi}^2 = \frac{M_{n-1}(4m_c^2)}{M_n(4m_c^2)} - 4m_c^2 \tag{20}$$

We choose the normalization point to be $Q^2 = 4m_c^2$. Then

$$\alpha_s = 0.21, \quad m_c = 1.24\,\text{GeV}. \tag{21}$$

For the matrix elements, we will use linear density approximation with $m_N = 0.93$ GeV and the nuclear matter density to be $\rho_0 = 0.17/\text{fm}^3$ [18].

In Fig.2(a)[18] , we plot the previous result, which includes the contribution only up to dimension 4. In Fig.2(b), we plot the present result which includes the total dimension 6 contribution. As can be seen from the comparison, the minimum occurs again at similar n value and the change from the vacuum results are also similar. We avoid fine tuning of the bare charm quark mass m_c to fit the vacuum J/ψ mass to its vacuum value, because we are only interested in the shift of the J/ψ mass, which is almost independent of this fine tuning.

Comparing the two graphs, one notes that the minimum occurs at the same n value and the graphs looks similar. Comparing the changes from the vacuum curve and the medium curve at the minimum point, we find

$$\Delta m_{J/\psi} \simeq -4\,\text{MeV}. \tag{22}$$

This mass shift is smaller than our previous calculation including dimension 4 only. The main reason for a smaller mass shift compared to including dimension 4 only is as follows. In the vacuum, the dimension 6 operators tend to cancel the dimension 4 operators[4]. This tendency is not only true but more effective in the medium. Therefore, including dimension 6 effects in medium would effectively correspond to a smaller change in the dimension 4 operators. This implies a smaller mass shift[18].

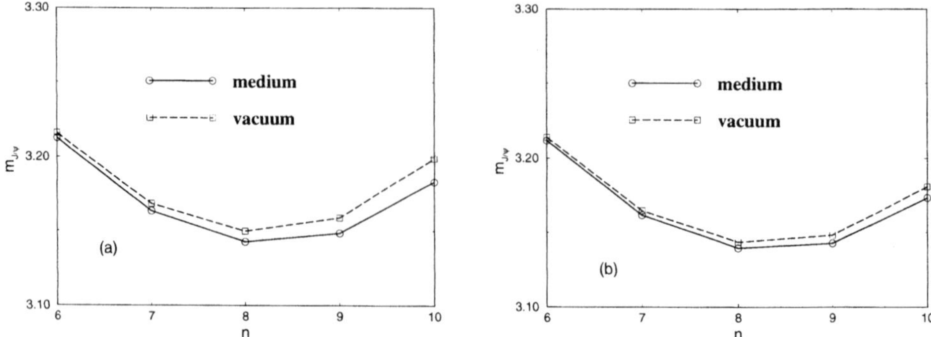

FIGURE 2. The mass of J/ψ in GeV determined from the plot of eq.(20) for different n at $\xi = 1$. We compare the result in normal nuclear matter (solid line) with the vacuum result (dashed line). (a) refers to the previous calculation, which includes only dimension 4 operators. (b) is the present calculation which includes all the dimension 6 operators.

CONCLUSION

In this paper we have summarised the OPE of the correlation function between two vector currents made of heavy quarks up to dimension 6 operators with any tensor structure. Using this result, we have applied our OPE to analyze the mass shift of J/ψ in nuclear medium using QCD moment sum rules for the heavy quark system. This is a generalization of our previous work, where we calculated the mass shift using the OPE only up to dimension 4 operators[16, 18]. Using an order of magnitude estimate for the matrix elements, we find that the mass of J/ψ would decrease by about 4 MeV in the nuclear medium. This shows that the dimension 6 effect is about 40% correction of the dimension 4 effects and goes in the opposite direction. This result seems consistent with the notion that the higher dimensional correction in the vacuum QCD sum rule for the heavy quark system goes like $\left(G^2/m_c^4\right)^K$ in the $r_n = \frac{M_n(\xi=0)}{M_{n-1}(\xi=0)}$ with alternating signs with K[20]. This also seems to be true in medium and the true mass shift is expected to lie between -4 and -7 MeV.

The resulting value of mass shift, is also consistent with the more recent estimates using a totally different approach[11, 12]. This is a result for a J/ψ at rest with respect to the nuclear medium. However, since we have calculated the OPE for a general external four momentum, our results can be easily and reliably generalized to study the moving J/ψ and also to finite temperature[21], which would be also interesting in relation to the ongoing discussion of J/ψ suppression in RHIC due to a comover model.

REFERENCES

1. M.A. Shifman, A.I. Vainstein and V.I. Zakharov, Nucl.Phys.B**147**(1979) 385
2. L.J. Reinders, H. Rubinstein and S. Yazaki, Phys.Rep.**127**(1985) 1
3. S.C. Generalis and D.J. Broadhurst, Phys. Lett.B**139**(1984) 88
4. S.N. Nikolaev and A.V. Radyushkin, Nucl.Phys.B**213**(1983) 285

5. J. Bartles, C. Bontus and H. Spiesberger, hep-ph/9908411
6. T. Matsui and H. Satz, Phys. Lett. B 178 (1986)416
7. R. Vogt, Phys. Rept. 310 (1999) 197
8. S.J. Brodsky and G.F. de Teramond, Phys.Rev.Lett **64**(1990) 1011
9. D.A. Wasson, Phys.Rev.Lett **67**(1991) 2237
10. M.Luke, A.V.Manohar and M.J.Savage, Phys.Lett.B**288**(1992) 355
11. A. B. Kaidalov and P.E. Volkovitsky, Phys.Rev.Lett. **69**(1992) 3155.
12. D. Kharzeev, nucl-th/9601029
13. S.J. Brodsky and G.A. Miller, Phys.lett.B**412**(1997) 125
14. G.F. de Teramond, R. Espinoza and M. Ortega-Rodriguez, Phys.Rev.D**58**(1998) 034012
15. V.A.Nivikov, M.A.Shifman, A.I.Vainstein and V.I.Zakharov, Fortschr.Phys. **32**(1984) 588
16. F. Klingl, S. Kim, S.H. Lee, P. Morath and W. Weise, Phys.Rev.Lett.**82**(1999) 3396
17. A. Hayashigaki, Prog.Theor.Phys.**101**(1999) 923
18. Sungsik Kim and Su Houng Lee, Nucl.Phys.A**679**(2001) 517
19. L.J. Reinders, H. Rubinstein and S. Yazaki, Nucl.Phys.B**186**(1981) 109
20. S.N. Nikolaev and A.V. Radyushkin, Phys.Lett.B**124**(1983) 243.
21. R. Furnstahl, T. Hatsuda, and Su H. Lee, Phys. Rev. D **42** (1990) 1744.

J/ψ at finite temperature - Lattice QCD result and potential model analysis

H. Matsufuru*, O. Miyamura†, H. Suganuma** and T. Umeda†

*Research Center for Nuclear Physics, Osaka University, Ibaraki 567-0047, Japan
(present address is YITP, Kyoto University, Kyoto 606-8502, Japan.)
†Department of Physics, Hiroshima University, Higashi-hiroshima 739-8526, Japan
**Faculty of Science, Tokyo Institute of Technology, Tokyo 152-8551, Japan

Abstract. We study the charmonium correlators in the deconfined phase, as well as at $T < T_c$, in two viewpoints. One is an analysis of correlation between quark and antiquark in the finite temperature lattice QCD at the quenched level. The result implies that the correlation is so strong that the quark and antiquark still tend to stay together even at $T \simeq 1.5T_c$. On the other hand, the potential model analysis, the second approach, results in the existence of no bound state at $T \simeq 1.2T_c$. The static quark potential obtained in lattice simulation is used in the latter analysis. The change of the bound state energy below T_c is also studied.

INTRODUCTION

The change of J/ψ properties near the QCD phase transition is one of most important signals of the formation of quark gluon plasma [1, 2]. Compared with the recent development of the heavy-ion collision experiment [3], theoretical understanding of the charmonium at $T > 0$ is not yet sufficient and its progress is desired. Although the lattice QCD simulation at $T \neq 0$ may give a stage of such an investigation directly based on QCD, at present only exploratory studies have been done [4]. Phenomenological approaches are thus important to understand the nature of hadronic states at finite temperature, as well as at finite density, where the lattice QCD is hardly applicable.

The most important ingredient of the potential model approaches is the static quark-antiquark potential, and it is numerically derived from the lattice QCD as a reliable quantity. On the other hand, both the lattice QCD and the potential model can provide the hadron properties. In this way, there is a mutual connection between lattice QCD simulations and phenomenological potential model analysis, and the consistent understanding between them would be significant.

In this paper, we discuss the fate of the charmonium above the critical temperature T_c, from the following two points of view.

(i) Finite temperature lattice QCD.
The charmonium correlator in the deconfined phase is studied using the finite temperature lattice QCD. Here we focus on the change of the q-$\bar{q}$ spatial correlation in terms of the Euclidean time t [4]. In the deconfined phase, q-$\bar{q}$ spatial correlation is naively expected to broaden as t, from a simple picture of almost free quarks. The static quark potential, which is an important ingredient of the potential model

CP594, *Hadrons and Nuclei: First International Symposium,* edited by Il-T. Cheon et al.

approach, is also measured [5].

(ii) Potential model analysis based on lattice QCD.
In the deconfined phase, the static quark potential is expected to be the Yukawa-type form characterized by its screening mass. This is indeed observed in the lattice QCD simulation [5, 6]. We employ the potential model approach together with the obtained static potential on the lattice for an investigation of the charmonium properties. First subject is the stationary state problem which give the condition of existence for a bound state. Below T_c, we observe the change of bound state energy accompanied by the change of the static quark potential as T. In the deconfined phase, we study whether the charmonium bound state is formed in the Yukawa potential. Then by solving the evolution of the wave function governed by the Schrödinger equation, we would estimate the decay distance of the created $c\bar{c}$ pair. (Even if the bound state is not formed, if the created $c\bar{c}$ pair keeps strong correlation during certain time period, $c\bar{c}$ pair may escape from the plasma phase without decaying and can form the charmonium state.)

The results of these two analyses are compared with each other. The present simulation is at the quenched level, which contains no dynamical quark effect. These results may give an important qualitative implications, and will be compared with the results on lattices with dynamical quark effect. At present, the last subject in (ii) is in progress, and we comment only on the technical point for a computation.

The next section describes the result of lattice QCD simulations, the charmonium correlator and the static Q-$\bar{Q}$ potential. In successive section, the potential model analysis is presented. The last section is devoted for summary and outlook.

LATTICE QCD RESULT

In this section, we summarize the results from lattice QCD simulations [4, 5].

Correlator analysis. Lattice QCD provides a nonperturbative procedure to investigate hadron correlators directly based on QCD. At finite temperature, however, an analysis of them must overcome several difficulties to obtain relevant result. One of them is a shortage of information in the temporal direction, since the temporal extent is restricted by the condition $T = 1/(N_t a_\tau)$, where a_τ is the temporal lattice spacing and N_t is the temporal lattice size. To increase N_t (therefore to decrease a_τ) keeping the spatial lattice size modest, we employ the anisotropic lattice [8]. Even on an anisotropic lattice, extraction of relevant information from the hadron correlators is a difficult task. QCD-TARO Collaboration measured the spatial q-$\bar{q}$ correlation and discussed their dependence on Euclidean time t to probe the possibility of the bound states [7]. This analysis has been applied also to the charmonium system [4], which we briefly summarize in the following.

The numerical simulation has been carried out on the lattices with sizes $16^2 \times 24 \times N_t$, at the quenched level. Gauge configurations have been generated with the anisotropic Symanzik (tree-level) action with parameters corresponding to the renormalized anisotropy $\xi \equiv a_\sigma/a_\tau = 3.95(2)$ and the lattice cutoffs a_σ^{-1}=1.61(1) GeV

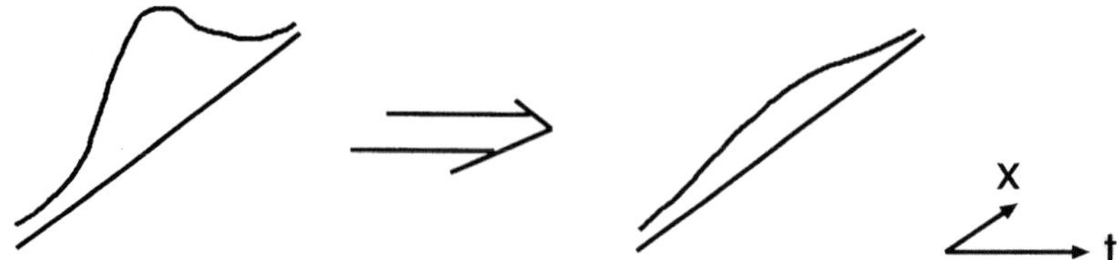

FIGURE 1. Schematic expected behavior of the q-$\bar{q}$ correlator without bound states.

($a_\sigma \simeq 0.125$ fm) and a_τ^{-1}=6.36(5) GeV. Used N_t's are as follows: N_t =96 ($T \simeq 0$), 28 (0.87 T_c), 26 (0.93 T_c), 20 (1.22 T_c) (for static potential) and 16 (1.52 T_c) (for q-$\bar{q}$ correlator analysis). The configurations are fixed to the Coulomb gauge. As the quark action, we adopt the $O(a)$ improved Wilson action [4, 9]. The quark parameters are tuned roughly so as to correspond to the charm quark mass.

The spatial correlation between quark and antiquark is represented by

$$w_\Gamma(r,t) = \sum_{\vec{x}} \langle \bar{q}(\vec{x}+\vec{r},t)\Gamma q(\vec{x},t) O_\Gamma^\dagger(0) \rangle, \tag{1}$$

where 4×4 matrix Γ specifies the spin quantum number. We focus on the vector channel in the following. As the operator $O_\Gamma(x)$, which overlaps with the meson states, we use the form $O_\Gamma(x) = \sum_{\vec{y}} \bar{q}(\vec{x}+\vec{y},t)\varphi(\vec{y})\Gamma q(\vec{x},t)$. The smearing function $\varphi(\vec{y})$ is tuned so that the overlap with the ground state is exclusively large. We observe the t-dependence of this correlator. The idea is rather simple: if there is no bound state (like free quark case), this q-$\bar{q}$ correlator becomes broader as t (Euclidean time), as in Figure 1. In this case, the normalized correlator at the spatial origin,

$$\phi_\Gamma(r,t) = w_\Gamma(r,t)/w_\Gamma(r=0,t), \tag{2}$$

increases with t.

Figure 2 shows the q-$\bar{q}$ correlation in the simulation, together with the case of free quarks smeared with the same function as in the simulation. In the left figure, at $T < T_c$, the $\phi_\Gamma(r,t)$ approaches to a stable shape with t. In contrast, the correlator with free quark and antiquark becomes broader with t, as expected. In the right figure, at $T \simeq 1.5T_c$, the correlator shows essentially the same feature as below T_c. This is qualitatively different behavior from the free quark case. This implies that up to this temperature the quark and the antiquark keeps strong correlation so that they still stay together. Although what the q-$\bar{q}$ correlator strictly means is not obvious than at $T = 0$, this result suggests the possibility of persistence of the bound state even in the deconfined phase.

As the next step, we examine this question in the potential model analysis. Before going into this subject, we obtain here the static quark potential used in the potential model on the same series of lattices.

Static quark potential. The static quark-antiquark potential at finite temperature is extracted from the correlation of the Polyakov loops,

$$P_2(\vec{r}) = \langle P(0)P^\dagger(\vec{r}) \rangle \simeq c \cdot \exp(-V_{Q\bar{Q}}(\vec{r})N_\tau), \tag{3}$$

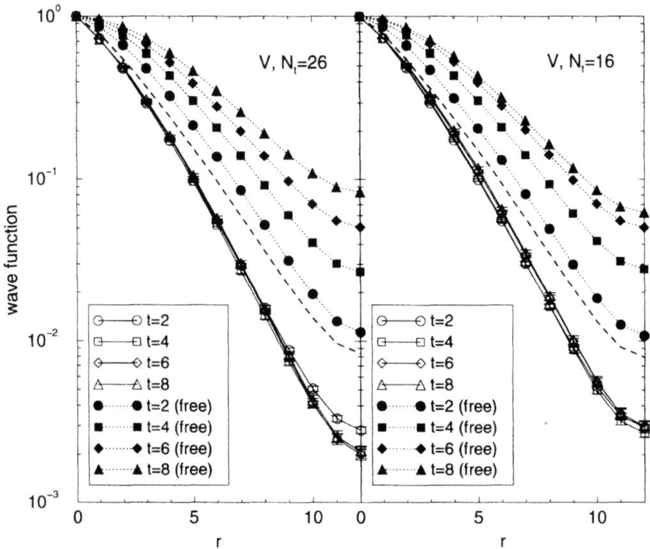

FIGURE 2. Observed q-$\bar{q}$ correlator, $\phi_\Gamma(r,t)$, below (left) and above (right) T_c. The dashed line is the smearing function.

where $P(\vec{x})$ is the Polyakov loop,

$$P(\vec{x}) = \mathrm{Tr} \prod_{t=0}^{N_\tau - 1} U_4(\vec{x},t). \tag{4}$$

We have measured the static quark potential on the same series of lattice used for the analysis of q-$\bar{q}$ spatial correlation.

The potentials at $T \neq 0$ are shown in Figure 3. Below T_c, we fit the obtained result to the form

$$V_{Q\bar{Q}}(r) = -\frac{A}{r} + \sigma r + C, \tag{5}$$

where the first term is the one-gluon-exchange Coulomb term, the second term is the confining linear term with the string tension σ, and the last term is constant due to the finiteness of lattice spacing. At $T = 0$, this form is known to well describe the potential obtained on the lattice. In Table 1, we summarize the result of fit at three values of T containing $T \simeq 0$. The large χ^2/N_{df} signals a large systematic error due to the lattice artifact for $T \simeq 0$. The values of the coefficient of the Coulomb term is not so reliable since the short range part is not appropriately incorporated in the fit. The string tension decreases with increasing T. In Table 1, we also quote the values of A and σ converted into the physical unit for the later use in the potential model analysis.

The right figure in Figure 3 shows the static potential above T_c ($T \simeq 1.2T_c$). In the deconfined phase, it is known that the potential is well described by the Yukawa-type form [6],

$$V_{Q\bar{Q}}(r) = -A\exp(-\mu r)/r + C, \tag{6}$$

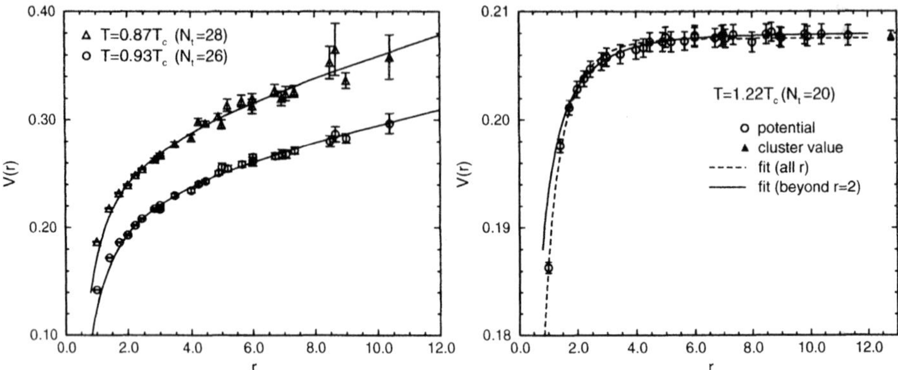

FIGURE 3. The lattice QCD results for the static quark potentials below (left) and above (right) T_c. In the left figure, upper data at $N_t = 28$ correspond to $T \simeq 0.87T_c$, and lower data at $N_t = 26$ correspond to $T \simeq 0.93T_c$. They are vertically slightly shifted. The solid lines are the fitted curves. In the right figure, the triangle symbol denotes the cluster value, which corresponds to the product of two independent Polyakov loops.

TABLE 1. Result of the fit for the potential below T_c. The last two columns are the same A and σ in the physical unit (the errors are omitted). A dimensionless quantity A is changed due to the anisotropy.

N_τ	T	fit range	A	σ	C	χ^2/N_{df}	$A_{(phys)}$	$\sigma_{(phys)}$
96	~ 0	$r \geq 2.8$	0.0587(49)	0.01781(29)	0.0587(49)	78.1/18	0.23	0.18 GeV^2
28	$0.87T_c$	$r \geq 2$	0.119 (11)	0.0089(11)	0.2817(70)	25.2/21	0.47	0.091 GeV^2
26	$0.93T_c$	$r \geq 2$	0.1253(63)	0.00638(7)	0.2934(41)	23.5/21	0.49	0.065 GeV^2

with the screening length as $1/\mu$. The extracted value of μ strongly depends on the fit range, and results in about 0.5 $\sim$ 1 GeV. One of our goal is to examine whether the bound state can survive with this screened potential.

POTENTIAL MODEL APPROACH

Stationary state analysis. For the charmonium system, the nonrelativistic approximation is applicable [10]. The subject we are interested in is essentially the T-dependence of the binding energy and the condition in which the bound state disappears. In this work, therefore, we treat the Schrödinger equation for a spin averaged states. For the stationary state, the equation to be solved is

$$\left(-\frac{\Delta}{2m_R} + V(x)\right)\psi(x) = E\psi(x), \tag{7}$$

TABLE 2. Result of fit for the potential at $T = 1.2T_c$. The last two columns are the same A and μ in the physical unit (the errors are omitted). A dimensionless quantity A is changed due to the anisotropy.

range	A	μ	C	χ^2 / N_{dof}	$A_{(phys)}$	$\mu_{(phys)}$
all r	0.0469(15)	0.803(40)	0.20755(88)	5.29 / 24	0.185	1.29 GeV
$r \geq 2$	0.0217(33)	0.385(73)	0.20794(87)	1.03 / 21	0.086	0.62 GeV

with the reduced mass, $m_R = m_Q/2$. For the S-state, this equation is reduced to

$$\left(-\frac{1}{2m_R}\frac{d^2}{dr^2} + V(r)\right)u(r) = Eu(r), \tag{8}$$

where $u(r) = r\psi_r(r)$ and $\psi_r(r)$ is the radial wave function. As the potential $V(r)$, we use the form in the previous section with parameters obtained in lattice QCD simulation.

Let us start with the change of the binding energy below T_c. As the potential $V(r)$, we use the form of eq. (5) without the constant term, which is due to the finite lattice spacing. The values of A and σ are set to ones determined in lattice QCD simulation (listed in the last two columns of Table 1). Eq. (8) is numerically solved and gives the binding energy as the function of m_Q as shown in the left figure of Figure 4. The binding energy with a fixed quark mass decreases as T approaches toward T_c. This is a natural tendency accompanied by the decrease of the string tension. Quantitatively, however, we may need the precise determination of the Coulomb coefficient A, because the charmonium state tends to have a small radius and then its precise spectroscopy requires the precise value of A. In fact, the present analysis of the static quark potential is to be developed with the accurate Coulomb coefficient A. With the quantitative determination of A, the potential model can provide the T-dependence of charmonium spectrum reflecting the T-dependence of the static potential based on QCD.

Above T_c, the potential is expressed with the screened form, eq. (6), without the constant term, which corresponds to the intrinsic energy concerning the static quark in the deconfined phase. In this case, we solve eq. (8) for fixed A and m_Q and obtain the binding energy as a function of μ. The right figure of Figure 4 shows the result for $A = 0.086$ and $m_Q = 1.3$GeV. Around $\mu \simeq 0.065\text{GeV} \equiv \mu_c$, the binding energy vanishes, and no bound state is formed for larger value of $\mu > \mu_c$. On the other hand, from Table 2, the value of μ at $T \simeq 1.2T_c$ is about 0.62 GeV, which is sufficiently larger than the critical value μ_c at which the bound state disappears. Similar result is also obtained for the case of $A = 0.185$. Therefore, we conclude that there exists no bound state at $T \simeq 1.2T_c$ from the analysis with the Q-$\bar{Q}$ potential obtained in quenched lattice QCD. This is a reproduction of the result by Karsch, Mehr and Satz [11]. On the lattice with dynamical configuration, the screening effect may be stronger, and this conclusion would not to be changed.

This result seems contradicting our previous result for q-$\bar{q}$ correlation at $T \simeq 1.5T_c$. In the latter case, the genuine dynamics of time evolution is not incorporated, and the argument concerning the bound state is based on a sort of conjecture. Since the temporal lattice extent is restricted to $1/T$, the dynamics of time scale larger than $1/T \sim 1$ fm is not probed. Then the next question is what is the time scale after which the quark

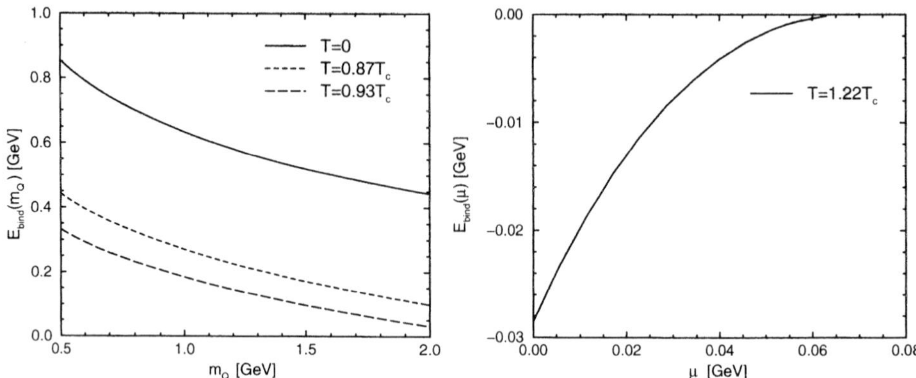

FIGURE 4. The potential model results for the binding energy below (left) and above (right) T_c. The left figure shows the binding energy as the function of m_Q at three values of T. The right figure shows the binding energy at $T > T_c$ as the function of μ. Here, $A = 0.086$ and $m_Q = 1.3$ GeV are used.

and antiquark are well separated. To answer this question, in the scope of the potential model, we treat the time evolution problem of the Schrödinger equation.

Toward non-stationary state analysis. To analyze the time evolution of the quark-antiquark system in the potential model, one needs to treat the time-dependent Schrödinger equation

$$i\frac{\partial}{\partial t}\psi(x) = \left(-\frac{\Delta}{2m_R} + V(x)\right)\psi(x). \tag{9}$$

With the Coulomb-like potential, one needs to manage the singularity in the vicinity of the spatial origin in the practical numerical computation. The most appropriate coordinate system is the modified cylindrical coordinate [12],

$$x = \xi^{2/3}\cos\phi, \qquad y = \xi^{2/3}\sin\phi, \qquad z = z. \tag{10}$$

The investigation based on this framework is now in progress, and would give an insight on the above-mentioned problem.

SUMMARY AND OUTLOOK

In this work, we have studied the properties of charmonium state in the lattice QCD and the potential model with the static potential measured in the lattice QCD.

In the lattice simulation, we have measured the q-$\bar{q}$ correlation in the deconfined phase as well as at $T < T_c$, and have found that the correlation is so strong that the quark and antiquark still stay together even at $T \simeq 1.5T_c$. The q-$\bar{q}$ correlator shows qualitatively different behavior from the free quark case, which may suggest the possibility of existence of the bound state even above T_c. The static quark potential is also measured on the same series of lattice. Another possible procedure to clarify the nature of hadronic correlators at $T > 0$ is the direct extraction of the spectral function [13].

In the potential model analysis for the stationary state, we have found contradicting result on the problem of the bound state in the deconfined phase: the bound state is not formed for the Yukawa-type static potential obtained in the lattice QCD at $T \simeq 1.2T_c$. To understand these two results, we are now preparing for the potential model analysis of non-stationary state problem. The change of the binding energy of the charmonium state below T_c is also an important result, and more quantitative analysis is desired.

Although present lattice calculation is at the quenched level, i.e. without dynamical quark effect, these analyses would give us an insight on the temperature effect on the charmonium state. The similar investigation on the charmonium at $T \neq 0$ is easily applicable to the simulation with dynamical quarks, in principle, and is expected to provide a novel information on the signal of quark-gluon-plasma formation.

We would like to thank to Profs. Il-Tong Cheon and Su Houng Lee for their warm hospitality at Yonsei University. The numerical simulation has been done on Intel Paragon XP/S and NEC HSP at INSAM, Hiroshima University and NEC SX-4 at Osaka University. H.M. is supported by center-of-excellence (COE) program at RCNP, Osaka University.

REFERENCES

1. T. Hashimoto, O. Miyamura, K. Hirose and T. Kanki, Phys. Rev. Lett. **57**, 2123 (1986).
2. T. Matsui and H. Satz, Phys. Lett. B **178**, 416 (1986).
3. NA50 Collaboration, Phys. Lett. B **477**, 28 (2000).
4. T. Umeda, R. Katayama, O. Miyamura and H. Matsufuru, hep-lat/0011085, to appear in Int. J. Mod. Phys. A.
5. H. Matsufuru, Y. Nemoto, H. Suganuma, T.T. Takahashi and T. Umeda, Nucl. Phys. B (Proc. Suppl.) **94**, 554 (2001).
6. M. Gao, Phys. Rev. D **41**, 626 (1990).
7. QCD-TARO Collaboration (Ph. de Forcrand et al.), Phys. Rev. D **63**, 054501 (2001).
8. F. Karsch, Nucl. Phys. **B205**, 285 (1982).
9. J. Harada, A.S. Kronfeld, H. Matsufuru, N. Nakajima and T. Onogi, hep-lat/0103026.
10. E. Eichten, K. Gottfried, T. Kinoshita, K.D. Lane and T.-M. Yan, Phys. Rev. D **17**, 3090 (1978), D **21**, 313 (1980) (E), D **21**, 203 (1980).
11. F. Karsch, M.T. Mehr and H. Satz, Z. Phys. C **37**, 617 (1988).
12. H. Kono, A. Kita, Y. Ohtsuki and Y. Fujimura, J. Comput. Phys. **130**, 148 (1997).
13. Y. Nakahara, M. Asakawa and T. Hatsuda, Phys. Rev. D **60**, 091503 (1999).

Perturbative aspects of the heavy ion collisions at RHIC and its measurement

J. H. Kang* and Y. Kwon†

*Department of Physics, Yonsei Univ.
†Institute of Physics & Applied Physics, Yonsei Univ.

Abstract. The physics related to the measurement of the hard probes at RHIC are surveyed through the processes of Drell-Yan pair production, direct γ production, heavy quark production, the inclusive production cross section of the hadrons resulting from the mini-jets, and the identified jets over the varing beam energy and the colliding nuclei. Some of the uncertainties in those measurements can be decided from the $p+p$ program. By performing a simple calculation, a few physics possibilities in the coming run are investigated with estimated uncertainties.

INTRODUCTION

With the operation of a new collider, new interesting measurements are feasible. The space time evolution of the collisions at the RHIC may be divided as pre-equilibrium, equilibrium, and hadronization stage. While the quantitative approaches can be made to all three stages through the perturbative QCD (Quantum ChromoDynamics), our main focus in this work will be in the initial stage (Pre-equilibrium). Some calculations based on the perturbative QCD for the $p+p$ reaction process such as the Drell-Yan lepton pair production, the direct γ production, the heavy flavour production and the high p_t particle and jet production are possible and have successfully reproduced the experimental results below and above the collision energy at RHIC to some degree [1] [2]. We survey the uncertainties in the calculation from the experimental side. While some of these can be an interesting issues for the $p+p$ program to decide, they form a building block to understand the heavy ion collisions. While measuring the initial evolution of the partonic system is interesting by itself, we also believe these measurements will significantly reduce the uncertainty in the study of the later stage of the evolution, namely the ones in the equilibrium stage and the hadronization stage, including the jet-quenching, the modification of the jet fragmentation, the J/ψ suppression, and the hydrodynamic evolution of the thermalized partonic matter.

PERTURBATIVE QCD CALCULATION & UNCERTAINTIES IN IT

The hard processes are important in the production of particles with $p_t \gg 1\ GeV/c$ or $M \gg 2\ GeV/c$. Since these processes involve large momentum transfers which are associated with a small coupling constant in QCD, the technique of perturbative QCD may be applicable. In the QCD improved parton model, we make use of the

CP594, *Hadrons and Nuclei: First International Symposium,* edited by Il-T. Cheon et al.

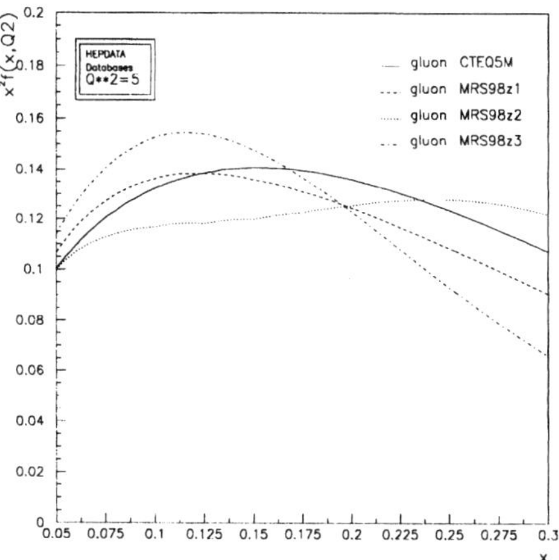

FIGURE 1. Comparison of the gluon distributions from MRST with those from CTEQ5HQ at the factorization scale Q = 5 GeV.

impulse approximation. The rule for calculating reaction rates for hadrons is as follows : one calculates the reaction rate for the basic process with free partons and then sums incoherently over the contributions of all partons in the hadrons. The picture behind the approximation is described in [2]. The hadron, say a nucleon, consists of a set of partons, in some virtual state of definite fractional momenta $\xi_i p$. In the center-of-mass system, the nucleon suffers both Lorentz contraction and time dilation, and the time it takes the partons cross each other vanishes in comparison to the life time of those virtual states as the center-of-mass energy goes to infinity. The incident parton must come as close to the colliding *frozen* parton as $O(1/Q), Q^2 = -q^2$ in the transverse direction to exchange a large momentum q^μ. The probability of finding an additional parton near enough to take part in the hard scattering is suppressed by the geometrical factor

$$\frac{1/Q^2}{\pi R_H^2}$$

The "initial-state" interactions between partons happen too early to affect the basic scattering, and hence the inclusive cross section, while the "final-state" interactions between fragments happen too late. Up to kinematic factors, the scattering is directly proportional to the density of partons

Uncertainty factors within this approximation exist in the parton distribution function, the parton fragmentation function, and the renormalization and factorization scale. While most of the distribution functions for the protons are well known or similar to each other within a few percent, accurate determination of the gluon distribution function is yet limited as shown in Fig 1. CTEQ use the inclusive jet data of CDF and D0 to complement the DIS constraints whereas MRST relies on direct photon production results of WA70, applying a range of k_t broadening corrections using the E706 data as a con-

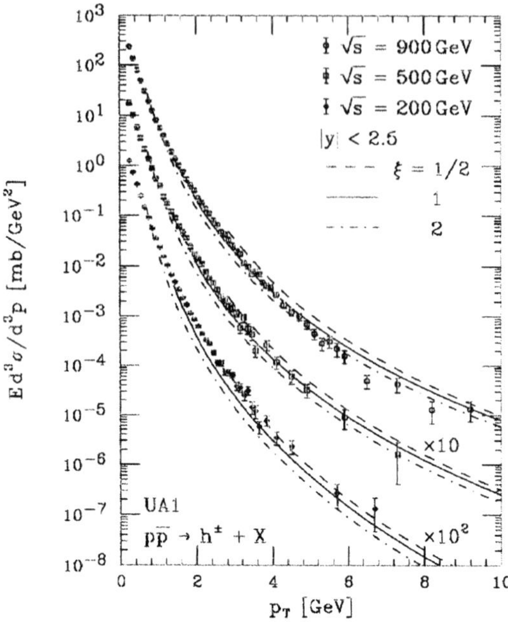

FIGURE 2. Figure taken from [5]. Differential cross section $E\frac{d^3\sigma}{dp^3}$(in mb/GeV^2) of inclusive charged-hadron hadroproduction in $p\bar{p}$ collisions $p\bar{p} \rightarrow h^{\pm} + X$ as a function of transverse momentum p_t at CM energies $\sqrt{s} = 200, 500, 900\ GeV$, averaged over rapidity interval $|y| < 2.5$. The NLO(Next to Leading Order) predictions for $\xi = 1/2, 1, 2$ are compared with data from UA1.

straint [4]. Hence the measurement of the direct photon production for the p + p reaction at the RHIC can be of interest. Fragmentation functions are obtained from the measurements of the e^+e^- annihilation. Fairly big uncertainties exist in the gluon fragmentation function into the charged π's [3], while a good description of inclusive charged-hadron hadroproduction in $\bar{p}p$ collisions is possible as in Fig. 2. The leading order estimates of cross-sections are only good to about 50% and often handled with the so called K-factor in the LO(Leading Order) prescription. This is mainly due to the uncertainty in the choice of renormalization or factorization scales. Ultimately the cross section should be independent of these scales and at next-to-leading order available presently, the uncertainty is significantly reduced to about 20% or so as can be seen in Fig. 2. Most of these uncertainties works on the parton level processes and appears in the processes of the $p+p$ reaction and the $A+A$ reaction commonly. Hence these uncertainties can be effectively handled for the $A+A$ program at the RHIC where simultaneous measurement of the $p+p$ reaction is made. Thickness function is often used to scale $p+p$ results to $A+A$ results and will be equivalent to the impulse approximation at the partonic level as long as the details of the parton interaction do not appear. The average number of inelastic nucleon-nucleon collisions averaged over the impact parameter ranges from about 200 for the minimum bias events to about 800 for the central events. Based on this estimation, *all* perturbative production will increase by factor of 4 from the minimum bias events to the central events.

Higher twist effects appears with the violation of the impulse approximation. The parton distribution function inside nuclei are modified due to this effect, and the so called shadowing can introduce modification of the distribution functions from the one by impulse approximation at the level of 10% in the small x region for the hadron collisions at RHIC [6]. Also this effect appears in the low p_t region of the produced particles at the low collision energy as in the intrinsic p_t broadening [11]. The higher twist effects are suppressed with the collision energy or Q variable. From the Fig. 2, we can observe the particle production at the RHIC energy and above follows the scaling by the QCD for $p+p$ reactions, and no significant higher twist effect is observed. We also expect the nuclear enhanced higher-twist effects. According to [7], the geometrical cross section for the $Pb+Pb$ reaction is saturated at $p_{sat} \sim 1.2\ GeV$, and there can be a fairly strong effect for the p_t region of a few GeV [11]. These higher twist effect can be estimated from the experimental measurement of the high p_t spectra at large $\sqrt{s}$ and their scaling at the lower collision energy.

MEASUREMENT

Here we study the possible measurements from the PHENIX experiment at RHIC which can measure various aspects of the heavy ion collisions, especially with penetrating probes. The acceptance for the γ, π^0 and charged particles in the central arm of the experiment is $-0.35 < \eta < 0.35$ in the pseudorapidity range with two almost back-to-back azimuthal coverages of $\pi/4$ each in ϕ. The acceptance for the muon pairs exist around the pair rapidity $Y \sim 2$ for a wide range of p_t [9]. This feature is quite effective especially when we desire a comprehensive understanding of the initial stage of the partonic system and its evolution/modification in the course of the heavy ion collisions. Typical interaction cross sections of the photons with the hadrons are smaller by two orders of magnitude [8], and the electromagnetic probes, the lepton pairs and photons, experience little interaction even in the hot and dense medium. Initially small number of charm pairs are produced even for the central heavy ion collisions. If thermal production is comparable or less than the primary production, the subsequent annihilation of the heavy flavoured quarks will not be strong.

The initial parton kinematics will appear

- in the direct γ's for the quark and gluon pairs
- in the lepton pair production and its pair mass and kinematics for the quark pairs
- in the total yield of the heavy flavoured quarks for the gluon pairs.

Hard partons with the kinematics equivalent to the produced and detected direct γ's at 10 GeV propagate through a macroscopic volume of a few fm. Depending on its momentum, the partons produced in the initial hard collisions will experience the perturbative and the non-perturbative corrections in their propagation through the matter and subsequent fragmentation which are good examples of the noble phenomena we want to study. Physics in this line is different from our initial scope, but we extend our survey somewhat since these modifications appear in the form of deviations from the standard calculation. The modification can appear

- in the E_t production (soft mini-jets).
- in the inclusive hadron production (mini-jets).
- in the identified jets, the leading particles and the jet variables.

We believe the identification of the individual jets can be feasible in the coming run if an efficient trigger can be achieved. The $p+p$ program can serve as a good starting point to define jets, and the high p_t partons in the heavy ion program will gain a large statistics.

The p_t spectrum of the prompt γ is affected by the shadowing of the parton distribution and the p_t broadening effect [10] [11]. The modification of the parton distribution by the shadowing is not big at $x \sim 0.1$ or greater [13], and these partons populate the direct γ spectrum at $p_t = 10\ GeV/c$ or above at midrapidity. The intrinsic p_t broadening for the basic $p+p$ reaction plays an important role when the collision energy is low, but is strongly suppressed with the beam energy. About 80% of the direct γ production is from the gluon compton process ($q+g \to q+\gamma$), and the measurement of the direct γ production in the mentioned momentum range will strongly constrain the gluon induced reaction rates. This includes the production rate of the gluons important to the particle production below 10 GeV/c in p_t.

The $pair-p_t$ spectrum of the Drell-Yan pairs also give valuable informations. The low p_t part of the $pair-p_t$ spectrum for the $p+p$ reaction comes from the primordial transverse momentum of the partons. The high p_t part of the $pair-p_t$ spectrum can be described by the NLO process given by $q+\bar{q} \to \gamma^* + g$ and $q+g \to \gamma^* + q$, and the cross section has a wide tail at high p_t [12]. If there is the p_t broadening in the direct γ production, the effect will be observed in the Drell-Yan process, and it will be the dominant factor for the p_t distribution of the DY pairs. By measuring the p_t broadening of the Drell-Yan pairs, we can estimate the strength of the rescattering if it exists in $A+A$ reaction.

The production cross section of the heavy quark pairs for the $p+p$ reactions at $\sqrt{s} = 200\ GeV$ can be estimated by the LO pQCD (Perturbative Quantum ChromoDynamics) and is about 0.2 % of the typical nucleon interaction cross section. The number of the binary nucleon collisions occuring in the central Au+Au collisions is about 800 and we expect about $2K$ pairs be produced (Here K stands for the K-factor for the binary nucleon collisions). These heavy quark pairs are relatively slow and will experience evolution of the fire ball up to the moment it hadronize. The produced pairs will be distributed over the transverse area of about 130 fm^2 and the amounts of initially produced quarks are still dilute to experience significant annihilation. Actual comparison between the yield measurements and the pQCD calculation shows a big discrepancy, and reference [14] is also scaling up the PYTHIA calculation by large factors. Considering the hugh discrepancy, the systematic study of the heavy quark production for the varying collision energy and colliding nuclei will be interesting. The impulse approximation is valid for the $p+A$ reactions induced by 800 GeV proton beam, which is the characteristics of perturbative production. In regard to the measurement of total yields, we note the large enhancement of the intermediate mass lepton pairs observed at CERN [15]. If those lepton pairs are from open charm as the pair p_t spectrum indicate, this means the violation of the impulse approximation and the failure of the conventional pQCD for central $Pb+Pb$ reaction at SPS energy. The yield of the coincident $e\mu$ pair within the PHENIX acceptance seems to be enough to clarify any difference by factor of 2 if there is no drastic modification in the meson kinematics or fragmentation [9].

The partons produced in the heavy ion collision experience the equilibrium and the hadronization stage of the evolution. Secondary collisions might change their distribution as pictured in [16]. These initial parton rate can be decided experimentally from the study of the direct γ production and Drell-Yan process. Additional possible modification is the jet quenching raised in QM01(Quark Matter 2001) [17]. The perturbatively(hence quickly) produced partons propagate through the macroscopic volume and experiences the modification in its energy and fragmentation. According to the picture in the previous section, hadronization process of the scattered partons occurs slowly when compared to the hard process. Since the partons resulting from A + A reaction propagate the macroscopic volume, actual estimation of the hadronization time is however needed to picture the modification. Taking the estimate based on the symmetric string break-up model [18], the proper particle production time [1] is

$$\tau_0 \sim 0.6\ fm/c.$$

In the picture, the high momentum particles are produced far away from the production point and at a later time due to the effect by Lorentz boost. The production point is estimated in this picture as

$$(t = \tau_0 \cosh y, x = \tau_0 \sinh y). \tag{1}$$

Taking $m_T \approx 0.4\ GeV$, the production time for the 8 GeV π's become about 12 fm/c. Taking this estimate, we can consider the fragmentation of the 12 GeV/c π's are free from the effect of the macroscopic medium. About 5 charged particles results from the fragmentation of 10 GeV gluons, and a large fraction of those charged particles will be below 2 GeV where non-perturbative component dominates. A large fraction of the captured particles will produce the hydrodynamic motion affecting $dN_{ch}/d\eta$ and $dE_t/d\eta$. The fragmentation of these soft jets still generate the high p_t particles above the exponential tail of the soft component. Using the estimate of the symmetric string break-up model, the hadronization point of these particles is given by

$$\tau_0 \cdot \frac{p_t}{m_t} = 0.6 \cdot \frac{p_t}{0.4\ GeV/c}\ fm. \tag{2}$$

The hadronization point of the π's with the transverse momentum of 4 GeV is 6 fm from the production point based on Eq. 2, and the modification of the fragmentation by the macroscopic medium will be modest. Inclusive hadron p_t spectrum will be useful if the parent parton production rate is determined from the measurement of the leptonic

[1] There are dependences on the energy of the string, but we take this number as a rough estimate.

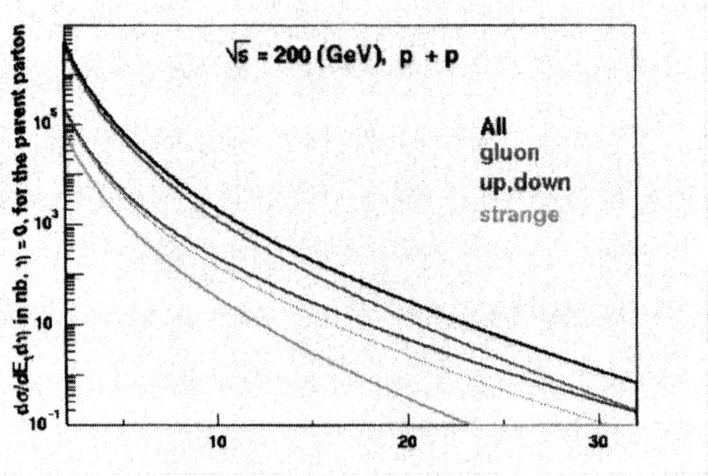

FIGURE 3. $d\sigma/dE_t d\eta(in\ nb)$ at $\eta = 0$ for the $p+p$ reaction at $\sqrt{s} = 200\ GeV$.

probe. When the energy of the hard parton reaches 20 GeV, the number of charged particles resulting from the gluon fragmentation reaches 8 and a significant portion of those particles will have $p_t > 2$, which will ease the identification of jet. Fig 3 shows the $d\sigma/dE_t d\eta$ at $\eta = 0$ calculated by a simple calculation program. It is obtained by

$$\frac{d\sigma}{dE_t d\eta} = 2\pi E_t \frac{d^3\sigma}{dyd^2p_t}.$$

Taking the acceptance and optimistic efficiency of the PHENIX for many factors as

$$\Delta\eta = 2 \cdot 0.35 = 0.7, \ \ \Delta\phi/2\pi = 1/4, \ \ \varepsilon = 0.5,$$

the effective luminousity of the RHIC

$$L = 0.25 \cdot 1200/\mu b = 300/\mu b,$$

and multiplying additional 200^2 for the number of binary collisions $< n_{NN} >$, We can read off the jet luminousity from Fig. 3 as

$$\Delta N = \ L \cdot A^2 \cdot \frac{d\sigma}{dE_t d\eta} \cdot \Delta\eta \cdot \Delta\phi/2\pi \cdot \varepsilon \cdot \Delta E_t \approx 1.1 \cdot 10^9 \cdot \Delta E_t \cdot \frac{d\sigma}{dE_t d\eta}(in\ mb) \tag{3}$$

Once enough number of jets are identified, many exciting possibilities appear. The p_t spectrum of the leading particles ($p_t > 12\ GeV$ and $E_{jet} > 12\ GeV$) for the identified jets will be closely related to the energy loss of the partons propagating through the medium. In regard to the evolution and modification of the parton fragmentation from the $p+p$ reaction to the $A+A$ reaction, we can study the difference in the modification of the hard particles ($p_t > 8\ GeV$) and soft particles ($p_t < 4\ GeV$) within the jet. Assuming the study on the global geometry including the simplified hydrodynamics are done ahead, effect on the fragmentation can be extracted from this study.

CONCLUSIONARY REMARKS AND OUTLOOK

We surveyed the physics issues to the hard probes described by the perturbative QCD. As the feasible exciting measurements, we note

- $dE_t/d\eta$ as the global observable,
- the invariant multiplicity for the produced hadrons, up to 20 GeV/c in p_t,
- the direct γ production up to 15 GeV/c in p_t,
- the identification of the hard jets and the study of its fragmentation,
- the Drell-Yan pair production $m_{ll} \geq 4\ GeV/c$,
- and the heavy quark production in $\mu + \mu$ and $e + \mu$ channel.

These measurement can be made for the $p+p$ and $A+A$ reaction simultaneously with a good impact parameter decision. It will reduce the uncertsinties in the generic pQCD, and help confirming the perturbative production. This proceeding is a short summary of a longer write-up and the original note can be obtained by request to the author.

ACKNOWLEDGMENTS

This efforts was started to express due respect to a retiring honourable professor by showing his disciples are struggling to follow the high standard he set up. While the work of the student is far from his high standard, the efforts of the disciples will be continued as long as they can.

REFERENCES

1. J. F. Owens, Rev.Mod.Phy. 59(1987)465,
2. G. Sterman and John Smith et. al., Rev.Mod.Phy. 67(1995)157
3. S. Kretzer, Phys. Rev. D62 054001
4. H. L. Lai, J. Huston, S. Kuhlmann, F. Olness, J. Owens, D. Soper, W. K. Tung, and H. Weerts, Phys. Rev. D55 (1997) 1281
5. B. A. Kniehl, G. Kramer, B. Pötter, Nucl. Phys. B582(2000) 514-536, $hep-ph/0011155$
6. N. Hammon, H. Stöcker and W. Greiner Phys. Lett. B448: 290-294, 1999, Jamal Jalilian-Marian and Xin-Nian Wang, $hep-ph/0005071$
7. K. J. Eskola, K. Tuominen, Phys. Lett. B489(2000) 329-336
8. Particle data group, EPJC15(2000)
9. The PHENIX Collaboration, "Phenix Conceptual Design Report", January, 1993
10. C. Y. Wong, H. Wang, Phys. Rev. C58(1998), 376-388
11. X. N. Wang, Phys.Rev. C61 (2000) 064910
12. Applications of Perturbative QCD, R. D. Field
13. G. Piller, W. Weise, Phys. Rept. 330(2000) 1-94.
14. P. Braun-Munzinger, D. Miskowiec, A. Drees, C. Lourenco, EPJC1(1998) 123-129
15. NA38 and NA50 Collaborations, CERN-EP-2000-012, Accepted by Euro. Phys. J. C
16. S. A. Bass and B. Müller, Phys. Lett. B471(1999) 108-112
17. M. Gyulassy, M. Plümer, Phys. Lett. B243(1990) 432-438, Xin-Nian Wang, Phys. Rept. 280(1997) 287-371
18. Introduction to High-Energy Heavy-Ion Collisions, C. Y. Wong

Issues of Elementary Matter: From Superheavies to Hypermatter and Antimatter

Walter Greiner

Institut für Theoretische Physik, J.W. Goethe-Universität,
D-60054 Frankfurt, Germany

Abstract. The extension of the periodic system into various new areas is investigated. Experiments for the synthesis of superheavy elements and the predictions of magic numbers are reviewed. Different ways of nuclear decay are discussed like cluster radioactivity, cold fission and cold multifragmentation, including the recent discovery of the tripple fission of ^{252}Cf. Furtheron, investigations on hypernuclei and the possible production of antimatter–clusters in heavy–ion collisions are reported. Various versions of the meson field theory serve as effective field theories at the basis of modern nuclear structure and suggest structure in the vacuum which might be important for the production of hyper– and antimatter. A perspective for future research is given.

There are fundamental questions in science, like e. g. "how did life emerge" or "how does our brain work" and others. However, the most fundamental of those questions is "how did the world originate?". The material world has to exist before life and thinking can develop. Of particular importance are the substances themselves, i. e. the particles the elements are made of (baryons, mesons, quarks, gluons), i. e. elementary matter. The vacuum and its structure is closely related to that. On this I want to report today. I begin with the discussion of modern issues in nuclear physics.

The elements existing in nature are ordered according to their atomic (chemical) properties in the **periodic system** which was developped by Mendeleev and Lothar Meyer. The heaviest element of natural origin is Uranium. Its nucleus is composed of $Z = 92$ protons and a certain number of neutrons ($N = 128 - 150$). They are called the different Uranium isotopes. The transuranium elements reach from Neptunium ($Z = 93$) via Californium ($Z = 98$) and Fermium ($Z = 100$) up to Lawrencium ($Z = 103$). The heavier the elements are, the larger are their radii and their number of protons. Thus, the Coulomb repulsion in their interior increases, and they undergo fission. In other words: the transuranium elements become more instable as they get bigger.

In the late sixties the dream of the superheavy elements arose. Theoretical nuclear physicists around S.G. Nilsson (Lund)[1] and from the Frankfurt school[2, 3, 4] predicted that so-called closed proton and neutron shells should counteract the repelling Coulomb forces. Atomic nuclei with these special **"magic" proton and neutron numbers** and their neighbours could again be rather stable. These magic proton (Z) and neutron (N) numbers were thought to be $Z = 114$ and $N = 184$ or 196. Typical predictions of their life times varied between seconds and many thousand years. Fig.1 summarizes the expectations at the time. One can see the islands of superheavy elements around

CP594, *Hadrons and Nuclei: First International Symposium,* edited by Il-T. Cheon et al.

$Z = 114, N = 184$ and 196, respectively, and the one around $Z = 164, N = 318$.

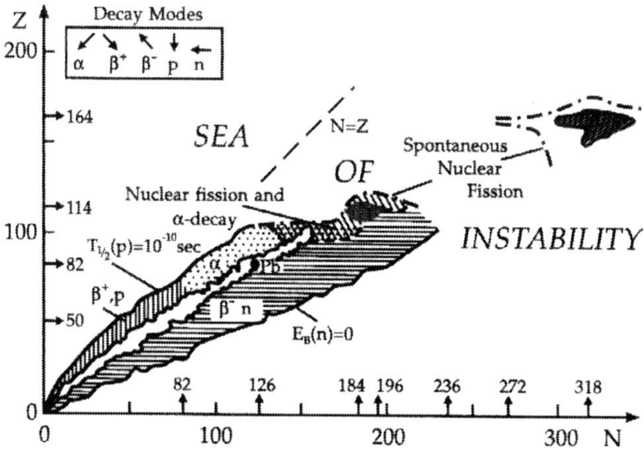

FIGURE 1. The periodic system of elements as conceived by the Frankfurt school in the late sixties. The islands of superheavy elements ($Z = 114, N = 184$, 196 and $Z = 164, N = 318$) are shown as dark hatched areas.

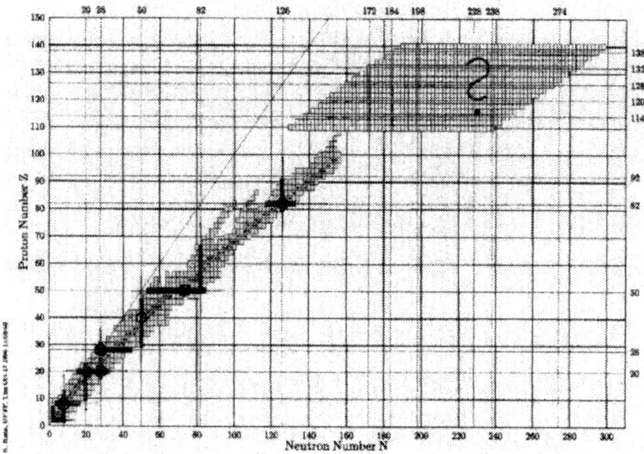

FIGURE 2. The shell structure in the superheavy region around $Z = 114$ is an open question. As will be discussed later, meson field theories suggest that $Z = 120, N = 172, 184$ are the magic numbers in this region.

The important question was how to produce these superheavy nuclei. There were many attempts, but only little progress was made. It was not until the middle of the seventies that the Frankfurt school of theoretical physics together with foreign guests (R.K. Gupta (India), A. Sandulescu (Romania))[6] theoretically understood and substantiated the concept of bombarding of double magic lead nuclei with suitable projectiles, which had been proposed intuitively by the russian nuclear physicist Y. Oganessian[7]. The two-center shell model, which is essential for the description of fission, fusion and nuclear molecules, was developped in 1969-1972 together with my then students

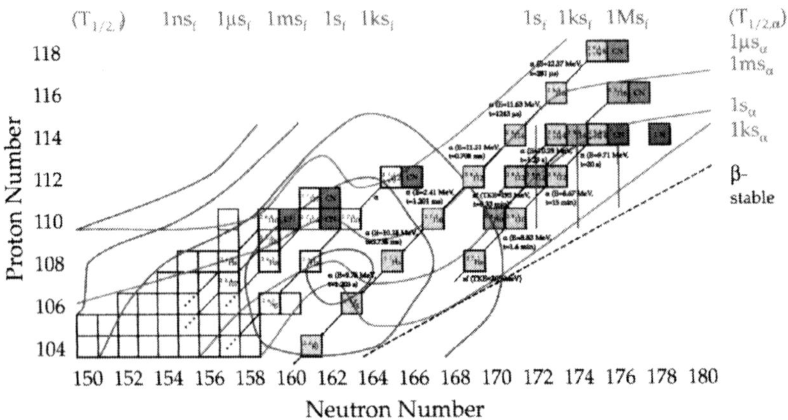

FIGURE 3. The $Z = 106 - 112$ isotopes were fused by the Hofmann–Münzenberg (GSI)–group. The two $Z = 114$ isotopes and the $Z = 116$ isotope were produced by the Dubna–Livermore group. It is claimed that three neutrons are evaporated. Obviously the lifetimes of the various decay products are rather long (because they are closer to the stable valley), in crude agreement with early predictions [3, 4] and in excellent agreement with the recent calculations of the Sobicevsky–group [12]. The recently fused $Z = 118$ isotope by V. Ninov et al. at Berkeley is the heaviest one so far, but needs confirmation.

U. Mosel and J. Maruhn[8]. It showed that the shell structure of the two final fragments was visible far beyond the barrier into the fusioning nucleus. The collective potential energy surfaces of heavy nuclei, as they were calculated in the framework of the two-center shell model, exhibit pronounced valleys, such that these valleys provide promising doorways to the fusion of superheavy nuclei for certain projectile-target combinations (Fig. 4). If projectile and target approach each other through those **"cold" valleys**, they get only minimally excited and the barrier which has to be overcome (fusion barrier) is lowest (as compared to neighbouring projectile-target combinations). In this way the correct projectile- and target-combinations for fusion were predicted. Indeed, Gottfried Münzenberg and Sigurd Hofmann and their group at GSI [9] have followed this approach. With the help of the SHIP mass-separator and the position sensitive detectors, which were especially developped by them, they produced the pre-superheavy elements $Z = 106, 107, \ldots 112$, each of them with the theoretically predicted projectile-target combinations, and only with these. Everything else failed. This is an impressive success, which crowned the laborious construction work of many years. The before last example of this success, the discovery of element 112 and its long α-decay chain, is shown in Fig. 6. Very recently the Dubna–Livermore–group produced two isotopes of $Z = 114$ element by bombarding ^{244}Pu with ^{48}Ca and also $Z = 116$ by ^{48}Ca + ^{248}C m.(Fig. 3). Also these are cold–valley reactions (in this case due to the combination of a spherical and a deformed nucleus), as predicted by Gupta, Sandulescu and Greiner [10] in 1977. There exist also cold valleys for which both fragments are deformed [11], but these have yet not been verified experimentally. The very recently reported $Z = 118$ isotope fused with the cold valley reaction [13] ^{58}Kr + ^{208}Pb by Ninov et al. [14] yields the latest support of the cold valley idea, but this one needs confirmation.

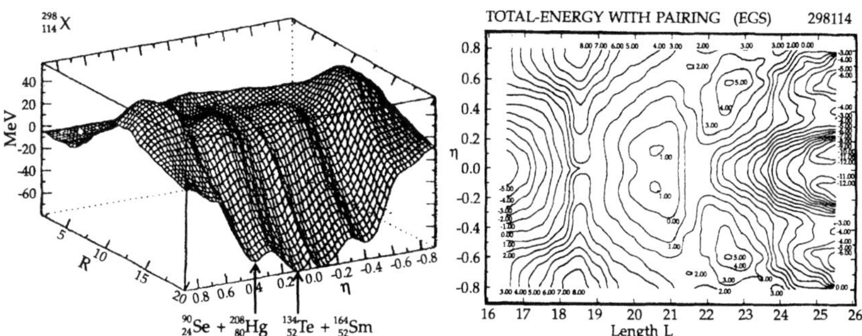

FIGURE 4. The collective potential energy surface of $^{264}108$ and $^{184}114$, calculated within the two center shell model by J. Maruhn et al., shows clearly the cold valleys which reach up to the barrier and beyond. Here R is the distance between the fragments and $\eta = \frac{A_1 - A_2}{A_1 + A_2}$ denotes the mass asymmetry: $\eta = 0$ corresponds to a symmetric, $\eta = \pm 1$ to an extremely asymmetric division of the nucleus into projectile and target. If projectile and target approach through a cold valley, they do not "constantly slide off" as it would be the case if they approach along the slopes at the sides of the valley. Constant sliding causes heating, so that the compound nucleus heats up and gets unstable. In the cold valley, on the other hand, the created heat is minimized. The colleagues from Freiburg should be familiar with that: they approach Titisee (in the Black Forest) most elegantly through the Höllental and not by climbing its slopes along the sides.

Studies of the shell structure of superheavy elements in the framework of the meson field theory and the Skyrme-Hartree-Fock approach have recently shown that the magic shells in the superheavy region are very isotope dependent [5, 15] (see Fig. 7). **According to these investigations** $Z = 120$ **being a magic proton number seems to be as probable as** $Z = 114$. Additionally, recent investigations in a chirally symmetric mean–field theory result also in the prediction of these two magic numbers[39, 41], see also below. The corresponding magic neutron numbers are predicted to be $N = 172$ and - as it seems to a lesser extend - $N = 184$. Thus, this region provides an open field of research. R.A. Gherghescu et al. have calculated the potential energy surface of the $Z = 120$ nucleus. It utilizes interesting isomeric and valley structures (Fig. 8). The charge distribution of the $Z = 120, N = 184$ nucleus indicates a hollow inside. This leads us to suggest that it might be essentially a fullerene consisting of 60 α-particles and one additional binding neutron per alpha. This is illustrated in Fig 5. The protons and neutrons of such a superheavy nucleus are distributed over 60 α particles and 60 neutrons (forgetting the last 4 neutrons).

The determination of the chemistry of superheavy elements, i. e. the calculation of the atomic structure — which is in the case of element 112 the shell structure of 112 electrons due to the Coulomb interaction of the electrons and in particular the calculation of the orbitals of the outer (valence) electrons — has been carried out as early as 1970 by B. Fricke and W. Greiner[16]. Hartree-Fock-Dirac calculations yield rather precise results.

The potential energy surfaces, which are shown prototypically for $Z = 114$ in Fig 4, contain even more remarkable information that I want to mention cursorily: if a given nucleus, e. g. Uranium, undergoes fission, it moves in its potential mountains from the

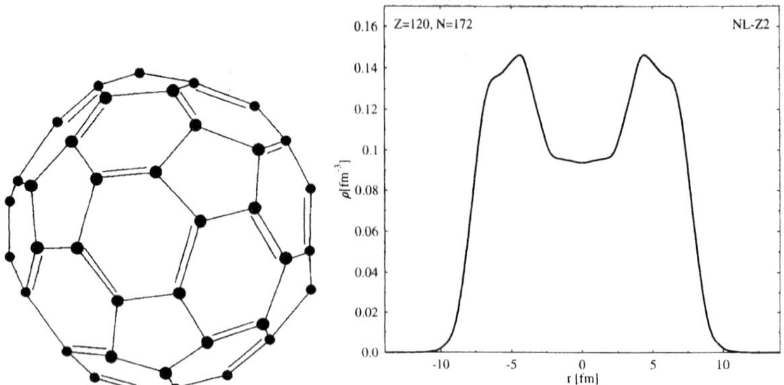

FIGURE 5. Typical structure of the fullerene ^{60}C. The double bindings are illutsrated by double lines. In the nuclear case the Carbon atoms are replaced by α particles and the double bindings by the additional neutrons. Such a structure would immediately explain the semi–hollowness of that superheavy nucleus, which is revealed in the mean–field calculations within meson–field theories. The radial density of the nucleus with 120 protons and 172 neutrons, as emerging from a meson-field calculation with the force NL-Z2 is shown on the right side. Note that the semi-bubble structure is mostly pronounced for this nucleus. When going to higher neutron numbers, this structures becomes less and less.

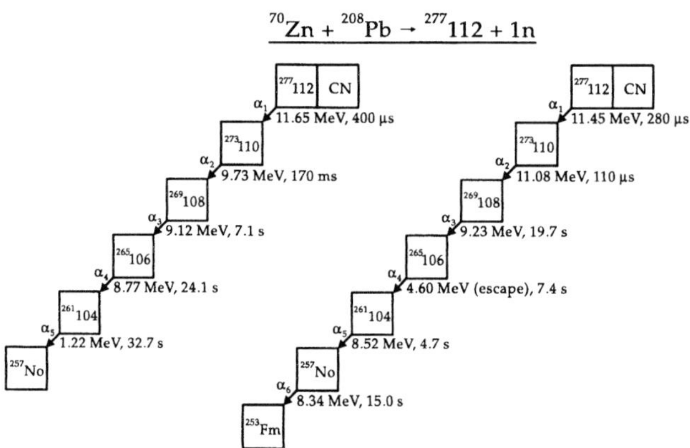

FIGURE 6. The fusion of element 112 with ^{70}Zn as projectile and ^{208}Pb as target nucleus has been accomplished for the first time in 1995/96 by S. Hofmann, G. Münzenberg and their collaborators. The colliding nuclei determine an entrance to a "cold valley" as predicted as early as 1976 by Gupta, Sandulescu and Greiner. The fused nucleus 112 decays successively via α emission until finally the quasi-stable nucleus ^{253}Fm is reached. The α particles as well as the final nucleus have been observed. Combined, this renders the definite proof of the existence of a $Z = 112$ nucleus.

interior to the outside. Of course, this happens quantum mechanically. The wave function of such a nucleus, which decays by tunneling through the barrier, has maxima where the potential is minimal and minima where it has maxima.

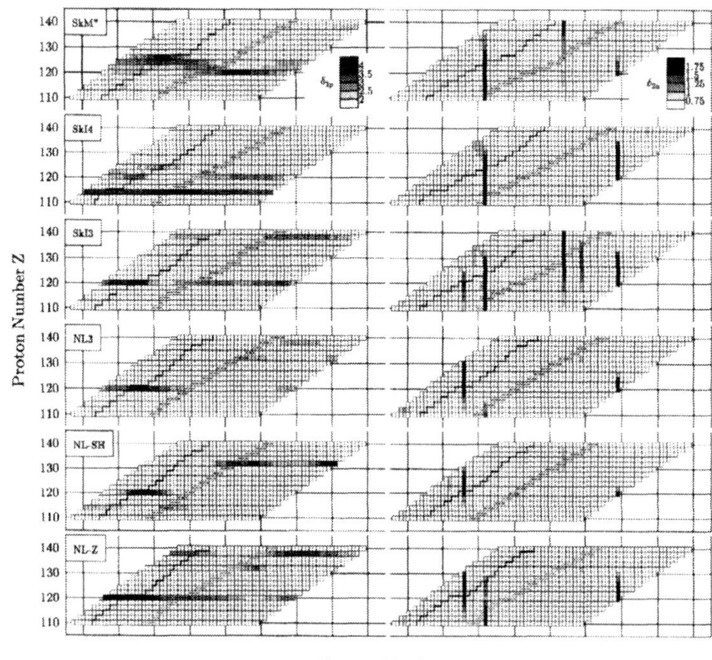

FIGURE 7. Grey scale plots of proton gaps (left column) and neutron gaps (right column) in the *N*-*Z* plane for spherical calculations with the forces as indicated. The assignment of scales differs for protons and neutrons, see the uppermost boxes where the scales are indicated in units of MeV. Nuclei that are stable with respect to β decay and the two-proton dripline are emphasized. The forces with parameter sets SkI4 and NL-Z reproduce the binding energy of $^{264}_{156}108$ (Hassium) best, i.e. $|\delta E/E| < 0.0024$. Thus one might assume that these parameter sets could give the best predictions for the superheavies. Nevertheless, it is noticed that NL-Z predicts only $Z = 120$ as a magic number while SkI4 predicts both $Z = 114$ and $Z = 120$ as magic numbers. The magicity depends — sometimes quite strongly — on the neutron number. These studies are due to Bender, Rutz, Bürvenich, Maruhn, P.G. Reinhard et al. [15].

The probability for finding a certain mass asymmetry $\eta = \dfrac{A_1 - A_2}{A_1 + A_2}$ of the fission is proportional to $\psi^*(\eta)\psi(\eta)d\eta$. Generally, this is complemented by a coordinate dependent scale factor for the volume element in this (curved) space, which I omit for the sake of clarity. Now it becomes clear how the so-called **asymmetric** and **superasymmetric** fission processes come into being. They result from the enhancement of the collective wave function in the cold valleys. And that is indeed, what one observes.

For large mass asymmetry ($\eta \approx 0.8$, 0.9) there exist very narrow valleys. They are not as clearly visible in Fig. 4, but they have interesting consequences. Through these narrow valleys nuclei can emit spontaneously not only α-particles (Helium nuclei) but also ^{14}C, ^{20}O, ^{24}Ne, ^{28}Mg, and other nuclei. Thus, we are lead to the **cluster radioactivity** (Poenaru, Sandulescu, Greiner [17]).

By now this process has been verified experimentally by research groups in Oxford,

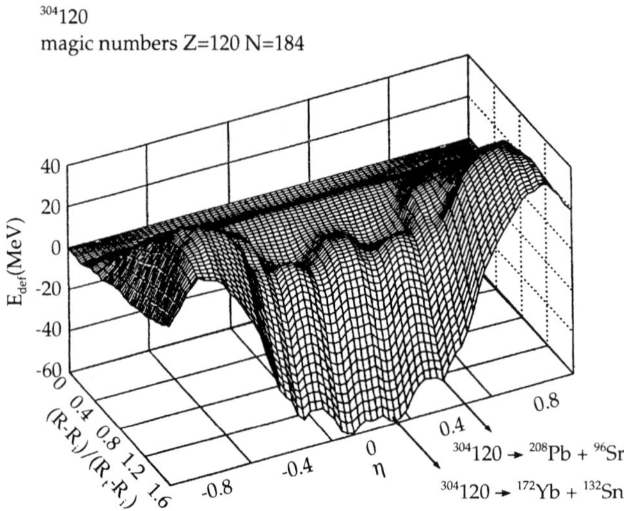

FIGURE 8. Potential energy surface as a function of reduced elongation $(R - R_i)/(R_t - R_i)$ and mass asymmetry η for the double magic nucleus $^{304}120$. $^{304}120_{184}$.

Moscow, Berkeley, Milan and other places. Accordingly, one has to revise what is learned in school: there are not only 3 types of radioactivity (α-, β-, γ-radioactivity), but many more. Atomic nuclei can also decay through spontaneous cluster emission (that is the "spitting out" of smaller nuclei like carbon, oxygen,...). Fig. 10 depicts some examples of these processes.

The knowledge of the collective potential energy surface and the collective masses $B_{ij}(R,\eta)$, all calculated within the Two-Center-Shell-Modell (TCSM), allowed H. Klein, D. Schnabel and J. A. Maruhn to calculate lifetimes against fission in an "ab initio" way [18]. The discussion of much more very interesting new physics cannot be persued here. We refer to the literature [19, 20, 21, 22, 23, 24, 25].

The "cold valleys" in the collective potential energy surface are basic for understanding this exciting area of nuclear physics! It is a master example for understanding the **structure of elementary matter**, which is so important for other fields, especially astrophysics, but even more so for enriching our "Weltbild", i.e. the status of our understanding of the world around us.

Nuclei that are found in nature consist of nucleons (protons and neutrons) which themselves are made of u (up) and d (down) quarks. However, there also exist s (strange) quarks and even heavier flavors, called charm, bottom, top. The latter has just recently been discovered. Let us stick to the s quarks. They are found in the 'strange' relatives of the nucleons, the so-called hyperons (Λ, Σ, Ξ, Ω). The Λ-particle, e. g., consists of one u, d and s quark, the Ξ-particle even of an u and two s quarks, while the Ω (sss) contains strange quarks only.

If such a hyperon is taken up by a nucleus, a **hyper-nucleus** is created. Hyper-nuclei with one hyperon have been known for 20 years now, and were extensively studied

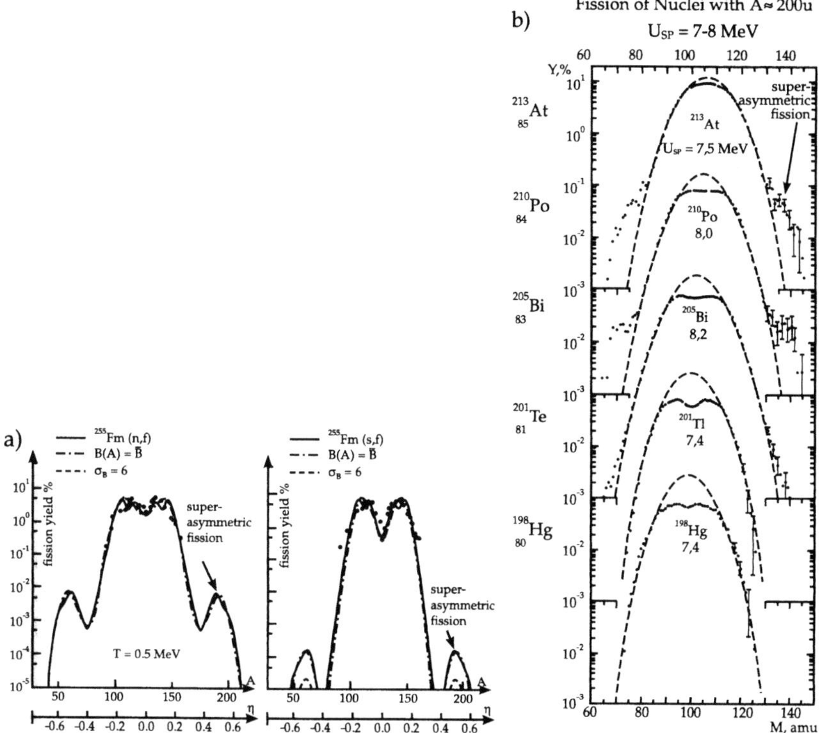

FIGURE 9. Asymmetric (a) and symmetric (b) fission. For domminantly symmetric fissioning nuclei, also superasymmetric fission is recognizable, as it has been observed only a few years ago by the russian physicist Itkis — just as expected theoretically.

by B. Povh (Heidelberg)[28]. Several years ago, Carsten Greiner, Jürgen Schaffner and Horst Stöcker[29] theoretically investigated nuclei with many hyperons, **hypermatter**, and found that the binding energy per baryon of strange matter is in many cases even higher than that of ordinary matter (composed only of u and d quarks). This leads to the idea of extending the periodic system of elements in the direction of strangeness.

One can also ask for the possibility of building atomic nuclei out of **antimatter**, that means searching e. g. for anti-helium, anti-carbon, anti-oxygen. Fig. 11 depicts this idea. Due to the charge conjugation symmetry antinuclei should have the same magic numbers and the same spectra as ordinary nuclei. However, as soon as they get in touch with ordinary matter, they annihilate with it and the system explodes.

Now the important question arises how these strange matter and antimatter clusters can be produced. First, one thinks of collisions of heavy nuclei, e. g. lead on lead, at high energies (energy per nucleon $\geq$ 200 GeV). Calculations with the URQMD-model of the Frankfurt school show that through **nuclear shock waves** [30, 31, 32] nuclear matter gets compressed to 5–10 times of its usual value, $\rho_0 \approx 0.17\ \mathrm{fm}^3$, and heated up to temperatures of $kT \approx 200$ MeV. As a consequence about 10000 pions, 100 Λ's, 40 Σ's

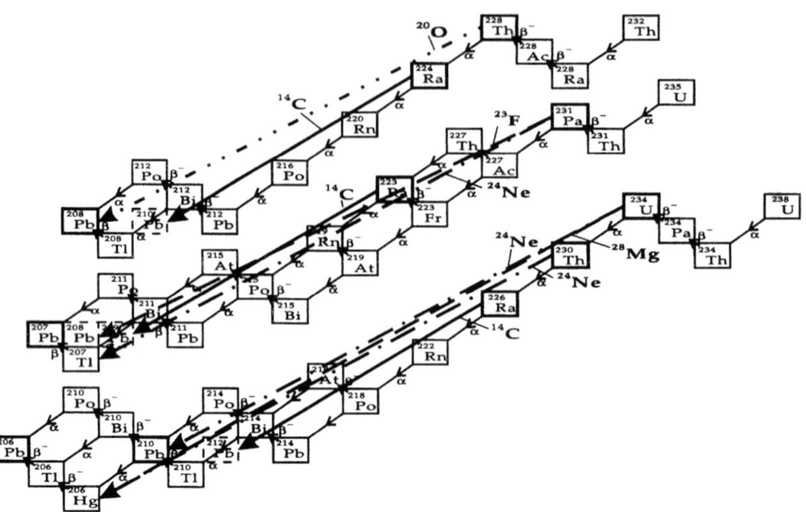

FIGURE 10. Cluster radioactivity of actinide nuclei. By emission of ^{14}C, ^{20}O,... "big leaps" in the periodic system can occur, just contrary to the known α, β, γ radioactivities, which are also partly shown in the figure.

and Ξ's and about as many antiprotons and many other particles are created in a single collision. It seems conceivable that it is possible in such a scenario for some Λ's to get captured by a nuclear cluster. This happens indeed rather frequently for one or two Λ-particles; however, more of them get built into nuclei with rapidly decreasing probability only. This is due to the low probability for finding the right conditions for such a capture in the phase space of the particles: the numerous particles travel with every possible momenta (velocities) in all directions. The chances for hyperons and antibaryons to meet gets rapidly worse with increasing number. In order to produce multi-Λ-nuclei and antimatter nuclei, one has to look for a different source.

In the framework of meson field theory within the mean-field approximation the energy spectrum of baryons in a nucleus has a peculiar structure, depicted in Fig. 12. It consists of an upper and a lower continuum, as it is known from the electrons (see e. g. [27]). The upper well represents the nuclear shell modell potential. It describes the overall structure throughout the nuclear table very well.

Of special interest in the case of the baryon spectrum is the potential well, built of the scalar and the vector potential, which rises from the lower continuum. It is known since P.A.M. Dirac (1930) that the negative energy states of the lower continuum have to be occupied by particles (electrons or, in our case, baryons). Otherwise our world would be unstable, because the "ordinary" particles are found in the upper states which can decay through the emission of photons into lower lying states. However, if the "underworld" is occupied, the Pauli-principle will prevent this decay. Holes in the occupied "underworld" (Dirac sea) are antiparticles. This has been extensively discussed in the context of QED of strong fields (overcritical fields, decay of the vacuum from a neutral one into a charged one [27]).

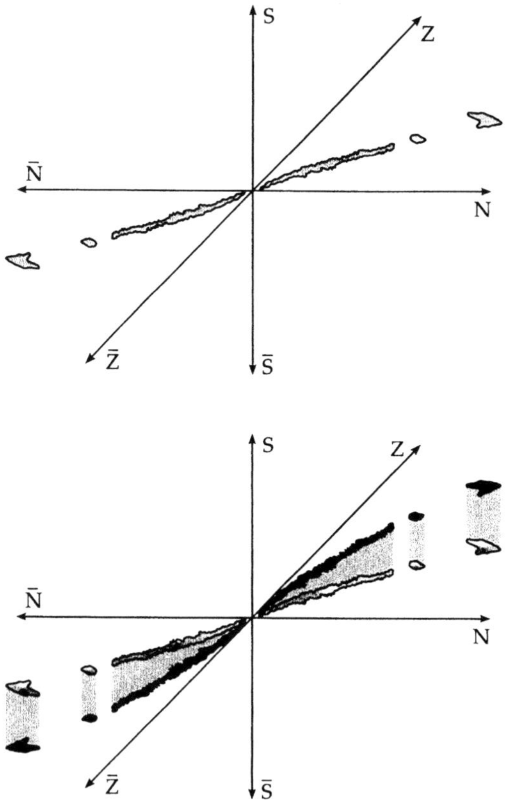

FIGURE 11. The extension of the periodic system into the sectors of strangeness (S, $\bar{S}$) and antimatter ($\bar{Z}$, $\bar{N}$). The stable valley winds out of the known proton (Z) and neutron (N) plane into the S and $\bar{S}$ sector, respectively. The same can be observed for the antimatter sector. In the upper part of the figure only the stable valley in the usual proton (Z) and neutron (N) plane is plotted, however, extended into the sector of antiprotons and antineutrons. In the second part of the figure it has been indicated, how the stable valley winds out of the Z-N-plane into the strangeness sector. This is due to an additional term proportional to $(\frac{A}{A} - \frac{S_0}{A})^2$ in the mass formula.

The occupied states of this underworld including up to 40000 occupied bound states of the lower potential well represent the **vacuum**. The peculiarity of this strongly correlated vacuum structure in the region of atomic nuclei is that — depending on the size of the nucleus — more than 20000 up to 40000 (occupied) bound nucleon states contribute to this polarization effect. Obviously, we are dealing here with a **highly correlated vacuum**. A pronounced shell structure can be recognized [33, 34, 35]. Holes in these states have to be interpreted as bound antinucleons (antiprotons, antineutrons). If the primary nuclear density rises due to compression, the lower well increases while the upper decreases and soon is converted into a repulsive barrier (Fig. 13). This compression of nuclear matter can only be carried out in relativistic nucleus-nucleus collision with the help of shock waves, which have been proposed by the Frankfurt school[30, 31] and

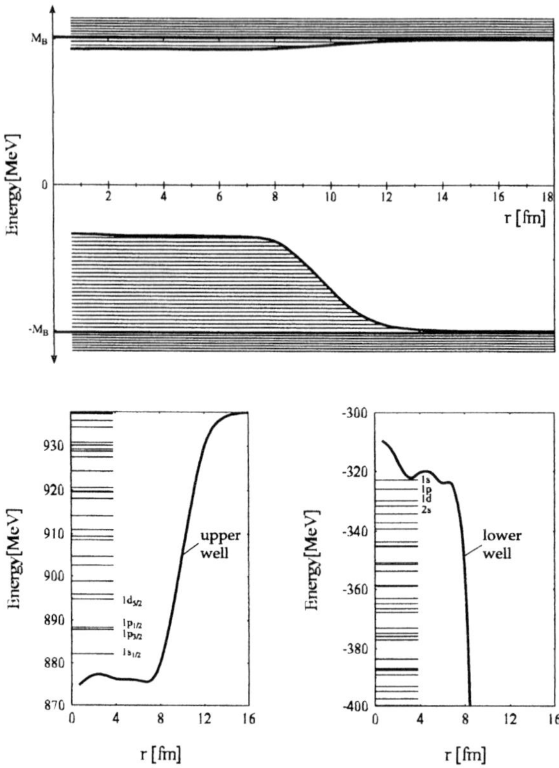

FIGURE 12. Baryon spectrum in a nucleus. Below the positive energy continuum exists the potential well of real nucleons. It has a depth of 50-60 MeV and shows the correct shell structure. The shell model of nuclei is realized here. However, from the negative continuum another potential well arises, in which about 40000 bound particles are found, belonging to the vacuum. A part of the shell structure of the upper well and the lower (vacuum) well is depicted in the lower figures.

which have since then been confirmed extensively (for references see e. g. [36]). These **nuclear shock waves** are accompanied by heating of the compressed nuclear matter. Indeed, density and temperature are correlated in terms of the hydrodynamic Rankine-Hugoniot-equations. Heating as well as the violent dynamics cause the creation of many holes in the very deep (measured from $-M_B c^2$) vacuum well and an equal number of particles (baryons) in the upper continuum. This is analogous to the dynamical $e^+ e^-$ pair creation in heavy ion collisions. [36]

These numerous bound holes resemble antimatter clusters which are bound in the medium; their wave functions have large overlap with antimatter clusters. When the primary matter density decreases during the expansion stage of the heavy ion collision, the potential wells, in particular the lower one, disappear.

The bound antinucleons are then pulled down into the (lower) continuum. In this way antimatter clusters may be set free. Of course, a large part of the antimatter will annihilate on ordinary matter present in the course of the expansion. However, it is important that this mechanism for the production of antimatter clusters out of the highly correlated vacuum does not proceed via the phase space. The required coalescence of many particles in phase space suppresses the production of clusters, while it is favoured by the direct production out of the highly correlated vacuum. In a certain sense, the highly correlated vacuum is a kind of cluster vacuum (vacuum with cluster structure). The shell structure of the vacuum levels (see Fig. 12) supports this latter suggestion. Fig. 14 illustrates this idea.

The mechanism is similar for the production of multi-hyper nuclei (Λ, Σ, Ξ, Ω). Meson field theory predicts also for the Λ energy spectrum at finite primary nucleon density the existence of upper and lower wells. The lower well belongs to the vacuum and is fully occupied by Λ's. Dynamics and temperature then induce transitions (e.g. $\Lambda\bar{\Lambda}$ creation) and deposit many Λ's in the upper well. These numerous bound Λ's (and similarly other hyperons) are sitting close to the primary baryons: in a certain sense a giant multi-Λ hypernucleus has been created. When the system disintegrates (expansion stage) the Λ's distribute over the nucleon clusters (which are most abundant in peripheral collisions). In this way multi-Λ hypernuclei can be formed. Also clusters of hyperons alone(Λ, Σ, ...) seem possible and quasistable [5, 29] and the Bethe-Weizsäcker mass formula requires at least one additional term proportional to $(f_S - f_{S_0})^2$, where f_S/A is the strangeness content in a hypernucleus.

Of course this vision has to be worked out and probably refined in many respects. This means much more and thorough investigation in the future. It is particularly important to gain more experimental information on the properties of the lower well by (e, e' p) or (e, e' p p') and also ($\bar{p}_c p_b$, $p_c \bar{p}_b$) reactions at high energy ($\bar{p}_c$ denotes an incident antiproton from the continuum, p_b is a proton in a bound state; for the reaction products the situation is just the opposite)[37]. Also the reaction (p, p' d), (p, p' ^{3}He), (p, p' ^{4}He) and others of similar type need to be investigated in this context. The systematic

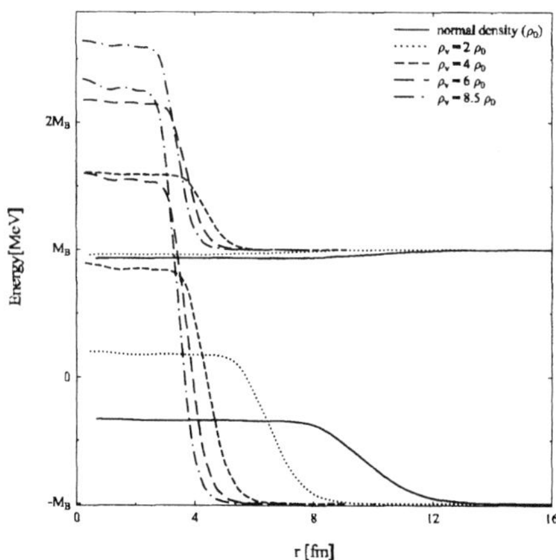

FIGURE 13. The lower well rises strongly with increasing primary nucleon density, and even gets supercritical (spontaneous nucleon emission and creation of bound antinucleons). Supercriticality denotes the situation, when the lower well enters the upper continuum.

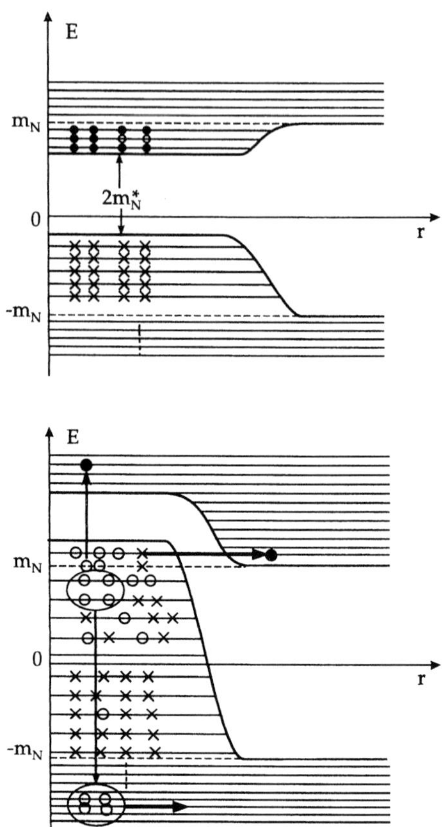

FIGURE 14.
Due to the high temperature and the violent dynamics, many bound holes (antinucleon clusters) are created in the highly correlated vacuum, which can be set free during the expansion stage into the lower continuum. In this way, antimatter clusters can be produced directly from the vacuum. The horizontal arrow in the lower part of the figure denotes the spontaneous creation of baryon-antibaryon pairs, while the antibaryons occupy bound states in the lower potential well. Such a situation, where the lower potential well reaches into the upper continuum, is called supercritical. Four of the bound holes states (bound antinucleons) are encircled to illustrate a "quasi-antihelium" formed. It may be set free (driven into the lower continuum) by the violent nuclear dynamics.

scattering of antiprotons on nuclei can contribute to clarify these questions: Time-like momentum transfer is required here! The Nambu-Jona-Lasigno (NJL) model seems to give much smaller lower wells, but does not describe the shell model potentials. Studies of I. Mishustin, L. Satarov et al. to improve the NJL model for applications to nuclear und baryon-meson sectors are on the way.

Problems of the meson field theory (e. g. Landau poles) can then be reconsidered. An effective meson field theory has to be constructed. Various effective theories, e. g. of Walecka-type on the one side and theories with chiral invariance on the other side, seem to give different strengths of the potential wells and also different dependence on the baryon density [38]. The Lagrangians of the Dürr-Teller-Walecka-type and of the chirally symmetric mean field theories look quantitatively quite differently. We exhibit them — without further discussion — in the following equations:

$$\mathcal{L} = \mathcal{L}_{\mathrm{kin}} + \mathcal{L}_{\mathrm{BM}} + \mathcal{L}_{\mathrm{vec}} + \mathcal{L}_0 + \mathcal{L}_{\mathrm{SB}}$$

Non-chiral Lagrangian:

$$\begin{aligned}
\mathcal{L}_{\mathrm{kin}} &= \frac{1}{2}\partial_\mu s \partial^\mu s + \frac{1}{2}\partial_\mu z \partial^\mu z - \frac{1}{4}B_{\mu\nu}B^{\mu\nu} - \frac{1}{4}G_{\mu\nu}G^{\mu\nu} - \frac{1}{4}F_{\mu\nu}F^{\mu\nu} \\
\mathcal{L}_{\mathrm{BM}} &= \Sigma_B \overline{\Psi}_B [i\gamma_\mu\partial^\mu - g_{\omega B}\gamma_\mu\omega^\mu - g_{\phi B}\gamma_\mu\phi^\mu - g_{\rho B}\gamma_\mu\tau_B\rho^\mu \\
&\qquad -e\gamma_\mu\frac{1}{2}(1+\tau_B)A^\mu - m_B^*]\psi_B \\
\mathcal{L}_{\mathrm{vec}} &= \frac{1}{2}m_\omega^2\omega_\mu\omega^\mu + \frac{1}{2}m_\rho^2\rho_\mu\rho^\mu + \frac{1}{2}m_\phi^2\phi_\mu\phi^\mu \\
\mathcal{L}_0 &= -\frac{1}{2}m_s^2 s^2 - \frac{1}{2}m_z^2 z^2 - \frac{1}{3}bs^3 - \frac{1}{4}cs^4
\end{aligned}$$

Chiral Lagrangian:

$$\begin{aligned}
\mathcal{L}_{\mathrm{kin}} &= \frac{1}{2}\partial_\mu\sigma\partial^\mu\sigma + \frac{1}{2}\partial_\mu\zeta\partial^\mu\zeta + \frac{1}{2}\partial_\mu\chi\partial^\mu\chi - \frac{1}{4}B_{\mu\nu}B^{\mu\nu} - \frac{1}{4}G_{\mu\nu}G^{\mu\nu} - \frac{1}{4}F_{\mu\nu}F^{\mu\nu} \\
\mathcal{L}_{\mathrm{BM}} &= \Sigma_B \overline{\Psi}_B [i\gamma_\mu\partial^\mu - g_{\omega B}\gamma_\mu\omega^\mu - g_{\phi B}\gamma_\mu\phi^\mu - g_{\rho B}\gamma_\mu\tau_B\rho^\mu \\
&\qquad -e\gamma_\mu\frac{1}{2}(1+\tau_B)A^\mu - m_B^*]\psi_B \\
\mathcal{L}_{\mathrm{vec}} &= \frac{1}{2}m_\omega^2\frac{\chi^2}{\chi_0^2}\omega_\mu\omega^\mu + \frac{1}{2}m_\rho^2\frac{\chi^2}{\chi_0^2}\rho_\mu\rho^\mu + \frac{1}{2}m_\phi^2\frac{\chi^2}{\chi_0^2}\phi_\mu\phi^\mu + g_4^4(\omega^4 + 6\omega^2\rho^2 + \rho^4) \\
\mathcal{L}_0 &= -\frac{1}{2}k_0\chi^2(\sigma^2+\zeta^2) + k_1(\sigma^2+\zeta^2)^2 + k_2(\frac{\sigma^4}{2}+\zeta^4) + k_3\chi\sigma^2\zeta - k_4\chi^4 \\
&\quad + \frac{1}{4}\chi^4\ln\frac{\chi^4}{\chi_0^4} + \frac{\delta}{3}\ln\frac{\sigma^2\zeta}{\sigma_0^2\zeta_0^2} \\
\mathcal{L}_{\mathrm{SB}} &= -\left(\frac{\chi}{\chi_0}\right)^2\left[m_\pi^2 f_\pi\sigma + \left(\sqrt{2}m_K^2 f_K - \frac{1}{\sqrt{2}}m_\pi^2 f_\pi\right)\zeta\right]
\end{aligned}$$

The non-chiral model contains the scalar-isoscalar field s and its strange counterpart z, the vector-isoscalar fields ω_μ and ϕ_μ, and the the ρ-meson ρ_μ as well as the photon A_μ. For more details see [38]. In contrast to the non-chiral model, the $SU(3)_L \times SU(3)_R$ Lagrangian contains the dilaton field χ introduced to mimic the trace anomaly of QCD in an effective Lagrangian at tree level (for an explanation of the chiral model see [38, 39]).

The connection of the chiral Lagrangian with the Walecka-type one can be established by the substitution $\sigma = \sigma_0 - s$ (and similarly for the strange condensate ζ). Then, e.g. the difference in the definition of the effective nucleon mass in both models (non-chiral:$m_N^* = m_N - g_s s$, chiral:$m_N^* = g_s\sigma$) can be removed, yielding:

$$m_N^* = g_s\sigma_0 - g_s s \equiv m_N - g_s s \tag{1}$$

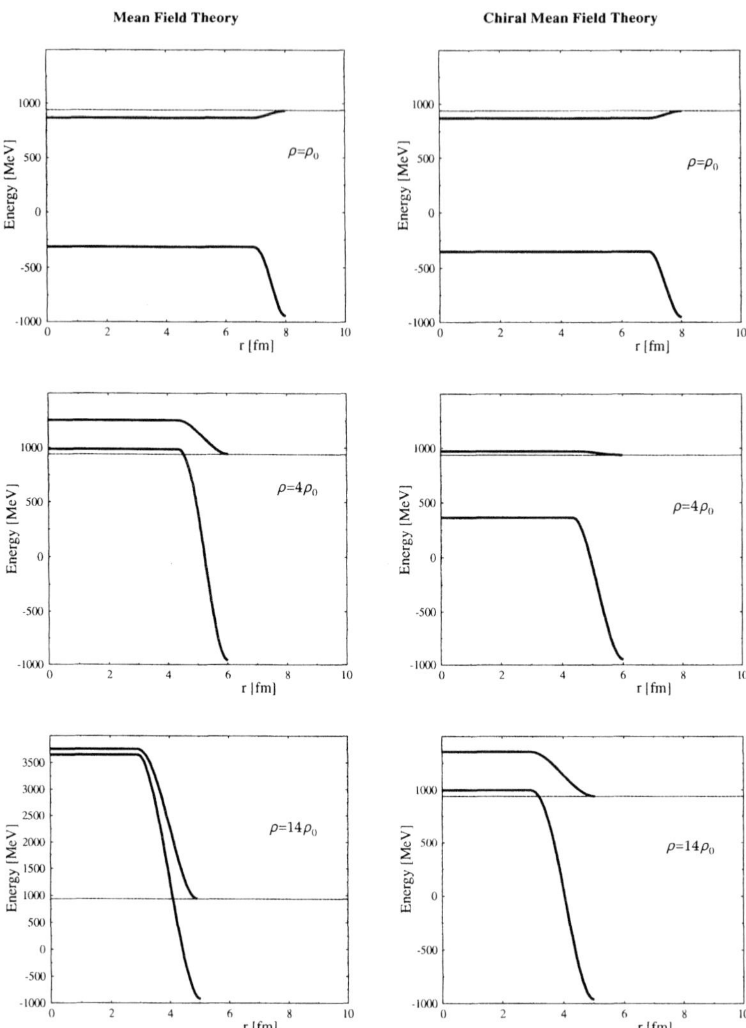

FIGURE 15. The potential structure of the shell model and the vacuum for various primary densities $\rho = \rho_0$, $4\rho_0$, $14\rho_0$. At left the predictions of ordinary Dürr-Teller-Walecka-type theories are shown; at right those for a chirally symmetric meson field theory as develloped by P. Papazoglu, S. Schramm et al. [38, 39]. Note however, that this particular chiral mean–field theory does contain ω^4 terms. If introduced in both effective models, they seem to predict quantitatively similar results.

for the nucleon mass in the chiral model.

Nevertheless, if the parameters in both cases (e.g. $g_s, g_\omega, g_\rho, m_s, b, c$ in the non-chiral case) are adjusted such that ordinary nuclei (binding energies, radii, shell structure,...) and properties of infinite nuclear matter (equilibrium density, compression constant K, binding energy) are well reproduced, the prediction of both effective Lagrangians for the dependence of the properties of the correlated vacuum on density and temperature is remarkably different. This is illustrated to some extend in Fig. 15. Accordingly, the chirally symmetric meson field theory predicts much higher primary densities (and temperatures) until the effects of the correlated vacuum are strong enough so that the mechanisms described here become effective. In other words, according to chirally symmetric meson field theories the antimatter-cluster-production and multi-hypermatter-cluster production out of the highly correlated vacuum takes place at considerably higher heavy ion energies as compared to the predictions of the Dürr-Teller-Walecka-type meson field theories. This in itself is an interesting, quasi-fundamental question to be clarified. Moreover, the question of the nucleonic substructure (form factors, quarks, gluons) and its influence on the highly correlated vacuum structure has to be studied. The nucleons are possibly strongly polarized in the correlated vacuum: the Δ resonance correlations in the vacuum are probably important. Is this highly correlated vacuum state, especially during the compression, a preliminary stage to the quark-gluon cluster plasma? To which extent is it similar or perhaps even identical with it? It is well known for more than 10 years that meson field theories predict a phase transition qualitatively and quantitatively similar to that of the quark-gluon plasma [40] — see Fig. 16.

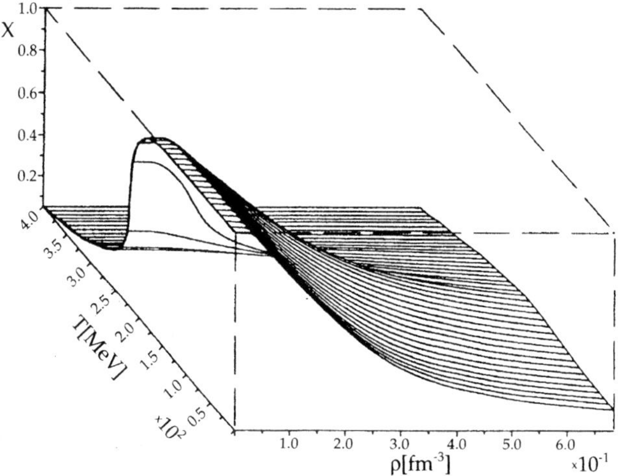

FIGURE 16. The strong phase transition inherent in Dürr-Teller-Walecka-type meson field theories, as predicted by J. Theis et al. [40]. Note that there is a first order transition along the ρ-axis (i.e. with density), but a simple transition along the temperature T-axis. Note also that this is very similar to the phase transition obtained recently from the Nambu-Jona-Lasinio-approximation of QCD [42].

The extension of the periodic system into the sectors hypermatter (strangeness) and antimatter is of general and astrophysical importance. Indeed, microseconds after the big bang the new dimensions of the periodic system we have touched upon, certainly have been populated in the course of the baryo- and nucleo-genesis. Of course, for the creation of the universe, even higher dimensional extensions (charm, bottom, top) come into play, which we did not pursue here. It is an open question, how the depopulation (the decay) of these sectors influences the distribution of elements of our world today. Our conception of the world will certainly gain a lot through the clarification of these questions.

For the Gesellschaft für Schwerionenforschung (GSI), which I helped initiating in the sixties, the questions raised here could point to the way ahead. Working groups have been instructed by the board of directors of GSI, to think about the future of the laboratory. On that occasion, very concrete (almost too concrete) suggestions are discussed — as far as it has been presented to the public. What is necessary, as it seems, is a **vision on a long term basis**. The ideas proposed here, the verification of which will need the **commitment for 2–4 decades of research**, could be such a vision with considerable attraction for the best young physicists. The new dimensions of the periodic system made of hyper- and antimatter cannot be examined in the "stand-by" mode at CERN (Geneva); a dedicated facility is necessary for this field of research, which can in future serve as a home for the universities. The GSI — which has unfortunately become much too self-sufficient — could be such a home for new generations of physicists, who are interested in the **structure of elementary matter**. GSI would then not develop just into a detector laboratory for CERN, and as such become obsolete. I can already see the enthusiasm in the eyes of young scientists, when I unfold these ideas to them — similarly as it was 30 years ago, when the nuclear physicists in the state of Hessen initiated the construction of GSI.

REFERENCES

1. S.G: Nilsson et al. Phys. Lett. 28 B (1969) 458
   Nucl. Phys. A 131 (1969) 1
   Nucl. Phys. A 115 (1968) 545
2. U. Mosel, B. Fink and W. Greiner, Contribution to "Memorandum Hessischer Kernphysiker" Darmstadt, Frankfurt, Marburg (1966).
3. U. Mosel and W. Greiner, Z. f. Physik 217 (1968) 256, 222 (1968) 261
4. a) J. Grumann, U. Mosel, B. Fink and W. Greiner, Z. f. Physik 228 (1969) 371
   b) J. Grumann, Th. Morovic, W. Greiner, Z. f. Naturforschung *26a* (1971) 643
5. W. Greiner, Int. Journal of Modern Physics E, Vol. 5 , No. 1 (1995) 1-90. This review article contains many of the subjects discussed here in an extended version, see also for a more complete list of references.
6. A. Sandulescu, R.K. Gupta, W. Scheid, W. Greiner, Phys. Lett. *60*B (1976) 225
   R.K. Gupta, A. Sandulescu, W. Greiner, Z. f. Naturforschung *32*a (1977) 704
   R.K. Gupta, A.Sandulescu and W. Greiner, Phys. Lett. *64*B (1977) 257
   R.K. Gupta, C. Parrulescu, A. Sandulescu, W. Greiner Z. f. Physik A283 (1977) 217
7. G. M. Ter-Akopian et al., Nucl. Phys. A*255* (1975) 509
   Yu.Ts. Oganessian et al., Nucl. Phys. A*239* (1975) 353 and 157
8. D. Scharnweber, U. Mosel and W. Greiner, Phys. Rev. Lett *24* (1970) 601
   U. Mosel, J. Maruhn and W. Greiner, Phys. Lett. *34*B (1971) 587
9. G. Münzenberg et al. Z. Physik A309 (1992) 89
   S.Hofmann et al. Z. Phys A*350* (1995) 277 and 288
10. R. K. Gupta, A. Sandulescu and Walter Greiner, Z. für Naturforschung *32*a (1977) 704
11. A. Sandulescu and Walter Greiner, Rep. Prog. Phys 55. 1423 (1992); A. Sandulescu, R. K. Gupta, W. Greiner, F. Carstoin and H. Horoi, Int. J. Mod. Phys. E1, 379 (1992)
12. A. Sobiczewski, Phys. of Part. and Nucl. 25, 295 (1994)
13. R. K. Gupta, G. Münzenberg and W. Greiner, J. Phys. G: Nucl. Part. Phys. 23 (1997) L13
14. V. Ninov, K. E. Gregorich, W. Loveland, A. Ghiorso, D. C. Hoffman, D. M. Lee, H. Nitsche, W. J. Swiatecki, U. W. Kirbach, C. A. Laue, J. L. Adams, J. B. Patin, D. A. Shaughnessy, D. A. Strellis

and P. A. Wilk, preprint
15. K. Rutz, M. Bender, T. Bürvenich, T. Schilling, P.-G. Reinhard, J.A. Maruhn, W. Greiner, Phys. Rev. C *56* (1997) 238.
16. B. Fricke and W. Greiner, Physics Lett *30*B (1969) 317
B. Fricke, W. Greiner, J.T. Waber, Theor. Chim. Acta (Berlin) *21* (1971) 235
17. A. Sandulescu, D.N. Poenaru, W. Greiner, Sov. J. Part. Nucl. 11(6) (1980) 528
18. Harold Klein, thesis, Inst. für Theoret. Physik, J.W. Goethe-Univ. Frankfurt a. M. (1992)
Dietmar Schnabel, thesis, Inst. für Theoret. Physik, J.W. Goethe-Univ. Frankfurt a.M. (1992)
19. D. Poenaru, J.A. Maruhn, W. Greiner, M. Ivascu, D. Mazilu and R. Gherghescu, Z. Physik A*328* (1987) 309, Z. Physik A*332* (1989) 291
20. E. K. Hulet, J. F. Wild, R. J. Dougan, R. W.Longheed, J. H. Landrum, A. D. Dougan, M. Schädel, R. L. Hahn, P. A. Baisden, C. M. Henderson, R. J. Dupzyk, K. Sümmerer, G. R. Bethune, Phys. Rev. Lett. *56* (1986) 313
21. K. Depta, W. Greiner, J. Maruhn, H.J. Wang, A. Sandulescu and R. Hermann, Intern. Journal of Modern Phys. A5, No. 20, (1990) 3901
K. Depta, R. Hermann, J.A. Maruhn and W. Greiner, in "Dynamics of Collective Phenomena", ed. P. David, World Scientific, Singapore (1987) 29
S. Cwiok, P. Rozmej, A. Sobiczewski, Z. Patyk, Nucl. Phys. A491 (1989) 281
22. A. Sandulescu and W. Greiner in discussions at Frankfurt with J. Hamilton (1992/1993)
23. J.H. Hamilton, A.V. Ramaya et al. Journ. Phys. G **20** (1994) L85 - L89
24. B. Burggraf, K. Farzin, J. Grabis, Th. Last, E. Manthey, H. P. Trautvetter, C. Rolfs, *Energy Shift of first excited state in* ^{10}Be *?*, accepted for publication in Journ. of. Phys. G
25. P. Hess et al., *Butterfly and Belly Dancer Modes in* $^{96}Sr + {}^{10}Be + {}^{146}Ba$, in preparation
26. E.K. Hulet et al. Phys Rev C *40* (1989) 770.
27. W. Greiner, B. Müller, J. Rafelski, QED of Strong Fields, Springer Verlag, Heidelberg (1985). For a more recent review see W. Greiner, J. Reinhardt, *Supercritical Fields in Heavy–Ion Physics*, Proceedings of the 15th Advanced ICFA Beam Dynamics Workshop on Quantum Aspects of Beam Physics, World Scientific (1998)
28. B. Povh, Rep. Progr. Phys. *39* (1976) 823; Ann. Rev. Nucl. Part. Sci. *28* (1978) 1; Nucl. Phys. A*335* (1980) 233; Progr. Part. Nucl. Phys. *5* (1981) 245; Phys. Blätter *40* (1984) 315
29. J. Schaffner, Carsten Greiner and H. Stöcker Phys. Rev. C*45* (1992) 322; Nucl. Phys. B*24B* (1991) 246; J. Schaffner, C.B. Dover, A. Gal, D.J. Millener, C. Greiner, H. Stöcker: Annals of Physics*235* (1994) 35; C. Greiner and J. Schaffner, *Physics of Strange Matter for Relativistic Heavy-Ion Collision* Int. J. Mod. Phys. E5, 239-300 (1996)
30. W. Scheid and W. Greiner, Ann. Phys. *48* (1968) 493; Z. Phys. *226* (1969) 364
31. W. Scheid, H. Müller and W. Greiner Phys. Rev. Lett. 13 (1974) 741
32. H. Stöcker, W. Greiner and W. Scheid Z. Phys. A 286 (1978) 121
33. I. Mishustin, L.M. Satarov, J. Schaffner, H. Stöcker and W.Greiner
Journal of Physics G (Nuclear and Particle Physics) *19* (1993) 1303
34. P.K. Panda, S.K. Patra, J. Reinhardt, J. Maruhn, H. Stöcker, W. Greiner, Int. J. Mod. Phys. E 6 (1997) 307
35. N. Auerbach, A. S. Goldhaber, M. B. Johnson, L. D. Miller and A. Picklesimer, Phys. Lett. B182 (1986) 221
36. H. Stöcker and W. Greiner, Phys. Rep. 137 (1986) 279.
37. J. Reinhardt and W. Greiner, to be published.
38. P. Papazoglou, D. Zschiesche, S. Schramm, H. Stöcker, W. Greiner, J. Phys. G 23 (1997) 2081; P. Papazoglou, S. Schramm, J. Schaffner-Bielich, H. Stöcker, W. Greiner, Phys. Rev. C 57 (1998) 2576.
39. P. Papazoglou, D. Zschiesche, S. Schramm, J. Schaffner–Bielich, H. Stöcker, W. Greiner, nucl–th/9806087, accepted for publication in Phys. Rev. C.
40. J. Theis, G. Graebner, G. Buchwald, J. Maruhn, W. Greiner, H. Stöcker and J. Polonyi, Phys. Rev. D 28 (1983) 2286
41. P. Papazoglou, PhD thesis, University of Frankfurt, 1998; C. Beckmann et al., in preparation
42. S. Klimt, M. Lutz, W. Weise, Phys. Lett. B*249* (1990) 386.

Chiral Transition and the Scalar and Vector Correlations

Teiji Kunihiro

Yukawa Institute for Theoretical Physics, Kyoto University, Sakyo-ku, Kyoto 606-8502, Japan

Abstract. The properties of the scalar and vector correlations in the hot and/or dense hadronic matter close to chiral transition are discussed. Presuming that the linear realization of chiral symmetry will become appropriate at least near the critical point, we argue that the strength function in the $I = J = 0$ channel will soften near the critical point, and the sigma meson will accordingly become a clearer resonance in hot and/or dense medium than in the free space. It is shown that the steep rise of the baryon-number susceptibility χ_B around the critical point seen in the lattice simulations may suggest that the interactions between quarks in the vector channel become week near the critical point; this implies that the peculiar behavior of the χ_B does not necessarily imply a proliferation of baryons and antibaryons (Skyrmions) which is though a natural scenario of chiral transition in the non-linear realization of chiral symmetry.

THE SIGMA MESON AND THE SOFT MODES

Chiral transition is a phase transition of the QCD vacuum with $\langle\bar{q}q\rangle$ being the order parameter. When exploring a phase transition in any physical system, the study of fluctuations of physical quantities, especially ones related to the order parameter is as important as that of the phase diagram for the system in equilibrium. The fluctuations of observables are also related with dynamical phenomena such as the transport properties of the system.

If a phase transition is of 2nd order or *weak* 1st order, there exist soft modes, which are the fluctuations of the order parameter as is well known in condensed matter physics[1] and nuclear physics[2]. For the chiral transition, the fluctuation of the order parameter $\langle(\bar{q}q)^2\rangle$ is a scalar-isoscalar meson which one calls the σ-meson. The σ meson may become the soft mode of chiral transition at $T \neq 0$ and/or $\rho_B \neq 0$[3, 4]. In fact, the existence of the sigma meson as the quantum fluctuation of the order parameter of the chiral transition is still controversial[5], though the recent active studies on the phase-shift analysis of the π-π scattering in the $I = J = 0$ channel, i.e., the sigma channel, have been confirming the existence of the pole deep in the second Riemann sheet with the real part ranging from 400 MeV to 800 MeV [6, 7]. If this pole is identified with the quantum fluctuation of the amplitude of the chiral condensate, it may imply that the linear realization of chiral symmetry is appropriate in the resonance energy region, although the chiral perturbation theory based on the non-linear chiral Lagrangians[8] works well in the rather low-energy region.

The first half of this report is concerned with the sigma meson and the precursory soft modes for chiral transition in nuclear medium, and is essentially a recapitulation

CP594, *Hadrons and Nuclei: First International Symposium,* edited by Il-T. Cheon et al.

of the talk presented at Tohoku university, a week before of the present workshop; the manuscript for the proceedings is now on the net[9]. Referring to the above proceedings for the detail of the content, here I only give a brief summary of the report and some comments:

(1) The existence of the σ meson as *the quantum fluctuation of the order parameter* of the chiral transition accounts for various phenomena in hadron physics which otherwise remain mysterious[10, 4, 11].

(2) There have been accumulation of experimental evidence of a low-mass pole in the σ channel in the pi-pi scattering matrix[6]. It should be emphasized that for obtaining this result, it is essential to respect chiral symmetry, analyticity and crossing symmetry even in an approximate way as in the N/D method[12].

(3)Partial restoration of chiral symmetry in hot and dense medium leads to an enhancement in the spectral function in the σ channel near the $2m_\pi$ threshold[13]. Such an enhancement has been observed in the reaction

$$\mathrm{A}(\pi^+, (\pi^+\pi^-)_{I=J=0})\mathrm{A}'$$

by CHAOS collaboration[14], which might be an experimental evidence of the partial restoration of chiral symmetry in heavy nuclei[16, 17]: The conventional approach without incorporating a partial restoration of chiral symmetry in the nuclear medium[18] failed to reproduce the CHAOS data.

(4) The spectral enhancement near the $2m_\pi$ threshold in the σ channel is predicted irrespective of the linear and nonlinear realization of chiral symmetry provided that the possible reduction of the quark condensate or f_π is taken into account[19].

(5) One should confirm that the near $2m_\pi$-threshold enhancement observed in the (π^+, $\pi^+\pi^-$) reactions by CHAOS collaboration is surely due to a partial restoration of chiral symmetry in nuclear medium by other means[16]. For this purpose, the strength function in the σ channel in the wider *two-dimensional* (ω, q) plane should be measured. To obtain such strength functions in the (ω, q) plane, various nuclear and electro-magnetic probes as well as heavy-ion collisions should can be utilized; for instance, photo- or electro-production[11] of the σ as well as the production by (d, ^{3}He) and (d, t) reactions are interesting [16, 20].

(6) Such experiments with nuclear targets for exploring the possible restoration of chiral symmetry in nuclear medium will automatically give clearer confirmation of the existence of the σ meson than is done in the free space.

(7) It should be emphasized that in any hadronic medium where the baryon density is finite, there arises a scalar-vector mixing[21, 11, 22],

$$\sigma \leftrightarrow \gamma, \omega,$$

as is familiar in the σ-ω model[23], which may cause a possible softening of the spectral functions at finite three momenta q, even in the vector channel, due to the softening of that in the sigma channel associated with the partial restoration of chiral symmetry in the hadronic medium. Such a softening may reflect in the spectral functions extracted in any experiment to try to see the spectral functions in the vector channel by seeing the lepton pairs from heavy-ion collisions such as CERES/NA45[24], proton-[25], electro- or *gamma*-nucleus reactions and so on.

(8) The formation of σ mesic nuclei by (d, ^{3}He) and (d, t) reactions are proposed as was done to produce the deeply-bound pionic atoms[20]. To identify the σ meson and the spectral function in that channel, detecting $2\pi^0$ and lepton pairs with $q \neq 0$ are interesting[11].

ENHANCEMENT OF THE BARYON-NUMBER SUSCEPTIBILITY AND THE DENSITY FLUCTUATIONS

The discussions in the preceding section put an emphasis on the sigma meson as a quantum fluctuation of the amplitude of the order parameter; this presumed that the linear realization of the chiral symmetry is appropriate at least near the critical point. I personally take it for granted that the linear realization is natural at least in the vicinity of the chiral critical point.

Nevertheless some stick to the nonlinear realization even in the vicinity of the critical point of the chiral transition. If the non-linear realization is appropriate even in the vicinity of the critical point, the chiral restoration might be associated with an anomalous proliferation of the baryons and the anti-baryons as Skyrmions and anti-Skyrmions near the critical point at finite temperature[26], which may account for a steep rise of the baryon-number susceptibility, χ_B [27] obtained in the lattice simulations[28, 29].

It has also been suggested that the vector-realization [30, 31], which is based on the non-linear sigma model with the vector mesons incorporated based on the ansatz of the hidden-local symmetry[32] and similar to the Georgi's vector limit[33], could be realized; then the decrease of the vector meson masses was conjectured[31].

In the second half of the present report, I showed that the baryon-number susceptibility give some information on the *vector* correlations at finite temperature[22], thereby the steep rise of χ_B seen in the lattice simulations may be accounted for without sticking to the non-linear realization of chiral symmetry.

The baryon-number susceptibility χ_B is the measure of the response of the baryon number density

$$\rho_B = \sum_{\beta=1\sim N_f} \rho_i$$

to infinitesimal changes in the quark chemical potentials μ_i [27, 22]:

$$\chi_B(T,\mu) = [\sum_{i=1}^{N_f} \frac{\partial}{\partial \mu_i}](\sum_{i=1}^{N_f} \rho_i) = \langle\langle N_B^2 \rangle\rangle / VT, \qquad (1)$$

where N_B is the baryon-number operator given by $N_B \equiv \sum_{i=1}^{N_f} N_i$, with

$$\rho_i = \mathrm{Tr} N_i \exp[-\beta(H - \sum_{i=u,d} \mu_i N_i)]/V \equiv \langle\langle N_i \rangle\rangle / V \qquad (2)$$

the i-th quark-number density, V the volume of the system and $\beta = 1/T$.

It is readily recognized that χ_B is the density-density correlation which is nothing but the 0-0 component of the vector-vector correlations[22];

$$\chi_B(T,\mu_q) = \beta \int d\mathbf{x} S_{00}(0,\mathbf{x}), \tag{3}$$

where

$$S_{\mu\nu}(t,\mathbf{x}) = \langle\langle j_\mu(t,\mathbf{x}) j_\nu(0,\mathbf{0})\rangle\rangle,$$

with

$$j_\mu(t,\mathbf{x}) = \bar{q}(t,\mathbf{x})\gamma_\mu q(t,\mathbf{x})$$

being the current operator. Using the fluctuation-dissipation theorem, one has

$$\chi_B(T,\mu_q) = -\lim_{k\to 0} L(0,\mathbf{k}), \tag{4}$$

where $L(\omega,\mathbf{k})$ is the longitudinal component of the retarded Green's function or the response function in the vector channel;

$$R_{\mu\nu}(\omega,k) = \text{F.T.}(-i\theta(t)\langle\langle[j_\mu(t,\mathbf{x}),\ j_\nu(0,\mathbf{0})]_-\rangle\rangle).$$

These formula clearly show the relevance of the vector correlations to the baryon-number susceptibility.

I also discussed the density fluctuations around the critical point of the chiral transition at finite temperature T and baryon density ρ_B. We notice that the baryon-number susceptibility at $\rho_B \neq 0$ is related with the (iso-thermal) compressibility of the system [22],

$$\kappa_T \equiv -N_B^{-1}(\partial V/\partial\mu)_{T,N_B} = \frac{\chi_B}{\rho^2}, \tag{5}$$

which tells that if χ_B is large and so is the density fluctuation, the system is easy to compress. One can then see that the stronger interaction in the vector channel suppress χ_B.

This part was largely based on a previous report by myself[34], so I only give a brief summary of this part here, referring to [34] for the details:

(1) The baryon-number susceptibility χ_B as an observable which reflects the confinement-deconfinement and the chiral phase transitions in hot and/or dense hadronic matter[27].

(2) The suppression of χ_B at low temperatures and steep rise around the critical temperature as shown in the lattice QCD may be roughly attributed to the confinement-deconfinement transition[29]. Nevertheless such a behavior of χ_B is also affected by the chiral transition[22].

(3) Noting that χ_B is a measure of the rate of the density fluctuation in the system, one can see that the chiral transition at finite chemical potential especially leads to an interesting phenomenological consequence to χ_B. When the vector coupling is small, the chiral transition at low temperatures is of first order in the density direction[35, 36, 37]. This implies a divergent behavior of χ_B, accordingly a huge density fluctuations[34].

(4) A large enhancement of the fluctuation can be also expected for the scalar density fluctuations due to the scalar-vector mixing at finite density mentioned above[22, 34]. Such a large enhancement may leads to an enhancement of the sigma-meson production due to the scalar-vector mixing at finite density. The above phenomena all have relevance to experiments to be done in RHIC and LHC[38, 39].

(5) The nature of the chiral transition as to the first order or not etc is sensitively dependent on the strength of the vector coupling[35]. An analysis of the lattice data suggests that the vector coupling is small in comparison with the scalar coupling at high temperature[22, 40].

(6) The susceptibility χ_B is nothing but the generalized susceptibility $\chi(\omega,k)$ at $\omega = k = 0$. One should examine $\chi(\omega,k)$ in the whole region of ω and k to get more information about the vector correlations and the density fluctuations theoretically and experimentally.

ACKNOWLEDGMENTS

I thank the organizers of this symposium for inviting me to the symposium which was organized to celebrate the retirement of Prof. I.-T. Cheon from Yonsei University. I was very much pleased that I had a chance to present a talk at such a memorable symposium. Some part of this report is based on the works done in collaboration with T. Hatsuda, D. Jido and H. Shimizu, to whom I am grateful. This work is partially supported by the Grants-in-Aid of the Japanese Ministry of Education, Science and Culture (No. 12640263 and 12640296).

REFERENCES

1. See e.g., P. W. Anderson, *Basic Notion of Condensed Matter Physics* (Benjamin, California, 1984).
2. See e.g., P. Ring and P. Schuck, *The Nuclear Many-Body Problem*, p. 450, (Springer-Verlag 1980); see also, T. Kunihiro, Prog. Theor. Phys. **65**, 1098 (1981), where it is suggested that spin-dependent isovector giant resonances with higher angular momenta could be precursory soft modes for the pion condensation in finite nuclei.
3. T. Hatsuda and T. Kunihiro , Phys. Lett. **B145**, 7 (1984); Prog. Theor. Phys. **74**, 765 (1985); Phys. Rev. Lett. **55**, 158 (1985); Phys. Lett. **B185**, 304 (1987).
4. T. Hatsuda and T. Kunihiro ,Phys. Rep. **247**, 221(1994).
5. See for example, T. Kunihiro, hep-ph/0009116, contained in [6].
6. The proceedings of Workshop at Yukawa Institute for Theoretical Physics, "Possible Existence of the σ-meson and Its Implications to Hadron Physics", KEK Proceedings 2000-4, (December, 2000), ed. by S. Ishida et al.
7. N. A. Törnqvist and M. Roos, Phys. Rev. Lett. **76**, 1575 (1996); M. Harada, F. Sannino and J. Schechter, Phys. Rev. **D54**, 1991 (1996); S. Ishida et al., Prog. Theor. Phys. **98**, 1005 (1997); J. A. Oller, E. Oset and J. R. Peláez, Phys. Rev. Lett. **80**, 3452 (1998); K. Igi and K. Hikasa, Phys. Rev. **D59**, 034005 (1999).
8. J. Gasser and H. Leutwyler, Nucl. Phys. **B250**, 465 (1985).
9. T. Kunihiro, the proceedings for LNS International Workshop on Physics with GeV electrons and γ-rays, Feb. 13 - 15, 2001, nucl-th/0103056;
see also [11].
10. V. Elias and M. D. Scadron, Phys. Rev. Lett. **53**, 1129 (1984).

11. T. Kunihiro, Prog. Theor. Phys. Supple. **120**, 75 (1995); nucl-th/0006035; nucl-th/9604019; hep-ph/9905262.
12. K. Igi and K. Hikasa, Phys. Rev. **D59**, 034005 (1999);
J. A. Oller, E. Oset and J. R. Peláez, Phys. Rev. Lett. **80**, 3452 (1998).
13. S. Chiku and T. Hatsuda, Phys. Rev. **D58**, 076001 (1998);
M. K. Volkov, E. A. Kuraev, D. Blaschke, G. Roepke and S. M. Schmidt, Phys. Lett. **B424**, 235 (1998).
14. F. Bonutti et al. (CHAOS Collaboration), Phys. Rev. Lett. **77**, 603 (1996); Nucl. Phys. **A677**, (2000), 213. The experminet was motivated to explore the possible strong pi-pi correlations in the nuclear medium[15].
15. P. Schuck, W. Norönberg and G. Chanfray, Z. Phys. **A330**, 119(1988); G. Chanfray, Z. Aouissat, P. Schuck and W. Nörenberg, Phys. Lett. **B256**, 325 (1991).
16. T. Hatsuda, T. Kunihiro and H. Shimizu, Phys. Rev. Lett. **82**, 2840 (1999).
17. Z. Aouissat, G. Chanfray, P. Schuck and J. Wambach, Phys. Rev. **C61**, 012202 (2000); D. Davesne, Y. J. Zhang and G. Chanfray, Phys. Rev. **C62**, 024604(2000).
18. R. Rapp et al, Phys. Rev. **C 59**, R1237 (1999); M. J. Vicente-Vacas and E. Oset, Phys. Rev. **C60**, 064621 (1999).
19. D. Jido , T. Hatsuda and T. Kunihiro, Phys.Rev. **D63**,011901 (2000).
20. S. Hirenzaki, T. Hatsuda, K. Kume, T. Kunihiro, H. Nagahiro, Y. Okumura, E. Oset, A. Ramos, H. Toki, Y.Umemoto Nucl.Phys. **A663**, 553 (2000).
21. H. A. Weldon, Phys. Lett. **B274**, 133 (1992)
22. T. Kunihiro, Phys. Lett. **B271**, 395 (1991).
23. B. D. Serot and J. D. Walecka, Adv. Nucl. Phys. **16**, 1 (1986).
24. CERES Collaboration, Phys. Rev. Lett. **75**, 1272 (1995); Phys. Lett. **B422**, 405 (1998); Nucl. Phys. **A638**, 159c, (1998).
25. K. Ozawa et al, nucl-ex/0011013.
26. C. DeTar, Phys.Rev. **D42**,224 (1990)
27. L. McLerran, Phys. Rev. **D36**, 3291(1987).
28. S. Gottlieb, W. Liu, D. Toussaint, R. L. Renken and R. L. Sugar, Phys. Rev. Lett. **59**, 2247 (1987); Phys. Rev. **D38**, 2888 (1988); R. V. Gavai, J. Potvin and S. Sanielevici, Phys. Rev. **D40**, 2743(1989); see [29] for recent lattice data.
29. C. DeTar, in *Quark Gluon Plasma 2* ed. by R.C. Hwa (World Scientific 1995).
30. M. Harada and K. Yamawaki, Phys. Rev. Lett. **86**, 757 (2001).
31. G.E. Brown and M. Rho, nucl-th/0101015, hep-ph/0103102.
32. M. Bando, T. Kugo, S. Uehara, K. Yamawaki and T. Yanagida, Phys. Rev. Lett. **54**, 1215 (1985); M. Bando, T. Kugo and K. Yamawaki, Phys. Rept. **164**,217 (1988).
33. H. Georgi, Phys.Rev.Lett. **63**, 1917 (1989); Nucl.Phys. **B331**, 311 (1990).
34. T. Kunihiro, the proceedings for International Symposium on Quantum Chromodynamics (QCD) and Color Confinement (Confinement 2000), Osaka, Japan, 7-10 Mar 2000, ed. by H. Suganuma, M. Fukushima and H. Toki, hep-ph/0007173.
35. Y. Asakawa and K. Yazaki, Nucl. Phys. **A504**, 668 (1989); M. Lutz, S. Klimt and W. Weise, Nucl. Phys. **A542**, 521 (1992) and references cited therein. See also, T.M. Schwarz, S.P. Klevansky, G. Papp, Phys.Rev.**C60**, 055205 (1999);
T. Kunihiro, Nucl. Phys. **B351**, 593 (1991).
36. T. Schaefer and E. Shuryak, Rev. Mod. Phys. **70**, 323 (1998), and the references cited therein.
37. As a review, J.I.M. Verbaaschot and T. Wettig, hep-ph/0003017, to be published in Ann. Rev. Nucl. Part. Sci.
38. K. Kumagai, O. Miyamura and T. Sugitate, the proceedings of the International Mini-symposium on High-Energy Heavy-Ion Reactions, 2-4 March, 1992, Hiroshima Univ., ed. by Y. Sumi.
39. M. Stephanov, K. Rajagopal and E. Shuryak, Phys. Rev. Lett. **81**, 4816 (1998);
G. Baym and H. Heiselberg, Phys. Lett.**B 469**, 7(1999);
S. Gavin, nucl-th/9908070.
40. G. Boyd, S. Gupta, F. Katsch and E. Laermann, hep-lat/9405006.

Signals of deconfinement? Strangeness and flow in heavy ion collisions

M. Bleicher, K. Paech, H. Weber, H. Stöcker, W. Greiner

Institut für Theoretische Physik, Goethe Universität, Frankfurt am Main, Germany

Abstract. Strangeness and flow in heavy ion collisions are discussed.

INTRODUCTION

One of the major goals of the relativistic heavy ion collider (RHIC) at Brookhaven National Laboratory is to explore the phase diagram of hot and dense matter near the quark gluon plasma (QGP) phase transition. The QGP is a state in which the individual hadrons dissolve into a gas of free (or almost free) quarks and gluons in strongly compressed and hot matter (for recent reviews on the topic, we refer to [1, 2]). The achievable energy- and baryon densities sensitively depend on the extend to which the nuclei are stopped during penetration; they also depend on mass number and bombarding energy.

Earlier RHIC estimates have been performed assuming boost-invariant hydrodynamics [3, 4, 5, 6, 7] and pQCD (Regge theory) motivated model [8, 9]: baryons are concentrated at projectile and target rapidity separated by a large region which is baryon free (in position and momentum space), i.e. the nuclei are transparent. The region between them is filled by the color fields which materialize, developing a plateau in the mesons' rapidity distribution. This scenario is supported experimentally for pp- and p$\bar{\text{p}}$-collisions at collider energies. It is the aim of the present work to examine whether this remains true also for the collision of large nuclei. From lower energy nucleus-nucleus collisions we know that only a small fraction of the total number of collisions takes place at the full incident energy while most of them take place at much lower energies. In fact, transport model studies show a fair amount of stopping at the RHIC energy with strong transverse expansion [10, 11] indicating that the collision of two nuclei is more than just the superposition of "A$\times$A" nucleon collisions at the same energy (i.e. that secondary interactions are very important at all investigated energies).

Similar to the RQMD model [10, 13], UrQMD is a microscopic transp ort approach based on the covariant propagation of constituent quarks and diquarks accompanied by mesonic and baryonic degrees of freedom. It simulates multiple interactions of ingoing and newly produced particles, the excitation and fragmentation of color strings and the formation and decay of hadronic resonances. At RHIC energies, the treatment of subhadronic degrees of freedom is of major importance. In the UrQMD model, these degrees of freedom enter via the introduction of a formation time for hadrons produced in the fragmentation of strings [14, 15, 16]. The leading hadrons of the fragmenting

CP594, *Hadrons and Nuclei: First International Symposium,* edited by Il-T. Cheon et al.

strings contain the valence-quarks of the original excited hadron. In UrQMD they are allowed to interact even during their formation time, with a reduced cross section defined by the additive quark model, thus accounting for the original valence quarks contained in that hadron [12]. Those leading hadrons therefore represent a simplified picture of the leading (di)quarks of the fragmenting string. Newly produced (di)quarks do, in the present model, not interact until they have coalesced into hadrons – however, they contribute to the energy density of the system. A more advanced treatment of the partonic degrees of freedom during the formation time ought to include soft and hard parton scattering [8] and the explicit time-dependence of the color interaction between the expanding quantum wave-packets [17]. However, such an improved treatment of the internal hadron dynamics has not been implemented for light quarks into the present model. For further details about the UrQMD model, the reader is referred to Ref. [12].

The UrQMD model has been applied successfully to explore heavy ion reactions from AGS energies ($E_{lab} = 1 - 10$ AGeV) up to the full CERN-SPS energy ($E_{lab} =$ 160 AGeV). This includes detailed studies of thermalization [18], particle abundancies and spectra [19], strangeness production [20], photonic and leptonic probes [21] , J/Ψ's [22] and event-by-event fluctuations [23].

The increasing importance of perturbative QCD effects (hard scattering)[8, 9, 24] and coherent parton dynamics [25] has lead to the speculations that transport models with string dynamics will fail to describe heavy ion collisions above a certain center of mass energy. Indeed, todays transport models are based on a probabilistic phase space approach, even in the earliest stage of the reaction. In this stage at RHIC energies, the protons and neutrons of the colliding nuclei should be described by coherent parton wave functions and should be modelled as such [25]. However, after initial parton or string production has taken place in the first 0.5 fm/c [26], this coherence is lost and the UrQMD ansatz may be applicable.

It is not known a priori at RHIC energies, whether pQCD effects (presumably taking place at the early stage of the collision, $t \sim 1$ fm/c) or the hadronic rescatterings dominate the evolution of the system and the hadronic spectra measured by the experiments after freeze-out. Models like UrQMD [12] or RQMD [13] can help to identify in the observables signals from different (early/late) stages of the collision dynamics.

FLOW AT RHIC

The exploration of the transverse collective flow is the earliest predicted observable to probe heated and compressed nuclear matter [27]. Its sensitivity to the equation of state (EoS) might be used to search for abnormal matter states and phase transitions [28, 29, 30].

Until now, the study of directed and anisotropic flow in high energy nuclear collisions is attracting large attention from both experimentalists and theorists [31, 32, 33, 34, 35]. Flow in general is sensitive to the equation of state [31, 32, 36, 37, 38] which governs the evolution of the system created in the nuclear collision. Elliptical flow [36, 39, 40, 41, 42, 43, 44, 45, 46] (i.e. squeeze-out except for a reversed sign in the observable) is especially sensitive to the early time scales of the reaction. It might serve as a keyhole to the (non-

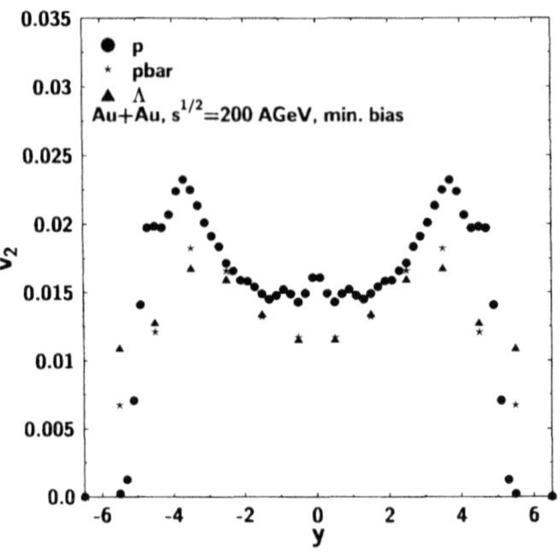

FIGURE 1. Elliptic flow parameter v_2 of protons, lambdas and anti-protons as a function of rapidity in Au+Au reactions at $\sqrt{s} = 200$ AGeV, min.bias.

)equilibrium dynamics of the strongly interacting matter even before hadronization.

Figs. 1 and 2 shows the anisotropic flow of nucleons, lambdas, anti-protons, pions and kaons as a function of rapidity for minimum biased Au+Au collisions at the full RHIC energy ($\sqrt{s} = 200$ AGeV). The elliptic flow of all inspected hadrons shows a prominent dip at central rapidities. This indicates a region of low 'pressure' (or small interaction strength, to be more specific). Comparing the elliptic flow parameters of baryons (Fig. 1) with those of the mesons (Fig. 2) shows that the mesons acquire 2/3 of the baryon elliptic flow. This scaling with the geometrical quark model cross section may indicate a connection between elliptic flow and in-medium cross section of the hadrons as will be discussed below.

A clear maximum of the elliptic flow in all particle species is observed for semi-peripheral collision. Our speculation about the formation of transverse flow in those ultra-relativistic collision is also supported by the transverse momentum dependence of v_2 (in min. bias Au+Au reactions) as depicted in Fig. 3: A strong increase of the ellipticity parameter with p_T is predicted which signals the existence of radial expansion.

The appearance of a dip in $v_2(y)$ might appear to be counter-intuitive. However, it points towards a distinct feature of the model dynamics in the early stage, namely the pre-equilibrium string dynamics and interactions on the parton level. Fig. 4 shows that the v_2 parameter is closely related to the formation time of particles in the string picture. A standard default setting of formation time results in an average formation time (in the string rest frame) of 1 fm/c. Consequently, the particles created in the initial stage of the collision are not allowed to interact during this time. With particle velocities near the speed of light, these particles do have a mean free path on the order of 1 fm at

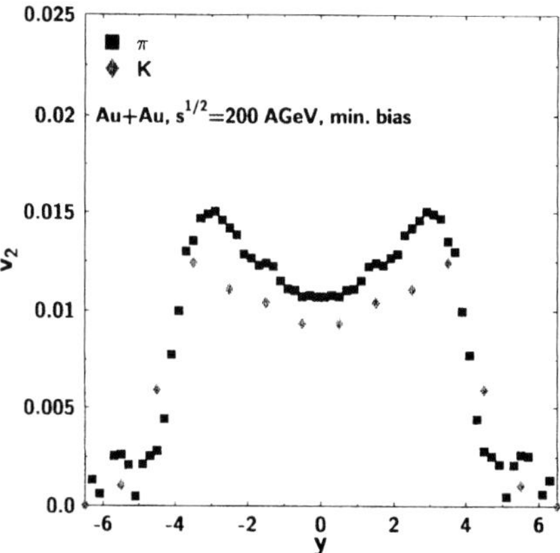

FIGURE 2. Elliptic flow parameter v_2 of kaons and pions as a function of rapidity in Au+Au reactions at $\sqrt{s} = 200$ AGeV, min.bias.

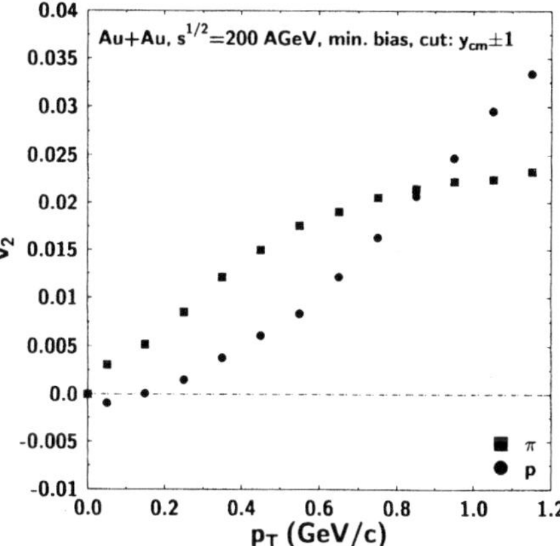

FIGURE 3. Elliptic flow parameter v_2 at midrapidity ($|y| \leq 1$) as a function of transverse momentum in Au+Au reactions at $\sqrt{s} = 200$ AGeV, min.bias.

midrapidity. If the formation time - and therefore the mean free path - is decreased, the elliptic flow parameter v_2 increases. In the limit of a vanishing mean free path (hydro limit) the elliptic flow in the present model becomes maximal and it is quantitatively consistent with hydrodynamical predictions [47]. In contrast, increasing the formation

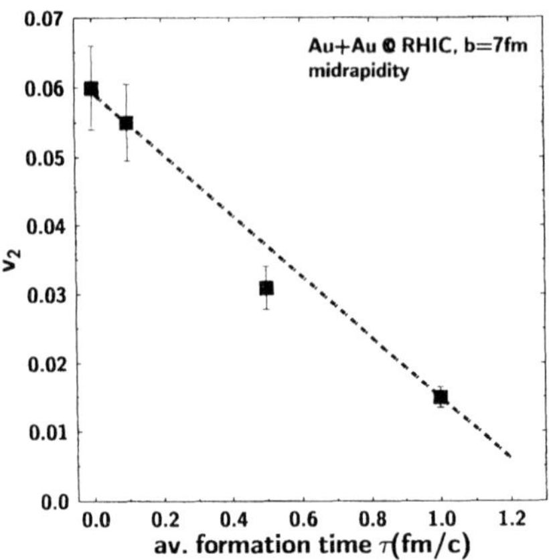

FIGURE 4. Relation between the elliptic flow parameter v_2 at midrapidity and the mean free path (formation time) of the particles in Au+Au reactions at $\sqrt{s} = 200$ AGeV, $b = 7$ fm.

time (mean free path) results in a vanishing v_2, in line with the limit given by Hijing calculations without quenching [48].

Thus, the strength of the anisotropic flow of pions is directly connected to the mean free path of the particles (partons, hadrons) forming the hot midrapidity region. The measurement of v_2 might therefore yield valuable information about the transport properties of QCD-matter, like the interaction frequencies and the viscosity of the excited partonic and hadronic matter at RHIC energies. Especially Ω baryons, with their small hadronic cross section, are supposed to measure QGP properties without any additional disturbance due to the hadronic phase.

STRANGENESS ENHANCEMENT AT RHIC

Strange particles, especially multi-strange baryons (which have more than one strange quark) carry vital information about the collision dynamics [49, 50, 51, 52, 53, 54, 55, 56]. Relative strangeness abundancies have been proposed as a powerful tool for searching the transition from hadronic matter to partonic matter in high energy nuclear collisions [49, 50, 51]. Indeed, strangeness enhancement has be observed in heavy ion collisions at all collision energies with different colliding systems.

A number of different mechanisms are under debate to understand this strong increase in strangeness production with centrality and beam energy:

- Equilibrated (gluon rich) plasma phase[49].
- Baryon-junctions[64].
- Diquark breaking and sea-diquarks[65].

• Strong color fields[67, 69, 70, 71, 72].

Two different scenarios will be explored in order to study the speculation of a string tension increase in heavy ion collision at RHIC energies: UrQMD calculations with the standard color flux tube break-up mechanism (i.e. a string tension $\kappa = 1$ GeV/fm) will be contrasted by UrQMD simulations with an in-medium κ increased to 3 GeV/fm. According to the implemented Schwinger mechanism for the string fragmentation (m denoting the ((di-)quark masses)

$$\gamma_X = \frac{P(X\overline{X})}{P(q\bar{q})} = \exp\left(-\frac{\pi(m_X^2 - m_q^2)}{\kappa}\right) \quad , \tag{1}$$

this results in an enhancement of the strangeness and diquark production probabilities to $\gamma_s = 0.72$ and $\gamma_{qq} = 0.46$ (for $\kappa = 3$ GeV/fm) compared to $\gamma_s = 0.37$ and $\gamma_{qq} = 0.093$ for $\kappa = 1$ GeV/fm. Note that a decrease of the (di-)quark masses due to chiral symmetry restoration may lead to a similar enhancement as the $\kappa = 3$ GeV/fm scenario.

Fig. 5 depicts the particle rapidity distributions (a, b) and ratios from Au+Au (b = 2 fm) to p+p collisions (c, d) at the RHIC full energy $\sqrt{s} = 200$AGeV. The symbols denote: $\Omega^- + \overline{\Omega^-}$ (full circles), $\Xi^- + \overline{\Xi^-}$ (full triangles), $\Sigma^0 + \overline{\Sigma^0}$ (reversed triangle), $\Lambda^0 + \overline{\Lambda^0}$ (full square) and anti-protons (open circles). (a) shows the rapidity distributions with string tension $\kappa = 1$GeV/fm; (b) shows the same as (a) with $\kappa = 3$GeV/fm. (c) depicts the ratio $R(y) = \frac{\frac{dN}{dy}^{\mathrm{Au}}(y)/A_{\mathrm{part}}}{\frac{dN}{dy}^{\mathrm{pp}}(y)/2}$ of the particle yields per participant from central AA collisions over that of the p+p collisions with $\kappa = 1$GeV/fm; (d) same as (c), however with $\kappa = 3$GeV/fm for Au+Au.

A strong enhancement of the (anti-)hyperon production compared to scaled pp interactions is predicted (cf. (c)). This enhancement is purely due to rescattering between constituent (di-)quarks and hadrons in the medium. If strong color fields, similar to Pb+Pb at SPS [71, 72], are present, the model predicts a dramatic enhancement in the multi-strange hadron production, up to a factor of 100 (in case of the Omega) at midrapidity.

The enhancement of the hyperon yields is strongly dependent on the centrality of the events as shown in Fig. 6: (a) shows the particle yields with string tension $\kappa = 1$GeV/fm, while (b) is same as (a) with $\kappa = 3$GeV/fm. Fig. 6(c) shows the ratio $R = \frac{N^{\mathrm{Au}}(A_{\mathrm{part}})/A_{\mathrm{part}}}{N^{\mathrm{pp}}/2}$ of the particle yields per participant[1] from central AA collisions over that of the inelastic p+p collisions with $\kappa = 1$GeV/fm. (d) depicts the same as (c) with $\kappa = 3$GeV/fm for Au+Au collisions. The symbols are: $\Omega^- + \overline{\Omega^-}$ (full circles), $\Xi^- + \overline{\Xi^-}$ (full triangles), $\Sigma^0 + \overline{\Sigma^0}$ (reversed triangle), $\Lambda^0 + \overline{\Lambda^0}$ (full square) and anti-protons (open circles).

The 4π yields and ratios as a function of number of participants from Au+Au at the RHIC full energy $\sqrt{s} = 200$AGeV increase rapidly with centrality (a, b) by two orders of magnitude when going from peripheral ($A_{part} \approx 25$) to central collision. The

---

[1] The number of participating nucleons in this paper is defined as: $A_{\mathrm{part}} = A_1 + A_2 - \Sigma$(Nucleons with $p_T \leq 270$ MeV). This prescription yields a reasonable parametrization of the experimental data on A_{part}.

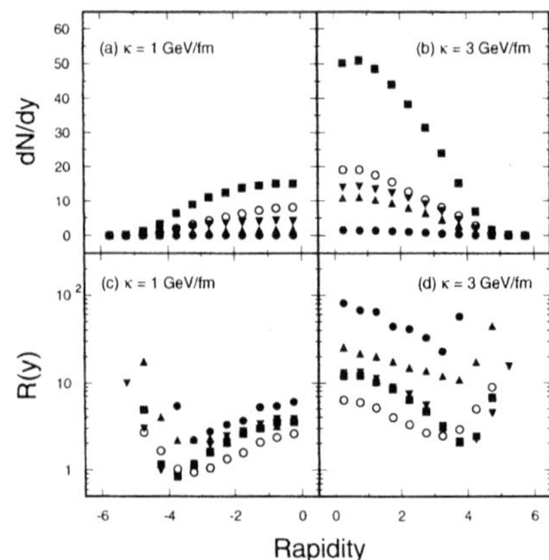

FIGURE 5. Particle rapidity distributions and ratios from Au+Au (b = 2 fm) and p+p collisions at the RHIC full energy $\sqrt{s} = 200$AGeV. (a) Rapidity distributions with string energy density $\kappa = 1\mathrm{GeV/fm}^3$; (b) same as (a) with $\kappa = 3\mathrm{GeV/fm}^3$; (c) ratio $R(y) = \frac{\frac{dN}{dy}^{\mathrm{Au}}(y)/A_{\mathrm{part}}}{\frac{dN}{dy}^{\mathrm{pp}}(y)/2}$ of the particle yields per participant from central nuclear collisions over that of the p+p collisions with $\kappa = 1$GeV/fm; (d) same as (c) with $\kappa = 3$GeV/fm. $\Omega^- + \overline{\Omega^-}$ (full circles), $\Xi^- + \overline{\Xi^-}$ (full triangles), $\Sigma^0 + \overline{\Sigma^0}$ (reversed triangles), $\Lambda^0 + \overline{\Lambda^0}$ (full squares) and anti-protons (open circles).

enhancement itself is most pronounced when scaled with the number of participating nucleons A_{part} and compared to p+p collisions at the RHIC full energy (c, d).

In conclusion, the UrQMD model has been applied to Au+Au reactions at RHIC energies. This model treats the dynamics of the hot and dense system by constituent (di-)quark and hadronic degrees of freedom. (Anti-)Hyperon yields in central Au+Au collisions at the full RHIC energy ($\sqrt{s} = 200$ AGeV) strongly enhanced in comparison to inelastic pp interactions at $\sqrt{s} = 200$ GeV or peripheral Au+Au collisions. This enhancement grows dramatically with the strangeness content of the hyperon. The present model predicts that strangeness enhancement occurs as a threshold effect already at rather small number of participants (≈ 25) due to rescattering. Increasing the string tension, similar to SPS energies (or reducing the effective masses of the constituent quarks to the current quark mass values yields large additional enhancement, which grows with the strangeness content even stronger: Λ's are enhancement by a factor of 7, while Ω's are enhanced by a factor of 60 compared to pp.

ACKNOWLEDGMENTS

This research used resources of the National Energy Research Scientific Computing Center (NERSC). Fruitful discussions with Drs. N. Xu, R. Snellings and P. Huovinen

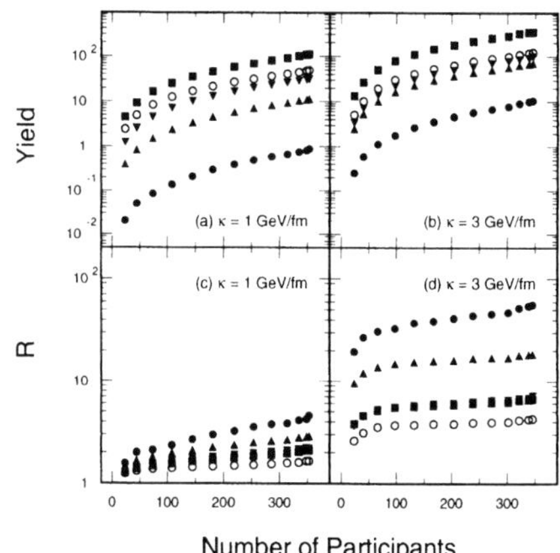

FIGURE 6. Particle 4π yields and ratios as a function of number of participants from Au+Au and inelastic p+p collisions at the RHIC full energy $\sqrt{s} = 200$AGeV. (a) yields with string energy density $\kappa = 1\text{GeV/fm}^3$; (b) same as (a) with $\kappa = 3\text{GeV/fm}^3$; (c) ratio $R = \frac{N^{Au}(A_{\text{part}})/A_{\text{part}}}{N^{pp}/2}$ of the particle yields per participant from central AA collisions over that of the p+p collisions with $\kappa = 1\text{GeV/fm}^3$; (d) same as (c) with $\kappa = 3\text{GeV/fm}^3$ for the Au+Au interactions. $\Omega^- + \overline{\Omega^-}$ (full circles), $\Xi^- + \overline{\Xi^-}$ (full triangles), $\Sigma^0 + \overline{\Sigma^0}$ (reversed triangles), $\Lambda^0 + \overline{\Lambda^0}$ (full squares) and anti-protons (open circles).

are gratefully acknowledged. This work is supported by GSI, DFG, BMBF.

REFERENCES

1. S.A. Bass, M. Gyulassy, H. Stöcker and W. Greiner, J. Phys. **G25** (1999) R1.
2. J. W. Harris, B. Müller, Ann. Rev. Nucl. Part. Sci. 46 (1996) 71
3. J.D. Bjorken, Phys. Rev. **D27** (1983) 140.
   K. Kajantie and L. McLerran, Nucl. Phys. **B214** (1983) 261.
   K.J. Eskola and M. Gyulassy, Phys. Rev. **C47** (1993) 2329.
4. R. B. Clare and D. Strottman, Phys. Reports **141** (1986) 179.
   N. S. Amelin, L. P. Csernai, E. F. Staubo and D. Strottman, Nucl. Phys. **A544** (1992) 463c.
   L.P. Csernai, J.I. Kapusta, G. Kluge and E.E. Zabrodin, Z. Phys. **C58** (1993) 453.
5. D.H. Rischke and M. Gyulassy, Nucl. Phys. **A597** (1996) 701, and **A608** (1996) 479.
   S. Bernard, D.H. Rischke, J.A. Maruhn, W. Greiner, Nucl. Phys. **A625** (1997) 473.
6. A. Dumitru, D. Rischke, Phys. Rev. C 59 (1999) 354.
7. C.M. Hung and E. Shuryak, Phys. Rev. **C57** (1998) 1891.
8. K. Geiger, Phys. Rev. **D46** (1992) 4965 and Phys. Rev. **D46** (1992) 4986.
   K. Geiger, Phys. Rev. **D47** (1993) 133.
   K. Geiger, Phys. Rept. 258 (1995) 237.
9. X. N. Wang, Phys. Rev. **D46** (1992) 1900.
   X. N. Wang, Phys. Rep. 280 (1997) 287.
10. T. Schönfeld, H. Stöcker, W. Greiner, H. Sorge, Mod. Phys. Lett. **A8** (1993) 2631,
    H. Stöcker, M. Hofmann, R. Mattiello, N.S. Amelin, A. Dumitru, A. Jahns, A. von Keitz, Y. Puersuen,

T. Schönfeld, C. Spieles, L.A. Winckelmann, J.A. Maruhn, W. Greiner, M. Berenguer, H. Sorge, Nucl. Phys. **A566** (1994) 15c.
11. B. Monreal *et al.*, Phys. Rev. **C60** (1999) 031901
12. M. Bleicher, E. Zabrodin, C. Spieles, S.A. Bass, C. Ernst, S. Soff, L. Bravina, M. Belkacem, H. Weber, H. Stöcker, W. Greiner, J. Phys. G 25 (1999) 1859;
S.A. Bass, M. Belkacem, M. Bleicher, M. Brandstetter, L. Bravina, C. Ernst, L. Gerland, M. Hofmann, S. Hofmann, J. Konopka, G. Mao, L. Neise, S. Soff, C. Spieles, H. Weber, L.A. Winckelmann, H. Stöcker, W. Greiner, C. Hartnack, J. Aichelin, N. Amelin,
Progr. Nucl. Phys. 41 (1998) 225
13. H. Sorge, H. Stocker and W. Greiner, Annals Phys. **192** (1989) 266.;
H. Sorge, Phys. Rev. C52 (1995) 3291
14. B. Andersson, G. Gustavson, and B. Nilsson-Almquist, Nucl. Phys. **B281**, 289 (1987).
15. B. Andersson *et al.*, Comp. Phys. Comm. **43**, 387 (1987).
16. T. Sjoestrand, Comp. Phys. Comm. **82**, 74 (1994).
17. L. Gerland, L. Frankfurt, M. Strikman, H. Stöcker, and W. Greiner, Phys. Rev. Lett. **81**, 762 (1998).
18. L.V. Bravina, M.I. Gorenstein, M. Belkacem, S.A. Bass, M. Bleicher, M. Brandstetter, M. Hofmann, S. Soff, C. Spieles, H. Weber, H. Stocker, W. Greiner, Phys. Lett. B434, 379 (1998) and Phys. Rev. C 60, 024904 (1999)
19. S.A. Bass, M. Belkacem, M. Brandstetter, M. Bleicher, L. Gerland, J. Konopka, L. Neise, C. Spieles, S. Soff, H. Weber, H. Stöcker, W. Greiner, Phys. Rev. Lett. **81** (1998) 4092;
M. Bleicher, C. Spieles, C. Ernst, L. Gerland, S. Soff, H. Stöcker, W. Greiner, S.A. Bass, Phys. Lett. B 447 (1999) 227
20. S. Soff, S.A. Bass, M. Bleicher, L. Bravina, E. Zabrodin, H. Stöcker, W. Greiner, nucl-th/9907026, Phys. Lett. B in print
21. C. Spieles, L. Gerland, M. Bleicher, S.A. Bass, W. Greiner, C. Lourenco, R. Vogt, Eur. Phys. J. C 5 (1998) 349;
C. Ernst, S.A. Bass, S. Soff, H. Stöcker and W. Greiner, nucl-th/9907118.
22. C. Spieles, R. Vogt, L. Gerland, S.A. Bass, M. Bleicher, H. Stöcker, W. Greiner, Phys. Rev. C 60 (1999) 054901
23. M. Bleicher, M. Belkacem, C. Ernst, H. Weber, L. Gerland, C. Spieles, S.A. Bass, H. Stöcker, W. Greiner, Phys. Lett. B 435 (1998) 9;
M. Bleicher, L. Gerland, C. Spieles, A. Dumitru, S. Bass, M. Belkacem, M. Brandstetter, C. Ernst, L. Neise, S. Soff, H. Weber, H. Stöcker, W. Greiner, Nucl. Phys. **A638** (1998) 391.
24. K. Werner, Phys. Rep. 232, 87 (1993)
25. L. McLerran and R. Venugopalan, Phys. Rev. **D49**, 2233 and 3352 (1994)
26. K. Eskola and X.N. Wang, Phys. Rev. **D49**, 1284 (1994)
27. W. Scheid, R. Ligensa and W. Greiner, Phys. Rev. Lett. **21**, 1479 (1968) and Phys. Rev. Lett. **32**, 741 (1974).
28. J. Hofmann et al., Phys. Rev. Lett. **36**, 88 (1976).
29. H. Stöcker and W. Greiner, Phys. Rep. **137**, 277 (1986).
30. H. Stocker, J. A. Maruhn and W. Greiner, ergy Heavy Ion Collisions," Z. Phys. **A290** (1979) 297.
31. L. P. Csernai, W. Greiner, H. Stocker, I. Tanihata, S. Nagamiya and J. Knoll, nergetic Collisions Of Heavy Nuclei," Phys. Rev. **C25** (1982) 2482.
32. W. Reisdorf and H.G. Ritter, Annu. Rev. Nucl. Part. Sci. **47**, 663 (1997); J.-Y. Ollitrault, Nucl. Phys. **A638**, 195c (1998); see also Quark Matter 99 proceedings.
33. E877 Collaboration, J. Barrette *et al.*, Phys. Rev. Lett. **73**, 2532 (1 994).
34. E895 Collaboration, H. Liu *et al.*, Nucl. Phys. **A638**, 451c (1998).
35. NA49 Collaboration, H. Appelshäuser *et al.*, Phys. Rev. Lett. **80**, 4 136 (1998).
36. H. Sorge, Phys. Rev. Lett. **82**, 2048 (1999).
37. S. Soff *et al.*, Phys. Rev. C **51**, 3320 (1995).
38. L. Bravina *et al.*, Phys. Lett. B **344**, 49 (1995).
39. J.-Y. Ollitrault, Phys. Rev. D **46**, 229 (1992).
40. C.M. Hung and E.V. Shuryak, Phys. Rev. Lett. **75**, 4003 (1995).
41. D.H. Rischke, Nucl. Phys. **A610**, 88c (1996).
42. H. Sorge, Phys. Rev. Lett. **78**, 2309 (1997).
43. H. Heiselberg and A.-M. Levy, Phys. Rev. C **59**, 2716 (1999).

44. L.P. Csernai and D. Röhrich, Phys. Lett. B **458**, 454 (1999).
45. J. Brachmann *et al.*, nucl-th/9908010.
46. B. Zhang, M. Gyulassy, C.M. Ko, nucl-th/9902016.
47. P. Huovinen, priv. comm.;
P. F. Kolb, J. Sollfrank and U. Heinz, Phys. Lett. **B459** (1999) 667 [nucl-th/9906003].
48. R. Snellings, priv. comm.
49. J. Rafelski, B. Müller, Phys. Rev. Lett. **48**, 1066 (1982); (E) **56** 2334 (1986).
50. P. Koch, B. Müller, J. Rafelski, Phys. Rep. **142**, 167 (1986).
51. P. Koch, B. Müller, H. Stöcker, W. Greiner, Mod. Phys. Lett. **A3**, 737 (1988).
52. P. Senger, H. Ströbele, Journal of Physics **G25**, R59 (1999).
53. Strangeness in Quark Matter (Padua, Italy, 1998), Journal of Physics **G25**, 143 (1999).
Strangeness in Quark Matter (Santorini, Greece, 1997), Journal of Physics **G23**, 1785 (1997).
Relativistic Aspects of Nuclear Physics (Rio, Brazil, 1995), T. Kodama et al., eds., World Scientific.
Strangeness in Hadronic Matter (Budapest, Hungary, 1996), Budapest, Akademiai Kiado.
Strangeness in Hadronic Matter (Tucson, AZ, 1995), AIP Conf. **340**.
Strange Quark Matter in Physics and Astrophysics (Aarhus, Denmark, 1991), Nucl. Phys. B **24B**, (1991).
54. R. Stock, Phys. Lett. **B456**, 277 (1999).
55. J. Rafelski, J. Letessier, A. Tounsi, Acta Phys. Pol. **B27**, 1037 (1996).
56. W. Cassing, E. L. Bratkovskaya, Phys. Rep. **308** , 65 (1999).
J. Geiss, W. Cassing, C. Greiner, Nucl. Phys. A **644**, 107 (1998).
57. E. Andersen et al. (WA97 collaboration), Phys. Lett **B433**, 209 (1998).
58. R. Lietava et al. (WA97 collaboration), Journal of Physics **G25**, 181 (1999).
59. R. Caliandro et al. (WA97 collaboration), Journal of Physics **G25**, 171 (1999).
60. S. Margetis et al. (NA49 collaboration), Journal of Physics **G25**, 189 (1999).
61. F. Gabler et al. (NA49 collaboration), Journal of Physics **G25**, 199 (1999).
62. D. Evans et al. (WA85 and WA94 collaborations), Journal of Physics **G25**, 209 (1999).
63. E. Shuryak, Phys. Rev. Lett. 68 (1992) 3270
64. S. E. Vance and M. Gyulassy, Phys. Rev. Lett. **83**, 1735 (1999) [nucl-th/9901009].
65. A. Capella and C. A. Salgado, Phys. Rev. **C60** (1999) 054906 [hep-ph/9903414].
66. J. Schwinger, Phys. Rev. **82**, 664 (1951).
67. T. S. Biro, H. B. Nielsen, J. Knoll, Nucl. Phys. **B245**, 449 (1984).
J. Knoll, Z. Phys. **C38**, 187 (1988).
68. H. Sorge, M. Berenguer, H. Stöcker, W. Greiner, Phys. Lett. **B289**, 6 (1992).
N. S. Amelin, M. A. Braun, C. Pajares, Phys. Lett. **B306**, 312 (1993).
H. Sorge, Nucl. Phys. **A630**, 522 (1998) and refs. therein.
69. M. Gyulassy, Quark Gluon Plasma, Advanced Series on Directions in High Energy Physics, Vol. 6, edited by R. C. Hwa, World Scientific, Singapore, 1990.
70. L. Gerland, C. Spieles, M. Bleicher, P. Papazoglou, J. Brachmann, A. Dumitru, H. Stöcker, W. Greiner, J. Schaffner, C. Greiner, Proc. of the 4th International Workshop, Relativistic Aspects of Nuclear Physics, Rio, Brazil, (1995), T. Kodama et al., eds., 437.
71. S. Soff, S. A. Bass, M. Bleicher, L. Bravina, E. Zabrodin, H. Stocker and W. Gre iner, tter?," Phys. Lett. **B471**, 89 (1999) [nucl-th/9907026].
72. M. Bleicher, M. Belkacem, S. A. Bass, S. Soff and H. Stocker, quark gluon matter?," Phys. Lett. **B485**, 133 (2000) [hep-ph/0004045].

Strangeness Production and Propagation in Relativistic Heavy Ion Collisions at SIS Energies

N. Herrmann

Physikalisches Institut der Universität Heidelberg, Heidelberg, Germany

Abstract. Strangeness production at incident energies close to the elementary production threshold is considered a tool to extract information about the properties of the hot and dense state of nuclear matter produced as a transient state in heavy ion collisions. The production yields and azimuthal anisotropies of strange particles available for nucleus nucleus collisions from the KaoS and FOPI experiments at the SIS accelerator of GSI, Darmstadt, are reviewed with respect to their significance of possible in-medium modifications and their sensitivity to the nuclear equation of state.

INTRODUCTION

Heavy ion collisions in the beam energy range 1-2 AGeV offer interesting possibilities to study the bulk properties of hot and dense hadronic matter. At these bombarding energies the density of the nuclear system created during the reaction is expected to reach up to three times normal nuclear matter density and the temperature is below 100 MeV [1, 2, 3]. Under those conditions the fireball consists of a composite system of strongly interacting nucleons, their resonances and produced mesons. Strange particles are an interesting probe of the properties of the transient high density state of the reaction especially at threshold energies, e.g. incident beam energies where the first chance nucleon nucleon collisions are not energetic enough to produce the strange particles. The energy necessary for the production can be accumulated in the nucleon resonances or the production could proceed via meson nucleon collisions. The number of produced mesons is thus determined by the number of collisions throughout the collision which in turn is sensitively dependent on the equation-of-state (EOS) of nuclear matter[4].

Recent theoretical developments suggest, however, that in the hot and dense nuclear medium additional effects might come in: strongly interacting particles might change their masses due to a partial restoration of the chiral symmetry of QCD [5, 6, 7, 8, 9, 10] or due to many body scattering effects with the surrounding medium [11]. These so called in-medium effects are visible in several ways: i) The masses of the various hadrons are changed and therefore the production yield is influenced. The mass of K^+ mesons is predicted to rise with the nuclear density, while the mass of K^- mesons is substantially dropping. K^+ (K^-) yields are therefore dropping (rising) with respect to the vacuum situation. ii) Once produced the propagation of the hadrons through the surrounding hadronic matter is modified due to the interaction. A rising (dropping) mass corresponds to a repulsive (attractive) potential that modifies the azimuthal correlation of strange particles with the main flow axis as defined by the nucleons.

To establish the fundamental mechanisms systematic studies have been performed

CP594, *Hadrons and Nuclei: First International Symposium,* edited by Il-T. Cheon et al.

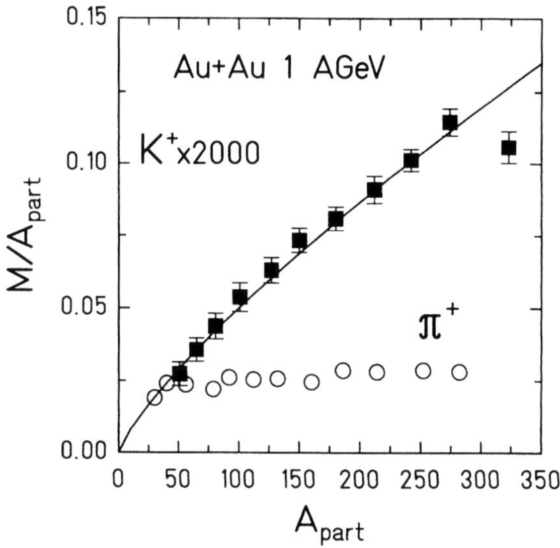

FIGURE 1. Kaon and pion multiplicities per participating nucleon as measured by the KaoS Collaboration [13]

during the last decade. In this respect, the study of sub-threshold or close-to-threshold kaon production (*i.e.* at SIS energies) offers several advantages: (i) K^+ and K^- show according to theoretical predictions opposite mass shifts. (ii) Having the same rest masses their phase space distribution due to kinematics is the same. (iii) Carrying the strangeness quantum number their yield is not changed due to the strong interaction. (iv) Heavy ion collisions at SIS energies exhibit strong collective flow effects that can be exploited to study the Kaon propagation through the hot and dense nuclear medium (see section).

The theoretical scenarios of dropping and/or melting hadron masses have attracted much interest lately since they are also used to describe the low mass dilepton enhancement observed by the CERES and HELIOS-3 collaborations at CERN SPS [12]. Resolving the question of in-medium modifications of hadron properties is in addition a prerequisite for the determination of the nuclear matter equation-of-state, a goal that is pending since 20 years.

In the following the results on the production and propagation of strange particles obtained with the KaoS and FOPI detector at SIS(GSI) are described. In the next section production yields are presented, followed by a discussion of the phase space distributions. After the presentation of the current status on flow observables of strange particles, finally the implications of the findings to the nuclear matter equation-of-state (EOS) will be discussed.

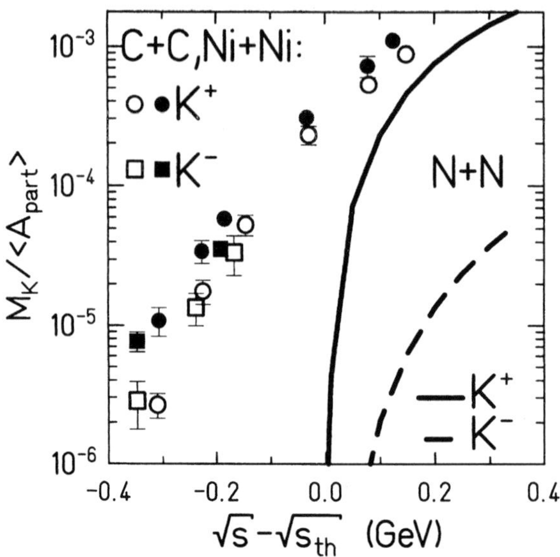

FIGURE 2. Kaon Multiplicities per participating nucleon for the systems C+C and Ni+Ni as function of the Q-value [14].

KAON YIELDS

Kaons are produced during the high density phase of the reactions. At subthreshold energies this property is most directly visible by the centrality dependence of K^+ production as shown in fig.1 [13]. While π^+ mesons are produced proportionally to the number of participating nucleons (A_{part}), i.e. proportionally to the volume of the fireball, kaons exhibit a stronger dependence on A_{part}. This indicates that more than one nucleon nucleon collision contributes to the production of one kaon. This can be achieved by multiple collisions or via intermediate resonances that are strongly enhanced during the high density phase of the reaction. In agreement with theoretical ana-lyses of transport model calculations kaons are therefore messengers of the highly compressed phase of the heavy ion collisions.

In order to investigate possible in-medium effects caused by the density of the hadronic system in fig.2 the production of K^+ and K^- in nucleus nucleus collisions in compared to proton-proton collisions [14]. To make the available data comparable the multiplicities of the strange mesons are normalized to the average number of participating nucleons and plotted as function of the excess energy (i.e. the Q-value) of the reaction. This is necessary since the threshold energy of K^- production in nucleon nucleon collisions of 2.5 GeV is much larger than the corresponding threshold for K^+ production (1.5 GeV) due to the different channels ($NNK^+ K^-$ or $NK^+ \Lambda(\Sigma)$,respectively). As compared to the nucleon-nucleon collision obtained from proton-proton data by averaging over the isospin antikaons are produced much more abundantly in nucleus-nucleus collisions. While their yield is suppressed by 1-2 orders of magnitude in the elementary reactions with respect to the K^+ production in nucleus-nucleus collisions the difference

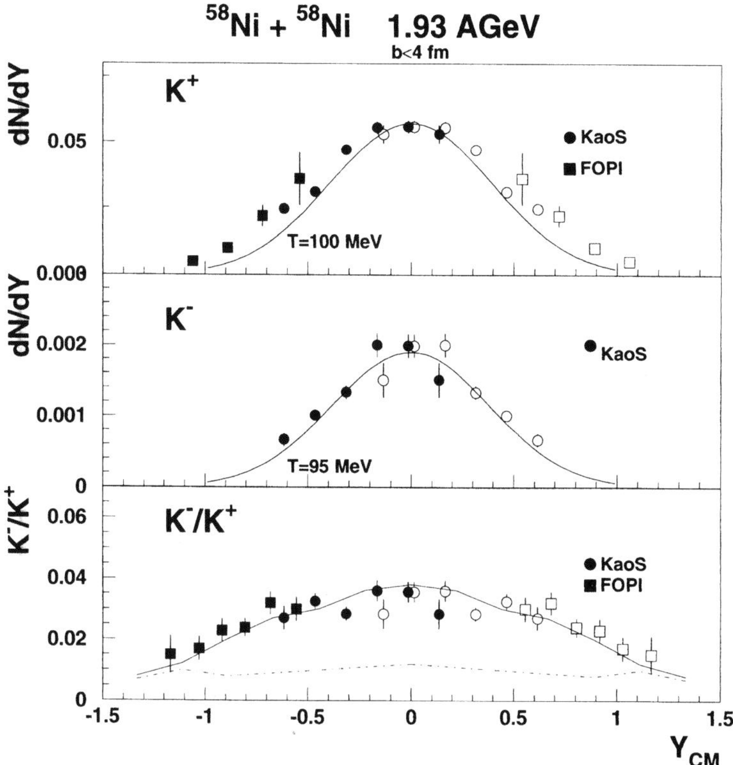

FIGURE 3. Rapidity density distributions for kaons and K^-/K^+ - ratiofor the system Ni+Ni at 1.93 AGeV combining data from the FOPI [16, 19] and KaoS [15] experiments.

in the production probability disappears. The enhancement of the K^- yield strongly supports the ideas of in-medium modifications of kaon masses.

KAON PHASE SPACE DISTRIBUTIONS

The measured longitudinal and transverse kaon phase space distributions can be used to extrapolate to the full solid angle. The extrapolation procedure is described in detail in [19, 20] and makes use of the close to thermal shape of the transverse mass spectra at the various rapidities.

In fig.3 the available data are plotted for the system Ni+Ni at 1.93 AGeV that allow for a comparison of the two different experiments, FOPI and KaoS. The extrapolated data are shown as function of the normalized rapidity $y_{CM} = y - \frac{1}{2} \cdot y_{proj}$ such that $y_{CM} = 0(0.89)$ corresponds to midrapidity (projectile rapidity). In the overlap region the data from both experiments are consistent despite their slightly different centrality selection. Both K^+ and K^- distributions are consistent with the expectations of an equilibrated thermal source, i.e. an gaussian shape of the rapidity density distribution. The parameters, however, to describe the width of the distribution are different for K^+ and K^-.

The lower panel in fig.3 presents the ratio of antikaon to kaon production. The observed different shapes of the rapidity distributions is highlighted in the ratio and is in contrast to the expectations for a fully equilibrated hadronic fireball scenario. Fig.3 also includes the K^- / K^+ ratio as obtained from RBUU calculation [21] (the same trends is obtained in [22]). Two different scenario are shown: Kaon production and phase space distributions with free (vacuum) cross sections (dashed line) and within a dropping mass scenario (solid line). Without in-medium effects the ratio is found to be almost independent of rapidity. Since the production thresholds favor the K^- production at center-of-mass rapidities the uniformity in the model distribution is non-trivial and is understood being due to the absorption of slow moving K^- mesons [23]. Oppositely, with in-medium effects, a pronounced maximum appears at mid-rapidity. Within the model assumptions of dropping K^- masses and slightly rising K^+ masses caused by the action of scalar and vector fields the phenomenon can be explained as the consequence of the interplay of three effects : *i)* the yield of K^- is enhanced more than the one of K^+ mesons is reduced, *ii)* this is the most pronounced at mid-rapidity where, due to the high baryon density, K^- are produced more abundantly, and *iii)* the effective kaon potential tends to push K^+ away from nucleons (*i.e.* to large rapidities) and to attract K^- towards nucleons (*i.e.* to small rapidities). The latter effect is also seen in the transverse momentum dependence of the K^-/K^+ - ratio[16].

SIDEFLOW OF STRANGE PARTICLES

Collective flow effects of various ejectiles emerging from a heavy ion reaction are of great interest since they are known as a sensitive probe of compression of nuclear matter [3, 17]. The dominating effect is the deflection of the incoming nucleons in the reaction plane * Quantitatively sideflow can be studied by a systematic expansion of azimuthal distributions with respect to the reaction plane [25, 26, 17] that allows to correct for the finite resolution due to finite particle number.

In this section the sideward flow not of baryons but of produced strange particles is investigated. K^+ and Λ flows were first predicted to be sensitive to their in-medium potential [27]. It should be noted that also the out-of-plane flow, as measured in [29] shows sensitivity to to in-medium potentials.

The reaction plane was experimentally reconstructed according to the method proposed in [24], with an accuracy of about 40° depending slightly on beam energy, system size and centrality.

---

* The reaction plane of a collision is defined by the beam axis and the impact parameter vector.

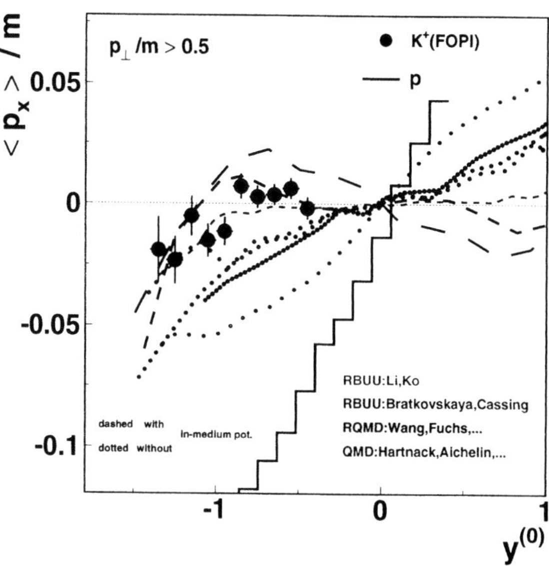

FIGURE 4. Sideflow of K^+ in the reaction Ni+Ni at 1.93 AGeV. The mean transverse momentum projected onto the reaction plane is shown for K^+ (circles) and protons (solid histogram) versus the normalized rapidity and compared to two classes of model calculations with (dashed lines) and without (dotted lines) in-medium potentials.

K^+ flow

Data on kaon sideflow were first published in [28] for the system Ni+Ni at 1.93AGeV and are updated in fig.4 with all the statistics available at the present date. The trend observed in [28], the vanishing sideflow signal of K^+ mesons, is reproduced with much smaller statistical error bars. With this improved statistical significance the discrepancies to some theoretical model calculations are much more clearly visible now.

As can be seen from the dotted lines fig.4 representing the predictions of various model calculations making use of vacuum kaon properties, all the models predict a finite kaon flow. Including medium modifications of the K^+ meson mass due to the presence of vector and scalar fields modifies the flow pattern and brings the theoretical model calculations close to the observed data. Although the magnitude of the predicted effect is slightly different from one model to another, the experimental trend clearly supports an in-medium potential scenario for K^+ thus implying the existence of a repulsive K^+-nucleon mean field. However this statement has to be taken carefully since *i)* an alternative description of the data was proposed in [33] invoking rescattering effects instead of in-medium kaon potentials, *ii)* the sensitivity of K^+ sideward flow to in-medium effects was found in [32] to be washed-out when a momentum dependence of the potential is included in the calculations, and *iii)* it was recently pointed-out in [33] that the lifetime of nuclear resonances (Δ's) used in the models is partially responsible for the magnitude of the K^+ sideward flow as it is a crucial ingredient for kaon production channels.

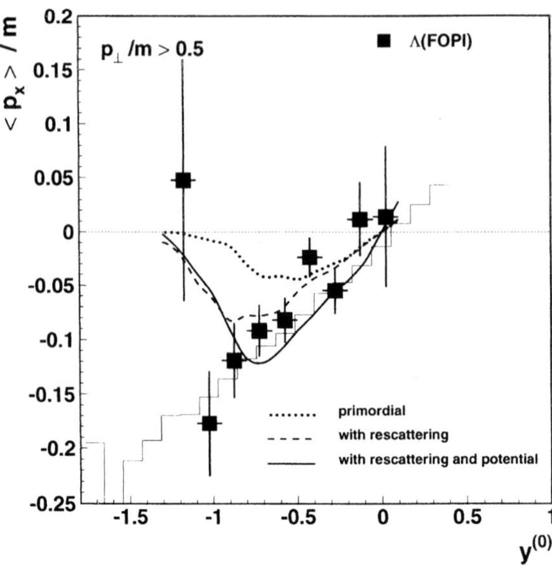

FIGURE 5. Sideflow of Λ in the reaction Ni+Ni at 1.93AGeV. Data are shown by squares for Λ baryons and by the solid histogram for protons. A p_t cut of $p_t/m > 0.5GeV/c$ was applied. Smooth lines represent the primordial (dotted), rescattered (dashed) and full (solid) distributions from the transport model calculations of G.Q.Li [34].

Λ flow

Experimentally an independent observation that Kaons experience the presence of the surrounding nuclear medium can be derived from the comparison to Λ flow distributions. The current status of the FOPI data for the system Ni+Ni is shown in fig.5. Λ baryons exhibit a very similar flow pattern as observed for protons. Since they are produced in an associate mode with the K^+ mesons the medium has to act differently on the K^+ meson and the Λ baryons in order to account for the final state differences. Indeed a theoretical analysis of the Λ interaction gives rise to an attractive potential. As can be seen from fig.5 the transport model calculation [34] show two contributions to the Λ flow: (i) the elastic scattering with the nucleon and (ii) the attractive Λ potential. The data are presently not good enough to make any quantitative statement about the relative contribution of both effects. Further analysis aiming on improving the statistics and studying the transverse momentum dependence of the Λ signal are presently undertaken.

Transverse momentum dependence of K^+ flow

With the statistics available by now more detailed analysis of the kaon flow patterns can be done. One interesting aspect is the systematic expansion of the azimuthal distribution in terms of Fourier coefficients [25, 26]. This method offers the advantage that

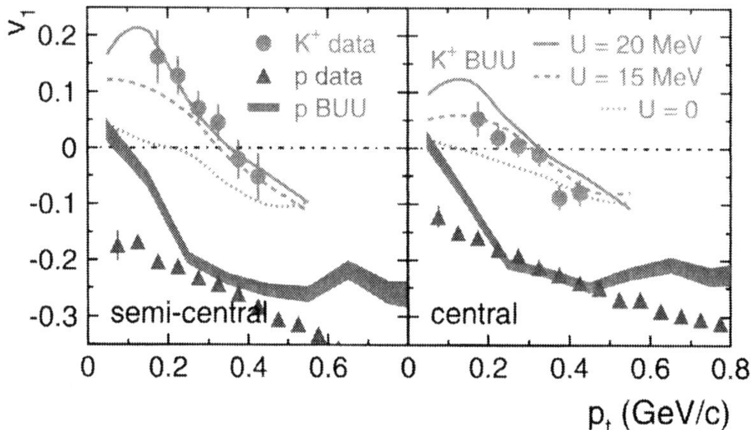

FIGURE 6. Kaon Sideflow in the reaction Ru+Ru at 1.69 AGeV. On the left hand side average transverse momenta projected onto the reaction plane are shown for Kaons (circles) and protons (triangles) versus the normalized rapidiy. The p_t dependence of the dipole fourier coefficient v_1 is shown on the right hand side. The upper (lower) figure is obtained for peripheral (central) collisions, respectively. Model predictions of the Giessen group [21] are plotted with (solid) and without (dashed) medium modifications.

the smearing due to a finite resolution of the reaction plane can be quantitatively taken into account. The second parameter influencing the strength of flow signals is the impact parameter that can be much better controlled with heavier system mass.

The first data along these lines are shown in fig.6 were the the first (dipole) fourier coefficients are shown for Ru+Ru collisions [30]. The different panels represent different centrality selections. The event samples correspond to average geometrical impact parameters of $4.5 fm$ (left side) and $2.5 fm$ (right side), respectively. Several important observations can be made from the figure: The vanishing sideflow of kaons is composed of low momentum kaons that flow opposite to the baryon direction and high momentum ones that are emitted into the direction of the baryon flow. The strength of the v_1 pattern changes with the centrality, resulting in a net Kaon antiflow for peripheral Ru+Ru collisions and a vanishing flow signal in central collisions. Systematic uncertainties do not allow, presently to fix the absolute scale to better than 20%, but the difference between central and peripheral collisions is unaffected by the systematic uncertainty.

For kaons the observations are in astonishingly good agreement with the prediction by RBUU calculations of the Giessen group [21]. The different lines in the figure correspond to kaon potentials at normal nuclear matter densities of 0., 15 and 20MeV and demonstrate the sensitivity of the differential flow observable to in-medium effects. Extending the comparison to protons, however, disturbs the picture. Although the integrated sideflow is reproduced nicely the p_t dependence of the dipole coefficient is completely missed. Before finally concluding on the interpretation of sideflow in terms of in-medium effects, a consistent description of the baryon data and the kaons must be achieved.

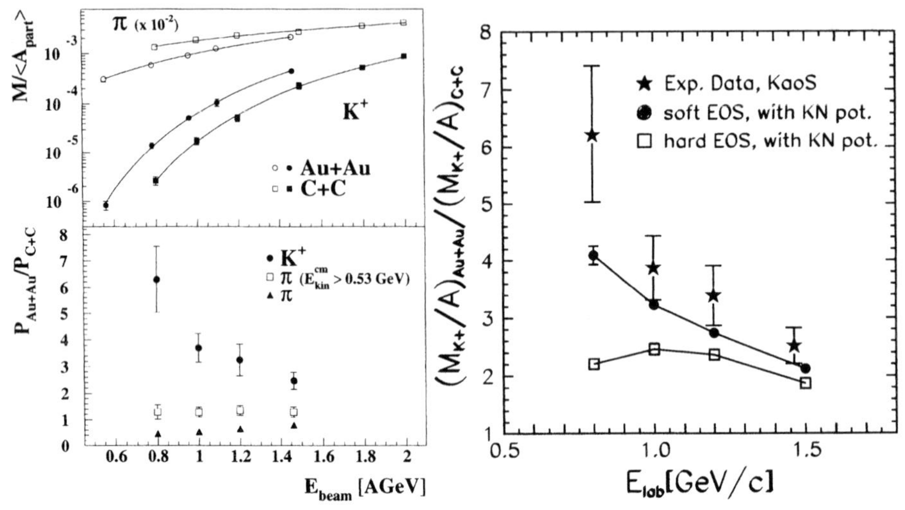

FIGURE 7. K^+ production from Au+Au compared to C+C [31]

IMPLICATIONS ON THE NUCLEAR EQUATION-OF-STATE

K^+ yields do not only depend on the in-medium mass of the potential but also probe the nuclear EOS [4]. A final conclusion about the magnitude of both effects can only be done after disentangling the different contributions. The problem is addressed in fig.7 were the K^+ production for a heavy and light collision system is compared as function of the beam energy [31]. As can be seen from the right side panel of the fig.7 K^+ are much more abundantly produced in heavy with respect to the light system. With larger densities produced in the heavier system, the increase of the effective mass is overcompensated by the additional collisions that occur at high densities. Therefore this ratio is sensitive to the EOS as well. Following the QMD calculations of [35] there is a clear distinction from a hard (K=380MeV) to a soft (K=200MeV) EOS. Before finally concluding about the nuclear EOS consistency is required, however, also for the dynamical flow observables. Even the baryons are found to be incompatible with the proposed Ansatz for the EOS in the QMD model [36]. Nevertheless, the available data provide the basis for reassessing the various contributions from medium effect, compressibility and momentum dependence of the interaction.

SUMMARY

Several observables of strange particles, e.g. production yields, phase space population and azimuthal asymmetries, available in heavy ion collisions at incident energies of 1.5 - 2 AGeV show an intriguing behavior that can be qualitatively reproduced by the concept of varying kaon masses with density. Even more dramatic effects are expected in the K^- phase space distributions, especially for the sideflow observables. The question whether the observations can be linked to low energy QCD properties is a challenging question that will be addressed in more detail by the next generation of experiments with the FOPI detector. To enhance the capabilities with respect to strange particles the experiment is currently being upgraded by a speed up of the data acquisition system and by a replacement of the TOF barrel. The final system should allow for kaon identification up to momenta of 0.9 GeV/c and event rates of 200 Hz.

Acknowledgments

Contributions from all members of the FOPI collaboration and of J.Aichelin, E.Bratkovskaya, W.Cassing, C.Fuchs, C.Hartnack and G.Q.Li are gratefully acknowledged.

REFERENCES

1. R. Stock, Phys.Rep. **135**, 259 (1986)
2. H.H. Gutbrod, A.M. Poskanzer, H.G. Ritter, Rep.Prog.Phys. **52**, 1267 (1989)
3. W. Reisdorf and H.G. Ritter, Annu. Rev. Nucl. Part. Sci. 47, 663 (1997)
4. J. Aichelin and C. M. Ko, PRL **55**, 2661 (1985)
5. D.B. Kaplan and A.E. Nelson, Phys.Lett. **B 175**, 57 (1986)
6. S.Klimt et al., Phys. Lett. B 249, 386 (1990)
7. G.E.Brown, M.Rho , PRL **66**, 2720 (1991)
8. J. Schaffner et al., Phys. Lett **B 334**, 268 (1994)
9. T. Hatsuda, S.H.Lee, Phys.Rev. **C46**, R34 (1992)
10. G.Q.Li, C.M. Ko, Phys. Lett. **B338**, 118 (1994)
11. M.Lutz, Phys. Lett. **B426**, 12 (1998)
12. J. Wambach, Nucl.Phys.**A638**, 171 (1998)
13. D. Miskowiec et al., KaoS Collaboration, PRL **72**, 3650 (1994)
14. F. Laue et al., KaoS Collaboration, PRL **82** 1640, (1999)
15. M. Menzel et al., KaoS Collaboration, Phys. Lett. **B495** 26, (2000)
16. K. Wisniewski et al., FOPI Collaboration, Euro. Phys. J. **A9**,515 (2000)
17. N. Herrmann, J. P. Wessels, T. Wienold, Ann. Rev. Nucl. Part. Sci. **49**, 581 (1999)
18. J. Ritmann et al., FOPI collaboration, Nucl. Pbys. **B44**, 708 (1995)
19. D. Best et al., FOPI Collaboration, Nucl.Phys. **A622**, 573 (1997)
20. B. Hong et al., FOPI Collaboration, PHys. Rev. **C57**, 244 (1998)
21. E.L. Bratkovskaya et al., Nucl. Phys. **A 622**, 593 (1997)
22. G.Q. Li and C.M. Ko, Nucl. Phys. **A 594**, 460 (1995)
23. G.Q. Li and G.E. Brown, Nucl. Phys. **A 636**,487 (1998)
24. P. Danielewicz and G. Odyniec, Phys. Lett. B 157 (1985) 146
25. S. Voloshin, Y. Zhang, *hep-ph/9407082*
26. J.Y. Ollitrault, *nucl-ex/9711003 v2*
27. G.Q.Li et al.: PRL **74**, 235 (1995)
28. J. Ritmann et al., FOPI collaboration, Z. Phys. **A 352**, 355 (1995)
29. Y. Shin et al., KaoS Collaboration, PRL **81** 1576, (1998)
30. P. Crochet et al., FOPI collaboration, Phys. Lett. **486**, 6 (2000)
31. C. Sturm et al., KaoS Collaboration, PRL **86**, 39 (2001)
32. C. Fuchs et al., Phys.Lett. **B434**, 245 (1998)
33. C. David et al., Nucl.Phys. A650, 358 (1999)
34. G.Q. Li, C.M. Ko, Phys. Rev. **C 54**, 1897 (1996)
35. C. Fuchs et al., PRL **86**, 1974 (2001)
36. P. Crochet et al., FOPI collaboration, Nucl. Phys. **628**, 687 (1997)

Conservation Laws and Particle Production in Heavy Ion Collisions

K. Redlich[†], J. Cleymans[**], H. Oeschler[‡] and A. Tounsi[§]

[*]Gesellschaft für Schwerionenforchung, D-64291 Darmstadt, Germany
[†]Institute of Theoretical Physics, University of Wroclaw, PL-50204 Wroclaw, Poland
[**]Department of Physics, University of Cape Town, Rondebosch 7701, Cape Town, South Africa
[‡]Institut für Kernphysik, Darmstadt University of Technology, D-64289 Darmstadt, Germany
[§]Laboratoire de Physique Théorique et Hautes Energies, Université Paris 7, F–75251 Cedex 05

Abstract. We discusse the role of the conservation laws related with U(1) internal symmetry group in the statistical model description of particle productions in ultrarelativistic heavy ion collisions. We derive and show the differences in particle multiplicities in the canonical and the grand canonical formulation of quantum number conservation. The time evolution and the approach to chemical equilibrium in the above ensembles is discussed in terms of kinetic master equation. The application of the statistical model to the description of (multi)strange particle yields at GSI/SIS and the SPS energies is also presented.

INTRODUCTION

There are in general two approaches to describe integrated particle yields measured in ultrarelativistic heavy ion collisions: (i) the microscopic transport models [1] and (ii) the macroscopic statistical thermal models [2, 3, 4, 5, 6]. In this article we will discuss the statistical approach and show that it provides a very satisfactory description of experimental data. We will emphasize the importance of the conservation laws in the particular strangeness conservation when modelling particle chemical freezeout conditions.

Within a statistical approach, the production of particles is commonly described using the grand canonical (GC) ensemble, where the charge conservation is controlled by the related chemical potential. In this description a net value of a given U(1) charge is conserved on the average. The (GC) approach can be only valid if the total number of particles carrying quantum number related with this symmetry is very large. In the opposite limit of a small particle multiplicities, conservation laws must be implemented exactly and locally, i.e., the canonical (C) ensemble for conservation laws must be used [7, 8]. The local conservation of quantum numbers in the canonical approach severely reduces the phase space available for particle productions. This treatment of charge conservation is of crucial importance in the description of particle multiplicities in proton induced processes [9, 10], in e^+e^- [9] as well as in central heavy ion collisions at low beam energies [11].

CP594, *Hadrons and Nuclei: First International Symposium,* edited by Il-T. Cheon et al.

In this article we describe the exact strangeness conservation in the context of relativistic statistical thermodynamics. A kinetic theory for the time evolution of particle production and the approach to the grand canonical and the canonical equilibrium distribution will be also introduced. Finally the example of the applications of the statistical model in (C) ensemble is presented in the context of low energy central as well as in high energy peripheral heavy ion collisions.

STATISTICAL MODEL AND PARTICLE MULTIPLICITY

The exact treatment of quantum numbers in statistical mechanics has been well established for some time now [7, 8, 11]. It is in general obtained by projecting the partition function onto the desired values of the conserved charges by using group theoretical methods. For our purpose we shall only consider the conservation laws related to the abelian U(1) symmetry group. In particular, we concentrate on strangeness conservation.

The basic quantity in the statistical mechanics describing a thermal properties of a system is the partition function $Z(T,V)$. In the (GC) ensemble,

$$Z^{GC}(T,V,\mu_Q) \equiv Tr[e^{-\beta(H-\mu_Q Q)}] \tag{1}$$

where Q is the conserved charge, H the hamiltonian of the system, μ_Q is the chemical potential which plays the role of the Lagrange multiplier which guarantees that the charge Q is conserved on the average in the whole system. Finally $\beta = 1/T$ is the inverse temperature.

In the (C) ensemble the charge Q is conserved exactly. Thus there is no more chemical potential under the trace and instead we calculate the partition function summing only these states which are carrying exactly the quantum number Q, that is

$$Z_Q^C(T,V) \equiv Tr_Q[e^{-\beta H}] \tag{2}$$

The canonical and the grand canonical partition functions are related through the following cluster decomposition,

$$Z^{GC}(T,V,\lambda) = \sum_{Q=-\infty}^{+\infty} \lambda^Q Z_Q^C(T,V) \tag{3}$$

where the fugacity $\lambda \equiv \exp(\beta\mu_Q)$ and the sum is taken over all possible values of the charge Q.

For the (GC) partition function, which is well behaving analytic function of the fugacity λ, the above relation can be inverted and the canonical partition function with a given value of the charge Q reads,

$$Z_Q = \frac{1}{2\pi}\int_0^{2\pi} d\phi e^{-iQ\phi}\tilde{Z}(T,V,\phi) \tag{4}$$

where the generating function $\tilde{Z}$ is obtained from the grand canonical partition function replacing the fugacity parameter λ by the factor $e^{i\phi}$,

$$\tilde{Z}(T,V,\phi) \equiv Z^{GC}(T,V,\lambda \to e^{i\phi}) \tag{5}$$

The form of the generating function $\tilde{Z}$ in the above equation is model dependent. Having in mind the application of the statistical description to particle production in heavy ion collisions we calculate $\tilde{Z}$ in the ideal gas approximation, however, including all particles and resonances [11]. This is not an essential restriction, because, describing the freeze-out conditions we are dealing with a dilute system where the interactions should not influence particle production anymore. We neglect any medium effects on particle properties. In general, however, already in the low-density limit, the modifications of resonance width or particle dispersion relations could be of importance [12]. For the sake of simplicity, we use classical statistics, i.e. we assume temperature and density regime so that all particles can be treated using Boltzmann distributions.

In nucleus-nucleus collisions the absolute values of the baryon number, electric charge and strangeness are fixed by the initial conditions. Modelling particle production in statistical thermodynamics would in general require the canonical formulation of all these quantum numbers. From the previous analysis [7, 11], however, it is clear, that in heavy ion collisions only strangeness should be treated exactly, whereas the conservation of baryon and electric charges can be described by the appropriate chemical potentials in the grand canonical ensemble.

Within the approximations described above and neglecting first the contributions from multi-strange baryons, the generating function in equation (4) has the following form,

$$\tilde{Z}(T,V,\mu_Q,\mu_B,\phi) = \exp(N_{s=0} + N_{s=1}e^{i\phi} + N_{s=-1}e^{-i\phi}) \tag{6}$$

where $N_{s=0,\pm1}$ is defined as the sum over all particles and resonances having strangeness $0, \pm 1$,

$$N_{s=0,\pm1} = \sum_k Z_k^1 \tag{7}$$

and Z_k^1 is the one-particle partition function defined as

$$Z_k^1 \equiv \frac{V g_k}{2\pi^2} m_k^2 T K_2(m_k/T) \exp(b_k \mu_B + q_k \mu_q) \tag{8}$$

with the mass m_k, spin-isospin degeneracy factor g_k, particle baryon number b_k and electric charge q_k. The volume of the system is V and the chemical potentials related with the electric charge and the baryon number are determined by μ_q and μ_B respectively.

With the particular form of the generating function equations (6,7,8) the ϕ-integration in equation (4) can be done analytically giving the canonical partition function for a gas with total strangeness S [11]:

$$Z_S(T,V,\mu_B,\mu_Q) = Z_0(T,V,\mu_B,\mu_Q) I_S(x) \tag{9}$$

where $Z_0 = \exp(N_{S=0})$ is the partition function of all particles having zero strangeness and the argument of the Bessel function

$$x \equiv 2\sqrt{S_1 S_{-1}}. \tag{10}$$

with $S_{\pm 1} \equiv N_{s=\pm 1}$. The parameter x thus measures the total number of strange particles in thermal fireball.

The calculation of the particle density n_k in the canonical formulation is straightforward. It amounts to the replacement

$$Z_k^1 \mapsto \lambda_k Z_k^1 \tag{11}$$

of the corresponding one-particle partition function in equation (7) and taking the derivative of the canonical partition function equation (4) with respect to λ_k

$$n_k^C \equiv [\lambda_k \frac{\partial}{\partial \lambda_k} \ln Z_Q(\lambda_k)]_{\lambda_k=1} \tag{12}$$

As an example, we quote the result for the density of thermal kaons in the canonical formulation assuming that the total strangeness of the system $S = 0$,

$$n_K^C = \frac{Z_K^1}{V} \frac{S_1}{\sqrt{S_1 S_{-1}}} \frac{I_1(x)}{I_0(x)} \tag{13}$$

Comparing the above formula with the result for thermal kaons density in the grand canonical ensemble, $n_K^{GC} = (Z_K^1/V)\exp(\mu_s/T)$, one can see that the canonical result can be obtained from the grand canonical one replacing the strangeness fugacity $\lambda_S \equiv \exp(\mu_S/T)$ in the following way:

$$n_K^C = n_K^{GC} \left(\lambda_S \mapsto \frac{S_1}{\sqrt{S_1 S_{-1}}} \frac{I_1(x)}{I_0(x)} \right) \tag{14}$$

In the limit of large x that is large volume and/or temperature the canonical and the grand canonical formulation are equivalent. For a small number of strange particles in a system, however, the differences are large. This can be seen in the most transparent way when comparing two limiting situations: the large and small x limit of the above equation. In the limit $x \to \infty$ we have

$$\lim_{x \to \infty} \frac{I_1(x)}{I_0(x)} \to 1 \tag{15}$$

and the kaon density is independent of the volume of the system as expected in the grand canonical ensemble. On the other hand in the limit of a small x we have

$$\lim_{x \to 0} \frac{I_1(x)}{I_0(x)} \to \frac{x}{2} \tag{16}$$

and the particle density is linearly dependent on the volume. It is thus clear, that the major difference between the canonical and the grand canonical treatment of the conservation laws appears through different volume dependence of strange particle densities as well as strong suppression of thermal particle phase space. The relevant parameter, F_S, which measures the suppression of particle multiplicities from their grand canonical result is determined by the ratio of the Bessel functions

$$F_S \equiv \frac{I_1(x)}{I_0(x)} \tag{17}$$

with the argument x defined in equation (10). In Fig. 1 we show the canonical suppres-

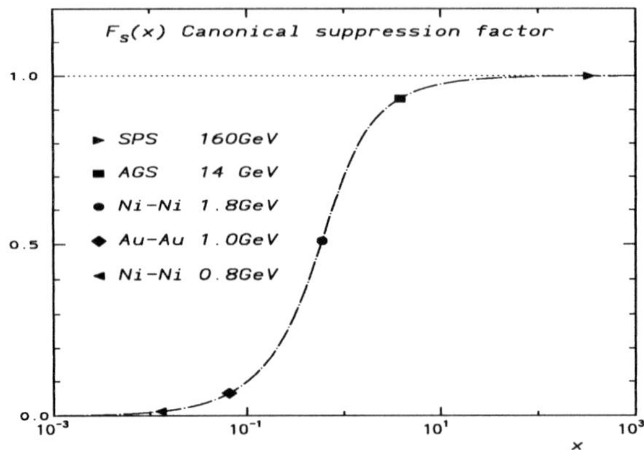

FIGURE 1. Canonical strangeness suppression factor (see text).

sion factor $F_S(x)$ as a function of the argument x. To relate the initial volume of the system with the number of participant in A-A collisions one uses the approximate relation $V \sim 1.9\pi A_{part}$. The corresponding value of x at SIS, AGS and SPS energy is calculated with the baryon chemical potential and temperature extracted from the measured particle multiplicity ratios [11]. The results in Fig. 1 shows the importance of the canonical treatment of strangeness conservation at SIS energy. Here, the canonical suppression factor can be even larger than an order of magnitude. For central Au-Au collisions at AGS or SPS energy this suppression is not relevant any more and the (GC)-formalism is adequate. In general, one expects, that the statistical interpretation of particle production in central heavy ion collisions requires the canonical treatment of strangeness conservation if the CMS collation energy $\sqrt{s} < 2 - 3$GeV. This is mainly because at these energies the freeze-out temperature is still too low to maintain large argument expansion of the Bessel functions in equation (15). The canonical description of strangeness conservation can be, however, also of importance at the SPS energy were $\sqrt{s} \sim 18$GeV when one considers the peripheral heavy ion collisions. This is particularly true for multistrange particle production since the canonical suppression of the thermal particle phase-space increases with strangeness content of the particle [10].

Multistrange particle multiplicities

The extension of the canonical description to multi-strange particle multiplicities is straightforward. One needs first to extend the generating functional in Eq. 6 by including the contributions of multistrange baryons. In this case the canonical partition function constraint by the strangeness neutrality condition reads [10, 13],

$$Z^C_{S=0} = \frac{1}{2\pi}\int_{-\pi}^{\pi} d\phi \, \exp\Big(\sum_{s=-3}^{3} S_s e^{is\phi}\Big), \tag{18}$$

where $S_s \equiv \sum_i Z_i^1$ and the sum is taken over all particles and resonances carrying strangeness s. The one-particle partition function Z_i^1 is defined in Eq.(8).

With the particular form of the partition function given by Eq. (18) the density n_s of particle i with strangeness s in volume V is obtained by the replacement $Z_i \mapsto \lambda_i Z_i$ in Eq. (18) and then taking an appropriate derivative [7, 10]:

$$\begin{aligned} n_{\pm s} &= \frac{\langle N_{\pm s}\rangle}{V} \equiv \Big[\frac{\lambda_i}{V}\frac{\partial \ln Z_0}{\partial \lambda_i}\Big]_{\lambda_i=1} \simeq Z_{\pm s}\frac{(S_{\pm 1})^s}{(S_{+1}S_{-1})^{s/2}} \\ &\quad \{I_s(x_1)I_0(x_2) + \sum_{m=1}^{\infty} I_m(x_2)[I_{2m+s}(x_1)A^{m/2} + I_{2m-s}(x_1)A^{-m/2}]\}/Z_{S=0} \end{aligned} \tag{19}$$

where

$$x_k \equiv 2\sqrt{S_k S_{-k}} \;, \; A \equiv \frac{S_{-1}^2 S_2}{S_1^2 S_{-2}}, \tag{20}$$

and the partition function

$$Z_{S=0} \simeq I_0(x_1)I_0(x_2) + \sum_{m=1}^{\infty} I_{2m}(x_1)I_m(x_2)[A^{m/2} + A^{-m/2}]. \tag{21}$$

In the derivation of Eq. (19) we have neglected, after differentiation over particle fugacity, the term $S_{\pm 3}$. This approximation, however, due to small value of $S_{\pm 3}$ coefficients in comparison with $S_{\pm 1}$ and $S_{\pm 2}$ is quite satisfactory. The complete expression for the partition function without any approximation can be found in reference [13].

In the large system like Pb-Pb and for large collision energy, required to reach high T, the density n_s of particle carrying strangeness s is V independent. In the opposite limit, however, this dependence is changed to $n_s \sim V^s$, which can be verified from Eq. (19). Indeed for small x_i we have approximately:

$$n_{\pm s} \simeq Z_{\pm s}\frac{(S_{\pm 1})^s}{(S_{+1}S_{-1})^{s/2}}\frac{I_s(x_1)}{I_0(x_1)}. \tag{22}$$

and when expanding the Bessel functions $I_n(x) \sim x^n$ in Eq. (22) we see that $n_s \sim V^s$.

From the above expression it is clear that strangeness suppression, which is measured by the ration I_s/I_0, is increasing with the strangeness content of the particle. Thus, there are two important ingredients of canonical modifications of multistrange particle density with respect to their (GC) value: (i) the density is volume dependent that is also centrality dependent and (ii) the thermal phase is suppressed and this suppression increases with strangeness content of the particle.

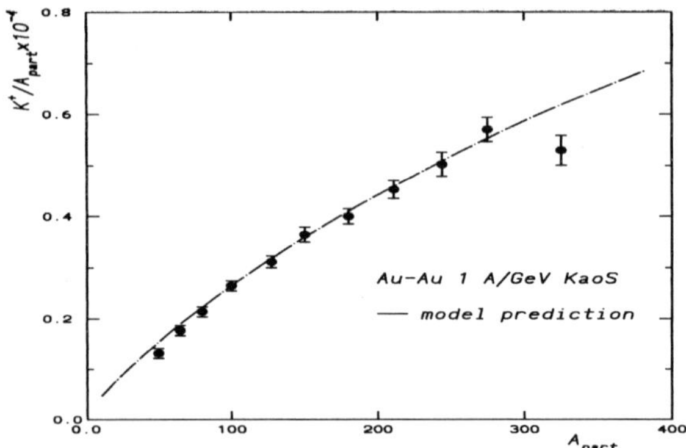

FIGURE 2. Measured K^+ multiplicity per participant A_{part} as a function of A_{part} for Au-Au collisions at 1 A/GeV [14] together with the canonical model results [11].

TIME EVOLUTION AND STRANGENESS EQUILIBRATION

In the last section we have formulated the statistical model for strange particle multiplicities $< N_S >$ assuming that the system is in thermal and chemical equilibrium. We have shown that dependently on the total number of strange particles there are two distinct equilibrium limits [8, 11]: if $< N_s >$ is small then we are in the canonical regime and in the opposite limit the canonical and grand canonical description coincides. In this section we consider the time evolution of the multiplicity of strange particles and formulate a kinetic master equations which distinguish between these two equilibrium limits. For a sake of illustration we consider a simple example of K^+K^- production in the environment of thermal pions in volume V and temperature T due to the the following binary process, $\pi^+\pi^- \to K^+K^-$. We formulate for this example the kinetics for the time evolution of kaon multiplicities and their approach to chemical equilibrium .

In the standard formulation [15], the rate equation for this binary process is described by the following population equation:

$$\frac{d < N_K >}{d\tau} = \frac{G}{V} < N_{\pi^+} >< N_{\pi^-} > - \frac{L}{V} < N_{K^+} >< N_{K^-} >, \tag{23}$$

where $G \equiv \langle \sigma_G v \rangle$ and $L \equiv \langle \sigma_L v \rangle$ give the momentum-averaged cross sections for the gain $\pi^+\pi^- \to K^+K^-$ and the loss $K^+K^- \to \pi^+\pi^-$ process respectively. The value of $< N_K >$ represents the total number of produced kaons.

To include the possible correlations between the production of K^+ and K^- [8], let us define $P_{i,j}$ as the probability to find i number of K^+ *and* j number of K^- in an event.

We also denote by P_i as the probability to find i number of K in an event. The average number of K per event is defined as:

$$\langle N_K \rangle = \sum_{i=0}^{\infty} i P_i. \tag{24}$$

We can now write the following general rate equation for the average kaon multiplicities:

$$\frac{d\langle N_K \rangle}{d\tau} = \frac{G}{V} \langle N_{\pi^+} \rangle \langle N_{\pi^-} \rangle - \frac{L}{V} \sum_{i,j} i j P_{i,j}. \tag{25}$$

Due to the local conservation of quantum numbers, we have:

$$\begin{aligned} P_{i,j} &= P_i \, \delta_{ij}, \\ \sum_{i,j} i j P_{i,j} &= \sum_i i^2 P_i \equiv \langle N^2 \rangle = \langle N \rangle^2 + \langle \delta N^2 \rangle, \end{aligned} \tag{26}$$

where $\langle \delta N^2 \rangle$ represents the event-by-event fluctuation of the number of K^+K^- pairs. Note that we always consider abundant π^+ and π^- so that we can neglect the number fluctuation of these particles and the change of their multiplicities due to the considered processes.

Following Eqs.(25-26) the general rate equation for the average number of K^+K^- pairs can be written as:

$$\frac{d\langle N_K \rangle}{d\tau} = \frac{G}{V} \langle N_{\pi^+} \rangle \langle N_{\pi^-} \rangle - \frac{L}{V} \langle N_K^2 \rangle. \tag{27}$$

For abundant production of K^+K^- pairs where $\langle N_K \rangle \gg 1$,

$$\langle N_K^2 \rangle \approx \langle N_K \rangle^2, \tag{28}$$

and Eq.(27) obviously reduces to the standard form:

$$\frac{d\langle N_K \rangle}{d\tau} \approx \frac{G}{V} \langle N_{\pi^+} \rangle \langle N_{\pi^-} \rangle - \frac{L}{V} \langle N_K \rangle^2. \tag{29}$$

However, for rare production of K^+K^- pairs where $\langle N_K \rangle \ll 1$, the rate equations (23) and (29) are no longer valid. We have instead

$$\langle N_K^2 \rangle \approx \langle N_K \rangle, \tag{30}$$

which reduces Eq.(27) to the following form [8]:

$$\frac{d\langle N_K \rangle}{d\tau} \approx \frac{G}{V} \langle N_{\pi^+} \rangle \langle N_{\pi^-} \rangle - \frac{L}{V} \langle N_K \rangle. \tag{31}$$

Thus, in the limit where $\langle N_K \rangle \ll 1$, the absorption term depends on the pair number only linearly, instead of quadratically for the limit of $\langle N_K \rangle \gg 1$. Thus, it is clear that the time evolutions and equilibrium values for kaon multiplicities are obviously different in the above limiting situations.

In the limit of large $< N_K >$ the equilibrium value for the number of K^+K^- pairs, which coincides with the multiplicity of K^+ and K^-, is obtained from Eq.(29) as,

$$< N_K >^{\rm GC}_{\rm eq} = \frac{V}{2\pi^2} m_K^2 T K_2(M_K/T) \tag{32}$$

thus, it is described by the (GC) result with vanishing chemical potential due to strangeness neutrality condition.

In the opposite limit where $\langle N_K \rangle \ll 1$, the time evolution is described by Eq.(31), which has the following equilibrium solution:

$$N^{\rm C}_{\rm eq} = \left[\frac{V}{2\pi^2} M^2_{K^+} T K_2(M_{K^+}/T)\right] \left[\frac{V}{2\pi^2} M^2_{K^-} T K_2(M_{K^-}/T)\right]. \tag{33}$$

The above equation demonstrates the locality of strangeness conservation. With each K^+ the K^- is produced in the same event in order to conserve strangeness exactly and locally. This is the result expected from the (C) formulation of the conservation laws as described in the previous section. We note that Eq.(33) is just the leading term in the expansion of the canonical result for multiplicities of particles which are carrying $U(1)$ charges. The general expression is given by Eq.(13) and Eq.(20).

Comparing Eq.(32) and Eq.(33), we first find that, for $\langle N_K \rangle \ll 1$, the equilibrium value in the canonical formulation is far smaller than what is expected from the grand canonical result as

$$< N_K >^{\rm C}_{\rm eq} = [< N_K >^{\rm GC}_{\rm eq}]^2 \ll < N_K >^{\rm GC}_{\rm eq}. \tag{34}$$

This shows the importance of the canonical description of quantum number conservation when the multiplicity of particles carrying non-zero $U(1)$ charges is small. We also note that the volume dependence in the two cases differs. The particle density in the GC limit is independent of V whereas in the opposite canonical limit the density scales linearly with V. Secondly, we note that the relaxation time for a canonical system is far shorter than what is expected from the grand canonical result [8]. It is also clear from Eq.(26) that (C) and (GC) limits are essentially determined by the size of $\langle \delta N^2 \rangle$, the event-by-event fluctuation of the number of K^+K^- pairs. The grand canonical results correspond to small fluctuations, i.e., $\langle \delta N^2 \rangle / \langle N \rangle^2 << 1$; while the canonical description is necessary in the opposite limit.

MODEL PREDICTIONS VERSUS EXPERIMENTAL DATA

In heavy ion collisions the number of produced strange particles depends on the collision energy and the centrality of these collisions. At low collision energies like eg. in GSI/SIS the freezeout temperature is relatively low being of the order of 50 – 80MeV. Consequently the number of strange particles in the final state is very small. Following our discussion in the last sections it is clear that the statistical description require here the canonical approach. In Fig. 2 we show the experimental data on K^+ yield per participant A_{part} as a function of A_{part} measured in Au-Au collisions at $E_{lab} \sim 1$ A/GeV [14]. The volume parameter in the statistical model scale with the number of participant. We can thus directly compare the model with data by fixing thermal parameters, the temperature and the baryon chemical potential, such that to reproduce the measured particle multiplicity ratios [11]. In Fig. 2 the results of the canonical model is shown by the full line. The results clearly indicates that strong almost quadratic dependence of kaon yield on the number of participant is well reproduced by the model. The quadratic dependence of particle multiplicity in the canonical regime is the basic property of the model. In Fig. 3 we calculate the ratio Ξ/K^+ multiplicities for central $Ni-Ni$ collisions

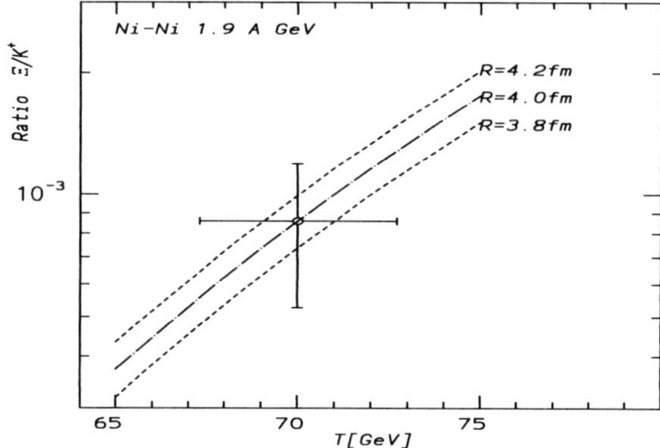

FIGURE 3. The ratio of Ξ to K^+ multiplicity as a function of temperature. The point with errors indicates the predictions of the thermal model for Ni-Ni collisions at 1.9 A/GeV. The lines corresponds to canonical calculations for different centrality measured by the initial radius R of the system.

in the temperature range which corresponds to the collision energy of $E_{lab} \sim 2$ A/GeV. The yield of Ξ is seen to be substantially smaller then the yield of K^+. This result is not only related with the differences in particle masses but particularly it appears due to the canonical suppression. Since Ξ carry strangeness minus two it has to be produced together with two K^+ to locally and exactly neutralized strangeness. It is also interesting to note that the Ξ/K^+ ratio is independent on the baryon chemical potential. This is because K^+ appears together with Λ, thus it contains the same dependence on μ_B as multistarnge baryons.

The importance of the canonical treatment of strangeness conservation is also seen in higher collision energies like at the SPS when considering centrality dependance of multistrange baryons. In peripheral collisions the yield of strange particles is small such that also here the canonical description should be applied. The canonical suppression of thermal particle phase-space increases with strangeness content of the particle. The exact conservation of strangeness requires that each particle carrying strangeness $\bar{s}$ has to appear e.g. with s other particles of strangeness one to satisfy strangeness neutrality condition.

In [10] the multiplicity/participant of Ω, Ξ, and Λ relative to its value in a small system with only two participants were calculated within canonical model. It was shown there that the statistical model in (C) ensemble reproduces the basic features of WA97 data [16]: the enhancement pattern and enhancement saturation for large A_{part} indicating here that (GC) limit is reached. The results obtained in [10] also demonstrate different A_{part} dependence of strange and multistrage baryons. For small A_{part} this dependence is power like as describe by Eq.(22). The quantitative comparison of the model with the experimental data would require an additional assumption on the variation of μ_B with centrality to account for larger value of $\bar{B}/B$ ratios in p+A than in Pb+Pb collisions [10, 16]. The most recent results of NA57 [17], showing an abrupt change of the enhancement for $\bar{\Xi}$ are, however, very unlikely to be reproducible in terms of the canonical approach.

SUMMARY AND CONCLUSIONS

We have discussed the importance of the conservation laws in the application of the statistical model to the description of strangeness production in heavy ion collisions. We have presented the arguments that the more general treatment of strangeness conservation based on the canonical ensemble is required if one compares the model with experimental data for particle yields obtained in central A-A collisions at SIS energies or peripheral collisions at the SPS. In both situations the number of produced strange particles per event is still too small to use the asymptotic grand canonical ensemble. The time evolution of strangeness production and the approach to chemical equilibrium limit was discussed in the context of a kinetic approach. We have shown on few examples that the statistical model predictions are consistent with the experimental data. A more complete presentation of the model versus data can be found in [2, 3, 4, 5, 6, 9, 10, 11].

ACKNOWLEDGMENTS

On of us (K.R.) acknowledge stimulating discussions with Su Houng Lee and the partial support of the Committee for Scientific Research (KBN-2P03B 03018). We also acknowledge stimulating discussions with P. Braun-Munzinger, B. Friman, V. Koch, Z. Lin, M. Stephanov, H. Satz and X.N. Wang.

REFERENCES

1. S. E. Vance, *et al.*, Phys. Rev. Lett. 83 (1999) 1735; J. Phys. G27 (2001) 627; M. Bleicher, W. Greiner, H. Stöcker and N. Xu, Phys. Rev.C62 (2000) 061901; A. Capella and C. A. Salgado, Phys. Rev.

C60 (1999) 054906; Z. Lin, *et al.*, nucl-th/0011059; W. Cassing, Nucl. Phys. A661 (1999) 468c; H. Drescher, J. Aichelin and K. Werner, Rapport, Subatech, 00-21.
2. R. Stock, Phys. Lett. 456 (1999) 277; Prog. Part. Nucl. Phys. 42 (1999) 295; J. Stachel, Nucl. Phys. A654 (1999) 119c; U. Heinz, Nucl. Phys. A685 (2001) 414; Nucl. Phys. A661 (1999) 349; J. Letessier and J. Rafelski, Int. J. of Mod. Phys. E9 (2000) 107.
3. P. Braun-Munzinger, J. Stachel, J. P. Wessels and N. Xu, Phys. Lett. B344, 43 (1995); Phys. Lett. B365, 1 (1996); P. Braun-Munzinger and J. Stachel, Nucl. Phys. A606, 320, 1996.
4. J. Cleymans and K. Redlich, Phys. Rev. Lett. 81, 5284 (1998) *and references therein.*
5. P. Braun-Munzinger, I. Heppe and J. Stachel, Phys. Lett. B465, 15 (1999).
6. F. Becattini, J. Cleymans, A. Keranen, E. Suhonen and K. Redlich, nucl-ph/0011322, Phys. Rev. C to appear.
7. R. Hagedorn, CERN yellow report 71-12, 101 (1971); E.V. Shuryak, Phys. Lett. B42, 357 (1972); K. Redlich and L. Turko, Z. Phys. B97, 279 (1980), L. Turko, Phys. Lett. B104, 153 (1981), H.-Th. Elze, W. Greiner and J. Rafelski, Phys. Lett. B124, 515 (1983); R. Hagedorn and K. Redlich, Z. Phys. A27, 541 (1985).
8. C.M. Ko, V. Koch, Z. Lin, K. Redlich, M. Stepanov and X.N. Wang, nuc-th/0010004, Phys. Rev. Lett. *in print.*
9. F. Becattini, Z. Phys. C69 (1996) 485; F. Becattini and U. Heinz, Z. Phys. C76 (1997) 269.
10. J. S. Hamieh, K. Redlich and A. Tounsi, Phys. Lett. B486 (2000) 61.
11. J. Cleymans and K. Redlich, Phys. Rev. C60, 054908 (1999); J. Cleymans, H. Oeschler and K. Redlich, Phys. Rev. C59, 1663 (1995); Phys. Let. B485, 27 (2000).
12. W. Weinhold, B. Friman, W. Nörenberg, Phys. Let. B433, 236 (1998); T. Hatsuda, S. Lee and H. Shiomi, Phys. Rev. C52, 3364 (1995), M. Lutz, B. Friman and G. Wolf, Nucl. Phys. A661, 526 (1999) and references therein.
13. J. Cleymans, K. Redlich and E. Suhonen, Z. Phys. C51, 137 (1991).
14. A. Wagner et al., (KaoS Collaboration), Phys. Lett. B420 (1998) 20, C. Muntz et al., (KaoS Collaboration) Z. Phys. C357 399 1997.
15. P. Koch, B. Müller and J. Rafelski, Phys. Rep. 142, 167 (1986); T. Matsui, B. Svetitsky and L.D. McLerran, Phys. Rev. D34, 783 (1986); T.S. Biro, E. van Doorn, B. Müller, M.H. Thoma and X.N. Wang, Phys. Rev. C48, 1275 (1993).
16. E. Andersen, *et al.*, WA97 Collaboration, Phys. Lett. B449 (1999) 401.
17. N. Carrer, NA57 Collaboration, In Proceedings of QM2001.

LATTICE QCD AND OTHER RELATED TOPICS

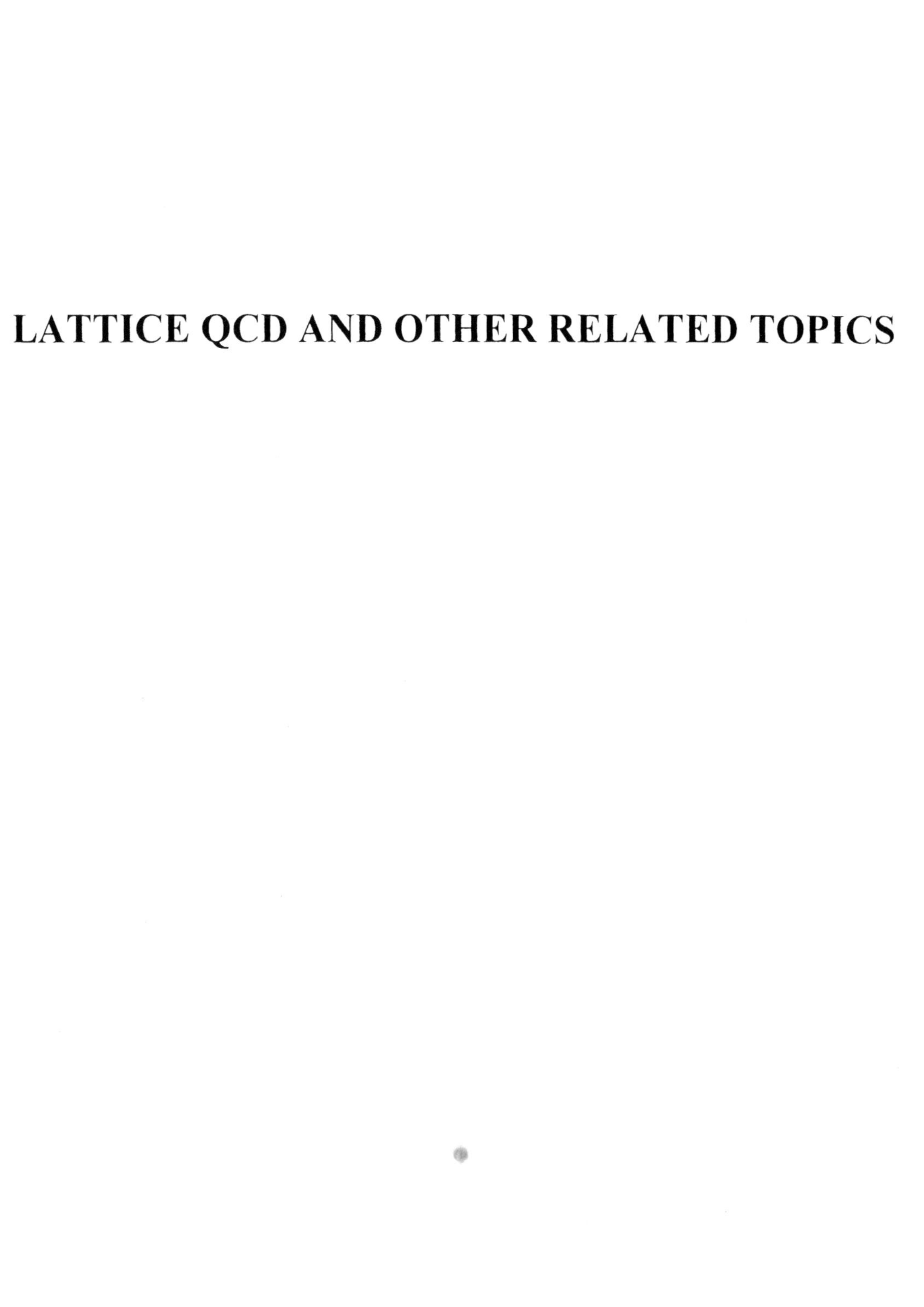

Weyl Symmetric Representation of SU(3) Gluodynamics in Abelian Projection

Y. Koma*, M. Takayama†, H. Toki† and D. Ebert†

*Institute for Theoretical Physics, Kanazawa University, Kanazawa 920-1192, Japan
†Research Center for Nuclear Physics (RCNP) Osaka University, Osaka 567-0047, Japan

Abstract. The dual Ginzburg-Landau (DGL) theory corresponding to the SU(3) gluodynamics in Abelian projection is formulated in a Weyl symmetric way. The Weyl symmetric DGL theory can be regarded as the sum of three types of the U(1) dual Abelian Higgs (DAH) model. As an application of this approach, the hadronic flux-tube solution corresponding to the baryonic state is investigated adopting the similar techniques used in the U(1) DAH model. The string representation of the DGL theory is also discussed in a Weyl symmetric way.

INTRODUCTION

The analysis of the nonperturbative properties of the SU(3) gluodynamics (QCD) is a very important subject in understanding not only the nontrivial vacuum structure but also the hadron properties observed in experiments. In this context, we consider that the so-called dual superconducting scenario [1, 2], described by the dual Ginzburg-Landau (DGL) theory [3], is a promising idea, where the quarks and gluons are confined into the inside of hadrons forming string-like structure. This is due to the dual Higgs mechanism (dual Meissner effect) caused by monopole condensation, which is numerically supported from recent studies of lattice QCD Monte-Carlo simulation in the maximally Abelian (MA) gauge. Such flux-tube (string) picture of hadrons reproduces not only the Regge trajectories, known from empirical data [4], but also the linear rising inter-quark potential, obtained as a result of area-law decay of the Wilson loop. Hence, the analysis of flux-tube dynamics in the dual superconducting vacuum would be useful for the purpose of systematic understanding of the observed hadrons in terms of quarks and gluons.

The DGL theory is obtained by performing the Abelian projection method [5] to the SU(3) gluodynamics, assuming the infrared Abelian dominance and monopole condensation. Its symmetry is the $U(1)\times U(1)$ dual gauge symmetry and the global Weyl symmetry, where $[U(1)]^2$ is directly related to the maximal torus subgroup of the SU(3) group. The Weyl symmetry guarantees the color-singlet criterion of the theory.

Recently, we have reformulated the DGL theory so as to make the Weyl symmetry manifest by extending the $[U(1)]^3$ dual gauge symmetry [6]. We have found the resulting one has very simple form like the U(1) dual Abelian Higgs (DAH) model, which means that the same techniques used in the U(1) DAH model can be adopted. As an example of the manifestly Weyl symmetric representation of the DGL theory, we study the hadronic flux-tube solution corresponding to the baryonic state [7]. We also discuss the

CP594, *Hadrons and Nuclei: First International Symposium,* edited by Il-T. Cheon et al.

string representation of the DGL theory as another application of this Weyl symmetric formulation.

WEYL SYMMETRIC FORMULATION OF THE DGL THEORY

We start from the DGL Lagrangian [3] in Euclidean metric, given by *

$$\mathcal{L}_{\rm DGL} = \frac{1}{4}\left((\partial\wedge\vec{B})_{\mu\nu} - e\vec{\Sigma}^{\rm open}_{\mu\nu}\right)^2 + \sum_{i=1}^{3}\left[\left|\left(\partial_\mu + ig\vec{\varepsilon}_i\cdot\vec{B}_\mu\right)\chi_i\right|^2 + \lambda\left(|\chi_i|^2 - v^2\right)^2\right], \quad (1)$$

where $\vec{B}_\mu$ and χ_i denote the dual gauge field with two components (B^3_μ, B^8_μ) and the complex scalar monopole field, respectively. The quark current $\vec{j}_\mu = \bar{q}\gamma_\mu\vec{H}q$, where $\vec{H} = (T_3, T_8)$, is represented by the boundary of a nonlocal string term $\vec{\Sigma}^{\rm open}_{\mu\nu}$, which expresses the color-electric (open) Dirac string singularity through the modified dual Bianchi identity $\partial^\nu {}^*\vec{\Sigma}^{\rm open}_{\mu\nu} = \vec{j}_\mu$. Note that $(\partial\wedge\vec{B})_{\mu\nu} \equiv \partial_\mu\vec{B}_\nu - \partial_\nu\vec{B}_\mu$ satisfies $\partial^{\nu *}(\partial\wedge\vec{B})_{\mu\nu} = 0$. Since the diagonal component of the matrix $\vec{H}$ gives the weight vector of the SU(3) algebra $\vec{w}_j$ ($j = 1,2,3$), where $\vec{w}_1 = \left(1/2, \sqrt{3}/6\right), \vec{w}_2 = \left(-1/2, \sqrt{3}/6\right), \vec{w}_3 = \left(0, -1/\sqrt{3}\right)$, one can define the color-electric charges of the quarks as $\vec{Q}^{\rm e}_j \equiv e\vec{w}_j$. Here, $j = 1,2,3$ correspond to the color-electric charges, red (R), blue (B), and green (G). Accordingly we can write the nonlocal term as $e\vec{\Sigma}^{\rm open}_{\mu\nu} = e\sum_{j=1}^{3}\vec{w}_j\Sigma^{\rm e:open}_{j\,\mu\nu}$ †. On the other hand, the root vectors of the SU(3) algebra $\vec{\varepsilon}_i$ are used to define the color-magnetic charges of the monopole field as $Q^{\rm m}_i \equiv g\vec{\varepsilon}_i$ ($i = 1,2,3$), where $\vec{\varepsilon}_1 = \left(-1/2, \sqrt{3}/2\right), \vec{\varepsilon}_2 = \left(-1/2, -\sqrt{3}/2\right), \vec{\varepsilon}_3 = (1,0)$. Both color-electric and color-magnetic charges satisfy the extended Dirac quantization condition $\vec{Q}^{\rm m}_i\cdot\vec{Q}^{\rm e}_j = 2\pi m_{ij}$ ($eg = 4\pi$). Here m_{ij} is an integer following the definition

$$m_{ij} = 2\vec{\varepsilon}_i\cdot\vec{w}_j = \sum_{k=1}^{3}\varepsilon_{ijk} = \{0, 1, -1\}, \quad (2)$$

where ε_{ijk} is the 3rd-rank antisymmetric tensor. Typical mass scales in the DGL theory are the mass of the dual gauge field $m_B = \sqrt{3}gv$ and of the monopole field $m_\chi = 2\sqrt{\lambda}v$. Their ratio, the so-called Ginzburg-Landau (GL) parameter $\kappa \equiv m_\chi/m_B$, characterizes the type of dual "superconductivity" of the vacuum. Like in real superconductive materials, the properties of the vacuum might be very different depending on the actual value of κ.

We make the Weyl symmetry of the DGL theory (1) manifest, with the help of an extended dual gauge field [6, 7], defined by

$$B_{i\,\mu} \equiv g\vec{\varepsilon}_i\cdot\vec{B}_\mu \qquad (i = 1,2,3). \quad (3)$$

---

* Throughout this paper, we use the following notations: Latin indices i, j express the labels 1,2,3, which are not to be summed over unless explicitly stated.

† We use the notation "e:···" and "m:···" denoting the color-electric (weight vector of SU(3) algebra) basis and the color-magnetic (root vector of SU(3) algebra) basis, respectively.

Here, a constraint $\sum_{i=1}^{3} B_{i\,\mu} = 0$ appears, since $\sum_{i=1}^{3} \vec{\varepsilon}_i = 0$. The DGL Lagrangian (1) is now written as

$$\mathcal{L}_{\rm DGL} = \sum_{i=1}^{3} \left[\frac{1}{4g_m^2} {}^{*}F_{i\,\mu\nu}{}^2 + \left| (\partial_\mu + iB_{i\,\mu}) \chi_i \right|^2 + \lambda \left(|\chi_i|^2 - v^2 \right)^2 \right], \tag{4}$$

$${}^{*}F_{i\,\mu\nu} \equiv (\partial \wedge B_i)_{\mu\nu} - 2\pi \Sigma_{i\,\mu\nu}^{\rm m:open} \qquad \left(\Sigma_{i\,\mu\nu}^{\rm m:open} \equiv \sum_{j=1}^{3} m_{ij} \Sigma_{j\,\mu\nu}^{\rm e:open} \right), \tag{5}$$

where the dual gauge coupling g is scaled as

$$g_m \equiv \sqrt{\frac{3}{2}} g. \tag{6}$$

The factor 2π in front of the Dirac string term is derived from the Dirac quantization condition. Clearly, the expression (4) is manifestly Weyl symmetric since all indices i and j are summed over. Apparently the dual gauge symmetry is extended to $[\mathrm{U}(1)]^3$, achieved by a set of transformation

$$\begin{aligned} &\chi_i \to \chi_i e^{if_i}, \quad \chi_i^* \to \chi_i^* e^{-if_i}, \\ &B_{i\,\mu}^{\rm reg} \to B_{i\,\mu}^{\rm reg} - \partial_\mu f_i, \qquad (i = 1,2,3). \end{aligned} \tag{7}$$

However, the number of gauge degrees of freedom is not enlarged because of the constraint $\sum_{i=1}^{3} B_{i\,\mu} = 0$. One finds that the expression of (4) is written as the sum of three types of the U(1) DAH model.

STRING SINGULARITIES

The interesting physics is hidden in the string singularities in the DGL theory. Here we find that the DGL theory has two kinds of string singularities. One is the open string singularity originating from the quark source. This becomes manifest when we dispose the behavior of the dual gauge field, which can be achieved by the decomposition of the dual gauge field into two parts, the regular (no Dirac string) part and the singular (Dirac string) part [6],

$$B_{i\,\mu} \equiv B_{i\,\mu}^{\rm reg} + B_{i\,\mu}^{\rm sing} \quad (i = 1,2,3), \tag{8}$$

where the singular part is determined so as to define the color-electric charge density $C_{i\,\mu\nu}^{\rm m}$ as

$$(\partial \wedge B_i^{\rm sing})_{\mu\nu} = 2\pi \left(\Sigma_{i\,\mu\nu}^{\rm m:open} + C_{i\,\mu\nu}^{\rm m} \right), \quad \text{or} \quad \varepsilon_{\mu\nu\lambda\rho} \partial_\lambda B_{i\,\rho}^{\rm sing} = 2\pi \left(\tilde{\Sigma}_{i\,\mu\nu}^{\rm m:open} + \tilde{C}_{i\,\mu\nu}^{\rm m} \right). \tag{9}$$

Here "~" denote dual quantities

$$\tilde{\Sigma}_{i\,\mu\nu}^{\rm m:open} = \frac{1}{2} \varepsilon_{\mu\nu\lambda\rho} \Sigma_{i\,\lambda\rho}^{\rm m:open}, \qquad \tilde{C}_{i\,\mu\nu}^{\rm m} = \frac{1}{2} \varepsilon_{\mu\nu\lambda\rho} C_{i\,\lambda\rho}^{\rm m}. \tag{10}$$

The explicit form of $C^{\rm m}_{i\,\mu\nu}$ is given by

$$C^{\rm m}_{i\,\mu\nu}(x) = \frac{1}{4\pi^2}\int d^4y \frac{1}{|x-y|^2}{}^*(\partial \wedge j^{\rm m}_i(y))_{\mu\nu}, \tag{11}$$

where $j^{\rm m}_{i\,\mu} = \partial^{\nu\,*}\Sigma^{\rm m:open}_{i\,\mu\nu}$. Note that if there is no quark source, we do not need to have a singular part $B^{\rm sing}_{i\,\mu}$. This means that the role of $\Sigma^{\rm m:open}_{i\,\mu\nu}$ is to cancel the string singularity in the dual gauge field. Then, the dual field strength tensor is rewritten as

$${}^*F_{i\,\mu\nu} = (\partial \wedge B^{\rm reg}_i)_{\mu\nu} + 2\pi C^{\rm m}_{i\,\mu\nu}. \tag{12}$$

In the static q-$\bar{q}$ system, $C^{\rm m}_{i\,\mu\nu}$ turns out to be the Coulombic color-electric field originating from the color-electric charge.

The other is the closed string singularity in the phase of the monopole field, which is extracted from the polar decomposition form of the monopole field $\chi_i = \phi_i \exp(i\theta_i)$ $(\phi, \theta \in \Re)$, as $\theta \equiv \theta^{\rm reg} + \theta^{\rm sing}$. Here $\theta^{\rm sing}$ is defined by the relation

$$[\partial_\mu, \partial_\nu]\theta^{\rm sing}_i = 2\pi\Sigma^{\rm m:closed}_{i\,\mu\nu}, \quad \text{or} \quad \varepsilon_{\mu\nu\lambda\rho}\partial_\lambda\partial_\rho\theta^{\rm sing}_i = 2\pi\tilde{\Sigma}^{\rm m:closed}_{i\,\mu\nu}. \tag{13}$$

Note that regular part of the phase satisfies $[\partial_\mu, \partial_\nu]\theta^{\rm reg}_i = 0$. Physically, such closed string singularity would correspond to the glueball state [8], since it does not contain any valence quarks.

BARYONIC FLUX-TUBE SOLUTION

As an interesting and an important application of the Weyl symmetric DGL theory, let us investigate the flux-tube system in the presence of three types of the color-electric charges. This system is determined by solving the field equations

$$\frac{1}{g_m^2}\partial^{\nu\,*}F_{i\,\mu\nu} = i\left(\chi_i^*\partial_\mu\chi_i - \chi_i\partial_\mu\chi_i^*\right) - 2B_{i\,\mu}\chi_i^*\chi_i, \tag{14}$$

$$\left(\partial_\mu + iB_{i\,\mu}\right)^2\chi_i = 2\lambda\chi_i(\chi_i^*\chi_i - v^2), \tag{15}$$

with proper boundary conditions. One finds that these expressions are exactly the same as the field equations in the U(1) DAH model, replicated with respect to the index $i=1$, 2, 3. Therefore, similar boundary conditions as in the U(1) DAH model can be adopted [6]. However, since these color-electric charges are defined in the weight vector diagram of SU(3) algebra as $\vec{Q}^{\rm e}_j = e\vec{w}_j$ $(j = 1,2,3)$, and the color-electric Dirac strings $\Sigma^{\rm e:open}_{j\,\mu\nu}$ which are attached to these charges carry the same quantities, respectively, these Dirac strings should join at a certain point to cancel each other $(\sum_{j=1}^3 e\vec{w}_j = 0)$, which we call a junction. Otherwise, the energy of the system cannot become finite. The non-vanishing Dirac string $\Sigma^{\rm e:open}_{1\,\mu\nu}$, $\Sigma^{\rm e:open}_{2\,\mu\nu}$, and $\Sigma^{\rm e:open}_{3\,\mu\nu}$ are properly included so as to minimize the length of the color-electric Dirac string, which corresponds to the energy minimization condition. Then, the position of the junction is given by the Fermat point [9]. As a result, we get a typical Y-shaped flux-tube object in the DGL theory, *i.e.* the baryonic flux tube.

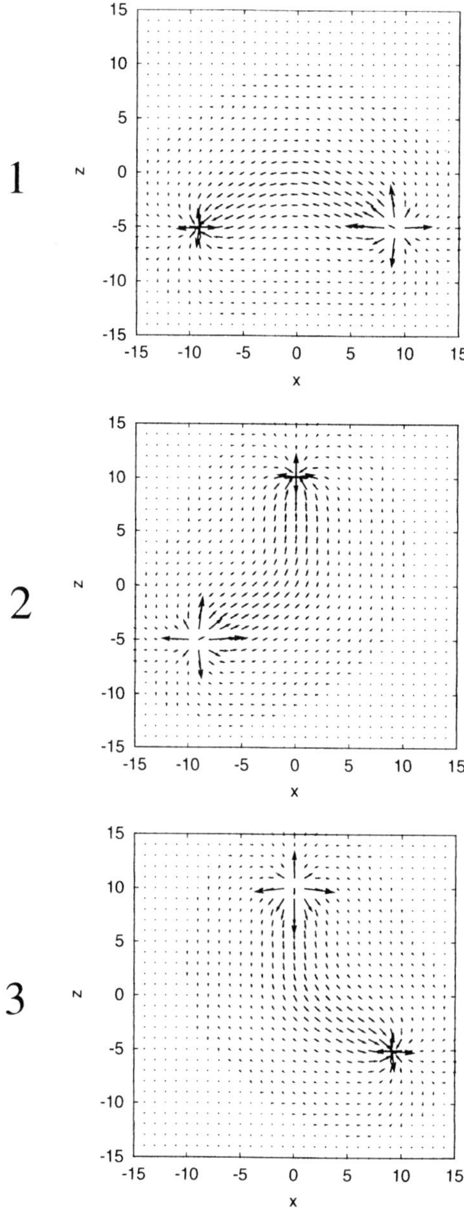

FIGURE 1. The profiles of the color-electric field in the Weyl symmetric representation, expressed on the root vectors of the SU(3) algebra, $\vec{\varepsilon}_1$ (upper), $\vec{\varepsilon}_2$ (middle), and $\vec{\varepsilon}_3$ (lower) in the baryonic flux-tube system in the x-z plane at $y = 0$. The junction and the quarks are located at $(x, y, z) = (0,0,0)$, and $R(0,0,9)$, $B(9,0,-5)$, $G(-9,0,-5)$, respectively.

In Fig. 1, we show the profiles of the color-electric field (spatial component of the dual field strength tensor ${}^*F_{i\,\mu\nu}$) corresponding to the Weyl symmetric representation of the dual gauge field, where the color-electric charges R, B, and G are placed on the corners of a regular triangle. One finds that in the Weyl symmetric formulation of the DGL theory, the baryonic flux-tube solution is obtained as the superposition of three types of bending flux tubes. The constraint $\sum_{i=1}^{3} {}^*F_{i\,\mu\nu} = 0$ which is originating from $\sum_{i=1}^{3} B_{i\,\mu} = 0$ is, of course, fulfilled. Here, we have adopted the dual lattice formulation in order to solve the field equations numerically, which is quite useful method to investigate various shapes of the flux-tube system [7].

STRING REPRESENTATION

As another application of the Weyl symmetric formulation of the DGL theory, it is interesting to discuss its string representation, which would be necessary when we discuss the flux-tube dynamics. The string representation is, in principle, obtained by the functional integration over the fields under keeping the string singularities in the theory, starting from the partition function of the DGL theory

$$\begin{aligned} \mathcal{Z} &= \int \prod_{i=1}^{3} (\mathcal{D}B_{i\,\mu}\mathcal{D}\chi_i\mathcal{D}\chi_i^*)\,\delta\left(\sum_{i=1}^{3} B_{i\,\mu}\right)\delta\left(\sum_{i=1}^{3}\arg\chi_i\right) \\ &\quad\times\exp\left[-\int d^4x\left\{\sum_{i=1}^{3}\left(\frac{1}{4g_m^2}\left((\partial\wedge B_i)_{\mu\nu} - 2\pi\Sigma_{i\,\mu\nu}^{\mathrm{m:open}}\right)^2\right.\right.\right. \\ &\qquad\left.\left.\left. +\left|(\partial_\mu + iB_{i\,\mu})\,\chi_i\right|^2 + \lambda\left(|\chi_i|^2 - v^2\right)^2\right)\right\}\right]. \end{aligned} \tag{16}$$

The main difference of the partition function between this form and the U(1) DAH model is only the δ-function constraint in the integration measure. However since this can be treated, for instance, as

$$\delta\left(\sum_{i=1}^{3}\arg\chi_i\right) = \int \mathcal{D}k\exp\left\{ik\left(\sum_{i=1}^{3}\arg\chi_i\right)\right\}, \tag{17}$$

and the Lagrange multiplier k is also integrated out, the same functional integration techniques as the U(1) DAH model can be adopted **. Then, the resulting string representation of the DGL theory is straightforwardly given by

$$\begin{aligned} \mathcal{Z} &= \int \prod_{i=1}^{3} (\mathcal{D}x_{i\,\mu}(\xi))\,\delta\left(\sum_{i=1}^{3}\tilde{\Sigma}_{i\,\mu\nu}^{\mathrm{m:open}}\right)\delta\left(\sum_{i=1}^{3}\tilde{\Sigma}_{i\,\mu\nu}^{\mathrm{m:closed}}\right) \\ &\quad\times\exp\left[-\int d^4x\int d^4y\sum_{i=1}^{3}\tilde{\Sigma}_{i\,\alpha\beta}^{\mathrm{m}}(x)D_{\alpha\beta,\gamma\delta}(x-y)\tilde{\Sigma}_{i\,\gamma\delta}^{\mathrm{m}}(y)\right], \end{aligned} \tag{18}$$

** In the derivation of the string representation of the U(1) DAH model, we follow similar manner as in Ref. [10] and take the London limit $\lambda \to \infty$ ($|\chi_i| \to v$).

where $\Sigma^{\rm m}_{i\,\mu\nu} \equiv \Sigma^{\rm m:open}_{i\,\mu\nu} + \Sigma^{\rm m:closed}_{i\,\mu\nu}$, and $x_{i\,\mu}(\xi)$ parametrizes the corresponding string world sheet. Note that $D_{\alpha\beta,\gamma\delta}(x-y)$ is the propagator of the Kalb-Ramond field, explicitly written as the combination of modified Bessel functions,

$$
\begin{aligned}
D_{\alpha\beta,\gamma\delta}(x-y) &= \left(\delta_{\alpha\delta}\delta_{\beta\gamma} - \delta_{\beta\gamma}\delta_{\alpha\delta}\right) \frac{m_B v^2}{4r} K_1(m_B r) \\
&+ \frac{v^2}{2 m_B r^2} \left[\left(\delta_{\alpha\gamma}\delta_{\beta\delta} - \delta_{\beta\gamma}\delta_{\alpha\delta}\right) \left(\frac{1}{r} K_1(m_B r) + \frac{m_B}{2}\left(K_0(m_B r) + K_2(m_B r)\right) \right) \right. \\
&- \frac{1}{2r} \left(\delta_{\alpha\gamma} r_\beta r_\delta + \delta_{\beta\delta} r_\alpha r_\gamma - \delta_{\alpha\delta} r_\beta r_\gamma - \delta_{\beta\gamma} r_\alpha r_\delta\right) \\
&\times \left. \left(3\left(\frac{m_B^2}{4} + \frac{1}{r^2} \right) K_1(m_B r) + \frac{3 m_B}{2r}\left(K_0(m_B r) + K_2(m_B r)\right) + \frac{m_B^2}{4} K_3(m_B r) \right) \right]. \quad (19)
\end{aligned}
$$

where $r \equiv |x-y|$. Expanding the propagator for small r, we get the so-called Nambu-Goto action and the rigidity term but extended to the SU(3) gluodynamics. More precise evaluation of the string representation of the DGL theory is still remaining as an interesting subject for the understanding of the string dynamics, which is in preparation.

CONCLUSION

We have studied the dual Ginzburg-Landau (DGL) theory corresponding to the SU(3) gluodynamics in Abelian projection in a Weyl symmetric way. The DGL theory describes the vacuum as the dual superconductor, which leads to the hadronic flux-tube configuration of the color-electric flux through the dual Meissner effect. The Weyl symmetric formulation of the DGL theory leads to the sum of three types of the U(1) dual Abelian Higgs (DAH) model, which makes it easy to handle the flux-tube solution in the DGL theory. As examples of the Weyl symmetric approach we have investigated the baryonic flux-tube solution and the string representation of the DGL theory, which are achieved by using the quite similar technique as in the U(1) DAH model. This means that the analysis of the U(1) DAH model is also helpful in understanding the SU(3) gluodynamics in Abelian projection.

It is interesting to note that as the further application of this approach, we can easily evaluate the string tension of flux tubes associated with static charges in various SU(3) representations. These results allow us to discuss the Casimir scaling problem in the dual superconducting scenario of confinement [11].

ACKNOWLEDGMENT

The authors acknowledge fruitful collaborations with E.-M. Ilgenfritz and T. Suzuki.

REFERENCES

1. Nambu, Y., *Phys. Rev.*, **D10**, 4262 (1974).
2. Mandelstam, S., *Phys. Rept.*, **23C**, 245 (1976).
3. Suzuki, T., *Prog. Theor. Phys.*, **80**, 929 (1988); Maedan, S., and Suzuki, T., *ibid.* **81**, 229 (1989).
4. Sailer, K., Schoenfeld, T., Schram, Z., Schaefer, A., and Greiner, W., *J. Phys.*, **G17**, 1005 (1991).
5. 't Hooft, G., *Nucl. Phys.*, **B190**, 455 (1981).
6. Koma, Y., and Toki, H., *Phys. Rev.*, **D62**, 054027 (2000), hep-ph/0004177.
7. Koma, Y., Ilgenfritz, E. M., Suzuki, T., and Toki, H., Weyl symmetric representation of hadronic flux tubes in the dual ginzburg-landau theory, to appear in *Phys. Rev.*, **D63** (2001), hep-ph/0011165.
8. Koma, Y., Suganuma, H., and Toki, H., *Phys. Rev.*, **D60**, 074024 (1999), hep-ph/9902441.
9. Kamizawa, S., Matsubara, Y., Shiba, H., and Suzuki, T., *Nucl. Phys.*, **B389**, 563–576 (1993).
10. Antonov, D., and Ebert, D., *Eur. Phys. J.*, **C8**, 343–351 (1999).
11. Koma, Y., Ilgenfritz, E. M., Toki, H., and Suzuki, T., Casimir scaling in a dual superconducting scenario of confinement, to appear in *Phys. Rev.*, **D63** (2001), hep-ph/0103162.

SU(3) Lattice QCD Study for Static Three-Quark Potential

T. T. Takahashi*, H. Suganuma†, H. Matsufuru** and Y. Nemoto**

*RCNP, Osaka University, Mihogaoka 10-1, Osaka 567-0047, Japan
†Faculty of Science, Tokyo Institute of Technology, Tokyo 152-8551, Japan
**YITP, Kyoto University, Kitashirakawa, Sakyo, Kyoto 606-8502, Japan

Abstract. We study the static three-quark (3Q) potential in detail using SU(3) lattice QCD with $12^3 \times 24$ at $\beta = 5.7$ and $16^3 \times 32$ at $\beta = 5.8, 6.0$ at the quenched level. For more than 200 patterns of the 3Q systems, we numerically derive 3Q ground-state potential V_{3Q} from the 3Q Wilson loop with the smearing technique, which reduces excited-state contaminations. The lattice QCD data of V_{3Q} are well reproduced within a few % deviation by a sum of a constant, the two-body Coulomb term and the three-body linear confinement term $\sigma_{3Q} L_{\rm min}$, with $L_{\rm min}$ the minimal value of total length of color flux tubes linking the three quarks. From the comparison with the Q-$\bar{\rm Q}$ potential, we find a universality of the string tension as $\sigma_{3Q} \simeq \sigma_{Q\bar{Q}}$ and the one-gluon-exchange result for Coulomb coefficients, $A_{3Q} \simeq \frac{1}{2} A_{Q\bar{Q}}$.

INTRODUCTION

Strong interaction in hadrons or nuclei is fundamentally governed by quantum chromodynamics (QCD). In a spirit of the elementary particle physics, it would be desirable to construct hadron physics and nuclear physics at quark level based on QCD. For instance, to bridge between QCD and hadron physics, the inter-quark potential[1, 2, 3] is one of the most important quantities, because it is directly responsible for the hadron structure at the quark level[4]. However, it still remains as a difficult problem to derive the inter-quark potential analytically from QCD, because of the strong-coupling nature of QCD in the infrared region.

Recently, the lattice QCD calculation has been adopted as a useful and reliable method for the nonperturbative analysis of QCD. For instance, quark-antiquark (Q$\bar{\rm Q}$) potential, which is responsible for the meson properties, has been well studied using lattice QCD[1, 5]. In Fig.1, we show the Q$\bar{\rm Q}$ static potential $V_{Q\bar{Q}}(r)$ as the function of the distance r between quark and antiquark in SU(3) lattice QCD at the quenched level. Here, the Q$\bar{\rm Q}$ potential can be well reproduced by a sum of the Coulomb term due to the perturbative one-gluon-exchange (OGE) process, a linear confinement term and a constant[1, 5],

$$V_{Q\bar{Q}} = -\frac{A_{Q\bar{Q}}}{r} + \sigma_{Q\bar{Q}} r + C_{Q\bar{Q}}, \tag{1}$$

where the string tension $\sigma_{Q\bar{Q}} \simeq 0.89$ GeV/fm represents the confinement force. The linear potential at the long distance can be physically interpreted with the flux-tube picture or the string picture for hadrons[4, 6, 7, 8]. In this picture, the quark and the

CP594, *Hadrons and Nuclei: First International Symposium,* edited by Il-T. Cheon et al.

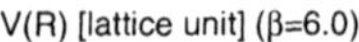

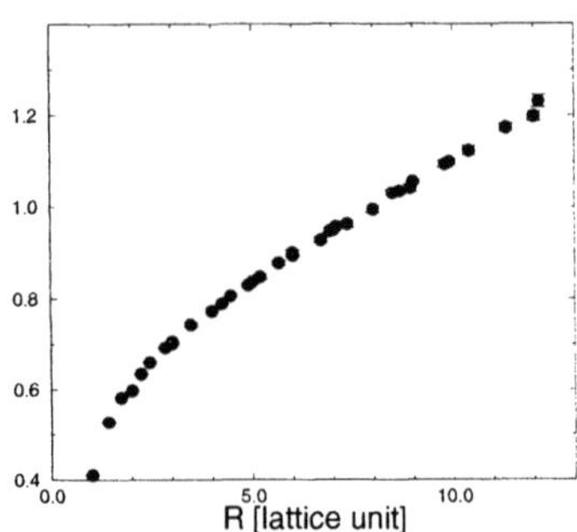

FIGURE 1. The $Q\bar{Q}$ static potential $V_{Q\bar{Q}}(r)$ obtained in SU(3) lattice QCD at the quenched level.

antiquark are linked with a one-dimensional flux-tube with the string tension $\sigma_{Q\bar{Q}}$. Hence, $Q\bar{Q}$ potential in the long-range region is proportional to the distance r between quark and antiquark. This flux-tube picture (or the string picture) in the infrared region is supported by the Regge trajectory of hadrons, empirical analysis of heavy quarkonia data, the strong-coupling expansion of QCD[7] and recent lattice QCD simulations.

However, there is almost no reliable formula to describe the three-quark (3Q) potential V_{3Q} directly based on QCD, besides the strong-coupling QCD[4, 7], although V_{3Q} is directly responsible for the baryon properties[4, 9, 10]. Up to now, the 3Q potential has been treated phenomenologically or hypothetically more than 20 years. We carry out the detailed study of the 3Q potential using SU(3) lattice QCD.

THEORETICAL CONSIDERATION

Theoretically, the 3Q potential[1, 3, 4] is expected to be expressed by a sum of a constant, the two-body Coulomb term from the perturbative OGE process at the short distance and the three-body linear confinement term at the long distance, similarly in the $Q\bar{Q}$ potential. Here, reflecting the $SU(3)_c$ gauge theory in QCD, the color flux tube has a junction which combines three different colors, R, B and G, in a color-singlet manner. For most 3Q systems, the flux-tube energy is minimized with the presence of the junction, and therefore a Y-type flux tube is expected to be formed among the three quarks [1, 4, 7, 9, 10]. We show in Fig.2 the flux-tube configuration which minimizes the total flux-tube length in the 3Q system.

Thus, the static 3Q potential is considered to take a form as

$$V_{3Q} = -A_{3Q} \sum_{i<j} \frac{1}{|\mathbf{r}_i - \mathbf{r}_j|} + \sigma_{3Q} L_{\min} + C_{3Q}, \tag{2}$$

where $L_{\min}$ denotes the miminal value of the total length of color flux tubes linking three quarks.

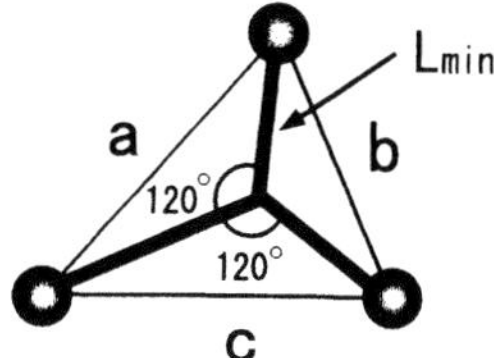

FIGURE 2. The flux-tube configuration of the 3Q system with the minimal value of the total flux-tube length $L_{\min}$.

Denoting by a, b and c the three side lengths of the 3Q triangle as shown in Fig.2, $L_{\min}$ is explicitly expressed as [1]

$$L_{\min} = \left[\frac{1}{2}(a^2+b^2+c^2) + \frac{\sqrt{3}}{2}\sqrt{(a+b+c)(-a+b+c)(a-b+c)(a+b-c)} \right]^{\frac{1}{2}}, \quad (3)$$

when any angle of the 3Q triangle does not exceed $2\pi/3$. For the case with one angle larger than $2\pi/3$, $L_{\min}$ is given as $L_{\min} = (a+b+c) - \max(a,b,c)$.

STATIC 3Q POTENTIAL IN SU(3) LATTICE QCD

The static 3Q potential can be extracted from the 3Q Wilson loop [1, 9, 10] in SU(3) lattice QCD calculations. As shown in Fig.3, the 3Q Wilson loop physically represents the creation of a gauge-invariant 3Q state at $t=0$, the system with spatially fixed three quarks in $0<t<T$ and the annihilation of the 3Q state at $t=T$.

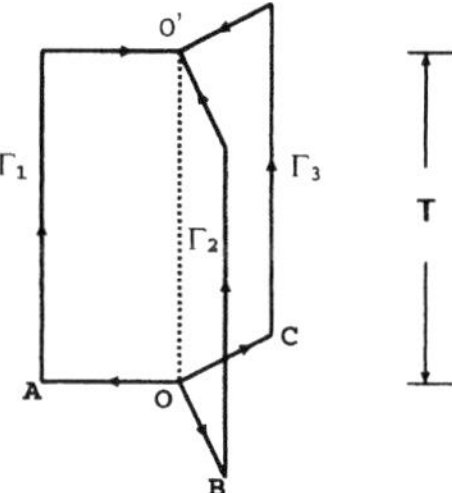

FIGURE 3. The 3Q Wilson loop in the Euclidean space-time. $\Gamma_k (k=1,2,3)$ denotes the path linking O and O'.

In four-dimensional Euclidean space-time, the 3Q Wilson loop is a color current defined in a gauge-invariant manner as

$$W_{3Q} \equiv \frac{1}{3!}\varepsilon_{abc}\varepsilon_{a'b'c'}U_1^{aa'}U_2^{bb'}U_3^{cc'}, \quad U_k = P\exp\{ig\int_{\Gamma_k} dx^\mu A_\mu(x)\} \in \mathrm{SU}(3). \quad (4)$$

Here, P denotes the path-ordered product along the path denoted by $\Gamma_k (k=1,2,3)$ linking O and O' in Fig.3. Similar to the derivation of the $Q\bar{Q}$ potential from the Wilson loop, the 3Q potential V_{3Q} can be obtained as $V_{3Q} = -\lim_{T\to\infty}\frac{1}{T}\ln\langle W_{3Q}\rangle$.

Next, let us consider the physical state of the 3Q system. The ground state of the 3Q system is expected to be composed by flux tubes rather than the strings[1, 8], and we denote by $|\text{g.s.};t\rangle$ the 3Q ground state at t. On the other hand, there are many excited states of the 3Q system corresponding to the flux-vibrational modes[1], and we denote by $|n\text{th e.s.};t\rangle$ the n-th excited 3Q state at t. In the 3Q Wilson loop, the gauge-invariant 3Q state created at $t=0$ and annihilated at $t=T$ can be expressed as

$$
\begin{aligned}
|3\text{Q};0\rangle &= c_0|\text{g.s.};0\rangle + c_1|\text{1st e.s.};0\rangle + c_2|\text{2nd e.s.};0\rangle + ..., \\
|3\text{Q};T\rangle &= c_0|\text{g.s.};T\rangle + c_1|\text{1st e.s.};T\rangle + c_2|\text{2nd e.s.};T\rangle + ...,
\end{aligned} \tag{5}
$$

where the coefficients are normalized as $\sum_{i=0}^{\infty}|c_i|^2 = 1$. (This normalization is found to be consistent with the definition of $W_{3\text{Q}}$ in Eq.(4).) Then, the expectation value of $W_{3\text{Q}}$ can be expressed as

$$
\begin{aligned}
\langle W_{3\text{Q}}\rangle &= |c_0|^2\langle \text{g.s.};T|\text{g.s.};0\rangle + |c_1|^2\langle \text{1st e.s.};T|\text{1st e.s.};0\rangle + ... \\
&= |c_0|^2\exp(-V_{\text{g.s.}}T) + |c_1|^2\exp(-V_{\text{1st e.s.}}T) + ...
\end{aligned} \tag{6}
$$

with the ground-state potential $V_{\text{g.s.}}$ of the 3Q system and the n-th excited-state potential $V_{n\text{-th e.s.}}$.

As increasing T, the excited-state components drop faster than the ground-state component in $\langle W_{3\text{Q}}\rangle$, however, the ground-state component $|c_0|^2\exp(-V_{\text{g.s.}}T)$ also decreases exponentially. Hence, we face a practical difficulty in extracting the numerical signal. To avoid this difficulty, we adopt the smearing technique[1, 5, 11] which enhances the ground-state overlap and removes the excited-state contamination efficiently. (There were a few lattice studies on the 3Q potential more than 13 years ago[12, 13], however, their results seem unreliable and misleading due to the fatal large excited-state contaminations. The smearing technique was mainly developed after their works.)

SMEARING TECHNIQUE TO REMOVE EXCITED MODES

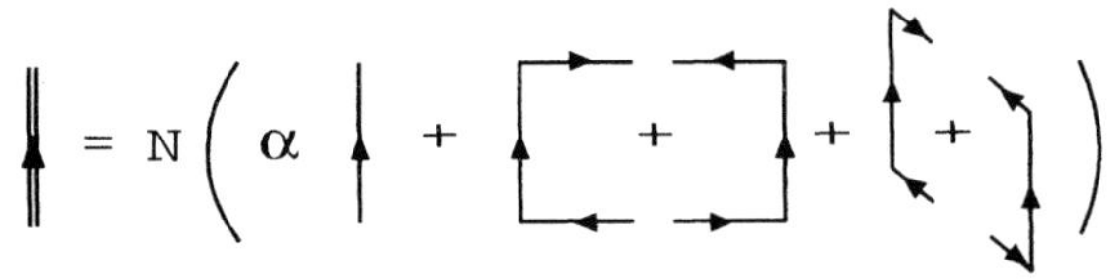

FIGURE 4. The schematic explanation of the smearing for the link-variables.

The smearing for the link-variable is expressed as the iterative replacement of the spatial link-variables by a obscured link-variables $\bar{U}_i(s) \in \text{SU}(3)$ which maximizes

$$
\text{Re tr}\left\{\bar{U}_i^\dagger(s)\left[\alpha U_i(s) + \sum_{j\neq i}\{U_j(s)U_i(s+\hat{j})U_j^\dagger(s+\hat{i}) + U_j^\dagger(s-\hat{j})U_i(s-\hat{j})U_j(s+\hat{i}-\hat{j})\}\right]\right\} \tag{7}
$$

with a real smearing parameter α. This can be visualized as in Fig.4. The obscured spatial line composed by the obscured link-variables physically corresponds to the flux tube with a finite cylindrical radius. The smearing parameter α and the iteration number of the smearing are suitably determined so as to maximize the ground-state overlap of the 3Q system at $t = 0$ and $t = T$ in the 3Q Wilson loop. In fact, after a suitable smearing, we expect saturation of the ground-state overlap as $|c_0|^2 \simeq 1$, i.e. strong reduction of the excited-state contaminations as $|c_i|^2 \simeq 0$ $(i = 1, 2, ..)$[1, 5, 11]. To see this in actual lattice

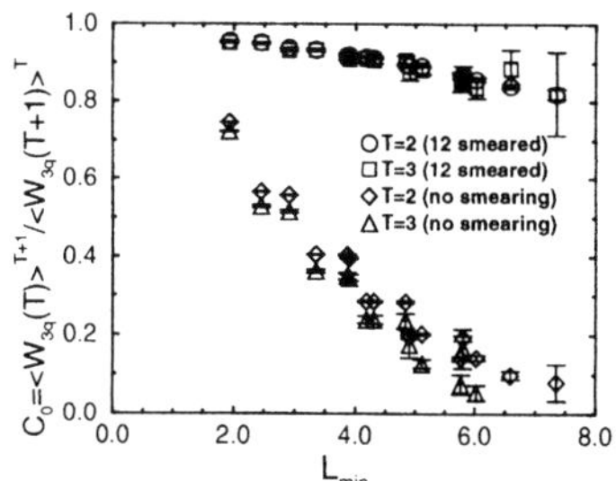

FIGURE 5. The ground-state overlap $C_0 \equiv \langle W_{3Q}(T)\rangle^{T+1}/\langle W_{3Q}(T+1)\rangle^T$ in SU(3) lattice QCD at $\beta = 5.7$. By the smearing, the ground-state overlap C_0 is largely enhanced from the lower data to the upper data as $0.8 < C_0 < 1$ for each 3Q system.

QCD calculations, we examine in Fig.5 the ground-state overlap

$$C_0 \equiv \langle W_{3Q}(T)\rangle^{T+1}/\langle W_{3Q}(T+1)\rangle^T \tag{8}$$

corresponding to $|c_0|^2$[1, 5], and we observe a large enhancement of the ground-state overlap as $0.8 < C_0 < 1$ by the smearing. To summarize here, owing to the smearing, we can set up the ground-state dominant 3Q state at $t = 0$ and $t = T$ in the 3Q Wilson loop, and therefore we enjoy accurate measurements for the 3Q potential without suffering from excited modes like the flux-tube vibration.

THE LATTICE QCD RESULTS FOR THE 3Q POTENTIAL

We study the 3Q potential V_{3Q} in detail by investigating more than 200 patterns of the 3Q systems using SU(3) lattice QCD with $12^3 \times 24$ at $\beta = 5.7$ and $16^3 \times 32$ at $\beta = 5.8, 6.0$ at the quenched level. Here, the lattice spacing a is determined so as to reproduce the Q$\bar{\text{Q}}$ string tension $\sigma_{Q\bar{Q}} = 0.89\text{GeV/fm}$ at each β. We show in Fig.6 the 3Q potential V_{3Q} in SU(3) lattice QCD at β=5.7 and β=5.8 as the function of $L_{\min}$, the minimal value of the total flux-tube length. At the large distance, where the Coulomb potential is negligible, we observe the linearity on $L_{\min}$, which is consistent with the expected form in Eq.(2). At the short distance, perturbative QCD would be applicable, and therefore the two-body Coulomb potential in V_{3Q} can be estimated as $V_{\text{Coul}} \equiv -\frac{A_{Q\bar{Q}}}{2}\sum_{i<j}\frac{1}{|\mathbf{r}_i-\mathbf{r}_j|}$, using the Coulomb coefficient $A_{Q\bar{Q}}$ in the lattice QCD result of the Q$\bar{\text{Q}}$ potential. (Here, the color factor $\frac{1}{2}$ appears in perturbative QCD.) To single out the confinement contribution, we subtract this perturbative Coulomb contribution V_{Coul} from the 3Q potential V_{3Q}. The

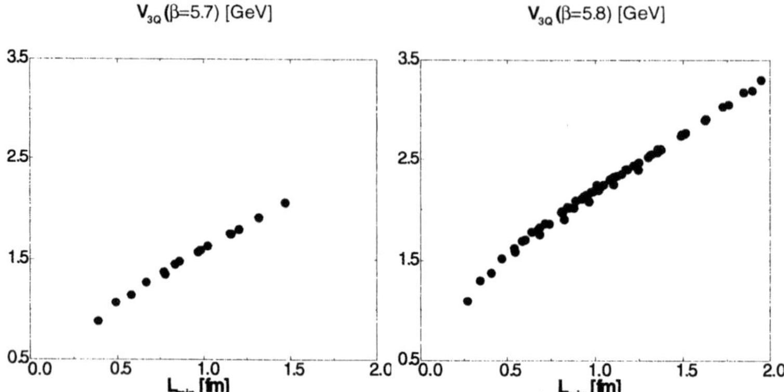

FIGURE 6. The 3Q potential V_{3Q} as the function of L_{min}, the minimal value of the total flux-tube length, in SU(3) lattice QCD at $\beta = 5.7$ (left) and at $\beta = 5.8$ (right).

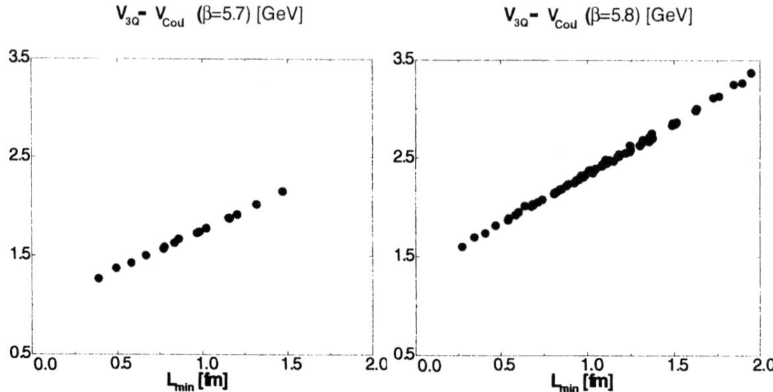

FIGURE 7. The SU(3) lattice QCD result for $V_{3Q} - V_{\rm Coul}$ as the function of L_{min}, the minimal value of the total flux-tube length, at $\beta = 5.7$ (left) and at $\beta = 5.8$ (right).

results are shown in Fig.7. The clear linearity on $L_{\rm min}$ in the whole region observed in Fig.7 means that the 3Q potential V_{3Q} can be well reproduced by a sum of the the perturbative Coulomb contribution as $V_{\rm Coul}$ and the linear confinement term proportional to $L_{\rm min}$ in Eq.(2).

Now, we consider the best fitting of the 3Q potential data with the form in Eq.(2). With three parameters A_{3Q}, σ_{3Q} and C_{3Q}, the 3Q potential data can be well reproduced within the accuracy better than a few %. We show the best fitting parameters of the 3Q potential V_{3Q} and the $Q\bar{Q}$ potential $V_{Q\bar{Q}}$ in Table 1 at various β.

First, we focus on the properties at the short distance, where the nonperturbative confinement force is almost negligible. We find $A_{3Q} \simeq \frac{1}{2} A_{Q\bar{Q}}$ for the Coulomb coefficient, and find the perturbative color factor $\frac{1}{2}$ originating from the OGE contribution.

Second, let us consider the constant term. The constant terms C_{3Q} and $C_{Q\bar{Q}}$ behave as $\frac{1}{a}$, and diverge in the continuum limit $a \to 0$, so that the constant term is not physical quantity. In the limit of $r \to 0$ in the $Q\bar{Q}$ system, the Wilson loop becomes trivial as $W_{Q\bar{Q}}(r=0)=1$ and then $V_{Q\bar{Q}}(r=0)=0$ holds, while one gets $V_{Q\bar{Q}}(r\simeq 0)\simeq -\frac{A_{Q\bar{Q}}}{\omega a}+C_{Q\bar{Q}}$ with $0<\omega<1$[1, 3] in the lattice regularization. Then, we find $C_{Q\bar{Q}}\simeq\frac{A_{Q\bar{Q}}}{\omega a}$. From the similar argument on the diquark limit, we find also $C_{3Q}-C_{Q\bar{Q}}\simeq\frac{A_{3Q}}{\omega a}$ [1, 3]. In this way, the constant term reflects the Coulomb singularity, and is kept to be finite in lattice QCD, because of the lattice reguralization[1, 3]. Physically, these constant terms C_{3Q} and $C_{Q\bar{Q}}$ represent the pairing energy at the limit when the quarks gathers at one spatial point. From the above two relations and $A_{3Q}\simeq\frac{1}{2}A_{Q\bar{Q}}$, we obtain $C_{3Q}\simeq\frac{3}{2}C_{Q\bar{Q}}$, which seems to hold as shown in Table 1.

Finally, we consider the infrared nonperturbative properties of the inter-quark potential. The universality of the string tension is observed as $\sigma_{3Q}\simeq\sigma_{Q\bar{Q}}$. In fact, the strength of the confinement force is universal for the $Q\bar{Q}$ system and the 3Q system, which supports the color flux-tube picture for hadrons [4, 8, 7, 9, 10].

SUMMARY AND CONCLUDING REMARKS

To bridge the gap between QCD and hadron physics, we have studied the 3Q potential responsible for the baryon properties. We have calculated the ground-state 3Q potential for more than 200 patterns of the 3Q systems using SU(3) lattice QCD at $\beta = 5.7, 5.8$ and 6.0 at the quenched level. In this calculation, we have adopted the smearing technique which enhances the ground-state overlap and reduces the excited-state contaminations like the flux vibrational modes. Within the accuracy better than a few %, the lattice QCD data for the 3Q potential have been well reproduced by a sum of a constant, the two-body Coulomb term, and the three-body linear confinement term proportional to $L_{\rm min}$, the minimal value of the total flux-tube length. We have observed the perturbative OGE result as $A_{3Q}\simeq\frac{1}{2}A_{Q\bar{Q}}$, and the universality of the string tension as $\sigma_{3Q}\simeq\sigma_{Q\bar{Q}}$.

We acknowledge Profs. Il-Tong Cheon and Su Hong Lee for their warm hospitality at Yonsei University. The lattice QCD simulations have been performed on NEC-SX4 at Osaka University and HITACHI-SR8000 at KEK.

TABLE 1. The best fitting parameters for the 3Q potential V_{3Q} at β=5.7, 5.8 and 6.0.

	σ (GeV/fm)	A	C (lattice unit)
3Q ($\beta = 5.7$)	0.832(15)	0.1331(66)	0.9182(213)
$Q\bar{Q}$ ($\beta = 5.7$)	0.890(25)	0.2793(116)	0.6203(161)
3Q ($\beta = 5.8$)	0.818(6)	0.1304(17)	0.9326(53)
$Q\bar{Q}$ ($\beta = 5.8$)	0.890(21)	0.2580(159)	0.6081(182)
3Q ($\beta = 6.0$)	0.811(7)	0.1363(11)	0.9590(35)
$Q\bar{Q}$ ($\beta = 6.0$)	0.890(12)	0.2768(24)	0.6374(30)

TABLE 2. Partial data for the 3Q potential in SU(3) lattice QCD at β=5.8. (i,j,k) labels the 3Q system where the three quarks put on $(i,0,0)$, $(0,j,0)$ and $(0,0,k)$ in $\mathbf{R}^3$ in the lattice unit.

(i,j,k)	V_{3Q}^{latt}	V_{3Q}^{fit}	$V_{3Q}^{latt}-V_{3Q}^{fit}$
(1,1,1)	0.9144	0.9008	0.0136
(1,1,2)	1.0657	1.0582	0.0075
(1,1,3)	1.1962	1.1883	0.0079
(1,1,4)	1.3049	1.3054	−0.0005
(1,1,5)	1.4134	1.4162	−0.0029
(1,1,6)	1.5339	1.5236	0.0103
(1,2,2)	1.1866	1.1877	−0.0011
(1,2,3)	1.3062	1.3076	−0.0014
(1,2,4)	1.4153	1.4203	−0.0050
(1,2,5)	1.5259	1.5287	−0.0029
(1,2,6)	1.6360	1.6347	0.0013
(1,3,3)	1.4178	1.4209	−0.0031
(1,3,4)	1.5216	1.5296	−0.0080
(1,3,5)	1.6276	1.6357	−0.0081
(1,3,6)	1.7296	1.7401	−0.0105
(1,4,4)	1.6291	1.6356	−0.0064
(1,4,5)	1.7190	1.7396	−0.0206
(1,4,6)	1.8203	1.8425	−0.0222
(2,2,2)	1.2793	1.2838	−0.0046
(2,2,3)	1.3815	1.3903	−0.0089
(2,2,4)	1.4904	1.4969	−0.0064
(2,3,3)	1.4751	1.4880	−0.0129
(2,3,5)	1.6814	1.6912	−0.0098

(i,j,k)	V_{3Q}^{latt}	V_{3Q}^{fit}	$V_{3Q}^{latt}-V_{3Q}^{fit}$
(4,5,6)	2.1263	2.1094	0.0169
(4,6,6)	2.2264	2.1979	0.0285
(5,6,6)	2.3105	2.2733	0.0372
(0,1,1)	0.7701	0.7727	−0.0027
(0,1,2)	0.9659	0.9694	−0.0035
(0,1,3)	1.1107	1.1071	0.0037
(0,1,4)	1.2323	1.2268	0.0055
(0,1,5)	1.3339	1.3388	−0.0049
(0,1,6)	1.4566	1.4469	0.0097
(0,1,7)	1.5730	1.5527	0.0203
(0,1,8)	1.6779	1.6570	0.0209
(0,2,2)	1.1375	1.1422	−0.0047
(0,2,3)	1.2660	1.2711	−0.0050
(0,2,4)	1.3815	1.3871	−0.0056
(0,2,5)	1.4928	1.4972	−0.0043
(0,2,6)	1.6069	1.6041	0.0028
(0,2,8)	1.8285	1.8128	0.0157
(0,3,3)	1.3893	1.3940	−0.0047
(0,3,4)	1.5007	1.5066	−0.0059
(0,3,5)	1.6135	1.6145	−0.0010
(0,4,4)	1.6079	1.6165	−0.0086
(0,4,5)	1.7085	1.7226	−0.0141
(0,4,6)	1.8263	1.8267	−0.0004

REFERENCES

1. T.T. Takahashi, H. Matsufuru, Y. Nemoto and H. Suganuma, Phys. Rev. Lett. **86**, 18 (2001).
2. T.T. Takahashi, H. Matsufuru, Y. Nemoto and H. Suganuma, Proc. of "Dynamics of Gauge Fields", Tokyo, 1999, A. Chodos et al. (eds.), Universal Academy Press, Tokyo, p.179 (2000).
3. H. Suganuma, H. Matsufuru, Y. Nemoto and T.T. Takahashi, Nucl. Phys. **A680**, 159 (2001).
4. S. Capstick and N. Isgur, Phys. Rev. **D34**, 2809 (1986).
5. G.S. Bali, C. Schlichter and K. Schilling, Phys. Rev. **D51**, 5165 (1995).
6. Y. Nambu, Phys. Rev. **D10**, 4262 (1974).
7. J. Kogut and L. Susskind, Phys. Rev. **D11**, 395 (1975).
8. H. Suganuma, S. Sasaki and H. Toki, Nucl. Phys. **B435**, 207 (1995).
9. N. Brambilla, G.M. Prosperi and A. Vairo, Phys. Lett. **B362**, 113 (1995).
10. M. Fable de la Ripelle and Yu. A. Simonov, Ann. Phys. **212**, 235 (1991).
11. APE Collaboration, M. Albanese et al., Phys. Lett. **B192**, 163 (1987).
12. R. Sommer and J. Wosiek, Phys. Lett. **149B**, 497 (1984); Nucl. Phys. **B267**, 531 (1986).
13. H.B. Thacker, E. Eichten and J.C. Sexton, Nucl. Phys. **B** (Proc. Suppl.) **4**, 234 (1988).

SU(3) lattice QCD study for octet and decuplet baryon spectra

N. Nakajima*, H. Matsufuru*, Y. Nemoto† and H. Suganuma**

*RCNP, Osaka University, Mihogaoka 10-1, Osaka 567-0047, Japan
†YITP, Kyoto University, Kitashirakawa, Sakyo, Kyoto 606-8502, Japan
**Faculty of Science, Tokyo Institute of Technology, Tokyo 152-8551, Japan

Abstract. The spectra of octet and decuplet baryons are studied using $SU(3)$ lattice QCD at the quenched level. As an implementation to reduce the statistical fluctuation, we employ the anisotropic lattice with $O(a)$ improved quark action. In relation to $\Lambda(1405)$, we measure also the mass of the $SU(3)$ flavor-singlet negative-parity baryon, which is described as a three quark state in the quenched lattice QCD, and its lowest mass is measured about 1.6 GeV. Since the experimentally observed negative-parity baryon $\Lambda(1405)$ is much lighter than 1.6 GeV, $\Lambda(1405)$ may include a large component of a $N\bar{K}$ bound state rather than the three quark state. The mass splitting between the octet and the decuplet baryons are also discussed in terms of the current quark mass.

INTRODUCTION

The lattice QCD simulation has become a powerful method to investigate hadron properties directly based on QCD. The hadron spectroscopy in the quenched approximation, i.e. without the dynamical quark effect, has been almost established, and reproduces experimental values of the low-lying hadron masses within 10 % deviation [1]. Detailed investigation with dynamical quarks will give us an insight on the dynamical quark effect on hadron properties [2]. With these situations, we now have a stage to proceed into more extensive and systematic studies of hadron structure in lattice QCD, and compare the lattice results with the model analysis or the phenomenological approaches such as the potential models [3]. The latter point of view would give us clearer physical picture of extracted information from lattice simulations, and may be useful to extract the physical quantities beyond the lattice QCD applicability. Furthermore, comparing hadron properties in the lattice QCD with the potential model analysis using the potential derived from lattice QCD, one can verify the applicability of these approaches in a self-consistent manner. This is our motivation of lattice calculation of the static three quark potential, which is responsible to the baryon properties [4].

Compared with the ground state hadrons, excited state hadrons are far from established. The purpose of our present study is to perform detailed investigation of excited states as well as the ground state hadrons. Among them, in this paper, we focus on two important subjects:

(a) Origin of Octet-Decuplet baryon mass splitting.
There have been proposed several models to explain the low-lying hadron spectrum. As one of celebrated models, the nonrelativistic quark model explains the N-Δ mass

CP594, *Hadrons and Nuclei: First International Symposium,* edited by Il-T. Cheon et al.

splitting with one-gluon-exchange (OGE) interaction and results in [5, 6]

$$M_\Delta - M_N \propto \sum_{i<j} \frac{1}{M_i M_j}, \tag{1}$$

where M_i is the *constituent* quark mass. This implies mass difference between octet and decuplet baryons decreases with increasing quark mass. In lattice QCD simulation, it is possible to change the *current* quark mass through the hopping parameter κ. Then we compare the obtained mass splitting with the form (1), although the relation between current and constituent quark masses is not so clear.

(b) Structure of $\Lambda(1405)$ and other negative parity baryons.
The detailed lattice study of the negative-parity baryons has been started rather recently [7]. Among the negative-parity baryons, we pay much attention to $\Lambda(1405)$ with $J^P = 1/2^-$, since its structure in terms of the constituent quark picture is not well understood. There are interesting two possibilities proposed for $\Lambda(1405)$, an $SU(3)_f$ singlet state (qqq) and a $N\bar{K}$ bound state as (qqq-$q\bar{q}$). In this paper, we investigate the flavor-singlet baryon spectrum with spin 1/2 and negative parity in lattice QCD, and compare it with $\Lambda(1405)$. The second possibility as the $N\bar{K}$ bound state is now in progress at the quenched level, where quark-antiquark pair creation (strictly speaking, dynamical quark loop effect) is absent and then the quark-level constitution in hadrons is definitely clear in the simulation.

In lattice QCD simulations, not only the quark mass, we can also change the number of dynamical quark flavor. This enables us to extract the quark loop effect on the hadron properties. However, the flavor-number dependence does not seem significant for the low-lying hadron spectrum. Then, we start with calculations at the quenched level (with no dynamical flavor). Before proceeding to the numerical calculation, however, we should describe a technical problem and equipment to circumvent it.

The difficulty of extracting the excited state masses lies in the rapid growth of the statistical fluctuations in the correlation functions. In the practical simulation for the correlation function, it is hard to identify the reliable range of its Euclidean temporal distance where the relevant information is kept without suffering from the large statistical noise. To overcome this problem, we adopt the anisotropic lattice, on which the temporal lattice spacing a_τ is finer than the spatial one, a_σ [8]. Detailed information in the temporal direction makes these analyses as the mass measurements extensively easy. This approach is especially efficient for the heavy particle correlators, such as excited states and glueballs, for which the noises grow rapidly against the signals. On the other hand, introduction of anisotropy requires us additional effort in the numerical simulations. Due to the quantum effect, the renormalized anisotropy $\xi \equiv a_\sigma/a_\tau$ differs from the bare one in general, and at the first stage of the simulation one need to tune the bare anisotropy so that the quark field retains the same renormalized anisotropy as the gluon field.

This paper is organized as follows. In the next section, we briefly summarize the anisotropic lattice, especially the quark action. Then, we describe the numerical simulation and discuss on the obtained result. The last section gives our summary and outlook.

ANISOTROPIC LATTICE

The anisotropic lattice has become an extensively useful tool in the lattice QCD simulations. In addition to aforementioned advantage, fine temporal resolution is particularly significant to extract the information of hadron correlators at finite temperature. Another advantage is that it can treat the relatively heavy quark without introducing the effective theoretical approaches. This feature is particularly suited for the study of charmonium and charmed hadrons. Although this is not a subject of this paper, this is one of reasons we adopt the anisotropic lattice. In this work, we treat the hadron correlators, and hence we focus here on the quark action on the anisotropic lattice.

As the quark action on the anisotropic lattice, we adopt the $O(a)$ improved Wilson action [9]. The construction of the action is along the program of Fermilab formulation [10]. Their treatment incorporates the full quark-mass dependence in the construction of lattice quark action, so that the quark mass region with $m_q \simeq a^{-1}$ is also available without large $O(ma)$ uncertainty. The Fermilab approach is naturally generalized to the anisotropic lattice. There is some arbitrariness in choosing the parameterization of action, and we use the form proposed in [11, 12]. In [12], the advantages of this form are discussed in detail. The quark action is written in the following form:

$$S_F = \sum_{x,y} \bar{\psi}(x) K(x,y) \psi(y), \tag{2}$$

$$\begin{aligned} K(x,y) = \delta_{x,y} &- \kappa_\tau \left\{ (1-\gamma_4) U_4(x) \delta_{x+\hat{4},y} (1+\gamma_4) U_4^\dagger(x-\hat{4}) \delta_{x-\hat{4},y} \right\} \\ &- \kappa_\sigma \sum_i \left\{ (r-\gamma_i) U_i(x) \delta_{x+\hat{i},y} (r+\gamma_i) U_i^\dagger(x-\hat{i}) \delta_{x-\hat{i},y} \right\} \\ &- \kappa_\sigma c_E \sum_i \sigma_{i4} F_{i4} \delta_{x,y} - r\kappa_\sigma c_B \sum_{ij} \frac{1}{2} \sigma_{ij} F_{ij} \delta_{x,y}. \end{aligned} \tag{3}$$

The gluon field is represented with the link variable $U_\mu \simeq \exp(-iga_\mu A_\mu)$, and ψ denotes the anticommuting quark field. The spatial and the temporal hopping parameters, κ_σ and κ_τ, respectively, are related to the bare quark mass m_0 and the bare anisotropy parameter γ_F as

$$\kappa_\sigma = 1/2(m_0 + \gamma_F + 3r), \qquad \kappa_\tau = \gamma_F \kappa_\sigma, \tag{4}$$

where m_0 is measured in the spatial lattice unit. The value of the Wilson parameter r is set as $r = 1/\xi$. The coefficients c_E and c_B in the clover terms are introduced to eliminate the $O(a)$ error induced by the Wilson term, and coincide with unity at the tree level.

We apply the mean-field improvement which reduces large contributions from the tadpole diagrams to the renormalization. This is achieved by replacing the link variable as $U \to U/u_0$, with the mean-field value of the link variable, u_0. On the anisotropic lattice, two mean-field values u_σ and u_τ are defined for the spatial and the temporal link variables, respectively. We employ the definition of the mean-field value through the average of U in the Landau gauge. With u_σ and u_τ, the mean-field improved values of the clover coefficients at the tree level are expressed as $c_E = 1/u_\tau^2 u_\sigma$ and $c_B = 1/u_\sigma^3$. The anisotropy parameter is related to the improved anisotropy $\tilde{\gamma}_F$ as $\gamma_F = \tilde{\gamma}_F \cdot u_\sigma / u_\tau$. It is convenient to define κ, as

$$\frac{1}{\kappa}=\frac{1}{\kappa_\sigma u_\sigma}-2(\tilde{\gamma}_F+3r-4) \qquad (=2(m_0+4)). \tag{5}$$

For the light quark systems, the extrapolation to the chiral limit is performed in $1/\kappa$.

In practical simulations, the anisotropy parameter γ_F should be tuned so that the fermionic anisotropy ξ_F defined with fermionic observable coincides with the gauge field anisotropy. This is called as "calibration". Several procedures have been used for the calibration. In this paper, we set the value of γ_F using the dispersion relation of the pseudoscalar and the vector mesons. We assume that the meson field is described with the lattice Klein-Gordon equation,

$$S=\frac{1}{2\xi_F}\sum_x \phi^\dagger(x)\left[-\xi_F^2 D_4^2-\vec{D}^2+m_0^2\right]\phi(x), \tag{6}$$

with D_μ the lattice covariant derivative. The free meson field satisfies the dispersion relation,

$$\cosh E(\vec{p})-\cosh E(0)=\vec{p}^2/2\xi_F^2. \tag{7}$$

This relation is used to define the fermionic anisotropy ξ_F [11].

NUMERICAL RESULTS

Lattice Setup. The SU(3) lattice QCD simulations are performed on an anisotropic lattice of the size $12^3\times 96$ with anisotropy $\xi=4$, at the quenched level. As the gauge field action, we adopt the anisotropic Wilson action with the parameters $(\beta,\gamma_G)=(5.75,3.072)$, determined by Klassen so as to give the renormalized anisotropy $\xi=4$ within 1 % uncertainty [13]. At these parameters, the lattice cutoff defined by setting the string tension $\sqrt{\sigma}$ to be 427 MeV is found to be $a_\sigma^{-1}(\sqrt{\sigma})\simeq 1.0$ GeV. The gauge configuration is fixed to the Coulomb gauge, which is convenient to smear the quark propagators by extending the quark source spatially on a time slice.

The mean-field values are determined on the lattice of half size in the temporal direction, $12^3\times 48$, at the same (β,γ_G). We fix these gauge configurations to the Landau gauge and determine the mean-field values u_σ and u_τ self-consistently as described in [11]. They result in $u_\sigma=0.7620(2)$ and $u_\sigma=0.9871$ (error is less than the last digit).

Calibration. The quark field calibration is performed along the course described in the last section. The pseudoscalar and the vector meson correlators are calculated with momentum $\vec{p}=0$, $2\pi/16$ and $2\cdot 2\pi/16$. We use the meson operators listed in Table 2 and a standard procedure to extract the meson energy.

Figure 1 shows the result of the calibration. The left three values of κ correspond to the quark masses around the strange quark mass. We use these three values of κ for the following analysis of hadron spectroscopy. As is clearly observed in Fig. 1, in the light quark region ($m_q\simeq m_s$), one can set the bare anisotropy $\tilde{\gamma}_F=4$. In Table 1, we list the values of κ, the pseudoscalar and the vector meson masses for the degenerate quark case.

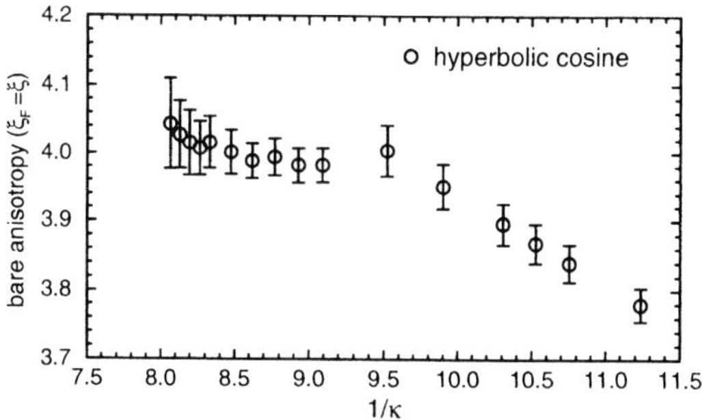

FIGURE 1. Results of the calibration in an anisotropic lattice. The value of $\tilde{\gamma}_F$ at which $\xi_F = \xi$ holds is determined for each value of κ.

TABLE 1. The hopping parameters used in the spectroscopy and the PS and V meson masses for the degenerate quark case.

κ	m_{PS}	m_V
0.124	0.1504(6)	0.2240(22)
0.123	0.2036(6)	0.2602(13)
0.122	0.2499(5)	0.2966(9)

The chiral extrapolation is carried out linearly in $1/\kappa$, and results in the critical hopping parameter as $\kappa_c = 0.12637(2)$.

Baryon spectrum. As listed in Table 2, we use the standard baryon operators which have the same quantum numbers as the corresponding baryons and survive in the non-relativistic limit. At large t (and large $N_t - t$), the baryon correlators are represented as

$$G_B(t) \equiv \sum_{\vec{x}} \langle B(\vec{x},t)\bar{B}(\vec{x},0)\rangle = (1+\gamma_4)\left[c_{B^+}\cdot e^{-tm_{B^+}} + bc_{B^-}\cdot e^{-(N_t-t)m_{B^-}}\right] + (1-\gamma_4)\left[bc_{B^+}\cdot e^{-(N_t-t)m_{B^+}} + c_{B^-}\cdot e^{-tm_{B^-}}\right], \quad (8)$$

where $b = +1$ and -1 for the periodic and antiperiodic temporal boundary conditions for the quark fields. Thus, combining the parity-projected correlators under two bound-

TABLE 2. Examples of the hadron operators. For baryon operators, the contraction with the color index is omitted.

Meson	Pseudoscalar	$M(K) = \bar{s}\gamma_5 u$
	Vector	$M_k(K^*) = \bar{s}\gamma_k u$
Baryon	Octet	$B_\alpha(\Sigma^0) = (C\gamma_5)_{\beta\gamma}[u_\alpha(d_\beta s_\gamma - s_\beta d_\gamma) - d_\alpha(s_\beta u_\gamma - u_\beta s_\gamma)]$
	Octet (Λ)	$B_\alpha(\Lambda) = (C\gamma_5)_{\beta\gamma}[u_\alpha(d_\beta s_\gamma - s_\beta d_\gamma) + d_\alpha(s_\beta u_\gamma - u_\beta s_\gamma) - 2s_\alpha(u_\beta d_\gamma - d_\beta u_\gamma)]$
	Singlet	$B_\alpha(\Lambda_1) = (C\gamma_5)_{\beta\gamma}[u_\alpha(d_\beta s_\gamma - s_\beta d_\gamma) + d_\alpha(s_\beta u_\gamma - u_\beta s_\gamma) + s_\alpha(u_\beta d_\gamma - d_\beta u_\gamma)]$
	Decuplet	$B_{\alpha k}(\Sigma^{*0}) = (C\gamma_k)_{\beta\gamma}[u_\alpha(d_\beta s_\gamma + s_\beta d_\gamma) + d_\alpha(s_\beta u_\gamma + u_\beta s_\gamma) + s_\alpha(u_\beta d_\gamma + d_\beta u_\gamma)]$

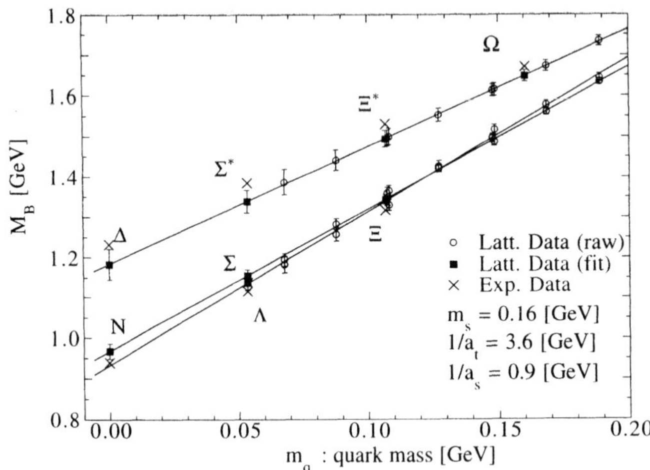

FIGURE 2. The spectra of octet and decuplet baryons with positive parity. The horizontal axis denotes the averaged current quark mass.

ary conditions, one can single out the positive and negative parity baryon states with corresponding masses m_{B^+} and m_{B^-}, respectively.

In our calculation, u,d current quark masses are taken to be the same value as $m_u = m_d$ or $\kappa_u = \kappa_d$, and the strange current quark mass m_s or κ_s is taken to be an independent value. Then, the baryon masses are expressed as the function of m_u and m_s like $M_B(m_u, m_s)$, and therefore the baryon masses $M_B(m_u, m_s)$ are to be depicted on the (m_u, m_s) plane. However, the lattice QCD result for the baryon masses M_B seem to be well described as a function of the averaged current quark mass over the three quarks, $m_q = (m_1 + m_2 + m_3)/3$, like $M_B(m_q)$ in each channel. Therefore, we will use the simplified figures as the function of m_q, although the actual lattice QCD calculations are performed with different quark masses on $m_u(= m_d)$ and m_s.

Let us start with the ground-state baryons shown in Figure 2, where the horizontal axis denotes the averaged current quark mass over the three quarks in the physical unit. We use the naive relation between the current quark mass and the hopping parameter as

$$m_q = \frac{1}{3}(m_1 + m_2 + m_3), \qquad m_i = \frac{1}{2}\left(\frac{1}{\kappa_i} - \frac{1}{\kappa_c}\right). \tag{9}$$

The baryon masses are linearly extrapolated to the chiral limit, $m_q = 0$. In the analysis of the baryon spectrum, we use the lattice scale determined by setting the averaged mass over the octet and the decuplet baryons at the chiral limit to the averaged mass of N and Δ. It results in $a_\sigma^{-1} = 0.9$ GeV. The strange quark mass is determined so that the octet and the decuplet baryons in Fig. 2 globally reproduce the experimentally measured masses. Thus, for the strange quark, we adopt $\kappa_s = 0.1229(1)$, which roughly corresponds to $m_s \simeq 0.16$ GeV as the current quark mass. This value seems consistent with the standard strange current-quark mass. Hereafter, we will fix $\kappa_s = 0.1229(1)$ for the strange quark.

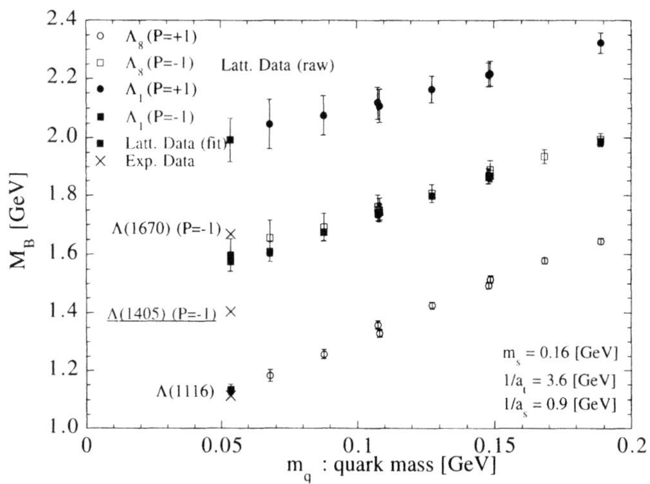

FIGURE 3. The positive and negative parity baryon masses in the flavor octet and singlet channels.

Figure 2 shows that the lattice result reproduces that the measured baryon spectrum within 5 % deviation.

Octet-Decuplet mass splitting. Let us consider the difference of octet and decuplet baryon masses. This mass splitting monotonously decreases as the current quark mass increases. This tendency is consistent with the one-gluon-exchange explanation in the nonrelativistic quark model, assuming the constituent quark mass M_q increases with the current quark mass m_q, as $M_q \simeq M_0 + m_q$.

Negative parity baryons. Figure 3 shows the mass spectra of the flavor octet and singlet baryons with positive and negative parities as the function of the averaged current quark mass m_q. In each channel, the baryon mass spectrum shows clear linear behavior on the averaged current quark mass m_q in the whole measured mass region. Then, we extrapolate the masses to the physical situation $m_u = m_d \simeq 0, m_s = 0.16\text{GeV}$, i.e., $m_q \simeq m_s/3$. One striking feature is that the singlet and octet negative parity baryons are almost degenerate. On the other hand, the lowest mass of the flavor-singlet positive-parity baryon is much larger than those in other channels. Extrapolating to the physical situation $m_u = m_d \simeq 0, m_s = 0.16\text{GeV}$ ($m_q \simeq m_s/3$), we obtain 1.6 GeV for the lowest mass of the flavor-singlet negative-parity baryon, which is described as a three quark state in the quenched lattice. Then, the experimentally observed negative-parity baryon $\Lambda(1405)$ is much lighter than the lowest mass of the flavor-singlet negative-parity baryon obtained in lattice QCD at the quenched level. Since the q-$\bar{q}$ pair creation is absent in the quenched QCD, this result implies that the simple three valence quark picture would not valid for $\Lambda(1405)$. One possible explanation is, as frequently suggested, that the $\Lambda(1405)$ is a mixture of a three quark state and an $N\bar{K}$ bound state. To clarify this subject, we are going to measure the $N\bar{K}$ state on the quenched lattice, where genuine bound state properties is apparent owing to the absence of the q-$\bar{q}$ pair creation.

SUMMARY AND OUTLOOK

In this paper, we report present status of our investigation of hadron properties using lattice QCD simulations. At this stage, systematic investigation of baryon spectrum is in progress at the quenched level. The result on the anisotropic lattice with $a_\sigma^{-1} \simeq 0.9$ GeV and anisotropy $\xi = 4$ is as follows: (a) The octet-decuplet baryon mass splitting decreases with increasing current quark mass. This tendency is consistent with the one-gluon-exchange explanation in the constituent quark model. (b) The negative-parity baryons in octet and singlet channels are measured in good statistical precision. However, the experimentally observed negative-parity baryon $\Lambda(1405)$ seems much lighter than the lattice QCD result for the lowest mass of the SU(3) flavor-singlet negative-parity baryon. This may suggest the possibility of that $\Lambda(1405)$ is a mixture of the $N\bar{K}$ bound state and the three quark state. This would be clarified by successive lattice calculations for qqq-$q\bar{q}$ system in the quenched approximation.

Our present results have been carried out on rather course lattice. Hence, we need to perform the simulations on finer lattices to remove lattice artifacts. Then, the anisotropic lattice will be a powerful device to study detailed properties of hadrons including excited states, exotics and glueballs. The simulation with dynamical quarks are also to be performed.

Another course of our program is a comparison with the potential model analysis using the static quark potential extracted from lattice QCD. This will give us important information on the quark wave function and novel insight on the quark structure of hadrons.

We are grateful to Profs. Il-Tong Cheon and Su Houng Lee for their warm hospitality at Yonsei University. The lattice QCD simulations have been performed on NEC SX4 at Osaka University, and Hitachi SR8000 at KEK.

REFERENCES

1. CP-PACS Collaboration (S.Aoki et al.), Phys. Rev. Lett. **84**, 238 (2000).
2. For recent review, see e.g. S.Aoki, Nucl. Phys. B (Proc. Suppl.) **94**, 3 (2001).
3. See also, H. Suganuma et al., these proceedings.
4. T.T. Takahashi, H. Suganuma, H. Matsufuru and Y. Nemoto, these proceedings.
5. N. Isgur and G. Karl, Phys. Rev. D **20**, 1191 (1979).
6. M. Oka and K. Yazaki, Prog. Theor. Phys. **66**, 556 (1981), *ibid.* 572.
7. S. Sasaki, T. Blum and S.Ohta, hep-lat/0102010, and references therein.
8. F. Karsch, Nucl. Phys. **B205**, 285 (1982).
9. T.R. Klassen, Nucl. Phys. B (Proc. Suppl.) **73**, 918 (1999).
10. A. El-Khadra, A.S. Kronfeld and P.B. Mackenzie, Phys. Rev. D **55**, 3933 (1997).
11. T. Umeda, R. Katayama, O. Miyamura and H. Matsufuru, hep-lat/0011085, to appear in Int. J. Mod. Phys. A.
12. J. Harada, A.S. Kronfeld, H. Matsufuru, N. Nakajima and T. Onogi, hep-lat/0103026.
13. T.R. Klassen, Nucl. Phys. **B533**, 557 (1998).

Current mass dependence of the quark condensate and the constituent quark mass

M.Musakhanov

Theoretical Physics Dept, Uzbekistan National University, Tashkent 7000174, Uzbekistan

Abstract. We discuss the current mass dependence of the basic quantities of the quark models – constituent quark mass M and quark condensate $i < \psi^\dagger\psi >$. The framework of the consideration is QCD instanton vacuum model.

INTRODUCTION

Among the quark models, instanton vacuum based Effective Action approach is a most promising since in this model the hadron properties and their interactions features are closely related to the properties of QCD vacuum.

Without any doubts instantons are a very important component of the QCD vacuum. Their properties are described by the average instanton size ρ and inter-instanton distance R. In 1982 Shuryak [1] fixed them phenomenologically as

$$\rho = 1/3\,fm, \;\; R = 1\,fm. \tag{1}$$

From that time the validity of such parameters was confirmed by theoretical variational calculations [2] and recent lattice simulations of the QCD vacuum (see recent review [3]). The presence of instantons in QCD vacuum very strongly affects light quark properties, owing consequent generation of quark-quark interactions. These effects lead to the formation of the massive constituent interacting quarks. This implies spontaneous breaking of chiral symmetry (SBCS), which leads to the collective massless excitations of the QCD vacuum–pions. The most important degrees of freedom in low-energy QCD are these quasiparticles. So instantons play a leading role in the formation of the lightest hadrons and their interactions, while the confinement forces are rather unimportant, probably. The properties of the hadrons and their interactions are concentrated in the QCD Effective Action in terms of quasiparticles. The features of light quarks placed into instanton vacuum are concentrated in the fermionic determinant $\det_N$ (in the field of N_+ instantons and N_- antiinstantons) calculated by Lee and Bardeen(LB) in 1979 [4]:

$$\det{}_N = \det B, \; B_{ij} = im\delta_{ij} + a_{ji}, \tag{2}$$

and a_{ij} is the overlapping matrix element of the quark zero-modes $\Phi_{\pm,0}$ generated by instantons(antiinstantons). This matrix element is nonzero only between instantons and antiinstantons (and vice versa) due to specific chiral properties of the zero-modes and equal to

$$a_{-+} = - < \Phi_{-,0}|i\hat{\partial}|\Phi_{+,0} > . \tag{3}$$

CP594, *Hadrons and Nuclei: First International Symposium,* edited by Il-T. Cheon et al.

The overlapping of the quark zero-modes provides the propagating of the quarks by jumping from one instanton to another one. So, the determinant of the infinite matrix was reduced to the determinant of the finite matrix in the space of only zero-modes. From Eqs. (3), (2) it is clear that for $N_+ \neq N_-$ $\det_N \sim m^{|N_+ - N_-|}$ which will strongly suppress the fluctuations of $|N_+ - N_-|$. Therefore in final formulas we will assume $N_+ = N_- = N/2$.

In (2) we observe the competition between current mass m and overlapping matrix element $a \sim \rho^2 R^{-3}$. With typical instanton sizes $\rho \sim 1/3 fm$ and inter-instanton distances $R \sim 1 fm$, a is of the order of the strange current quark mass, $m_s = 150 MeV$. So in this case it is very important to take properly into account the current quark mass.

The fermionic determinant $\det_N$ averaged over instanton/anti-instanton positions, orientations and sizes can be considered as a partition function of light quarks Z_N. Then the properties of the hadrons and their interactions are concentrated in the QCD Effective Action. We calculate this one via fermionic representation of $\det_N$ which provide easy way for the averaging over instanton collective coordinates – positions and orientations [5]. This approach leads to the Diakonov-Petrov(DP) Effective Action [6] with a specific choice of the degrees of freedom [7]. It was shown that DP Effective Action is a good tool in the chiral limit but failed beyond this limit, checked by the calculations of the axial-anomaly low energy theorems [7]. The solution of this puzzle related with the observation that the fermionisation of $\det_N$ is not unique procedure and another fermionic representation of $\det_N$ leads to a different choice of the degrees of freedom in the Effective Action. Within this approach it was proposed so called Improved Effective Action(IA) which is more properly taken into account current quark masses and satisfied axial-anomaly low energy theorems also beyond the chiral limit [8] at least at $O(m)$ order.

Completely another approach to the same problem was developed by Pobylitsa [9]. He directly summed up planar diagrams for the propagator in the instanton medium in large N_c limit for two extreme cases: $N/VN_c -> 0$ and $N/VN_c -> \infty$. We will compare his result for constituent mass with our one and will calculate quark condensate within this approach too.

In the present case we concentrate on the calculation of the current mass dependence of the quark condensate $i < \psi^\dagger \psi >$. As a byproduct we find also current mass dependence of the constituent quark mass M. Since the quark condensate does not dependent on the specific choice of the degrees of freedom in the Effective Action and entirely is defined by the current mass dependence of the partition function Z_N, it gives important information on the accuracy of the fundamental LB result by comparison with phenomenological data. LB result (2) itself has accuracy which is $O(m^2)$ order. We consider here the m dependencies of quark condensate $i < \psi^\dagger \psi >$ and constituent quark mass M within DP and IA and compare with results of slightly modified version of Improved Action(MIA), which has difference from IA on of $O(m^2)$ order terms. So, both DP and IA approaches must lead to the same m-dependence of the quark condensate and must coincide with MIA approach, at least within $O(m^2)$ accuracy. We consider first modification of Improved Action, and further calculate the current mass dependencies of abovementioned quantities within all variants of Effective Action. The comparison of these results with the result of the calculations within Pobylitsa approach provide independent test of the calculations. Another test of the results is provided by heavy quark limit, under the assumption that the gluon field strength $G^a_{\mu\nu}$ is much less than the square of quark mass m^2. We find that the quark condensate in all approaches based on instanton vacuum model has almost the same rather strong m dependence and in the region $m > 0.3\ GeV$ they are in accordance with heavy quark approximation. As example, the strange quarks condensate $< s^\dagger s > \sim 0.5 < u^\dagger u >$ at $m_s \sim 0.15\ GeV$. We find also rather strong m dependence of the constituent quark mass M. Since in Pobylitsa, IA and MIA cases the total quark mass is $m+M$, the total mass is almost constant in the region $m < 0.2\ GeV$. This dependencies contrasted very much naive expectations and need detailed phenomenological analysis.

MODIFICATION OF IMPROVED EFFECTIVE ACTION

First, accordingly Eqs. (2) and (3) and by introducing the Grassmanian (N_+, N_-) variables $\Omega_i\ \bar{\Omega}_j$ we represent

$$\det B = \int d\Omega d\bar{\Omega} \exp(\bar{\Omega} B \Omega), \tag{4}$$

where

$$\bar{\Omega} B \Omega = \bar{\Omega}_i (im + a^T)_{ij} \Omega_j = -\Omega_j (\Phi^+_{j,0}(-i\hat{\partial} + im)_{ji} \Phi_{i,0}) \bar{\Omega}_i \tag{5}$$

This formula is transformed to:

$$\begin{aligned} \bar{\Omega} B \Omega &= \Omega_j (\Phi^+_{j,0}(i\hat{\partial}(i\hat{\partial} + im)^{-1} i\hat{\partial} + m^2 (i\hat{\partial} + im)^{-1})_{ji} \Phi_{i,0}) \bar{\Omega}_i \\ &= \Omega_j (\Phi^+_{j,0}(i\hat{\partial}(i\hat{\partial} + im)^{-1} i\hat{\partial})^{-1})_{ji} \Phi_{i,0}) \bar{\Omega}_i + \ldots \end{aligned} \tag{6}$$

where we are neglecting by $O(m^2)$ term, since Lee-Bardeen result for det_N itself was derived within this accuracy. The next step is to introduce N_+, N_- sources η_i and $\bar{\eta}_j$ defined as:

$$\bar{\eta}_i = -\Phi^+_{i,0} \Omega_i i\hat{\partial}, \eta_j = i\hat{\partial} \Phi_{j,0} \bar{\Omega}_j \tag{7}$$

Then $(\bar{\Omega} B \Omega)$ can be rewritten as

$$(\bar{\Omega} B \Omega) = -\bar{\eta}(i\hat{\partial} + im)^{-1} \eta = -\sum_{ij} \bar{\eta}_j (i\hat{\partial} + im)^{-1} \eta_i \tag{8}$$

and $\det B$ can be rewritten as

$$\begin{aligned} \det B &= \int d\Omega d\bar{\Omega} \exp(\bar{\Omega} B \Omega) \\ &= \left(\det(i\hat{\partial} + im)\right)^{-1} \int d\Omega d\bar{\Omega} D\psi D\psi^\dagger \exp \int dx (\psi^\dagger(x)(i\hat{\partial} + im)\psi(x) \\ &+ \sum_i (\bar{\eta}_i(x)\psi(x) + \psi^\dagger(x)\eta_i(x))) \end{aligned} \tag{9}$$

The integration over Grassmanian variables Ω and $\bar{\Omega}$ (with the account of the N_f flavors $\det_N = \prod_f \det B_f$) provides finally the fermionized representation of fermionic determinant (6) in the form:

$$\det B = \int D\psi D\psi^\dagger \exp\left(\left(\int d^4x \sum_f \psi_f^\dagger (i\hat{\partial} + im_f)\psi_f\right)\right.$$
$$\left.\times \prod_f \left\{ \prod_+^{N_+} V_+[\psi_f^\dagger, \psi_f] \prod_-^{N_-} V_-[\psi_f^\dagger, \psi_f] \right\}\right., \tag{10}$$

where

$$V_\pm[\psi_f^\dagger, \psi_f] = \int d^4x \left(\psi_f^\dagger(x) i\hat{\partial}\Phi_{\pm,0}(x;\xi_\pm)\right) \int d^4y \left(\Phi_{\pm,0}^\dagger(y;\xi_\pm)(i\hat{\partial}\psi_f(y)\right). \tag{11}$$

Now the averaging over collective coordinates $\xi_\pm$ become trivial problem. The further steps are the exponentiation and the bosonization of the integrand [6]. Finally, the corresponding partition function in terms of constituent quarks has a form [8]:

$$Z_N = \int d\lambda_+ d\lambda_- D\Phi_+ D\Phi_- \exp\left(-S[\lambda_+, \Phi_+; \lambda_-, \Phi_-]\right), \tag{12}$$

where

$$S[\lambda_+, \Phi_+; \lambda_-, \Phi_-] = -\sum_\pm \left(N_\pm \ln[\left(\frac{4\pi^2\rho^2}{N_c}\right)^{N_f} \frac{N_\pm}{V\lambda_\pm}] - N_\pm \right) + S_\Phi + S_\psi, \tag{13}$$

$$S_\Phi = \int d^4x \sum_\pm (N_f - 1)\lambda_\pm^{-\frac{1}{N_f-1}} (\det \Phi_\pm)^{\frac{1}{N_f-1}},$$

$$S_\psi = -\mathrm{Tr}\ln((-\hat{k} + im_f\delta_{fg} + iF(k_1)F(k_2)\sum_\pm \Phi_{\pm,fg}(k_1 - k_2)\frac{1\pm\gamma_5}{2})(-\hat{k} + im_f)^{-1}).$$

Variation of the total action $S[\lambda_+, \Phi_+; \lambda_- \Phi_-]$ over $\lambda_\pm$ leads to the saddle-point:

$$\lambda_\pm = (N_\pm^{-1} \int d^4x (\det \Phi_\pm)^{\frac{1}{N_f-1}})^{(N_f-1)},$$

The additional variation over $\Phi_\pm$ must vanish in the common saddle-point. Since we take $N_+ = N_- = N/2$, this one is

$$\Phi_{\pm,fg} = \Phi_{\pm,fg}(0) = M_f \delta_{fg},$$

and

$$\lambda_\pm = \lambda = \frac{2V}{N} \prod_f M_f,$$

This condition leads to the the saddle-point equation for the momentum dependent constituent mass $M_f(k)$, i. e.,

$$M_f(k) = M_f F^2(k). \tag{14}$$

The contribution of the quark loop to the saddle-point equation is

$$\mathrm{Tr}\ln[(-\hat{k}+im+iF^2(k)\sum_{\pm}\Phi_{\pm}\frac{1\pm\gamma_5}{2})(-\hat{k}+im)^{-1}]. \tag{15}$$

Then we get the saddle-point equation

$$N/V=4N_c\int\frac{d^4k}{(2\pi)^4}\frac{M_fF^2(k)(m_f+M_fF^2(k))}{k^2+(m_f+M_fF^2(k))^2} \tag{16}$$

The form-factor $F(k)$ is related to the zero–mode wave function in momentum space $\Phi_{\pm}(k;\xi_{\pm})$ [6]. We use simplified expression for this form-factor:

$$F(k)=\frac{L^2}{L^2+k^2}, \tag{17}$$

where $L^2\sim 2/\rho^2=0.72\,GeV^2$. which was proposed in [11].

The important steps in the derivation of these formulas (12) and (13) are:
1. The fermionisation of (6) (which is in fact not unique procedure);
2. Independent averaging over positions and orientations of the instantons, due to the small packing parameter of the instanton media – $(\rho/R)^4\sim(1/3)^4$;
3. The exponentiation and the bosonization of the partition function, described in [6].

The matrices $\Phi_{\pm}$, whose usual decomposition is $\Phi_{\pm}=\exp(\pm\frac{i}{2}\phi)M\sigma\exp(\pm\frac{i}{2}\phi)$, ϕ and σ being $N_f\times N_f$ matrices, describes mesons and $M_{fg}=M_f\delta_{fg}$. At the saddle-point $\sigma=1,\phi=0$. The usual decomposition for the pseudoscalar fields $\phi=\sum_0^8\lambda_i\phi_i$ may be used. These mesons are considered as a small fluctuation near the saddle point.

The account of the fluctuations of number of instantons N can be easily done. Let $N_{\pm}=0.5(N\pm\Delta)$, $\Delta<<N$. Then assuming the saddle-points $\Phi_{\pm fg}=\delta_{fg}M_{f\pm}$, $M_{f\pm}=M_f(1\pm\delta_f)$, $\delta_f<<1$ we find additional to (16) another saddle-point equation

$$\Delta/V=4N_c\delta_f\int\frac{d^4k}{(2\pi)^4}\frac{m_fM_fF^2(k)}{k^2+(m_f+M_fF^2(k))^2} \tag{18}$$

Taking into account the definition of the condensate $i<\psi^\dagger\psi>$ (27) we find

$$m_f\delta_f=\frac{\Delta}{Vi<\psi_f^\dagger\psi_f>}(1+O(m_f^2)) \tag{19}$$

This formula leads to the Δ–distribution, which is in accordance with general theorems [6].

CURRENT MASS DEPENDENCE OF CONSTITUENT MASS

The saddle-point equation (16) leads to the momentum dependent constituent mass $M_f(k)=M_fF^2(k)$. The constituent quark propagator, as it is follows from (12), has a form:

$$S=(-\hat{k}+i(m_f+M_fF^2(k)))^{-1}, \tag{20}$$

In IA approach analogous saddle-point condition:

$$N/V = 4N_c \int \frac{d^4k}{(2\pi)^4} \frac{M_f^{IA} F^2(k)(1+m_f^2/k^2)(m_f + M_f^{IA} F^2(k)(1+m_f^2/k^2))}{k^2 + (m_f + M_f^{IA} F^2(k)(1+m_f^2/k^2))^2} \tag{21}$$

also define M_f^{IA} and the constituent quark propagator has a form:

$$S^{IA} = (-\hat{k} + i(m_f + M_f^{IA} F^2(k)(1+m_f^2/k^2)))^{-1}, \tag{22}$$

On the other hand, DP Effective Action leads to the propagator:

$$S^{DP} = (-\hat{k} + M_f^{DP}(k)))^{-1}, \tag{23}$$

where $M_f^{DP}(k) = M_f^{DP} F^2(k)$ is followed from analogous saddle-point equation:

$$\frac{4VN_c}{N} \int \frac{d^4k}{(2\pi)^4} \frac{M_f^{DP2}(k)}{k^2 + M_f^{DP2}(k)} = 1 - \frac{M_f^{DP} m_f V N_c}{2\pi^2 \rho^2}, \tag{24}$$

As was mentioned in the Introduction, Pobylitsa [9] in quenched approximation directly summed up planar diagrams for the propagator in the instanton medium in large N_c limit for two extreme cases: $N/VN_c -> 0$ and $N/VN_c -> \infty$. In the first (and most interesting) case his result can be summarized in the form:

$$S_P = (-\hat{k} + i(m + M^P(k)))^{-1}, \tag{25}$$

where

$$\begin{aligned} M^P(k) &= M_0 F^2(k)[(1+m^2/d^2)^{1/2} - m/d], \\ d &= \left(\frac{0.08385}{2N_c}\right)^{0.5} \frac{8\pi\rho}{R^2} = 0.198\,GeV. \end{aligned} \tag{26}$$

Fig.1 represent different versions of the current mass dependence of constituent mass derived from saddle-point equations (16), (21), (24) and Pobylitsa result (26).

CURRENT MASS DEPENDENCE OF THE QUARK CONDENSATE

First, we calculate the quark condensate by using the evident formula

$$\begin{aligned} i < \psi_f^\dagger \psi_f > &= V^{-1} Z_N^{-1} \frac{\partial Z_N}{\partial m_f} \\ &= \mathrm{Tr} i[(-\hat{k} + im_f + iM_f(k))^{-1} - (-\hat{k} + im_f)^{-1}]. \end{aligned} \tag{27}$$

In (27) the saddle-point condition was taken into account. It is evident, the condensates in IA and Pobylitsa approaches are calculated with similar formula.

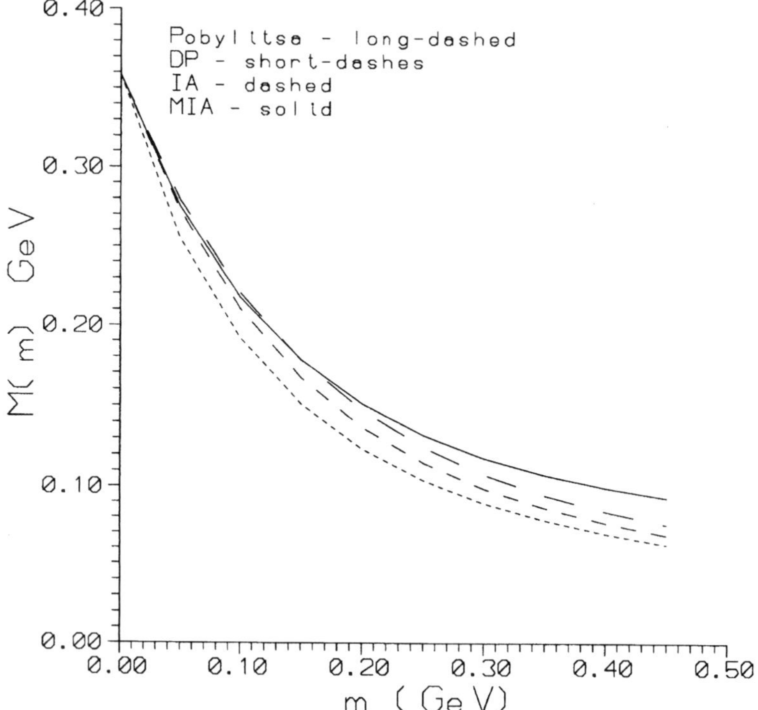

FIGURE 1. *Current mass dependence of the constituent mass: Solid line – Modified Improved Action calculations. Long dashed line – Pobylitsa approach calculations. Dashed line – Improved Action calculations. Short dashed line – DP Action calculations.*

With DP Action [6] the condensate is

$$i < \psi_f^\dagger \psi_f >^{DPW} = \frac{N_c M_f^{DP}}{2\pi^2 \rho^2}. \tag{28}$$

Simple numerical calculations leads to the condensate as a function of current mass presented in Fig.2. We present here also heavy quark approximation result for the quark condensate. In this limit, for heavy quarks we have to use the expansion over small parameter G/m^2 under the assumption that the gluon field strength $G^a_{\mu\nu}$ is much less than the square of quark mass m^2 [12, 13]. Then

$$\begin{aligned} \int d^4xi < \psi^\dagger \psi > &= \mathrm{Tr}\left(\frac{i}{\hat{\mathcal{P}} + im_c} - \frac{i}{\hat{p} + im}\right) = m\mathrm{Tr}\left(\frac{1}{\mathcal{P}^2 + m_c^2 + \frac{g}{2}\sigma G} - \frac{1}{p^2 + m^2}\right) \\ &= \int d^4x\left(\frac{g^2}{24\pi^2 m}\mathrm{tr}_c G^2_{\alpha\beta} + \frac{ig^3}{360\pi^2 m^3}\mathrm{tr}_c G_{\alpha\beta}G_{\alpha\gamma}G_{\beta\gamma} + \ldots\right) \end{aligned} \tag{29}$$

Here $\sigma G \equiv \sigma_{\mu\nu}G_{\mu\nu}$, $\sigma_{\mu\nu} \equiv \frac{i}{2}[\gamma_\mu, \gamma_\nu]$, $\hat{\mathcal{P}} \equiv \mathcal{P}_\mu \gamma_\mu$ and $\mathcal{P}_\mu = iD_\mu = i(\partial_\mu - igA^a_\mu t^a)$.

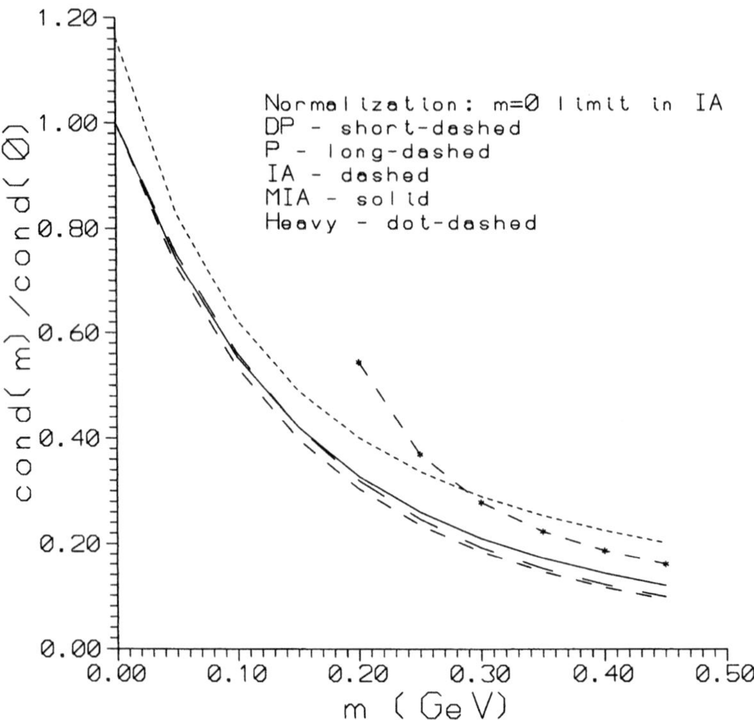

FIGURE 2. *Current mass dependence of the quark condensate: Solid line – Modified Improved Action calculations. Long dashed line – Pobylitsa approach calculations. Dashed line – Improved Action calculations. Short dashed line – DP Action calculations. Dashed-dot line – Heavy quark approximation (30).*

In the instanton vacuum:

$$\int d^4x \mathrm{tr}_c g^2 G^2_{\alpha\beta} = (4\pi)^2 N, \quad \int d^4x \mathrm{tr}_c g^3 G_{\alpha\beta} G_{\alpha\gamma} G_{\beta\gamma} = -i(4\pi)^2 \frac{6}{5\rho^2} N$$

and

$$i < \psi^\dagger \psi >= \frac{2}{3mR^4} + \frac{4}{75\rho^2 m^3 R^4} = \frac{2}{3mR^4}(1 + \frac{12}{150\rho^2 m^2}) \tag{30}$$

We see that MIA,IA, DP and Pobylitsa results almost coincide with each other and in the good correspondence with heavy quark approximation at $m > 0.3\,GeV$.

CONCLUSION

It were investigated the current quark mass dependencies of quark condensate and constituent quark mass in QCD instanton vacuum model. It were considered different

approaches:
1. Diakonov&Petrov effective action (see recent papers [6]);
2. Improved Action [8] and presented here Modified Improved Action.
They are essentially based on Lee& Bardeen result for the quark determinant in instanton vacuum background [4].
3. Direct summation of planar diagrams for the propagator in the instanton medium in quenched approximation [9].
All of these approaches leads to rather fast dependencies of quark condensate and constituent quark mass on current quark mass, as were demonstrated in Figs 1,2. Then, the strange quarks condensate $< s^{\dagger}s > \sim 0.5 < u^{\dagger}u >$ at $m_s \sim 0.15\, GeV$ and total quark mass $m + M$ is almost constant in the region $m < 0.2\, GeV$.

We conclude that strange quark physics might be very different from usual expectation based on old-fashioned quark model and demand careful phenomenological reanalysis.

ACKNOWLEDGMENTS

I am very grateful to M.Birse, D.Diakonov, V.Petrov, P.Pobylitsa and M.Polyakov for useful discussions.

REFERENCES

1. E. V. Shuryak, *Nucl. Phys.* **B 203**, 93, 116 (1982)
2. D. Diakonov and V. Petrov, *Nucl. Phys.* **B 245**, 259 (1984)
3. T. De Grand, A. Hasenfratz, T. Kovacs, *Progr.Theor. Phys. Suppl.* **131**, 573 (1998)
4. C. Lee, W. A. Bardeen, *Nucl. Phys.* **B 153**, 210 (1979)
5. M. M. Musakhanov, F. C. Khanna, *Phys. Lett.* **B 395**, 298 (1997)
6. D.I. Diakonov, M.V. Polyakov, C. Weiss, *Nucl. Phys.* **B 461**, 539 (1996); Dmitri Diakonov, hep-ph/9602375, hep-ph/9802298
7. E. Di Salvo, M.M. Musakhanov, *Europ.Phys.J.* **C 5**, 501 (1998)
8. M.M. Musakhanov, *Europ. Phys. J.* **C9**, 235 (1999), hep-ph/9810295
9. P. Pobylitsa, *Phys.Lett.* **B 226**, 387 (1989); Quarks in the instanton vacuum I. Quark propagator, preprint 1598, LINP, 1990
10. V.F. Tokarev, preprint INP P-0406, Moscow 1985; *Soviet J. Teor. Math. Phys.*, **73**, 223 (1987)
11. V.Yu. Petrov, M.V. Polyakov, R. Ruskov, C. Weiss and K. Goeke, hep-ph/9807229
12. A.I.Vainshtein, V.I.Zakharov, V.A.Novikov, M.A.Shifman, Sov.J.Nucl.Phys. **39**, 77 (1984).
13. F. Araki, M. Musakhanov, H. Toki, Axial Currents of Virtual Charm in Light Quark Processes, hep-ph/9808290.

Spectral Change of Hadrons and Chiral Symmetry

T. Hatsuda

Physics Department, University of Tokyo, Tokyo 113-0033, Japan

Abstract. After a brief summary of the QCD phase structure with light quarks, we discuss two recent developments on in-medium hadrons. First topic is the σ-meson which is a fluctuation of the chiral order parameter $\bar{q}q$. Although σ is at best a broad resonance in the vacuum, it may suffer a substantial red-shift and show a characteristic spectral enhancement at the $2m_\pi$ threshold at finite temperature and baryon density. Possible experimental signatures of this phenomenon are also discussed. Another topic is the first principle lattice QCD calculation of the hadronic spectral functions using the maximum entropy method (MEM). The basic idea and a successful example of MEM are presented. Possible applications of MEM to study the in-medium hadrons in lattice QCD simulations are discussed.

INTRODUCTION

One of the most intriguing phenomena in quantum chromo dynamics (QCD) is the dynamical breaking of chiral symmetry. This explains the existence of the pion and dictates most of the low energy phenomena in hadron physics. The dynamical breaking of chiral symmetry is associated with the condensation of quark - anti-quark pairs in the QCD vacuum, $\langle\bar{q}q\rangle$, which is analogous to the condensation of Cooper pairs in the theory of superconductivity [1]. As the temperature (T) and/or the baryon density (n_B) increase, the QCD vacuum undergoes a phase transition to the chirally symmetric phase where $\langle\bar{q}q\rangle$ vanishes. Studying what are the possible phase structure and how the phase transition takes place are one of the main aims of the modern hadron physics [2].

When we talk about the phase structure, we need to fix appropriate variables to define the phases. Such variables in QCD are current masses of light quarks ($m_{u,d,s}$), T, and n_B. Since the following relation holds, neglecting $m_{u,d}$ in the first approximation and studying the phase structure in the m_s-T-n_B space would give us a good insight into the real world:

$$m_{u,d} \ll m_s \sim T_c \sim n_c^{1/3} \sim \Lambda_{QCD}. \tag{1}$$

Here T_c (n_c) are the critical temperature (density). In Fig.1, possible phase structures in the m_s-T plane (with n_B=0) and m_s-n_B plane (with T=0) are shown. m_s=0 (m_s=∞) corresponds to the limit of SU(3)⊗SU(3) (SU(2)⊗SU(2)) chiral symmetry. Recent lattice simulations show $T_c \simeq 175$ (155) MeV for 3 (2) flavors [3]. Furthermore, the first order transition for small m_s turns into the second order one for large m_s at the tricritical point [4]. The precise value of m_s at the tricritical point is not known yet.

The phase transition at finite baryon density is not well understood mostly because the information from lattice QCD is poor. Naive analysis based on the bag model shows

CP594, *Hadrons and Nuclei: First International Symposium,* edited by Il-T. Cheon et al.

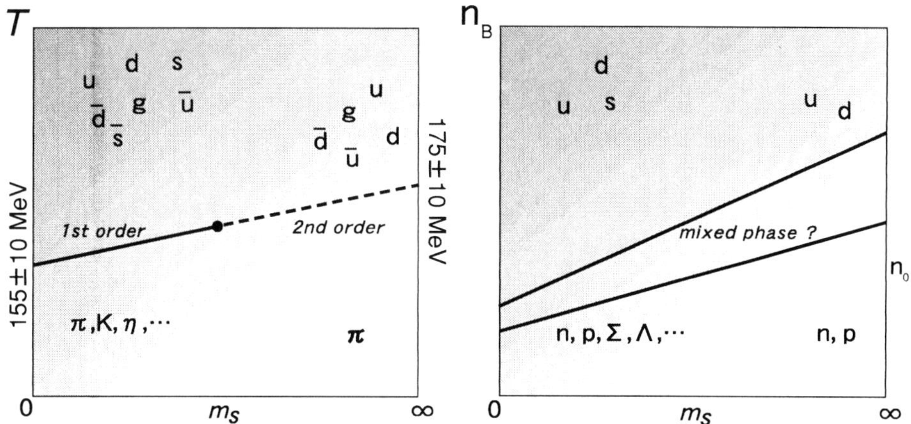

FIGURE 1. Possible phases in the m_s-T plane at n_B=0 (left panel) and the m_s-n_B plane at T=0. $m_{u,d}$=0 is assumed. Relevant degrees of freedom in each phase are also shown. $n_0 = 0.17\text{fm}^{-3}$ is the normal nuclear matter density.

the first order transition [5], which implies the existence of a mixed phase in the m_s-n_B plane as shown in Fig.1 (right panel). However, the possibility of a smooth hadron-quark transition is not excluded. The bag model analysis also suggests the strange quark matter as the true ground state of matter for sufficiently small m_s [6]. The system at finite baryon density, which is intrinsically quantum, is obviously richer in physics than that at finite T and has various phases such as the color superconductivity, meson condensations, and baryon superfluidity (see the reviews, [7]).

Now, what are the observables associated with the chiral transition? Model independent statement is that hadronic correlations in a same chiral multiplet (such as S (scalar)$-P^a$ (pseudo-scalar) and V_μ^a (vector)$-A_\mu^a$ (axial-vector)) should be degenerate when $\langle \bar{q}q \rangle \to 0$, namely

$$\langle S(x)S(y)\rangle \to \langle P^a(x)P^a(y)\rangle, \quad \langle A_\mu^a(x)A_\nu^b(y)\rangle \to \langle V_\mu^a(x)V_\nu^b(y)\rangle. \tag{2}$$

Thermal susceptibilities for hadronic operators O defined in the Euclidian space,

$$\chi_H = \int_0^{1/T} d\tau \int d^3x \langle O^\dagger(\tau,\vec{x})O(0,\vec{0})\rangle, \tag{3}$$

may be used to check such chiral degeneracy. Shown in Fig.2 is the full QCD simulation of $\sqrt{1/\chi_H}$ on the lattice with 2 light flavors [8]. One can see the degeneracy between the σ-channel (I=0, J^P=0^+) and the π-channel (I=1, J^P=0^-) above the critical point. There remains a splitting between the δ-channel (I=1, J^P=0^+) and the π-channel above the critical point, which reflects the breaking of $U_A(1)$ symmetry.

To make a direct connection with the experimental observables, however, one needs to go further and calculate the Minkowski (real-time) correlation and the hadronic spectral

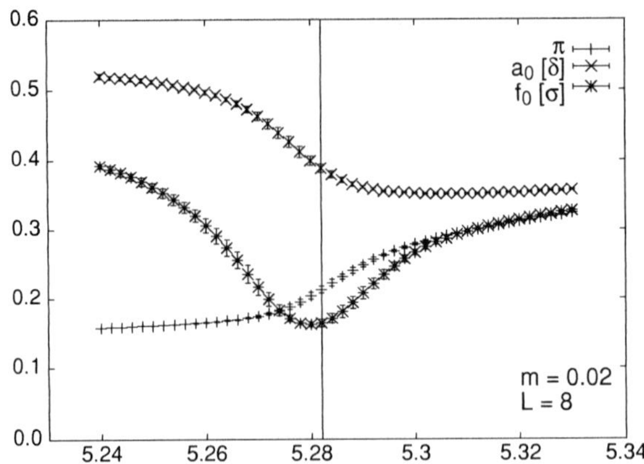

FIGURE 2. Thermal susceptibilities in three different channels (π, σ, and δ) for two-flavor QCD on the $8^3 \times 4$ lattice with $m_{u,d}\ a = 0.02$ [8]. The vertical (horizontal) axis denotes $\sqrt{1/\chi_H}$ (the lattice coupling $\beta = 6/g^2$). Courtesy of Nucl. Phys. B.

functions at finite T and n_B [9]. The aim of this talk is to show some recent attempts toward this goal. In Sec.2 and Sec.3, the spectral function in the σ-channel will be discussed near 2π threshold where almost model-independent argument is possible.

Experimentally, the resonances in the scalar and axial-vector channels have large width and are difficult to measure with possible exception discussed in Sec.2,3. For this reason, in-medium neutral vector-mesons which couple directly to the virtual photon have been studied both theoretically and experimentally as a probe of the chiral transition [10]. Unfortunately, theoretical calculations of the spectral function in the vector channel on the basis of the effective Lagrangians are rather model dependent, and the first principle lattice QCD study has been called for. Recently, a new promising approach was proposed: it utilizes the maximum entropy method for extracting the spectral functions from lattice QCD data. We will discuss this new development in Sec.4.

SPECTRAL ENHANCEMENT IN HOT PLASMA

The fluctuation of the order parameter becomes large as the system approaches to the critical point of a second order or weak first order phase transition. In QCD, the fluctuation of the phase (amplitude) of the chiral order parameter $\langle \bar{q}q \rangle$ corresponds to π (σ). The vital role of such fluctuations in the *dynamical* phenomena near the critical point at finite T was originally studied in [11, 12].

It was suggested that the chiral restoration gives rise to a softening (the red-shift) of the σ, which in turn leads to a small σ-width due to the suppression of the phase-space of the decay $\sigma \to 2\pi$ [11]. Therefore, σ may appear as a narrow resonance at finite T although it is at best a very broad resonance in the free space with a width comparable to its mass [13]. (For the phenomenological applications of this idea in relation to the relativistic heavy ion collisions, see [14].)

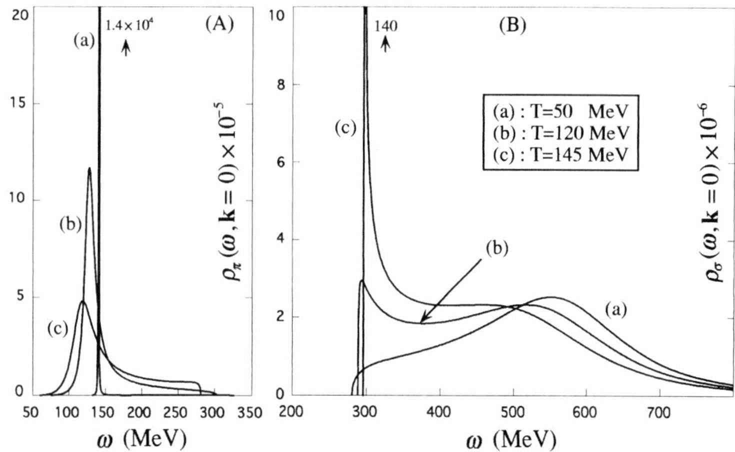

FIGURE 3. Spectral functions in the π channel (A) and in the σ channel (B) for T=50, 120, and 145 MeV. [15] $\rho_\sigma(\omega)$ in (B) shows only a broad bump at low T (a), while the spectral concentration is developed as T increases, (a)→(b)→(c). Courtesy of Phys. Rev. D.

Further theoretical analysis by taking account the coupling $\sigma \leftrightarrow 2\pi$ shows that (i) the spectral function is the most relevant quantity for studying the true nature of σ, and (ii) the spectral function of σ has a characteristic enhancement at the 2π threshold near T_c [15]. In Fig.3, shown are the spectral functions $\rho_{\pi(\sigma)}$ in the π (σ)-channel at finite T calculated in the $O(4)$ linear σ model: Two characteristic features are the broadening of the pion peak (Fig.3(A)) and the spectral concentration at the 2π threshold ($\omega \simeq 2m_\pi$) in the σ-channel (Fig.3(B)). The latter may be measured by the 2γ spectrum from the hot plasma created in the relativistic heavy ion collisions [15].

SPECTRAL ENHANCEMENT IN NUCLEAR MATTER

$\langle \bar{q}q \rangle$ at finite baryon density obeys an exact theorem in QCD [16]:

$$\frac{\langle \bar{q}q \rangle}{\langle \bar{q}q \rangle_0} = 1 - \frac{n_B}{f_\pi^2 m_\pi^2}\left[\Sigma_{\pi N} + \hat{m}\frac{d}{d\hat{m}}\left(\frac{E(n_B)}{A}\right)\right], \tag{4}$$

where $\Sigma_{\pi N} = 45 \pm 10$ MeV is the pion-nucleon sigma term and $E(n_B)/A$ is the nuclear binding energy per particle with $\hat{m} = (m_u + m_d)/2$. $\langle \bar{q}q \rangle_0$ denotes the chiral condensate in the vacuum. The density-expansion of the right hand side of (4) gives a reduction of almost 35 % of $\langle \bar{q}q \rangle$ already at the nuclear matter density $n_0 = 0.17\text{fm}^{-3}$.

The near-threshold enhancement discussed in Sec.3 has been also studied at finite baryon density [17]. Because of the decrease of the chiral condensate discussed above, the spectral function in the σ-channel, ρ_σ, has a following generic behavior at densities

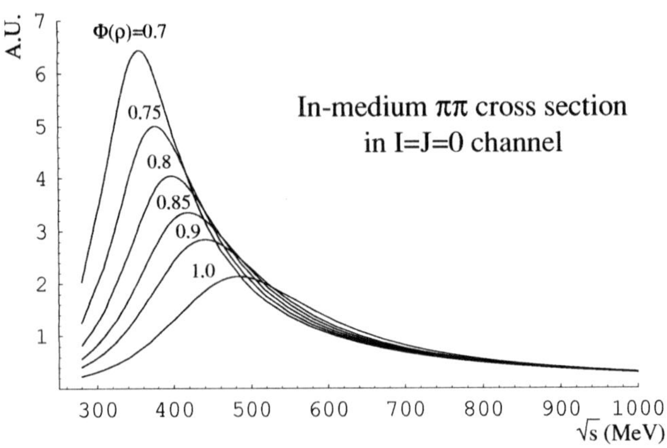

FIGURE 4. In-medium $\pi\pi$ cross section in the I=J=0 channel as a function of the c.m. energy $\sqrt{s}$ for several different values of the condensate ratio $\Phi = \langle\sigma\rangle/\langle\sigma\rangle_0$ in nuclear matter [23].

not far from n_0;

$$\rho_\sigma(\omega \simeq 2m_\pi) \propto \frac{\theta(\omega - 2m_\pi)}{\sqrt{1 - \frac{4m_\pi^2}{\omega^2}}}, \tag{5}$$

which shows a spectral concentration at the 2π threshold.

Recently CHAOS collaboration [19] reported the data on the $\pi^+\pi^\pm$ invariant mass distribution $M^A_{\pi^+\pi^\pm}$ in the reaction $A(\pi^+,\pi^+\pi^\pm)A'$ with the mass number A ranging from 2 to 208: They observed that the yield for $M^A_{\pi^+\pi^-}$ near the 2π threshold is close to zero for $A = 2$, but increases dramatically with increasing A. They identified that the $\pi^+\pi^-$ pairs in this range of $M^A_{\pi^+\pi^-}$ is in the I=J=0 state. This experiment was originally motivated by a possibility of strong $\pi\pi$ correlations in nuclear matter [20]. However, the state-of-the-art calculations using the nonlinear chiral Lagrangian together with $\pi N\Delta$ many-body dynamics do not reproduce the cross sections in the I=0 and I=2 channels simultaneously [21]; the final state interactions of the emitted two pions in nuclei give rise to a slight enhancement of the cross section in the I=0 channel, but is not sufficient to reproduce the experimental data. This indicates that an additional mechanism such as the partial restoration of chiral symmetry may be relevant for explaining the data [17, 22].

To make a close connection between the idea of the spectral enhancement and the CHAOS data, the in-medium π-π cross section has been calculated in the linear and non-linear σ models [23]. It was shown that, in both cases, substantial enhancement of the π-π correlation in the I=J=0 channel near the threshold can be seen due to the partial restoration of chiral symmetry. Furthermore, an effective 4π-N-N vertex responsible for the enhancement in the non-linear chiral Lagrangian is identified. In Fig.4, the in-medium π-π cross section in the σ-channel calculated in the O(4) linear σ model is shown.

Theoretically, it is of great importance to make an extensive analysis of the π-π interaction in heavy nuclei with the effect of partial chiral restoration to understand the

CHAOS data. Experimentally, it is definitely necessary to confirm the CHAOS result not only in the same reaction but also in other reactions. Measuring $\sigma \to 2\pi^0 \to 4\gamma$ is one of the clean experiments, since it is free from the ρ meson background inherent in the $\pi^+\pi^-$ measurement. Measuring $\sigma \to 2\gamma$ is interesting because of the small final state interactions. Dilepton detection through the scalar-vector mixing in matter, $\sigma \to \gamma^* \to e^+e^-$ and the formation of the σ mesic nuclei through the nuclear reactions such as (d, t) are other possible experiments [18, 17].

SPECTRAL FUNCTIONS FROM LATTICE QCD

The lattice QCD simulations have remarkable progress in recent years for calculating the properties of hadrons as well as the properties of QCD phase transition [24]. In particular, masses of light mesons and baryons in the quenched QCD simulation agree within 5-10 % with the experimental spectra [25]. However, the lattice QCD had difficulties in accessing the dynamical quantities in the Minkowski space, because measurements on the lattice can only be carried out for discrete points in imaginary time. The analytic continuation from the imaginary time to the real time using the noisy lattice data is highly non-trivial and is even classified as an ill-posed problem.

Recently a new approach to extract spectral functions of hadrons from lattice QCD data by using the maximum entropy method (MEM) was started [26]. MEM has been successfully applied for similar problems in image reconstruction in crystallography and astrophysics, and in quantum Monte Carlo simulations in condensed matter physics [27]. There are three important aspects of MEM: (i) it does not require a priori assumptions or parametrizations of SPFs, (ii) for given data, a unique solution is obtained if it exists, and (iii) the statistical significance of the solution can be quantitatively analyzed.

Basic idea of MEM

The Euclidean correlation function $D(\tau)$ of an operator $O(\tau,\vec{x})$ and its spectral decomposition at zero three-momentum read

$$D(\tau > 0) = \int \langle O^\dagger(\tau,\vec{x}) O(0,\vec{0}) \rangle d^3x \equiv \int_0^\infty K(\tau,\omega) A(\omega) d\omega, \qquad (6)$$

where ω is a real frequency, and $A(\omega)$ is the spectral function (or sometimes called the *image*), which is positive semi-definite. $K(\tau,\omega)$ is a known integral kernel (it reduces to Laplace kernel $e^{-\tau\omega}$ at zero T.)

Monte Carlo simulation provides $D(\tau_i)$ on the discrete set of temporal points $0 \leq \tau_i/a \leq N_\tau$. From this data with statistical noise, we need to reconstruct the spectral function $A(\omega)$ with continuous variable ω. This is a typical ill-posed problem, where the number of data is much smaller than the number of degrees of freedom to be reconstructed. This makes the standard likelihood analysis and its variants inapplicable unless strong assumptions on the spectral shape are made [28]. MEM is a method to circumvent this difficulty through Bayesian statistical inference of the most probable *image* together with its reliability.

Let's start with the Bayes' theorem: $P[X|Y] = P[Y|X]P[X]/P[Y]$, where $P[X|Y]$ is the conditional probability of X given Y. The most probable image $A(\omega)$ for given lattice data D is obtained by maximizing the conditional probability

$$P[A|D] \propto P[D|A]P[A]. \tag{7}$$

The reliability of the obtained result can be checked by the second variation of $P[A|D]$ with respect to $A(\omega)$.

$P[D|A]$ in (7) is a standard χ^2, namely $P[D|AH] = Z_L^{-1}\exp(-L)$ with a likelihood function $L = (1/2)\sum_{i,j}(D(\tau_i) - D^A(\tau_i))C_{ij}^{-1}(D(\tau_j) - D^A(\tau_j))$ and a normalization factor Z_L. $D(\tau_i)$ is the lattice data averaged over gauge configurations and $D^A(\tau_i)$ is the correlation function obtained by given A. C is a covariance matrix of the data.

$P[D|A]$ (the prior probability) in (7), can be written with parameters α and m as $P[A] = Z_S^{-1}\exp(\alpha S)$ by the combinatorial argument. Here S is the generalized information entropy,

$$S = \int_0^\infty \left[A(\omega) - m(\omega) - A(\omega)\log\left(\frac{A(\omega)}{m(\omega)}\right)\right] d\omega, \tag{8}$$

with Z_S being a normalization factor. α is a real and positive parameter and $m(\omega)$ is a real and positive function called the default model.

The output image A_{out} is given by a weighted average over A and α:

$$A_{out}(\omega) = \int A_\alpha(\omega)\, P[\alpha|Dm]\, d\alpha, \tag{9}$$

where $A_\alpha(\omega)$ is obtained by minimizing the "free-energy" $F \equiv L - \alpha S$. α dictates the relative weight of the entropy S (which tends to fit A to the default model m) and the likelihood function L (which tends to fit A to the lattice data). Note that α appears only in the intermediate step and is integrated out in the final result.

One can prove that the solution of $\delta F = 0$ is unique if it exits. The error analysis of the reconstructed image can be studied by evaluating the second derivative of the free energy $\delta^2 F/\delta A(\omega)\delta A(\omega')$. The default model m is chosen such that the error becomes minimum.

MEM with lattice data

In Fig.5, shown is an example of the spectral function in the pion and the rho meson channels at $T = 0$ extracted from the quenched lattice QCD data [26]. The lattice size is $20^3 \times 24$ with $\beta = 6.0$, which corresponds to $a = 0.0847$ fm ($a^{-1} = 2.33$ GeV), and the spatial size of the lattice $L_s a = 1.69$ fm. Hopping parameters are chosen to be $\kappa = 0.153$, 0.1545, and 0.1557 with $N_{conf} = 161$ for each κ. For the quark propagator, the Dirichlet (periodic) boundary condition is employed for the temporal (spatial) direction. We use data at $1 \leq \tau_i/a \leq 12$.

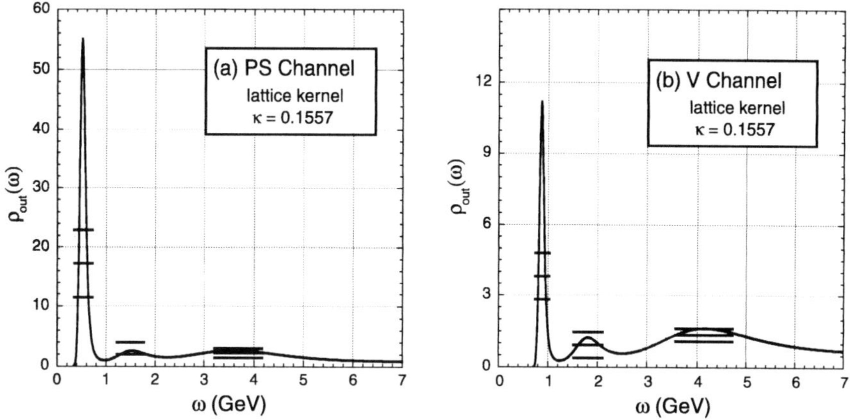

FIGURE 5. Spectral functions $\rho_{out}(\omega) \equiv A_{out}(\omega)/\omega^2$ obtained by MEM using the quenched lattice QCD data. The lattice size is $20^3 \times 24$ with $a = 0.0847$ fm. 12 data points in the temporal direction ($1 \leq \tau_i/a \leq 12$) are used for the MEM analysis. (a) is for the pion channel and (b) is for the rho-meson channel. The figures are based on the calculation in ref. [26].

The obtained images have the low-energy peaks corresponding to π and ρ, and the broad structure in the high-energy region. The mass of the ρ-meson in the chiral limit extracted from the peak in Fig.5 is consistent with that determined by the asymptotic behavior of $D(\tau)$. Although the maximum value of the fitting range $\tau_{max}/a = 12$ marginally covers the asymptotic limit in τ, we can extract reasonable masses for π and ρ. The width of π and ρ in Fig.5 is an artifact due to the statistical errors of the lattice data. In fact, in the quenched approximation, there is no room for the ρ-meson to decay into two pions.

As for the second peaks, the error analysis shows that their spectral "shape" does not have much statistical significance, although the existence of the non-vanishing spectral strength is significant. Under this reservation, we fit the position of the second peaks and made linear extrapolation to the chiral limit and obtain results consistent with experimental data on π' and ρ'.

The error in Fig.5 indicates the uncertainty of the spectrum averaged over the interval in the frequency space. Better data with smaller a and larger lattice volume will be helpful for obtaining spectral functions with smaller errors.

The future of MEM

MEM introduces a new way of extracting physical information from the lattice QCD data. It is now time to study how hadrons are modified in the medium using this new approach. The long-standing problem of in-medium spectral functions of vector mesons (ρ, ω, ϕ, J/ψ, Υ, $\cdots$, etc) and scalar/pseudo-scalar mesons (σ, π, $\cdots$, etc) can be studied using MEM combined with finite T lattice simulations. The in-medium behavior of the light vector mesons [29] and scalar mesons [11] is intimately related to the chiral restoration in hot/dense matter, while that of the heavy vector mesons is related to deconfinement [30]. Simulations with an anisotropic lattice is necessary for this purpose to have enough data points in the temporal direction at finite T, which is now under way [31]. (See also an attempt on the basis of NRQCD simulation [32].) It is interesting to study the spectral functions with finite three-momentum $\mathbf{k}$ for treating moving-hadrons in the medium.

Non-perturbative collective modes above the critical temperature of the QCD phase transition speculated in [11, 33, 34] may be studied efficiently by using MEM, since the method does not require any specific ansätze for the spectral shape. Correlations in the diquark channels in the vacuum and in medium are interesting to be explored in relation to the q-q correlation in baryon spectroscopy and to color superconductivity at high n_B [35].

SUMMARY

Although the light σ-meson does not show up clearly in the free space because of its large width due to the strong coupling with two pions, it may appear as a soft and narrow collective mode in the hadronic medium when the chiral symmetry is (partially) restored. A characteristic signal is the enhancement of the spectral function and the π-π cross section near the 2π threshold. They could be observed in the hadronic reactions with heavy nuclear targets as well as in the heavy ion collisions through the detection of pion pairs and photon pairs. Finding such signal provides us with a better understanding of the non-perturbative structure of the QCD vacuum and its quantum fluctuations.

The maximum entropy method (MEM) opens the door to extract the spectral functions from the first principle lattice QCD data without making a priori assumptions on the spectral shape. The method has been tested at T=0 with reasonable success. The uniqueness of the solution and the quantitative error analysis make MEM superior to any other approaches adopted previously. The method has a promising future for analyzing in-medium properties of hadrons and its relation to chiral structure of matter.

REFERENCES

1. Y. Nambu, Nucl. Phys. **A638** (1998) 35c.
2. T. Hatsuda and T. Kunihiro, Phys. Rep. **247** (1994) 221.
3. F. Karsch, hep-ph/0103314, these proceedings.
4. R. D. Pisarski and F. Wilczek, Phys. Rev. **D29** (1984) 338. F. Wilczek, Int. J. Mod. Phys. **A7** (1992) 3911, 6951; Int. J. Mod. Phys. **D3** (1994) 63.
5. H. Heiselberg and M. Hjorth-Jensen, Phys. Rep. **328** (2000) 237.
6. J. Madsen, astro-ph/9809032.
7. K. Rajagopal and F. Wilczek, hep-ph/0011333.
   Brown and Rho, hep-ph/0103102 (to appear in Phys. Rep).
8. F. Karsch, Nucl. Phys. B (Proc. Suppl.) **83** (2000) 14.
9. See the review, E. V. Shuryak, Rev. Mod. Phys. **65** (1993) 1.
10. See the reviews, R. Rapp and J. Wambach, Adv. Nucl. Phys. **25** (2000) 1.
    J. Alam, S. Sarkar, P. Roy, T. Hatsuda and B. Sinha, Ann. Phys. (N.Y.), **286** (2000) 159. G. Chanfray, nucl-th/0012068.
11. T. Hatsuda and T. Kunihiro, Phys. Rev. Lett. **55** (1985) 158; Phys. Lett. **B185** (1987) 304.

12. K. Rajagopal and F. Wilczek, Nucl. Phys. **B399** (1993) 395 .
13. D.E. Groom et al. (Particle Data Group), Eur. Phys. J. **C15** (2000) 1.
Proceedings of "Possible Existence of the σ-Meson and its Implications to Hadron Physics", KEK-Proceedings/2000-4, (Proc. site,
http://amaterasu.kek.jp/YITPws/online/index.html).
14. H. A. Weldon, Phys. Lett. **B274** (1992) 133. C. Song and V. Koch, Phys. Lett. **B404** (1997) 1. M. Stephanov, K. Rajagopal and E. Shuryak, Phys. Rev. Lett. **81** (1998) 4816. J. Schaffner-Bielich, Phys. Rev. Lett. **84** (2000) 3261.
15. S. Chiku and T. Hatsuda, Phys. Rev. **D57** (1998) R6, ibid. **D58** (1998) 076001.
M. K. Volkov et al., Phys. Lett. **B424** (1998) 235.
16. E. G. Drukarev and E. M. Levin, Prog. Paet. Nucl. Phys. **A556** (1991) 467.
R. Brockmann and W. Weise, Phys. Lett. **B367** (1996) 40.
17. T. Hatsuda, T. Kunihiro and H. Shimizu, Phys. Rev. Lett. **82** (1999) 2840.
18. T. Kunihiro, Prog. Theor. Phys. Supplement **120** (1995) 75.
19. F. Bonutti *et al.* (CHAOS Collaboration), Nucl. Phys. **A677** (2000) 123.
20. See e.g., G. Chanfray, Z. Aouissat, P. Schuck and W. Nörenberg, Phys. Lett. **B256** (1991) 325.
21. R. Rapp *et al.*, Phys. Rev. **C59** (1999) R1237.
M. J. Vicente-Vacas and E. Oset, Phys. Rev. **C60** (1999) 064621.
22. Z. Aouissat, G. Chanfray, P. Schuck and J. Wambach, Phys. Rev. **C61**, 012202 (2000). D. Davesne, Y. J. Zhang and G. Chanfray, Phys. Rev. **C62** (2000) 024604 .
23. D. Jido, T. Hatsuda and T. Kunihiro, Phys. Rev. **D63** (2001) 011901.
24. Lattice 2000 Proceedings, Nucl. Phys. B (Proc. Suppl.) **94** (2000) 1.
25. S. Aoki et al., Phys. Rev. Lett. **84** (2000) 238.
26. Y. Nakahara, M. Asakawa and T. Hatsuda, Phys. Rev. **D60** (Rapid Comm.) (1999) 091503; Nucl. Phys. B (Proc. Suppl.) **83** (2000) 191; hep-lat/0011040 to appear in Prog. Part. Nucl. Phys. **47** (2001). For earlier attempts to aim a similar goal, Ph. de Forcrand et al., Nucl. Phys. B (Proc. Suppl.) **63A-C** (1998) 460; E. G. Klepfish, C. E. Creffield and E. P. Pike, Nucl. Phys. B (Proc. Suppl.) **63A-C** (1998) 655.
27. B. R. Frieden, J. Opt. Soc. Am., **62** (1972) 511.
M. Jarrell and J. E. Gubernatis, Phys. Rep. **269** (1996) 133.
N. Wu, *The Maximum Entropy Method*, (Springer-Verlag, Berlin, 1997).
28. T. Hashimoto, A. Nakamura and I. O. Stamatescu, Nucl. Phys. **B400** (1993) 267; ibid. **B406** (1993) 325. M. -C. Chu, J. M. Grandy, S. Huang and J. W. Negele, Phys. Rev. D **48** (1993) 3340. D. B. Leinweber, Phys. Rev. D **51** (1995) 6369. D. Makovoz and G. A. Miller, Nucl. Phys. **B468** (1996) 293. C. Allton and S. Capitani, Nucl. Phys. **B526** (1998) 463. Ph. de Forcrand et al. (QCD-TARO Collaboration), hep-lat/0008005.
29. R. D. Pisarski, Phys. Lett. **B110** (1982) 155. G. E. Brown and M. Rho, Phys. Rev. Lett. **66** (1991) 2720. T. Hatsuda and S. Lee, Phys. Rev. **C46** (1992) R34.
30. T. Hashimoto et al., Phys. Rev. Lett. **57** (1986) 2123.
T. Matsui and H. Satz, Phys. Lett. **B178** (1986) 416.
31. M. Asakawa and T. Hatsuda, in progress.
32. M. Oevers, C. Davies and J. Shigemitsu, Nucl. Phys. B (Proc. Suppl.) **94** (2001) 363.
33. C. DeTar, Phys. Rev. **D32** (1985) 276.
34. A. Peshier and M. Thoma, Phys. Rev. Lett. **84** (2000) 841.
F. Karsch, M. G. Mustafa and M. H. Thoma, Phys. Lett. **B 497** (2001) 249.
35. I. Wetzorke and F. Karsch, hep-lat/0008008.

Hadron Physics and Confinement Physics in Lattice QCD

H. Suganuma*, K. Amemiya*, H. Ichie†, N. Ishii*, H. Matsufuru†, N. Nakajima†, Y. Nemoto**, M. Oka* and T.T. Takahashi†

*Faculty of Science, Tokyo Institute of Technology, Meguro, Tokyo 152-8551, Japan
†RCNP, Osaka University, Mihogaoka 10-1, Ibaraki, Osaka 567-0047, Japan
**Yukawa Institute for Theoretical Physics, Kyoto University, Kyoto 606-8502, Japan

Abstract. We are aiming to construct Quark Hadron Physics and Confinement Physics based on QCD. Using $SU(3)_c$ lattice QCD, we are investigating the three-quark potential at $T=0$ and $T\neq 0$, mass spectra of positive and negative-parity baryons in the octet and the decuplet representations of the SU(3) flavor, glueball properties at $T=0$ and $T\neq 0$. We study also Confinement Physics using lattice QCD. In the maximally abelian (MA) gauge, the off-diagonal gluon amplitude is strongly suppressed, and then the off-diagonal gluon phase shows strong randomness, which leads to a large effective off-diagonal gluon mass, $M_{\rm off}\simeq 1.2$GeV. Due to the large off-diagonal gluon mass in the MA gauge, infrared QCD is abelianized like nonabelian Higgs theories. In the MA gauge, there appears a macroscopic network of the monopole world-line covering the whole system. From the monopole current, we extract the dual gluon field B_μ, and examine the longitudinal magnetic screening. We obtain $m_B\simeq 0.5$ GeV in the infrared region, which indicates the dual Higgs mechanism by monopole condensation. From infrared abelian dominance and infrared monopole condensation, low-energy QCD in the MA gauge is described with the dual Ginzburg-Landau (DGL) theory.

QUARK HADRON PHYSICS FROM LATTICE QCD

Quantum chromodynamics (QCD) established as the fundamental theory of the strong interaction takes a simple form [1, 2],

$$\mathcal{L}_{\rm QCD} = -\frac{1}{2}{\rm tr}G_{\mu\nu}G^{\mu\nu} + \bar{q}(i\not{D} - m_q)q, \tag{1}$$

however, it is still hard to understand nonperturbative phenomena, such as color confinement and dynamical chiral-symmetry breaking, due to the infrared strong-coupling feature [1, 2]. In this decade, lattice QCD Monte Carlo calculations have been developed and have been mainly applied to (i) *hadron spectroscopy* and (ii) *finite temperature QCD phase transition*, with a great success. However, lattice QCD is applicable also to (iii) *Quark Hadron Physics* and (iv) *Confinement Physics*, as a useful and reliable method.

Our group is aiming to construct Quark Hadron Physics and Confinement Physics based on lattice QCD. Our strategy is to adopt lattice QCD calculations to the relevant quantities pointed out in Quark Hadron Physics or Confinement Physics, in order to obtain the reliable physical picture based on QCD.

CP594, *Hadrons and Nuclei: First International Symposium*, edited by Il-T. Cheon et al.

In relation to Quark Hadron Physics, we are investigating several important subjects in Quark Hadron Physics using SU(3)$_c$ lattice QCD at the quenched level as follows.

1. We numerically derive the static three-quark (3Q) potential V_{3Q} at $T = 0$, which is responsible to the baryon properties, and find that V_{3Q} is well reproduced by the sum of a constant, the Coulomb term and the linear confinement term proportional to the total flux-tube length, with the accuracy better than a few % [3-5].
2. From the Polyakov-loop correction, we obtain the Q-$\bar{Q}$ potential and the 3Q potential at $T \neq 0$, which characterize the hadron structure at $T \neq 0$.
3. We measure the lowest mass of the positive and the negative parity baryons in singlet, octet and decuplet representations of the SU(3) flavor, respectively [6]. The experimentally observed $\Lambda(1405)$ is much lighter than the corresponding baryon with the mass of 1.6GeV on the lattice, which may suggest the $\bar{K}N$ molecule picture for $\Lambda(1405)$. The octet-decuplet baryon mass splitting is also investigated.
4. We are investigating the glueball properties both at $T = 0$ and at $T \neq 0$.

Here, for the accurate measurement of the 3Q potential at $T = 0$ and the lowest hadron mass in each channel, we adopt the *smearing technique* which reduces the excited-state contamination. Furthermore, to get maximal information on the temporal correlation, we use *anisotropic lattices* where the temporal lattice spacing a_t is finer than the spatial lattice spacing a_s [6]. In particular, the anisotropic lattice is quite useful for the finite temperature QCD, because of the limitation of the temporal distance [7].

Since these studies are introduced in [5-7], we mainly present the recent progress of our studies on Confinement Physics based on SU(2) lattice QCD at the quenched level.

CONFINEMENT PHYSICS FROM LATTICE QCD

To understand the confinement mechanism is one of the most difficult problems remaining in the particle physics. As the Regge trajectories and lattice QCD calculations indicate, quark confinement is characterized by *one-dimensional squeezing* of the color-electric flux and the *string tension* $\sigma \simeq 1\text{GeV/fm}$, which is the key quantity of confinement. On the confinement mechanism, Nambu first proposed the *dual superconductor theory* for quark confinement [8], based on the electro-magnetic duality in 1974. In this theory, there occurs the one-dimensional squeezing of the color-electric flux by the *dual Meissner effect* due to condensation of bosonic color-magnetic monopoles. However, there are *two large gaps* between QCD and the dual superconductor theory [2,9-11]

1. The dual superconductor theory is based on the *abelian gauge theory* subject to the Maxwell-type equations, where electro-magnetic duality is manifest, while QCD is a nonabelian gauge theory.
2. The dual superconductor theory requires color-magnetic monopole condensation as the key concept, while QCD does not have color magnetic monopoles as the elementary degrees of freedom.

These gaps may be simultaneously fulfilled by taking *MA gauge fixing,* which reduces QCD to an abelian gauge theory including color-magnetic monopoles.

In Euclidean QCD, the maximally abelian (MA) gauge is defined so as to minimize the total amount of the off-diagonal gluons [2,9-11]

$$R_{\rm off}[A_\mu(\cdot)] \equiv \int d^4x \, {\rm tr}\left\{[\hat{D}_\mu,\vec{H}][\hat{D}_\mu,\vec{H}]^\dagger\right\} = \frac{e^2}{2}\int d^4x \sum_\alpha |A_\mu^\alpha(x)|^2 \qquad (2)$$

by the SU(N_c) gauge transformation. Here, we have used the Cartan decomposition, $A_\mu(x) = \vec{A}_\mu(x)\cdot\vec{H} + \sum_\alpha A_\mu^\alpha(x)E^\alpha$. Since the SU($N_c$) covariant derivative operator $\hat{D}_\mu \equiv \hat{\partial}_\mu + ieA_\mu$ obeys the adjoint gauge transformation, the local form of the MA gauge condition is easily derived as $[\vec{H},[\hat{D}_\mu,[\hat{D}_\mu,\vec{H}]]] = 0$. In the MA gauge, the gauge symmetry $G \equiv {\rm SU}(N_c)_{\rm local}$ is reduced into $H \equiv {\rm U}(1)_{\rm local}^{N_c-1} \times {\rm Weyl}_{N_c}^{\rm global}$, where the global Weyl symmetry is a subgroup of SU(N_c) relating the permutation of N_c bases in the fundamental representation. In the MA gauge, off-diagonal gluons behave as charged matter fields like $W_\mu^\pm$ in the Standard Model, and provide the color-electric current in terms of the residual abelian gauge symmetry. In addition, color-magnetic monopoles appear as topological objects reflecting the nontrivial homotopy group [2,9-14]

$$\Pi_2({\rm SU}(N_c)/{\rm U}(1)^{N_c-1}) = \Pi_1({\rm U}(1)^{N_c-1}) = \mathbf{Z}_\infty^{N_c-1} \qquad (3)$$

in a similar manner to similarly in the GUT monopole. Here, the global Weyl symmetry and color-magnetic monopoles are relics of nonabelian nature of QCD.

In this way, in the MA gauge, QCD is reduced into an abelian gauge theory including color-magnetic monopoles, which is expected to provide a theoretical basis of the dual superconductor theory for quark confinement. Furthermore, recent lattice QCD studies show remarkable features of *abelian dominance* and *monopole dominance* for nonperturbative QCD (NP-QCD) in the MA gauge.

1. Without gauge fixing, all the gluon components equally contribute to NP-QCD, and it is difficult to extract relevant degrees of freedom for NP-QCD.
2. In the MA gauge, QCD is reduced into an abelian gauge theory including the electric current j_μ and the magnetic current k_μ, which forms a *global network of the monopole world-line covering the whole system.* (See Fig.3(a).) In the MA gauge, lattice QCD shows *abelian dominance* for NP-QCD (confinement [9], gluon propagators [15], chiral symmetry breaking [17]): only the diagonal gluon is relevant for NP-QCD, while off-diagonal gluons do not contribute to NP-QCD.
3. By the Hodge decomposition, the diagonal gluon is decomposed into the "photon part" ($j_\mu \neq 0, k_\mu = 0$) and the "monopole part" ($k_\mu \neq 0$, $j_\mu = 0$), corresponding to the separation of j_μ and k_μ. In the MA gauge, lattice QCD shows *monopole dominance* [17-19] for NP-QCD: the monopole part leads to NP-QCD, while the photon part seems trivial like QED and does not contribute to NP-QCD. For example, on the Q-$\bar{\rm Q}$ potential, the purely linear potential appears in the monopole part, while the Coulomb potential appears in the photon part like QED [16].

Thus, in the MA gauge, QCD is reduced into an abelian gauge theory with color-magnetic monopoles, *keeping essence of infrared nonperturbative features*, and the *relevant collective mode for NP-QCD* can be extracted as the color-magnetic monopole.

Strong Randomness of Off-diagonal Gluon Phase, Abelian Dominance and Large Mass of Off-diagonal Gluons in MA Gauge

To find out essence of the MA gauge, we study the off-diagonal gluon field $A_\mu^\pm \equiv \frac{1}{\sqrt{2}}(A_\mu^1 \pm iA_\mu^2)$ in the MA gauge in SU(2) lattice QCD. There are two remarkable features in the off-diagonal gluon field $A_\mu^\pm(x) = e^{\pm\chi_\mu(x)}|A_\mu^\pm(x)|$ in the MA gauge [2,9-11].

1. The off-diagonal gluon amplitude $|A_\mu^\pm(x)|$ is strongly suppressed by SU(N_c) gauge transformation in the MA gauge.
2. The off-diagonal gluon phase $\chi_\mu(x)$ tends to be random, because $\chi_\mu(x)$ is not constrained by MA gauge fixing at all, and only the constraint from the QCD action is weak due to a small accompanying factor $|A_\mu^\pm|$.

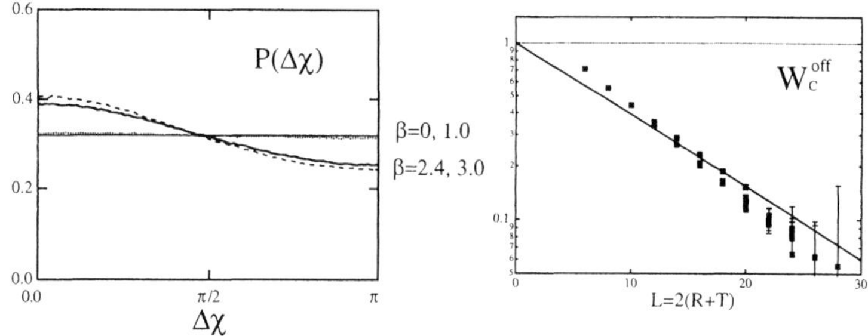

FIGURE 1. (a) The probability distribution $P(\Delta\chi)$ of the difference $\Delta\chi \equiv |\chi_\mu(s) - \chi_\mu(s+\hat{\nu})|(\mathrm{mod}\pi)$ in the MA gauge plus U(1)$_3$ Landau gauge at β=0 ($a = \infty$, thin line), β=1.0 ($a \simeq 0.57$fm, dotted curve), β=2.4 ($a \simeq 0.127$fm, solid curve), β=3.0 ($a \simeq 0.04$fm, dashed curve). (b) The off-diagonal gluon contribution to the Wilson loop W_C^{off} v.s. the perimeter length $L \equiv 2(R+T)$ in the MA gauge on 16^4 lattice with β=2.4. The thick line denotes the theoretical estimation in Eq.(4).

Now, we consider the difference $\Delta\chi \equiv |\chi_\mu(s) - \chi_\mu(s+\hat{\nu})|(\mathrm{mod}\pi)$ in the MA gauge. If the off-diagonal gluon phase $\chi_\mu(x)$ is a continuum variable, as the lattice spacing a goes to 0, $\Delta\chi \simeq a|\partial_\nu\chi_\mu|$ must go to zero, and hence $P(\Delta\chi)$ approaches to the δ-function like $\delta(\Delta\chi)$. However, as shown in Fig.1(a), $P(\Delta\chi)$ is almost a-independent and almost flat. These features indicate the *strong randomness of the off-diagonal gluon phase* $\chi_\mu(x)$ in the MA gauge. Then, $\chi_\mu(x)$ behaves as a random angle variable in the MA gauge.[1]

Within the random-variable approximation for the off-diagonal gluon phase $\chi_\mu(s)$ in the MA gauge, we analytically prove abelian dominance of the string tension [2,9-11] from the analysis of the SU(2) Wilson loop $\langle W_C[A_\mu^a]\rangle$, the abelian Wilson loop $\langle W_C[A_\mu^\pm \equiv 0, A_\mu^3]\rangle_{\mathrm{MA}}$ and the off-diagonal gluon contribution to the Wilson loop $W_C^{\mathrm{off}} \equiv \langle W_C[A_\mu^a]\rangle / \langle W_C[A_\mu^\pm \equiv 0, A_\mu^3]\rangle_{\mathrm{MA}}$. The point is the cancellation of the off-diagonal gluon contribution due to the random phase as $\langle e^{i\chi_\mu(s)}\rangle_{\mathrm{MA}} \simeq \frac{1}{2\pi}\int_0^{2\pi} d\chi_\mu(s) e^{i\chi_\mu(s)} = 0$.

[1] Near the monopole, a large amplitude of $|A_\mu^\pm(s)|$ remains even in the MA gauge, and $\chi_\mu(s)$ is constrained so as to reduce the QCD action. Hence, $\chi_\mu(s)$ cannot be regarded as a random variable near monopoles.

Near the continuum limit $a \simeq 0$, we find a relation between the *macroscopic* quantity $W_C^{\rm off}$ and the *microscopic* quantity of the off-diagonal gluon amplitude $\langle |eA_\mu^\pm|^2 \rangle_{\rm MA}$ as

$$W_C^{\rm off} \equiv \langle W_C[A_\mu^a] \rangle / \langle W_C[A_\mu^\pm \equiv 0, A_\mu^3] \rangle_{\rm MA} \simeq \exp\{L_{\rm phys} a \langle |eA_\mu^\pm|^2 \rangle_{\rm MA}/4\} \tag{4}$$

with the perimeter $L_{\rm phys} \equiv La$ of the Wilson loop. This relation is checked in lattice QCD as shown in Fig.1(b). In fact, the off-diagonal gluon contribution $W_C^{\rm off}$ obeys the *perimeter law*,[2] and then *perfect abelian dominance for the string tension*, $\sigma_{\rm SU(2)} = \sigma_{\rm Abel}$, is derived by regarding off-diagonal gluon phases to be random in the MA gauge.

As another remarkable fact, *strong randomness of off-diagonal gluon phases leads to rapid reduction of off-diagonal gluon correlations.* In fact, if $\chi_\mu(x)$ is a complete random phase, Euclidean off-diagonal gluon propagators exhibit the δ-functional reduction as

$$\langle A_\mu^+(x) A_\nu^-(y) \rangle_{\rm MA} = \langle |A_\mu^+(x)||A_\nu^-(y)| e^{i\{\chi_\mu(x)-\chi_\nu(y)\}} \rangle_{\rm MA} = \langle |A_\mu^\pm(x)|^2 \rangle_{\rm MA} \delta_{\mu\nu} \delta^4(x-y), \tag{5}$$

which means the infinitely large mass of off-diagonal gluons. Of course, the real off-diagonal gluon phases are not complete but approximate random phases. Then, the off-diagonal gluon mass would be large but finite. Thus, *strong randomness of off-diagonal gluon phases is expected to provide a large effective mass of off-diagonal gluons.*

Large Mass Generation of Off-diagonal Gluons in MA Gauge : Essence of Infrared Abelianization of QCD

We quantitatively study the Euclidean gluon propagator $G_{\mu\nu}^{ab}(x-y) \equiv \langle A_\mu^a(x) A_\nu^b(y) \rangle$ $(a,b = 1,2,3)$ and the off-diagonal gluon mass $M_{\rm off}$ in the MA gauge, using SU(2) lattice QCD with $2.2 \le \beta \le 2.4$ and various sizes ($12^3 \times 24$, 16^4, 20^4). As for the residual U(1)$_3$ gauge symmetry, we take U(1)$_3$ Landau gauge, to extract most continuous gluon configuration under the MA gauge constraint, for the comparison with the continuum theory. The continuum gluon field $A_\mu^a(x)$ is derived from the link variable as $U_\mu(s) = \exp\{iaeA_\mu^a(s)\frac{\tau^a}{2}\}$. We show in Fig.2(a) the scalar-type gluon propagators $G_{\mu\mu}^3(r)$ and $G_{\mu\mu}^{+-}(r) \equiv \langle A_\mu^+(x) A_\mu^-(y) \rangle = \frac{1}{2}\{G_{\mu\mu}^1(r) + G_{\mu\mu}^2(r)\}$, which depend only on the four-dimensional Euclidean distance $r \equiv \sqrt{(x_\mu - y_\mu)^2}$. We find *infrared abelian dominance for the gluon propagator in the MA gauge*: only the abelian gluon $A_\mu^3(x)$ propagates over the long distance and can influence the infrared physics [10, 11, 15].

Since the four-dimensional Euclidean propagator of the massive vector boson with the mass M takes a Yukawa-type asymptotic form as

$$G_{\mu\mu}(r) = \frac{3}{4\pi^2}\frac{M}{r} K_1(Mr) + \frac{1}{M^2}\delta^4(x-y) \simeq \frac{3M^{1/2}}{2(2\pi)^{3/2}} \frac{e^{-Mr}}{r^{3/2}}, \tag{6}$$

the infrared effective mass $M_{\rm off}$ of the off-diagonal gluon $A_\mu^\pm(x)$ can be extracted from the slope in the logarithmic plot of $r^{3/2} G_{\mu\mu}^{+-}(r) \sim \exp(-M_{\rm off} r)$ in Fig.2(b). From the

---

[2] The off-diagonal contribution $W_C^{\rm off}$ becomes trivial as $W_C^{\rm off} \to 1$ in the continuum limit $a \to 0$.

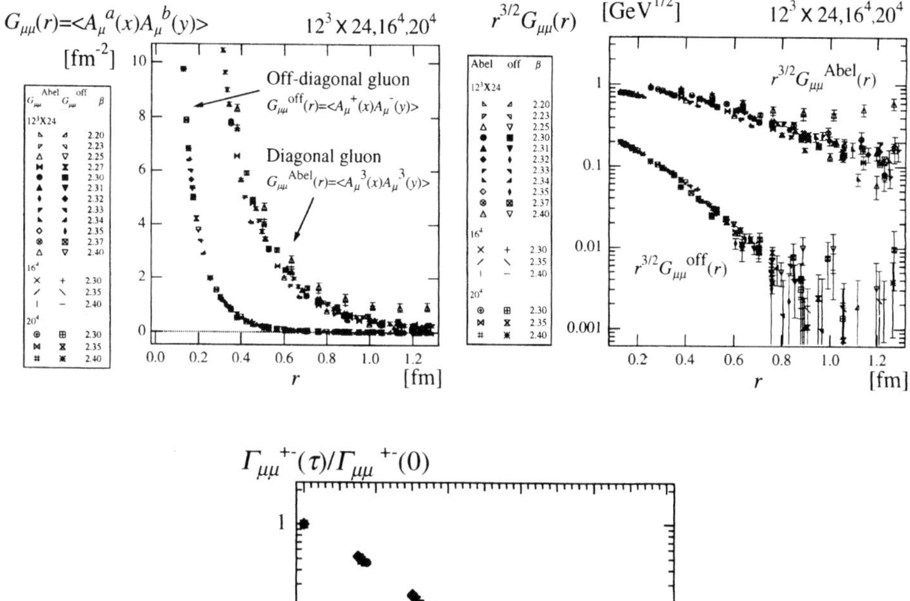

FIGURE 2. (a) The scalar-type gluon propagator $G^a_{\mu\mu}(r)$ as the function of the four-dimensional distance r in the MA gauge in SU(2) lattice QCD with $2.2 \le \beta \le 2.4$ with various sizes ($12^3 \times 24$, 16^4, 20^4). (b) The logarithmic plot of $r^{3/2}G^a_{\mu\mu}(r)$ v.s. r. The off-diagonal gluon propagator behaves as the Yukawa-type function, $G_{\mu\mu} \sim \frac{\exp(-M_{\rm off}r)}{r^{3/2}}$. (c) The logarithmic plot of the temporal correlation $\Gamma^{+-}_{\mu\mu}(\tau) \equiv \langle O^+_\mu(\tau)O^-_\mu(0)\rangle$ as the function of the temporal distance τ in SU(2) lattice QCD with $2.3 \le \beta \le 2.35$ with $16^3 \times 32$ and $12^3 \times 24$. From the slope in (b) and (c), the effective mass of the off-diagonal gluon $A^\pm_\mu$ is estimated as $M_{\rm off} \simeq 1.2$GeV.

slope analysis of the lattice QCD data with $r \ge 0.2$fm, we obtain the off-diagonal gluon mass as $M_{\rm off} \simeq 1.2$ GeV in the MA gauge.

We carry out also the *mass measurement of off-diagonal gluons from the temporal correlation of the zero-momentum projected operator* $O_\mu^\pm(\tau)$,

$$\Gamma_{\mu\mu}^{+-}(\tau) \equiv \langle O_\mu^+(\tau) O_\mu^-(0)\rangle, \quad O_\mu^\pm(\tau) \equiv \int d\mathbf{x}\, A_\mu^\pm(\mathbf{x},\tau), \tag{7}$$

in the MA gauge plus $U(1)_3$ Landau gauge. in SU(2) lattice QCD with $2.3 \leq \beta \leq 2.35$ with $16^3 \times 32$ and $12^3 \times 24$. Again, we find the off-diagonal gluon mass $M_{\rm off} \simeq 1.2$GeV in the MA gauge from the slope of the logarithmic plot of $\Gamma_{\mu\mu}^{+-}(\tau)$ in Fig.2(c) [2, 10].

Thus, the *off-diagonal gluon* $A_\mu^\pm$ *acquires a large effective mass* $M_{\rm off} \simeq 1.2$GeV *in the MA gauge*, which is *essence of infrared abelian dominance* [10, 11, 15]. In the MA gauge, due to the large effective mass $M_{\rm off} \simeq 1.2$GeV, off-diagonal gluons $A_\mu^\pm$ can propagate only within a short range as $r < M_{\rm off}^{-1} \simeq 0.2$fm, and becomes *infrared inactive* like weak bosons in the Standard Model. Then, in the MA gauge, off-diagonal gluons $A_\mu^\pm$ cannot contribute to the infrared NP-QCD, which leads to infrared abelian dominance.

QCD-Monopole Structure in terms of the Off-diagonal Gluon and Infrared Monopole Condensation

In the MA gauge, there appears a global network of monopole world-lines covering the whole system as shown in Fig.3(a), and this monopole-current system (the monopole part) holds essence of NP-QCD. We examine the dual Higgs mechanism by monopole condensation in this NP-QCD vacuum in the MA gauge using SU(2) lattice QCD [10,11]. So far, the "electric sector" in QCD has been well studied with the Wilson loop, since QCD is described by the "electric variable" such as quarks and gluons. To investigate the hidden "magnetic sector", it is useful to introduce the "dual (magnetic) variable" such as the *dual gluon field* B_μ, which is the dual partner of the diagonal gluon A_μ^3 and directly couples with the magnetic current k_μ. Owing to the absence of the electric current j_μ in the monopole part, the dual gluon B_μ can be introduced as the regular field satisfying $(\partial\wedge B)_{\mu\nu} = {}^*F_{\mu\nu}$ and the dual Bianchi identity, $\partial^\mu{}^*(\partial\wedge B)_{\mu\nu} = j_\nu = 0$. In the dual Landau gauge $\partial_\mu B^\mu = 0$, the field equation is simplified as $\partial^2 B_\mu = \partial^\alpha{}^*F_{\alpha\mu} = k_\mu$, and the dual gluon field B_μ can be obtained from the monopole current k_μ as

$$B_\mu(x) = (\partial^{-2}k_\mu)(x) = -\frac{1}{4\pi^2}\int d^4y \frac{k_\mu(y)}{(x-y)^2}. \tag{8}$$

Here, the mass generation of the dual gluon B_μ physically means the dual Higgs mechanism by monopole condensation, and leads to the *longitudinal magnetic screening*, which can be observed in the following phenomena [2, 10, 11].

1. Due to the longitudinal screening effect on the magnetic flux, the inter-monopole potential $V_M(r)$ is screened and behaves as a short-range Yukawa potential.
2. The Euclidean dual gluon propagator $\langle B_\mu(x)B_\nu(y)\rangle_{\rm MA}$ is exponentially reduced as in Eq.(6).

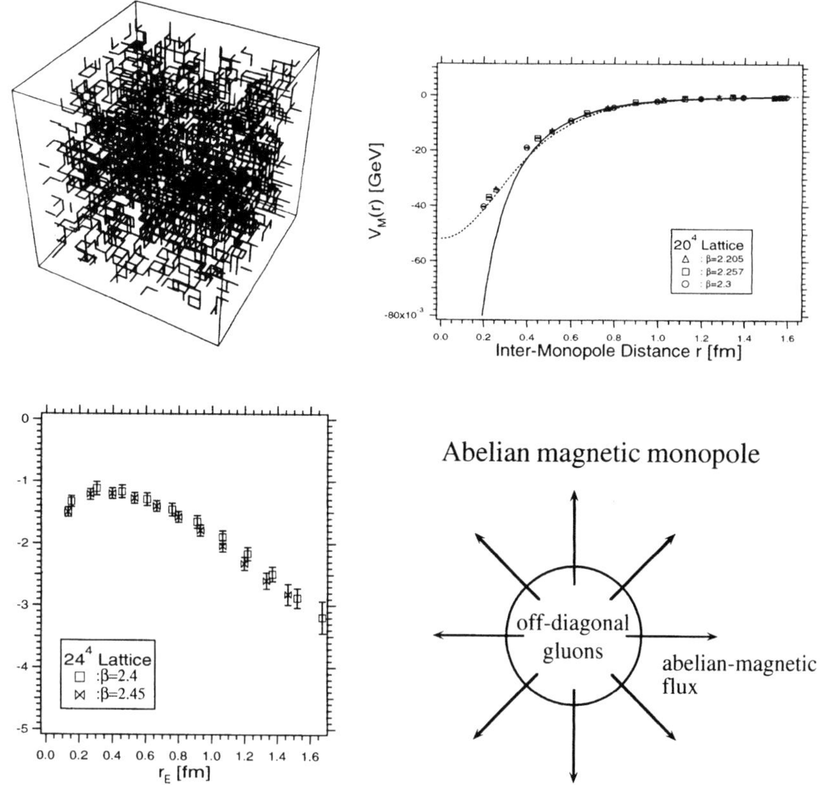

FIGURE 3. The SU(2) lattice-QCD results in the MA gauge. (a) The monopole world-line projected into $\mathbf{R}^3$ on the $16^3 \times 4$ lattice with $\beta = 2.2$ (the confinement phase). There appears a global network of monopole currents covering the whole system. (b) The inter-monopole potential $V_M(r)$ v.s. the three-dimensional distance r in the monopole-current system on the 20^4 lattice. The solid curve denotes the Yukawa potential with $m_B = 0.5$GeV. The dotted curve denotes the Yukawa-type potential including the monopole-size effect. (c) The scalar-type dual gluon correlation $\ln(r_E^{3/2} \langle B_\mu(x) B_\mu(y) \rangle_{\rm MA})$ as the function of the four-dimensional Euclidean distance r_E on the 24^4 lattice. The slope corresponds to the dual gluon mass m_B. (d) The schematic figure of the QCD-monopole structure in the MA gauge. The QCD-monopole includes a large amount of off-diagonal gluons around its center as well as the diagonal gluon.

Through these tests using lattice QCD, we investigate the dual gluon mass m_B.

First, by putting test magnetic charges in the monopole-current system in the MA gauge in SU(2) lattice QCD, we define the *dual Wilson loop* $W_D(C)$ [10, 11] as

$$W_D(C) \equiv \exp\{i\frac{e}{2}\oint_C dx_\mu B^\mu\} = \exp\{i\frac{e}{2}\iint d\sigma_{\mu\nu}{}^*F^{\mu\nu}\}, \tag{9}$$

which is the *dual version of the abelian Wilson loop* $W_{\rm Abel}(C) \equiv \exp\{i\frac{e}{2}\oint_C dx_\mu A_3^\mu\} = \exp\{i\frac{e}{2}\iint d\sigma_{\mu\nu} F^{\mu\nu}\}$. The monopole-antimonopole potential $V_M(r)$ can be derived as $V_M(r) = -\lim_{T\to\infty}\frac{1}{T}\ln\langle W_D(r,t)\rangle$ from the dual Wilson loop. In the monopole part in the MA gauge, $V_M(r)$ behaves as the Yukawa potential $V_M(r) \simeq -\frac{(e/2)^2}{4\pi}\frac{e^{-m_B r}}{r}$ at the long distance, as shown in Fig.3(b). From the infrared behavior of $V_M(r)$, the dual gluon mass is estimated as $m_B \simeq 0.5$GeV in the MA gauge.

Second, we investigate also the Euclidean scalar-type dual gluon propagator $\langle B_\mu(x)B_\mu(y)\rangle_{\rm MA}$ as shown in Fig.3(c), and estimate the dual gluon mass as $m_B \simeq 0.5$ GeV from its long-distance behavior [10, 11].

From these two tests, we obtain the dual gluon mass $m_B \simeq 0.5$ GeV in the infrared region, as the direct evidence of the dual Higgs mechanism by monopole condensation.

To conclude, lattice QCD in the MA gauge exhibits *infrared abelian dominance* and *infrared monopole condensation*, and hence the dual Ginzburg-Landau (DGL) theory [14,22-30] can be constructed as the infrared effective theory based on QCD.

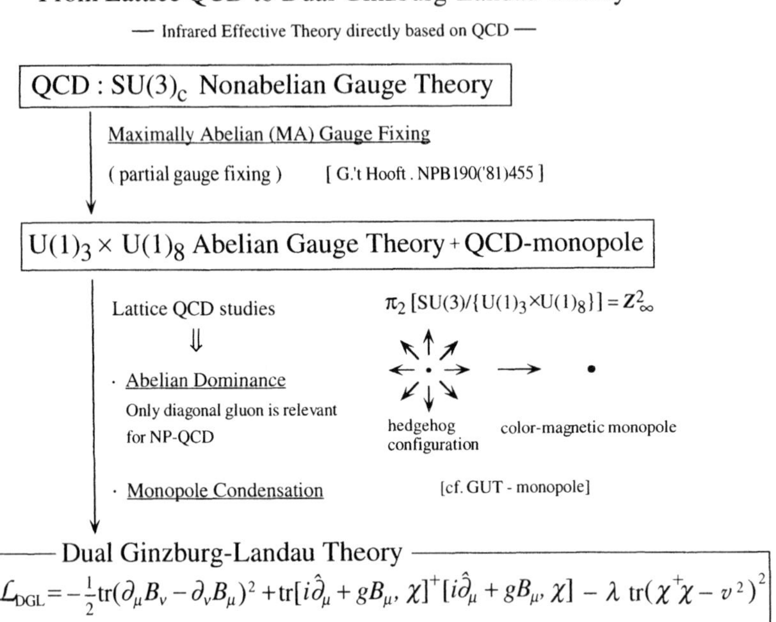

FIGURE 4. The dual Ginzburg-Landau (DGL) theory from lattice QCD in the MA gauge.

Here, we compare the QCD-monopole with the point-like Dirac monopole. In QED, there is no point-like monopole, because the QED action diverges around the point-like monopole. *The QCD-monopole also accompanies a large abelian action density, however, the total QCD action is kept finite even around the QCD-monopole, owing to cancellation with the off-diagonal gluon contribution.* This is the reason why monopoles can appear in QCD. In [2,9-11], using SU(2) lattice QCD, we investigate the QCD-monopole structure in the MA gauge in terms of the SU(2) action density $S_{\rm SU(2)}$, the abelian action density $S_{\rm Abel}$, and the off-diagonal contribution $S_{\rm off} \equiv S_{\rm SU(2)} - S_{\rm Abel}$. We summarize the results on the QCD-monopole structure and its related topics as follows.

1. Around the QCD-monopole, the abelian action density $S_{\rm Abel}$ takes a large value, but the off-diagonal gluon contribution $S_{\rm off}$ cancels with the abelian fluctuation and keeps the total QCD-action density $S_{\rm SU(2)}$ small.
2. The QCD-monopole has an *intrinsic structure relating to the large amount of off-diagonal gluons* around its center like the 't Hooft-Polyakov monopole. (See Fig.3(d).) At a large scale, off-diagonal gluons inside the QCD-monopole become invisible, and QCD-monopoles can be regarded as point-like Dirac monopoles.
3. From the concentration of off-diagonal gluons around QCD-monopoles in the MA gauge, we can predict a *local correlation between monopoles and instantons*: instantons tend to appear around the monopole world-line in the MA gauge, because instantons need full SU(2) gluon components for existence [9-11,18-21].

Gluonic Higgs and Gauge Invariant Description of MA Projection

In the MA gauge, the gauge group $G \equiv {\rm SU}(N_c)$ is partially fixed into its subgroup $H \equiv {\rm U}_{\rm local}^{N_c-1} \times {\rm Weyl}_{N_c}^{\rm global}$, and then the gauge invariance becomes unclear.[3] In this section, we propose a *gauge invariant description of the MA projection in QCD*. Even without explicit use of gauge fixing, we can define the MA projection by introducing a "gluonic Higgs scalar field" $\vec{\phi}(x)$. For a given gluon field configuration $\{A_\mu(x)\}$, we define a gluonic Higgs scalar $\vec{\phi}(x) \equiv \Omega(x)\vec{H}\Omega^\dagger(x)$ with $\Omega(x) \in {\rm SU}(N_c)$ so as to minimize

$$R[\vec{\phi}(\cdot)] \equiv \int d^4x \, {\rm tr}\left\{[\hat{D}_\mu, \vec{\phi}(x)][\hat{D}_\mu, \vec{\phi}(x)]^\dagger\right\}. \tag{10}$$

We summarize the features of this description as follows [2, 9].

1. The gluonic Higgs scalar $\vec{\phi}(x)$ does not have amplitude degrees of freedom but has only color-direction degrees of freedom, and $\vec{\phi}(x)$ corresponds to a "color-direction" of the nonabelian gauge connection $\hat{D}_\mu$ averaged over μ at each x.
2. Through the projection along $\vec{\phi}(x)$, we can extract the abelian ${\rm U}(1)^{N_c-1}$ sub-gauge-manifold which is most close to the original ${\rm SU}(N_c)$ gauge manifold. This projection is manifestly gauge invariant, and is mathematically equivalent to the ordinary MA projection [2, 9].
3. Similar to $\hat{D}_\mu$, the gluonic Higgs scalar $\vec{\phi}(x)$ obeys the adjoint gauge transformation, and $\vec{\phi}(x)$ is diagonalized in the MA gauge. Then, monopoles appear at the hedgehog singularities of $\vec{\phi}(x)$ as shown in Fig.5 [2, 9, 14].
4. In this description, infrared abelian dominance is interpreted as infrared relevance of the gluon mode along the color-direction $\vec{\phi}(x)$, and QCD seems similar to a nonabelian Higgs theory.

---

[3] In Refs.[2, 9, 19], we show a useful *gauge-invariance criterion* on the operator $O_{\rm MA}$: *If $O_{\rm MA}$ defined in the MA gauge is H-invariant, $O_{\rm MA}$ is also invariant under the whole gauge transformation of G.*

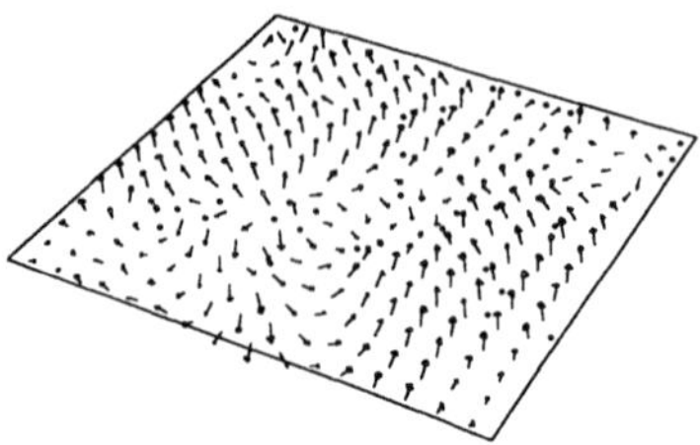

FIGURE 5. The correlation between the gluonic Higgs scalar field $\phi(x) = \phi^a(x)\frac{\tau^a}{2}$ and monopoles (dots) in SU(2) lattice QCD with $\beta = 2.4$ and 16^4. The arrow denotes the SU(2) color direction of $(\phi^1(x), \phi^2(x), \phi^3(x))$. The monopole appears at the hedgehog singularity of the gluonic Higgs scalar $\phi(x)$.

ACKNOWLEDGMENTS

We would like to thank Professor Yoichiro Nambu for his useful suggestions. We are grateful to Professor Il Tong Cheon for his encouragement. The lattice calculations have been performed on NEC-SX4 at Osaka University.

REFERENCES

1. W. Greiner and A. Schäfer, "Quantum Chromodynamics", (Springer-Verlag, Berlin), p.1 (1994).
2. H. Suganuma, K. Amemiya, H. Ichie, H. Matsufuru, Y. Nemoto, T.T. Takahashi, "Quantum Chromodynamics and Color Confinement", H. Suganuma et al. (eds.) (World Scientific), p.103 (2001).
3. T.T. Takahashi, H. Matsufuru, Y. Nemoto and H. Suganuma, Phys. Rev. Lett. **86**, 18 (2001).
4. H. Suganuma, H. Matsufuru, Y. Nemoto and T.T. Takahashi, Nucl. Phys. **A680**, 159 (2001).
5. T.T. Takahashi, H. Suganuma, H. Matsufuru and Y. Nemoto, this Proceedings (2001).
6. N. Nakajima, H. Matsufuru, Y. Nemoto and H. Suganuma, this Proceedings (2001).
7. H. Matsufuru, O. Miyamura, H. Suganuma and T. Umeda, this Proceedings (2001).
8. Y. Nambu, Phys. Rev. **D10**, 4262 (1974).
9. H. Ichie, H. Suganuma, Phys.Rev.**D60**, 77501 (1999); Nucl.Phys.**B548**, 365 (1999) ;**B574**, 70 (2000).
10. H. Suganuma, K. Amemiya, H. Ichie and A. Tanaka, Nucl. Phys. **A670**, 40 (2000).
11. H. Suganuma, H. Ichie, A. Tanaka and K. Amemiya, Prog. Theor. Phys. Suppl. **131**, 559 (1998).
12. G. 't Hooft, Nucl. Phys. **B190**, 455 (1981).
13. Z.F. Ezawa and A. Iwazaki, Phys. Rev. **D25**, 2681 (1982); **D26**, 631 (1982).
14. H. Suganuma, S. Sasaki and H. Toki, Nucl. Phys. **B435**, 207 (1995).
15. K. Amemiya and H. Suganuma, Phys. Rev. **D60**, 114509 (1999).
16. M.I. Polikarpov, Nucl. Phys. **B** (Proc. Suppl.) **53**, 134 (1997) and its references.
17. O. Miyamura, Phys. Lett. **B353**, 91 (1995).
18. H. Suganuma, S. Umisedo, S. Sasaki, H. Toki and O. Miyamura, Aust. J. Phys. **50**, 233 (1997).
19. H. Suganuma et al., Nucl. Phys. **B** (Proc. Suppl.) **47**, 302 (1996); **53**, 528 (1997); **65**, 29 (1998).
20. M. Fukushima et al., Phys. Lett. **B399**, 141 (1997).
21. M. Fukushima, H. Suganuma and H. Toki, Phys. Rev. **D60**, 94504 (1999).
22. H. Suganuma, S. Sasaki, H. Toki and H. Ichie, Prog. Theor. Phys. Suppl. **120**, 57 (1995).
23. S. Sasaki, H. Suganuma, H. Toki, Prog. Theor. Phys. **94**, 373 (1995); Phys. Lett. **B387**, 145 (1996).
24. H. Ichie, H. Suganuma and H. Toki, Phys. Rev. **D52**, 2994 (1995); **D54**, 3382 (1996).
25. S. Umisedo, H. Suganuma and H. Toki, Phys. Rev. **D57**, 1605 (1998).
26. H. Monden, H. Ichie, H. Suganuma and H. Toki, Phys. Rev. **C57**, 2564 (1998).
27. Y. Koma, H. Suganuma and H. Toki, Phys. Rev. **D60**, 74024 (1999).
28. H. Toki and H. Suganuma, Prog. Part. Nucl. Phys. **45**, 397 (2000).
29. K.-I. Kondo, Phys. Rev. **D57**, 7467 (1998); **D58**, 105016 (1998); Phys. Lett. **B455**, 251 (1999).
30. Y.M. Cho, Phys. Rev. **D21**, 1080 (1980); **D62**, 074009 (2000).

Accelerator Mass Spectrometry

J.C. Kim

Seoul National University

Abstract. A brief description for the basic principle of accelerator mass spectrometry is given followed by an introduction of Seoul National University ^{14}C-AMS system. Some results of our recent measurements are also presented.

INTRODUCTION

Accelerator mass spectrometry is an ultra sensitive mass spectrometry which has been evolved from nuclear physics laboratories about twenty years ago [1, 2] by using a tandem accelerator, the properties of negative ions and atom counting. When used in conjunction with mass spectrometry, these additional capabilities removed the traditional limitations of molecular interferences and the presence of isobars.

This has made possible the unique identification of trace elements and rare radioisotopes with long half life. Using these methods, a significant reduction in sample size and increased detection sensitivities, compared to beta ray counting, were possible for the study of the long lived radionuclides in nature.

In this paper, the basic principle of accelerator mass spectrometry will be described ,and one of the modern dedicated AMS facilities, the Seoul National University AMS will be introduced. Also, typical examples applied to archaeology and geoscience will be presented.

THE BASIC PRINCIPLE OF AMS

Long-lived radionuclides in nature such as ^{10}Be, ^{14}C, ^{26}Al, ^{36}Cl, and ^{129}I are rare in abundances, but these minute trace elements are extremely useful for dating purposes and also for environmental tracings [3, 4, 5, 6]. However, measuring these nuclides by radiation detection is difficult due to its long half life and small abundances, thereby, limiting its capacity of application considerably. However, now with the advent of a new detection method, which called accelerator mass spectrometry[AMS], it become possible to measure them directly by atom through the atom counting.

The advantage of the direct atom-counting over the conventional radiation detection is illustrated in **Fig. 1.** In this figure, counting rates of both radioactive measurement and mass spectrometry method are shown for 1g[dotted line] and 1mg[solid line] samples of various long-lived nuclides in nature, whose half lives are ranging from 10^{-2} to $\sim 10^6$ years. For example, while one gram of 55500 yr. old carbon sample, in which the ^{14}C /

CP594, *Hadrons and Nuclei: First International Symposium,* edited by Il-T. Cheon et al.

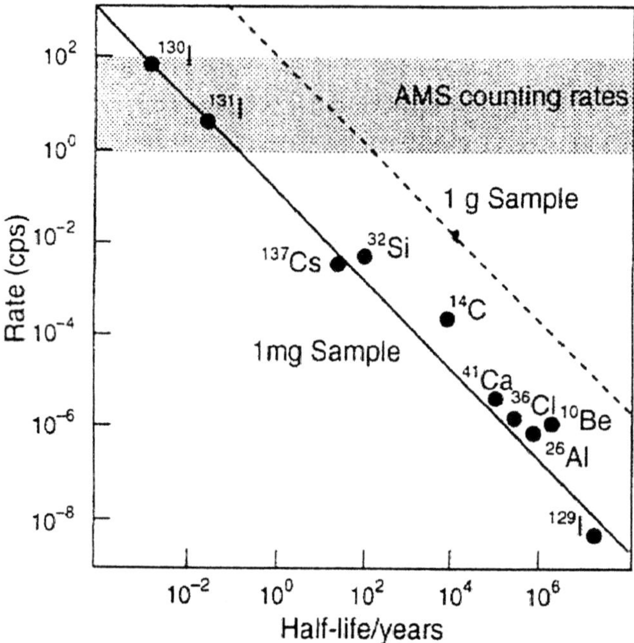

FIGURE 1. Countrates for samples of 1g[dotted line] and 1mg[solid line] are plotted *vs* half-lives of radioactive isotopes

^{12}C ratio is 1.2×10^{-15}, will give decay rate of 1 cts/hr., in AMS 1mg carbon sample of the same age, 1 cts/sec count rate will be obtained by using 100μA C^- beam in the ion source.

To isolate and identify rare long-life radionuclides by conventional mass specrometry such as SIMS [secondary ion mass spectrometry], is hampered by difficulties associated with molecular interferences and isobar problems. Conventional mass spectrometers, utilizing magnet spectrometer and electric analyser, select ions of a certain values $\{E/q, M/q\}$, where E, M and q are energy, mass and charge of the ion respectively. In measuring radiocarbon ^{14}C using low resolving spectrometry, for example, molecules such as $^7Li^+$, $^{12}CH_2{}^+$, $^{13}CH^+$, $^{12}C^{16}O^{++}$, $^{12}C^{13}CH_3{}^{++}$ will have the same $\{E/q, M/q = 14/1\}$, and these molecules interferes the ^{14}C measurement.[see, for example, **Fig. 2**]

For rare ^{14}C measurement, the abundant isobar, ^{14}N, should also be discriminated. There are isobars for other long life radionuclides as shown in **table I.** In stead of building a prohibitedly large [high resoloution] analysing system, the problem of molecules and isobars will be more easily solved by introducing a tandem accelerator. In tandem accelerator, negative ions are injected in the low energy side of the accelerator and goes through charge changing process at the termnal of the accelerator.

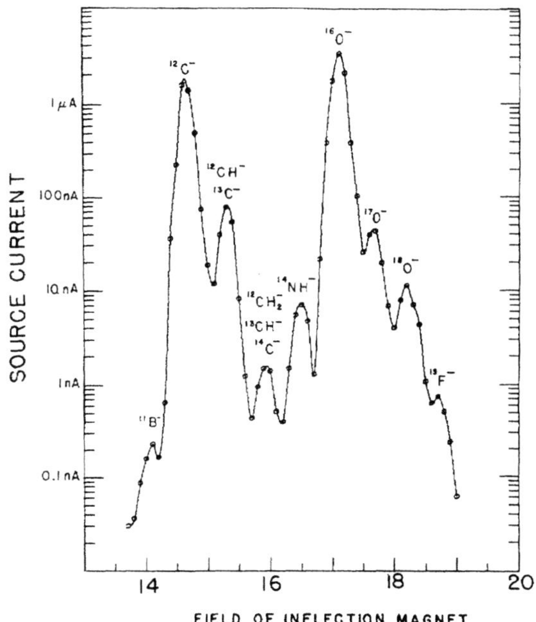

FIGURE 2. Mass spectrum to be obtained by conventional mass spectrometry following ion source to measure ^{14}C. ^{14}C is swamped by $^{12}CH_2^-$, $^{13}CH^-$ molecules.

TABLE 1. Isobar : Case by case studies

	$T_{1/2}$	isobar	solution	application
^{14}C	5730y	^{14}N	Neg. ion	Dating
^{26}Al	7.4×10^6y	^{26}Al	dE/dx	Nervous system
^{36}Cl	3.01×10^5y	^{36}S	dE/dx, $q=5^+$	Hydrology
^{41}Ca	1.03×10^5y	^{41}K	Coulomb Dissociation	Archaeology
^{129}I	1.57×10^7y	^{129}Xe	dE/dx, TOF	TRACER, Nuclear safe guard
^{81}Kr	2.13×10^5y	^{81}Br	Neutral injection	Hydrology
^{194}Pt		stable	$Q=5^+$	Microminerалogy

After charge changing, molecules will be positive ions of various charge states. However, molecules of charge state higher than 3^+ are unstable due to large Coulomb repulsion and will be dissociated into component elements, while elemental ions remain stable. Therefore, the molecules will be removed by selecting charge state 3^+ ions once again following the tandem acceleration. Population of charge state 3^+ in the charge changing process depends on the velocity of ions, which dictates terminal voltage of the tandem accelerator. For ^{14}C measurements, terminal voltage 3MV gives the

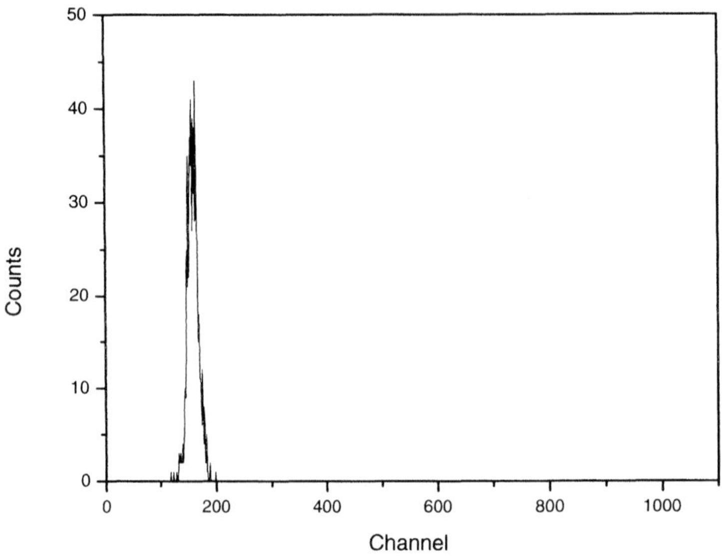

FIGURE 3. shows the $^{14}C^{3+}$ peak obtained by AMS ^{14}C detector.

maximum stripping efficiency of *ca* 50%. The energy, tens of MeV, which the ions obtained after tandem acceleration are usefully incorporated for further selections of ions using standard nuclear detection technologies such as time of flight, energy loss discrimination etc.. These techniques have been used to solve isobar problems in case of other cosmogenic nuclides. For radiocarbon ^{14}C, the isobar problem has been naturally solved by the use of negative ions in conjunction with tandem acceleration [3]. It is observed that negative ion $^{14}N^{-}$ is not exist, or unstable.

SNU-AMS SYSTEM

The Seoul National University AMS is a dedicated radiocarbon AMS system, a new generation type, of which the full description was already given elsewhere [7]. The most distinguished feature of the new generation radiocarbon AMS is that it measures three carbon isotopes; $^{12,13,14}C$ simultaneously [8].

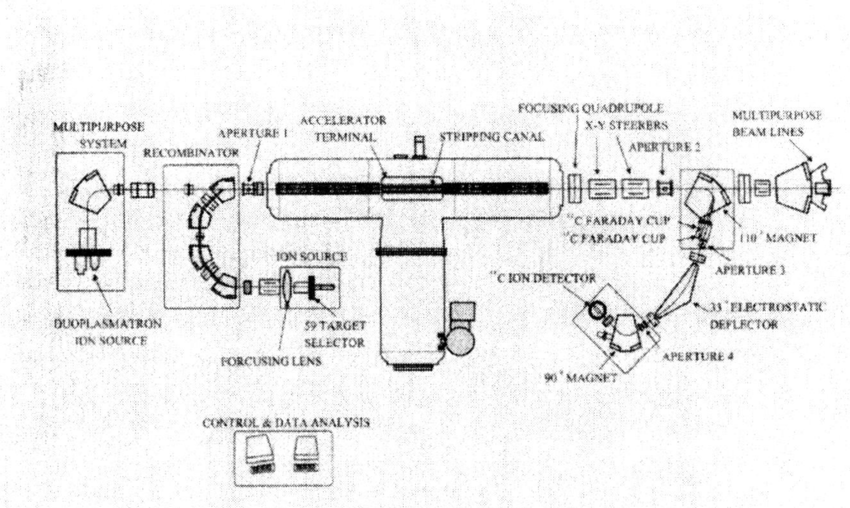

FIGURE 4. The Tandetron 4130 AMS/MPS* system at SEOUL National University. * : Product of High Voltage Engineering Europa co.

The ^{14}C AMS ion source

Our ^{14}C AMS system is equipped with HVEE [High Voltage Engineering Europa co.] model 846B Cs sputter ion source. This ion source has a carousol which can hold up to 59 targets. The targets are loaded (unloaded) by a computerized robot system. Being a high current ion source, unanlysed beams of negative ions C^- of 30μA are easily produced by this ion source.

Injector

The simultaneous injection of beams of three carbon isotopes is made possible by introducing the separator/ recombinator system. It is consisted of four 45° magnets(ρ=31cm) as main components, and vertical steerers. For proper function of this recombinator system precise allignments are required. For proper function of this recombinator system precise allignments are required. A chopper is placed at the center of the recombinator, e.g. the azimuthal symmetry point of the optical path. Here the 99% of ^{12}C beam is chopped and only 1% allowed to enter the accelerating tube.

Tandetron Accelerator

The terminal is charged by the Cockcroft-Walton type solid state power supply. The maximum terminal voltage of 3.3MV for over 4 hours has been demonstrated during the installation period. The conditionning time up to this terminal voltage was about 4 hours. and the stability was within 2kV. This terminal voltage stability is obtained by ^{13}C slit feedback system and GVM is used for voltage reading of the terminal. The stripper gas canal is 90cm long and has a diameter of 15mm, which is 3mm larger than those of its predecessors. A single turbo-molecular pump (300 *l*/s) located inside the terminal shell circulates Ar stripping gas. An immersion lense at the entrance of the accelerating tube, called Q-snout, allows low injection voltages and focuses the injection beam into the stripper canal. The radiation level just outside the pressure vessel is observed as 0.156 μSv/hr. This low radiation level is attributed to permanents magnets helically distributed around the acceleration tube and also to 3mm thick Pb shields clad on the pressure vessel. The vacuum of the accelerator tube is kept better than 10^{-7}mb when there is no stripper gas in the tube and 10^{-6}mb when there is stripper gas by two cryogenic pumps(800*l*/s) located at the both ends of the tube. SF6 gas is used as insulating gas in the pressure vessel, and pressurized to 8 atm. It also requires drying to a dew point of below -40°C for the spark free conditioning.

High energy beam line

The high energy section consists, sequentially in the beam travel direction, of an electrostatic quadrupole doublet, X-Y elecrostatic steerers, 110° magnet(ME/Z^2=15MeV amu), 33° electrostatic analiser(ρ=163 cm), 90° magnet and an inoization detector. The $^{12,13}C$ beam currents are measured by Faraday cups placed in the focal plane of 110° magnet. ^{14}C are analysed in further down stream , by 33° ESA[Electro Static Analyser] and 90° magnet, and detected in the gas ionization chamber. The ionization chamber is filled with isobutane to a pressure of 30 mb. The window of gas ionization chamber is 5 m thick Mylar foil of rectangular shape, 5 mm wide and 10 mm long. There is a slit in front of the ^{13}C Faraday cup,and offset signals from the pair of electrode of this slit are used to stabilize terminal voltage.

APPLICATIONS OF SNU-AMS

The AMS facility of the Seoul National University was operating since early 1999 just after completion of the installation work in December 1998 and extensive testing of the accuracy and reproducibility of the system has been carried out, and *ca.* 500 unknown samples have been measured so far. We obtained a precision of 4 $^{o}/_{oo}$ for modern ages, and an accuracy of *ca.* 40 yr. was demonstrated by analyzing samples which were previously dated with a conventional technique and by other AMS laboratories. **Fig 5** shows a typical batch results.

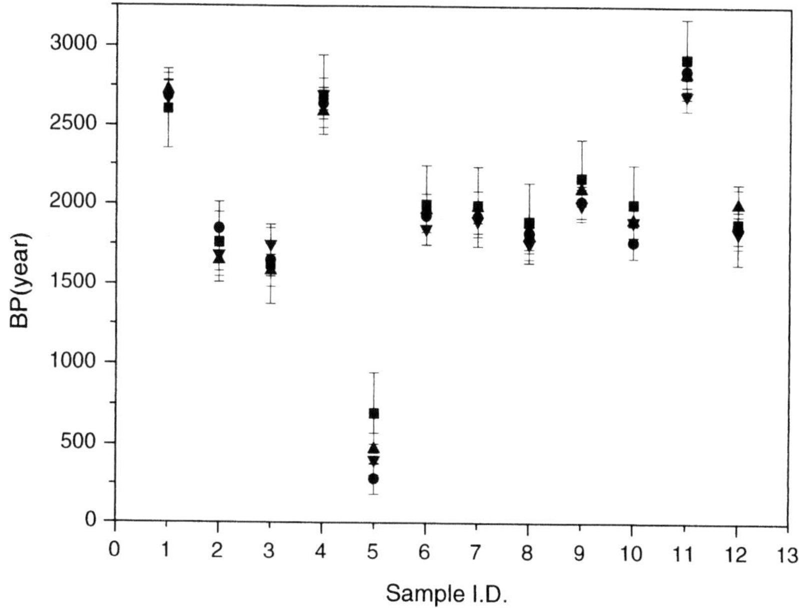

FIGURE 5. BATCH(00305) RESULTS : The result from an examplary batch run. For each sample, of which the I.D. is denoted in the abscissa, measurements were made 3 times and the resulting BP values are shown in the ordinate. BP error of ca. 40 yr is easily obtained from standard measurement procedures described in the text.

In the following three examples of our recent measurements: 1 geological, 2 archaeological are given to illustrate the usefulness of the AMS measurement.

Dating of Siberian permafrost [A Case of Geology]

Ice-wedges in the permafrost region have a vertical stratification of ice like glaciers and ice sheets in arctic region or high mountains, and dating the strata of this ice-wedges will be invaluable for the study of paleo-climate or paleo-environments [9]. We have been dating samples from ice-wedges of Siberian permafrost in collaboration with Russian group [10]. Earlier datings of ice-wedge samples in the conventional radiocarbon laboratories have been carried out using bulk samples, which are mixed with contaminated, re-deposited organic materials. Our main task in this collaboration is to measure various fractions of small samples and to establish a method to distinguish later intruded organic materials from the real one, autochthonous organic materials which are correlated with the growth age of ice-wedges.

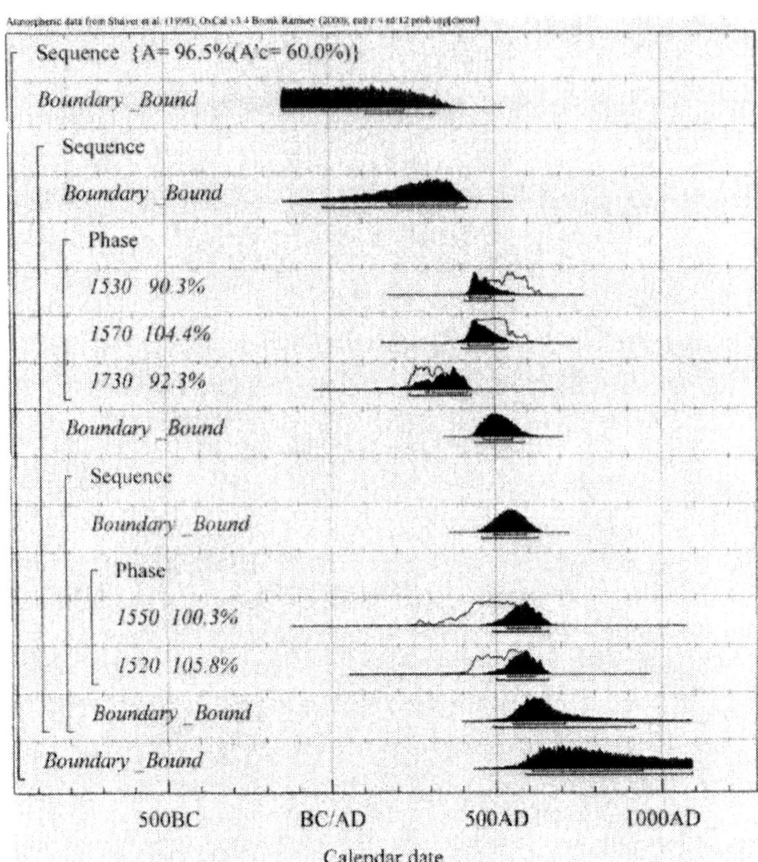

FIGURE 6. Result of Oxcal calculation for the determination of calendar year age of Wangnam-daechong and Cheonma-Chong.In this calculation, a priori condition that Whangnam-daechong is precedent to Cheonma-chong was used.

Dating King's tombs of the Kingdom of Old Silla Korea

WhangNam dae-chong, the largest king's tomb in Korea, which is located in WhangNam-ri, KyungJu-city, the capital of the ancient dynasty, Silla Korea has not been dated yet, neither the well known CheonMa-chong [Heavenly Horse Tomb]. These two tombs were excavated in 1973 [11], but there are much disputes among scholars about the burial date and ownership of the tomb . There are wide discrepancies, ranging 200-300 years, among scholars for the estimations of the burial date [12], which is based on pottery typology. Fortunately, however, there seems to be only two choices, among scholars, about the ownership of the tomb, i.e., the King Nulgi or the King Naemul. King Nulgi founded a state scale nation from the kingdom of old Silla Korea and Naemul is his son.

Therefore, the determination of ages of these kings' tomb is very important for the

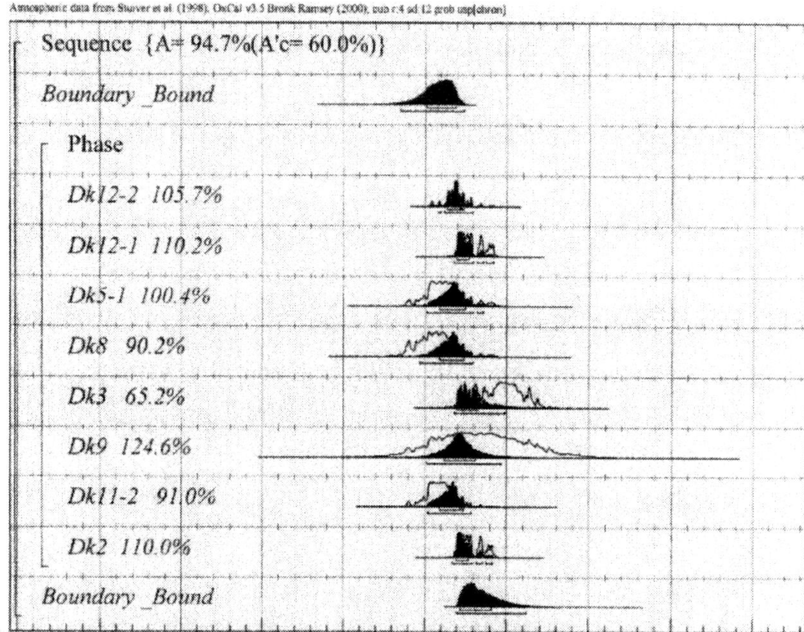

FIGURE 7. Result of Oxcal calculation for the determination of calendar year age of the Sejuk Neolithic site. In this calculation, all events are assumed to be a single phase events.

study of developement of the old Silla Korea as well as for disclosure of the true ownership of the tomb. Figure shows the result of AMS dating. The calendar year[AD] were obtained using a calibration program Oxcal, which uses Bayesian Statistical approach [13]. The samples which we used in this AMS measurement are textiles and tree barks, which was available in scares quantities. More measurements, which uses rare samples such as coffin resins are in progress, and the results meets the accuracy for the identification of the owner of the tomb.

Pottery dating of Sejuk-ri: the oldest Neolithic Site in Korea

Traditionally most of pottery sherd datings were indirectly made by radiocarbon measurements of organic materials collected from the same layer where the pottery sherds were found. As a direct method, TL[thermo-luminiscence] has been employed with less accuracy [14]. Now, the fact that AMS measurement requires only very small quantities of sample makes direct radiocarbon dating of pottery sherds possible.

Pottery sherds are often coated with organic rich charred food remains or soots on their surfaces. These coatings provide fairly reliable samples. Also temper straws in the pot sherds could be picked by crashing and used for the AMS dating. Sejuk-ri[longitude-lattidude] is located at coastal area of the city Ulsan, which is in the south-east part of Korean peninsula, and was excavated ,for the first time, in August-November 2000 by the archaeolgical team of Donguk University [15]. We have dated 12 samples obtained by scraping the surfaces of pottery sherds from this site. The results are shown in **Fig 7.** calibrated using OXcal. It turned out that this site is the oldest Neolithic site in Korea.

ACKNOWLEDGMENTS

This work is supported by KOSEF, grant no. 98-0702-06-01-3.

REFERENCES

1. K.H. Purser, R.B. Liebert, A.E. Litherland, R.P. Beukens, H.E. Gove, C.L. Bennet, M.R. Clover, and W.E. Sondheim, *Revue de Physique* ,(12)1977, 1487-1492.
2. C.L. Bennet, R.P. Beukens, M.R. Clover, H.E. Gove, A.E. Litherland, K.H. Purser, and W.E. Sondheim, *Science*, 198, 508-510.
3. A.E. Litherland, *Ann. Rev. Nucl. Part. Sci.* 30(1980) 437-473.
4. E. Elmore and F. M. Phillips, *Science*, 236(1987) 543-550.
5. W. Kutschera and M. Paul, *Ann. Rev. Nucl. Part. Sci.* 40(190) 411-438.
6. H.E. Gove, The Development and Applications of Accelerator Mass Spectrometry [Institute of Physics Publishing, London, 1999]
7. J.C. Kim, C.H. Lee, I.C. Kim, J.H. Park. J. Kang, M.K. Cheoun, Y.D. Kim, C.B. Moon, *Nucl. Instr. and Meth. In Phys. Res.* B172(2000) 13-17.
8. A.E. Litherland and L.R. Kilius, *Nucl. Instr. and Meth.* B52(1190)375.
9. Y.K. Vasil'chuk and V.M. Kotlyakov, Principle of Isotope Geocryology and Glaciology, 2000, Moscow University Press
10. Y.K. Vasil'chuk and A.C. Vasil'chuk, *Radiocarbon*, 40(1998) 883-893.
11. Excavation Report[in Korea], Kyongju National Research Institute of Cultural Properties.
12. W.R. Kim, Introduction to Korea Archaeology, 1986, [in Korean], 3rd edition, Il-Zi publishing Co.
13. Bronk Ramsey C, *Radiocarbon* 37(1995) 425-430.
14. M.J. Aitken, *Thermoluminescence Dating*, 1985, Academic Press.
15. J.H. Ahn, Excavation Report of Sejuk-ri Site 2000, Research Institute of Buried Artifacts of Kyongju Campus, Dongkuk University.

Three Quarters of a Century in Nuclear Physics

Il-Tong Cheon

Yonsei University, Seoul 120-749, Korea

Nuclear physics has a long history of almost three quarters of a century. There has been a tremendous amount of discoveries made in this field and various models have been proposed to explain specific phenomena revealed by experiments. Recent research works are increasingly concerned with hadron systems and various symmetries in the frame of lagrangian formalism. Since it seems timely to review novel topics and new research directions at the dawn of the twenty-first century, we have organized an International Symposium on Hadrons and Nuclei. Intensive discussions and new attempts of interpretation of nuclear properties were presented and discussed in this symposium along this flow of development. Since all the contents of each talk will be included in the proceedings, I would like to go through roughly the history of developments in nuclear physics rather than summarizing all talks.

First of all, let me list some of the main discoveries in physics before nuclear force was established.

1900	Max Planck	Black body radiation
1901	W. Thomson (Kelvin)	Atomic model
1902	E. Rutherford	Atomic scattering
1903	J. J. Thomson	Atomic model
1904	H. A. Lorentz	Lorentz transformation
1905	A. Einstein	Special theory of relativity
1909	R. A. Millikan	Measurement of electron charge
1911	M. Kamerlingh-Onnes	Superconductivity
1913	N. Bohr	Atomic structure (Bohr model)
	J. Stark	Stark effect
1915	A. Einstein	General theory of relativity
1921	O. Stern & W. Gerlach	Measurement of electron magnetic moment
1923	L. de Broglie	Electron wave (matter wave)
	A. H. Compton	Compton effect
1924	S. Bose	Bose-Einstein statistics
1925	W. Pauli	Exclusion principle
	W. Heisenberg	Quantum mechanics
1926	E. Schrödinger	Wave mechanics
	M. Born	Interpretation of wave function
1928	P. A. M. Dirac	Relativistic quantum mechanics
		Dirac equation, positron
	C. V. Raman	Raman effect

CP594, *Hadrons and Nuclei: First International Symposium,* edited by Il-T. Cheon et al.

1929	Heisenberg-Pauli	Quantum Electrodynamics
	A. Einstein	Unified field theory
1931	E. D. Lawrence	Cyclotron
	W. Pauli	Neutrino hypothesis
1932	C. D. Anderson	Positron (experimentally discovered)
	J. Chadwick	Neutron
1935	H. Yukawa	Meson theory
1937	C. D. Anderson	Meson
1938	O. Hahn	Nuclear fission
1945		Atomic bomb
1949		$\pi-$ meson discovered

Now, let us look at the of developments in nuclear physics. To see some correlations between nuclear physics and particle physics, we list up current topics in each era.

	Particle Physics	Nuclear Physics
1930	QED	**Nuclear force**
	Field Theory	
1950	Baryons	**Nuclear structure**
	Mesons	Shell model
	Resonances	Liquid drop model
		Collective motion
		(Many body system)
		Nuclear reaction
		Direct reaction
		Compound states
		Optical model
		Resonances
1960	Dispersion theory	Short range of Nuclear force
	S-matrix theory	Nucleon-Nucleon correlation
	Regge pole theory	E & M properties of nuclei
	Analyticity	Deformation
	Causality	Meson-nucleus interaction
	Current algebra	Mesonic atom
	Group theory $SU(2) \to SU(3)$	Electron scattering
	Quark hypothesis (p,n,Λ)	Meson exchange current
1970	Renormalization	Quark degrees of freedom
	Electroweak theory	MIT bag model
	Weinberg-Salam theory	

1980	Confinement	Baryons & mesons
	String theory	Chiral bag model
	Gravity	Skyrmion model
	Grand unified theory	Solition
		Q.C.D. lagrangian
		Chiral invariance
1990	High energy phenomena	Quark-gluon plasma (QGP)
	Standard model	(phase transition)
	Neutrino oscillation	Hyperons
	Weak bosons	Strangeness in nuclei
	Higgs	Chiral perturbation theory
	Top quark	Relativistic mean field theory
	QGP	Meson condensation
2000	Supersymmetry	CP violation
	Super String	Astronuclear physics
		Unstable nuclei
		Superheavy nuclei
		Anti-nuclei
		Nuclear waste

As can be seen in this list, some parts of nuclear physics run in parallel with particle physics, but the former follows the latter with a certain time difference. When problems are handed over to nuclear physicists, they are investigated in great details and complicated ways. There is obviously a drastic phase transition from nuclear physics based on nucleons to new approaches with quark degrees of freedom. However, as Abdus Salam once pointed out, nuclear physics has not revealed fundamental concept as particle physics has achieved so far. In this sense, nuclear physics was devoted to the investigation of nuclear properties rather than fundamental principles. We, nuclear physicists may need to meditate on our philosophy and attitude towards nature.

In this twenty-first century, nuclear physics might move to a new direction. It will not remain merely as a field of studying the physical properties of nuclei. An example is the creation of anti-nuclei in the vacuum of superheavy nuclei as suggested by Greiner. Another example is to change nuclear energy levels and their widths by means of artificial controls. If nuclear life-time could be freely controlled, it would be a great achievement equivalent to nuclear energy gripped by our hands.

Anyway, this symposium held in the first year of the twenty-first century is really meaningful in presenting recent research works and in looking for future directions. Finally, I sincerely thank all the speakers and participants for their elaborated presentations and fruitful discussions.

APPENDIX

SCIENTIFIC PROGRAM

8:30 am **Registration**

9:00 am **Opening**

Plenary Sessions 1 – 4 (Feb. 20, Tues)

Plenary Session 1 *(Chairman; B. T. Kim, Sungkyunkwan University)*

Welcoming address; **W. S. Kim (President of Yonsei University)**

Welcoming address; **H. S. Song (President of the Korean Physical Society)**

Opening address; **Il-T. Cheon (Yonsei University)**

9:10 am **H. Toki (RCNP, Osaka)**
Surface Pion Condensation in Relativistic Mean Field Theory
for Finite Nuclei

9:50 am **A. I. Titov (JINR, Dubna)**
Baryon Resonances and the Validity of OZI-rule in πN Reactions

10:30 am Break

Plenary Session 2 *(Chairman; W. Kim, Kyungpook University)*

11:00 am **F. Iazzi (INFN-Sezione di Torinoand Politecnico di Torino)**
Antineutron-Nucleus Interaction at Low Energy

11:30 am **B. Saghai (Saclay)**
From Known to Undiscovered Nucleonic Resonances

12:10 am **B. A. Mecking (JLAB)**
The CEBAF Electron Accelerator Physics Program and First Results

12:40 am Lunch

Plenary Session 3 *(Chairman; N. Herrmann, University of Heidelberg)*

2:00 pm **J. K. Ahn (RCNP, Osaka)**
First Results from SPring-8/LEPS and Prospects

2:30 pm **B. Stella (University and INFN)**
Proton Structure Functions at HERA

3:00 pm **H. C. Kim (Pusan National Univ.)**
Why is the Nucleon Strange?

3:30 pm Break

Plenary Session 4 *(Chairman; M. Fujiwara, RCNP, Osaka University)*

3:50 pm **T. Motoba (Osaka E-C Univ.)**
Prospects of Hypernuclear Structure

4:30 pm **H. Tamura (Tohoku Univ.)**
Hypernuclear γ Spectroscopy and ΛN Interactions

5:00 pm **Y. Akaishi (IPNS, KEK)**
Nuclear $\bar{K}$ Bound States in Light Nuclei

5:30 pm **J. C. Kim (Seoul National Univ.)**
Accelerator Mass Spectrometry : An Application of Nuclear Physics to the Interdisciplinary Sciences

Plenary Session 5 (Feb. 21, Wed)

Plenary Session 5 *(Chairman; H. Stöcker, Goethe Universität)*

9:00 am **M. Oka (Tokyo Institute of Tech.)**
Roles of Σ in Weak and Electromagnetic Interactions of Hypernuclei

9:30 am **H. Bhang (Seoul National Univ.)**
Non-Mesonic Weak Decay of Λ Hypernuclei and the Final State Interaction

10:00 am **B. S. Hong (Korea Univ.)**
Proton and Pion Distributions in Heavy-Ion Collisions at SIS Energies

10:30 am Photo Session

10:40 am Break

Parallel Sessions 1 – 3 (Feb. 21, Wed)

Parallel Session 1 *(Chairman; J. Kasagi, Tohoku University)*

11:00 am **T. Suzuki (Fukui Univ., RIKEN)**
Mesonic Nuclear Collective States

11:30 am **S. Hirenzaki (Nara Women's Univ.)**
Deeply Bound Pionic Atoms

11:50 am **A. Hosaka (Numazu College of Technology)**
Chiral Symmetry of Baryons and Its Observation

12:10 am **V. Dmitrašinović (RCNP, Osaka)**
Discriminating Between Models of $U_A(1)$ Symmetry Breaking

12:30 am **K. S. Kim (Sung Kyun Kwan University)**
Coulomb Distortion Effects for Electron or Positron Induced (e,e') Reactions in the Quasielastic Region

12:50 am Lunch

Parallel Session 2 *(Chairman; H. Bhang, Seoul National University)*

2:00 pm **M. K. Cheoun (Seoul National Univ.)**
Pion and Kaon Electromagnetic Form Factors by an $SU(3) \otimes SU(3)$ Effective Lagrangian and Their Related Decays

2:20 pm **M. Takayama (RCNP, Osaka)**
Spin of a Relativistic Composite System and Nucleon Magnetic Moment

2:40 pm **K. Nakazawa (Gifu Univ.)**
Doubly Strange Nuclei by a Hybrid-Emulsion Experiment E373 at KEK

3:10 pm **E. Hiyama (IPNS, KEK)**
Four-body Calculation of ${}^4_\Lambda$H and ${}^4_\Lambda$He with Realistic YN and NN Interactions

3:30 pm **H. Nemura (IPNS, KEK)**
Quark Pauli Principle in Λ-Hypernuclear Systems

3:50 pm Break

Parallel Session 3 *(Chairman; Y. Akaishi, KEK)*

4:10 pm **S. Fujii (RIKEN)**
Shell-Model Calculations for $^{17}_{\Lambda}$O and $^{16}_{\Lambda}$O) Using Microscopic ΛN and ΣN Effective Interactions

4:30 pm **J. H. Kim (Seoul National Univ.)**
Neutron Spectra from the NMWD of $^{12}_{\Lambda}$C and $^{89}_{\Lambda}$Y

4:50 pm **I. Kumagai-Fuse (CIMS, Hokkaido University)**
Pionic Weak-Decay of Lightest double-Λ Hypernucleus $^{4}_{\Lambda\Lambda}$H

5:10 pm **N. Sawado (FRCCC, Science University of Tokyo)**
Strange Dibaryon States in the Chiral Quark Soliton Model

Parallel Sessions 4 – 6 (Feb. 21, Wed)

Parallel Session 4 *(Chairman; T. Hatsuda, Uinversity of Tokyo)*

11:00 am **M. Maruyama (Tohoku Univ.)**
Time Evolution of Quantum Many Body Systems
in the Projection Operator Method

11:30 am **S. Choe (Hiroshima Univ.)**
$\frac{\partial m}{\partial \mu}$ in the Nambu–Jona-Lasinio model

11:50 am **S. Kim (Yonsei Univ.)**
QCD Sum Rules for J/ψ in the Nuclear Medium

12:10 pm **H. Matsufuru (RCNP, Osaka)**
J/ψ at Finite Temperature - Lattice QCD Result and Potential Model Analysis

12:30 am **Y. Kwon (Yonsei Univ.)**
Perturbative Aspects of the Heavy Ion Collisions at RHIC and
Its Measurement

12:50 am Lunch

Parallel Session 5 *(Chairman; K. Redlich, University of Wroclaw)*

2:00 pm **O. Miyamura (Hiroshima Univ.)**
Recent Results on Hadron Spectroscopy at Finite Temperature from Lattice Gauge Theory

2:30 pm **H. S. Roh (Sung Kyun Kwan University)**
Quantum Nucleardynamics as an $SU(2)_N \times U(1)_Z$ Gauge Theory

2:50 pm **Y. Koma (RCNP, Osaka)**
Weyl Symmetric Representation of SU(3) Gluodynamics in Abelian Projection

3:10 pm **T. T. Takahashi (RCNP, Osaka)**
SU(3) Lattice QCD Study for Static Three-Quark Potential

3:30 pm **N. Nakajima (RCNP, Osaka)**
Systematical Study for Octet and Decuplet Baryon Spectra in $SU(3)$ Lattice QCD

3:50 pm Break

Parallel Session 6 *(Chairman; K. Redlich, Uinversity of Wroclaw)*

4:10 pm **M. Musakhanov (Uzbekistan National Univ.)**
Current Mass Dependence of the Quark Condensate and the Constituent Quark Mass

Plenary Sessions 6 – 8 (Feb. 22, Thur)

Plenary Session 6 *(Chairman; H. Toki, RCNP, Osaka University)*

9:00 am **W. Greiner (Frankfurt Univ.)**
Nuclear Matter, Hypermatter, Antimatter and Strong Correlations in the Vacuum

9:35 am **T. Kunihiro (YITP, Kyoto University)**
Chiral Transition and the Scalar and Vector Correlations

10:10 am **H. Stöcker (Goethe Univ.)**
Probing QCD matter at RHIC

10:45 am Break

Plenary Session 7 *(Chairman; B. A. Mecking, Jefferson Laboratory)*

11:05 am **N. Herrmann (GSI/Heidelberg)**
Production and Propagation of Strange Hadrons in Dense Nuclear Matter

11:40 am **K. Redlich (Univ. of Wroclaw)**
Production and Propagation of Strange Hadrons in Dense Nuclear Matter

12:15 am Lunch

Plenary Session 8 *(Chairman; A. I. Titov, Joint Institute for Nuclear Research)*

1:30 pm **T. Hatsuda (Univ. of Tokyo)**
Maximum Entropy Analysis of the Spectral Functions in Lattice QCD

2:05 pm **H. Suganuma (Tokyo Inst. of Technology)**
Confinement Physics and Hadron Physics in SU(3) Lattice QCD

2:40 pm **Il-T. Cheon (Yonsei University)**
Closing Remarks

LIST OF PARTICIPANTS

Jung Keun Ahn
Research Center for Nuclear Physics
Osaka University
Osaka 567-0047, Japan
jkahn@rcnp.osaka-u.ac.jp

Yoshinori Akaishi
Institute of Particle and Nuclear Studies
KEK
Tsukuba, Ibaraki 305-0801, Japan
akaishi@post.kek.jp

H. Bhang
Department of Physics
Seoul National University
Seoul 151-747, Korea
bhang@phya.snu.ac.kr

Il-Tong Cheon
Department of Physics
Yonsei University
Seoul 120-749, Korea
itcheon@phya.yonsei.ac.kr

Myung Ki Cheoun
AMS Group, IUCNSF
Seoul National University
Seoul 151-742, Korea
cheoun@phya.snu.ac.kr

Seungho Choe
Department of Physics
Hiroshima University
Higashi-Hiroshima 739-8526, Japan
schoe@hiroh2.hepl.hiroshima-u.ac.jp

Hyunah Choi
Department of Physics
Pusan National University
Pusan 609-735, Korea
hapeace@hanmail.net

Taekeun Choi
Institute of Physics and Applied Physics
Yonsei University
Seoul 120-749, Korea
tkchoi@phya.yonsei.ac.kr

Veljko Dmitrasinovic
Research Center for Nuclear Physics
Osaka University
Osaka 567-0047, Japan
dmitra@rcnp.osaka-u.ac.jp

Shinichiro Fujii
RI Beam Project Offic
RIKEN
Saitama 351-0198, JAPAN
sfujii@postman.riken.go.jp

Mamoru Fujiwara
Research Center for Nuclear Physics
Osaka University
Osaka 567-0047, Japan
fujiwara@rcnp.osaka-u.ac.kr

Walter Greiner
Institut fuer Theoretische Physik
Goethe Universitaet
60054 Frankfurt am Main, Germany
greiner@th.physik.uni-frankfurt.de

Tetsuo Hatsuda
Department of Physics
University of Tokyo
Tokyo 113-0033, Japan
hatsuda@phys.s.u-tokyo.ac.jp

Arata Hayashigaki
Department of Physics
University of Tokyo
Tokyo 113-0033, Japan
arata@nt.phys.s.u-tokyo.ac.jp

Norbert Herrmann
Physikalisches Institut
University of Heidelberg
D-69120 Heidelberg, Germany
N.Herrmann@gsi.de

Satoru Hirenzaki
Department of Physics
Nara Women's University
Nara 630-8506, Japan
zaki@phys.nara-wu.ac.jp

Emiko Hiyama
Institute of Particle and Nuclear Studies
KEK
Tsukuba, Ibaraki 305-0801, Japan
hiyama@post.kek.jp

Byungsik Hong
Department of Physics
Korea University
Seoul 130-701, Korea
hong@mail.korea.ac.kr

S. T. Hong
Department of Physics
Sogang University
Seoul 121-742, Korea
sthong@ccs.sogang.ac.kr

Seung-Woo Hong
Department of Physics
Sung Kyun Kwan University
Suwon 440-746, Korea
swhong@skku.ac.kr

Atsushi Hosaka
Department of Physics
Numazu College of Technology
Numazu 410-8501, Japan
hosaka@la.numazu-ct.ac.jp

Felice Iazzi
Dipartimento di Fisica
Politecnico di Torino
10129 Torino, Italy
iazzi@to.infn.it

Moon Taek Jeong
Department of Physics
Dongshin University
Naju 520-714, Korea
drwoman@hanmail.net

Jirohta Kasagi
Laboratory of Nuclear Science
Tohoku University
Sendai 982-0826, Japan
kasagi@lns.tohoku.ac.jp

Byoung-Chul Kim
Department of Physics
Pusan National University
Pusan 609-735, Korea
kipperch@chollian.net

B. T. Kim
Department of Physics
Sung Kyun Kwan University
Suwon 440-746, Korea
btkim@skku.ac.kr

B. S. Kim
Department of Physics
Yonsei University
Seoul 120-749, Korea
sozin76@netian.com

C. K. Kim
Department of Physics
Yonsei University
Seoul 120-749, Korea
kjk@phya.yonsei.ac.kr

Gui Nyun Kim
Pohang Accelerator Laboratory
POSTECH
Pohang Kyungpook 790-784, Korea
gnkim@postech.ac.kr

Hyun-Chul Kim
Department of Physics
Pusan National University
Pusan 609-735, Korea
hchkim@hyowon.pusan.ac.kr

Hungchong Kim
Department of Physics
Yonsei University
Seoul 120-749, Korea
hung@phya.yonsei.ac.kr

I. T. Kim
Department of Physics
Yonsei University
Seoul 120-749, Korea
greatkit@phya.yonsei.ac.kr

Ji-Young Kim
Department of Physics
Pusan National University
Pusan 609-735, Korea
wooyada@yahoo.co.kr

Jungho Kim
Department of Physics
Seoul National University
Seoul 151-742, Korea
jungho@ieplab.snu.ac.kr

J. C. Kim
Department of Physics
Seoul National University
Seoul 151-742, Korea
jckim@phya.snu.ac.kr

K. H. Kim
Department of Physics
Inha University
Incheon 402-751, Korea

Kyungsik Kim
Department of Physics
Sung Kyun Kwan University
Suwon 440-746, Korea
kyungsik@color.skku.ac.kr

M. H. Kim
Department of Physics
Sung Kyun Kwan University
Suwon 440-746, Korea
creepha@color.skku.ac.kr

Sungsik Kim
Department of Physics
Yonsei University
Seoul 120-749, Korea
sskim@phya.yonsei.ac.kr

Wooyoung Kim
Department of Physics
Kyungpook National University
Taegu 702-701, Korea
wooyoung@bh.knu.ac.kr

Y. K. Ko
Department of Physics
Yonsei University
Seoul 120-749, Korea
yongkyu@phya.yonsei.ac.kr

Yoshiaki Koma
Research Center for Nuclear Physics
Osaka University
Ibaraki 567-0047, Japan
koma@rcnp.osaka-u.ac.jp

Izumi Kumagai-Fuse
Center for Information and Multimedia Studies
Hokkaido University
Sapporo 060-0811, Japan
ifuse@cims.hokudai.ac.jp

T. Kunihiro
Yukawa Institute for Theoretical Physics
Kyoto University
Kyoto 606-8502, Japan
kunihiro@yukawa.kyoto-u.ac.jp

Yongil Kwon
Institute of Physics and Applied Physics
Yonsei University
Seoul 120-749, Korea
ykwon@phya.yonsei.ac.kr

Jeonghan Lee
Department of Physics
Pusan National University
Pusan 609-735, Korea
citadel@cheerful.com

Su Houng Lee
Department of Physics
Yonsei University
Seoul 120-749, Korea
suhoung@phya.yonsei.ac.kr

Masahiro Maruyama
Department of Physics
Tohoku University
Sendai 980-8578, Japan
maruyama@nucl.phys.tohoku.ac.jp

Hideo Matsufuru
Research Center for Nuclear Physics
Osaka University
Osaka 567-0047, Japan
matufuru@rcnp.osaka-u.ac.jp

Toru Matsumura
Synchrotron Radiation Research Center
Japan Atomic Energy Research Institute
Hyogo 679-5148, Japan
toru@rcnp.osaka-u.ac.jp

Bernhard A. Mecking
Physics Division
Jefferson Laboratory
Virginia 23606, U.S.A
mecking@jlab.org

Osamu Miyamura
Department of Physis
Hiroshima University
Higashi-Hiroshima 839-8526, Japan
miyamura@sci.hiroshima-u.ac.jp

T. Motoba
Department of Physics
Osaka E-C University
Osaka 572-8530, Japan
motoba@isc.osakac.ac.jp

C. B. Moon
Department of Physics
Hoseo University
Chungnam Asan 336-795, Korea
cbmoon@dogsuri.hoseo.ac.kr

Mirzayusuf Musakhanov
Theor. Phys. Dept.
Uzbekistan National University
Tashkent 700174, Uzbekistan
yousuf@iaph.silk.org

Noriaki Nakajima
Research Center for Nuclear Physics
Osaka University
Osaka 567-0047, Japan
nakaji@rcnp.osaka-u.ac.jp

Kazuma Nakazawa
Physics Department
Gifu University
Gifu 501-1193, Japan
nakazawa@cc.gifu-u.ac.jp

Seng-Il Nam
Department of Physics
Pusan National University
Pusan 609-735, Korea
sinam@instanton.phys.pusan.ac.kr

Hidekatsu Nemura
Institute of Particle and Nuclear Studies
KEK
Tsukuba 305-0801, Japan
hidekatsu.nemura@kek.jp

Makoto Oka
Deapartment of physics
Tokyo Institute of Technology
Tokyo 151-8551, Japan
oka@th.phys.titech.ac.jp

Sang-Hyun Park
Department of Physics
Pusan National University
Pusan 609-735, Korea
upanishad@instanton.phys.pusan.ac.kr

Ki Jun Park
Department of Physics
Kyungpook National University
Taegu 702-701, Korea
parkkj@fermi.knu.ac.kr

Y. J. Park
Department of Physics
Yonsei University
Seoul 120-749, Korea
skynever@netian.com

Krzysztof Redlich
Department of Physics
University of Wroclaw
PL-50204 Wroclaw, Poland
redlich@rose.ift.uni.wroc.pl

H. S. Roh
Department of Physics
Sung Kyun Kwan University
Suwon 440-746, Korea
hroh@nature.skku.ac.kr

C. Y. Ryu
Department of Physics
Sung Kyun Kwan University
Suwon 440-749, Korea
cyryu@color.skku.ac.kr

Bijan Saghai
Service de Physique Nucleaire, Bat. 703
CEA/Saclay
91191 Gif-sur-Yvette Cedex, France
bsaghai@cea.fr

S. Sato
Department of Physics
University of Tsukuba
Tsukkuba 305-8577, Japan
ssato@bnl.gov

Nobuyuki Sawado
Department of Physics and Technonogy
Science University of Tokyo
Chiba 278-8510, Japan
a6293701@rs.noda.sut.ac.jp

Hajime Shimizu
Research Center for Nuclear Physics
Osaka University
Osaka 567-0047, Japan
hshimizu@rcnp.osaka-u.ac.jp

Y. M. Shin
Department of Physics
Yonsei University
Seoul 120-749, Korea
shin@netscape.net

S. A. Shin
Department of Physics
Ewha Woman's University
Seoul 120-750,Korea
sashin@mm.ewha.ac.kr

W. Y. So
Department of Physics
Sung Kyun Kwan University
Suwon 440-746, Korea
wyso@mail.skku.ac.kr

H. S. Song
Department of Phsics
Seoul National University
Seoul 151-742, Korea
hssong@physs.snu.ac.kr

Bruno Stella
III Universitá degli Studi di Roma and INFN
Sezione Roma 3
V. della Vasca Navale 89, I-001. Roma, Italy
bruno.stella@roma1.infn.it

H. Stoecker
Institut fuer Theoretische Physik
Goethe Universitaet
60054 Frankfurt am Main, Germany
stoecker@uni-frankfurt.de

Hideo Suganuma
Faculty of Science
Tokyo Institute of Technology
Tokyo 152-8551, Japan
suganuma@th.phys.titech.ac.jp

Toshio Suzuki
Department of Applied Physics
Fukui University
Fukui 910-8507, Japan
suzuki@quantum.apphy.fukui-u.ac.jp

Toru T. Takahashi
Research Center for Nuclear Physics
Osaka University
Osaka 567-0047, Japan
ttoru@rcnp.osaka-u.ac.jp

Miho Takayama
Research Center for Nuclear Physics
Osaka University
Osaka 567-0047, Japan
takayama@rcnp.osaka-u.ac.jp

Hirokazu Tamura
Laboratory of Nuclear Science
Tohoku University
Sendai 980-8578, Japan
tamura@lambda.phys.tohoku.ac.jp

Alexander I. Titov
Bogolyubov Lab. Theor. Physics
Jiont Instiute for Nuclear Research
Dubna, Moscow Region 141980, Russia
atitov@thsun1.jinr.ru

Hiroshi Toki
Research Center for Nuclear Physics
Osaka University
Osaka 567-0047, Japan
toki@rcnp.osaka-u.ac.jp

Jae Hyung Yee
Department of Physics
Yonsei University
Seoul 120-749, Korea
jhyee@phya.yonsei.ac.kr

Masaru Yosoi
Research Center for Nuclear Physics
Osaka University
Osaka 567-0047, Japan
yosoi@rcnp.osaka-u.ac.jp

Byung Geel Yu
Department of Physics
Hankuk Aviation University
Koyang 412-791, Korea
bgyu@mail.hangkong.ac.kr

J. H. Yoon
Department of Physics
Inha University
Incheon 402-751, Korea
jinyoon@inha.ac.kr

AUTHOR INDEX

A

B

C

D

E

F

G

H

I

J

K

L

M

N

O

P

R